ISBN 978-0-265-91390-1
PIBN 11146433

1 MONTH OF
FREE
READING

at

www.ForgottenBooks.com

By purchasing this book you are eligible for one month membership to ForgottenBooks.com, giving you unlimited access to our entire collection of over 1,000,000 titles via our web site and mobile apps.

To claim your free month visit:

www.forgottenbooks.com/free1146433

JAMES WATT

AND THE

STEAM ENGINE

THE MEMORIAL VOLUME
PREPARED FOR THE COMMITTEE OF THE
WATT CENTENARY COMMEMORATION
AT BIRMINGHAM 1919

BY

H. W. DICKINSON & RHYS JENKINS

OXFORD
AT THE CLARENDON PRESS
1927

JAMES WATT

AND THE

STEAM ENGINE

THE MEMORIAL VOLUME
PREPARED FOR THE COMMITTEE OF THE
WATT CENTENARY COMMEMORATION
AT BIRMINGHAM 1919

BY

H. W. DICKINSON & RHYS JENKINS

OXFORD
AT THE CLARENDON PRESS
1927

Printed in Great Britain

PREFACE

THE publication of a Memorial Volume to mark in a lasting manner the centenary of the death of James Watt was, from the first, one of the aims of the Committee set up in Birmingham in 1919 to commemorate that event. The genesis of the movement, the proceedings during the Commemoration week, and the measure of success in attaining the objects contemplated will be found set forth in their appropriate places in the pages that follow.

The preparation of the material for the Memorial Volume was entrusted to the present Authors with wide discretionary powers and with sufficient funds to carry out the work. They take this opportunity of thanking the Committee for their confidence and of stating that for the matter contained in the volume they take entire responsibility. They would add that its preparation has been to them a work of very great interest.

James Watt was one of the few really great inventors who have been recognized as such during their lifetimes. Even to-day, over a hundred years after his death, his name is perhaps the best and most widely known among inventors. The object of the present work is not to glorify further the inventor, but rather to discuss critically the origin of the invention, the steps by which it was reduced to a practical form, and the development of the steam engine by him and under his direction, as well as the conditions and environment in which the development was effected. Accordingly it stops with the retirement of Watt from active business in the year 1800. From some points of view this is to be regretted, since the succeeding years up to the time of his death present features of much interest in the development of the steam engine, of mechanical engineering, and of the engine works at Soho.

The life story of James Watt has indeed been so fully dealt with by previous writers that it was felt that this aspect of the subject might be disposed of in a briefer fashion than would otherwise have been necessary; at the same time it will be found that fresh light has been shed on his life and that some erroneous views previously held have been corrected—yet, while the inventions and discoveries of Watt in other directions have been touched upon, the main theme of the book is the steam engine. In following his work, after dealing with the experimental stage, the Authors have confined the earlier sections to the broad lines of engine construction and relegated the consideration of the evolution of the parts, details, or accessories of the engine to separate chapters. This has involved a certain amount of repetition, but the advantages appear to justify this; it is possible now to follow out the development of, for instance, the condenser, in a consecutive narrative.

The investigation of the records placed at the Authors' disposal—the Boulton Papers, the Doldowlod Papers, and the Boulton and Watt Collection at Birmingham —has shown that they are of great historical value in directions apart from the constructional side of the steam engine, as, for instance, in the history of the early days of mechanical engineering, and the Authors feel that the sections devoted to the description of the manner in which the engines were made, of the engine yard at the Soho Manufactory and of the Soho Foundry will be of wide interest.

In one sense the task may be said to have been a light one, because the material had already been accumulated in years gone by and was available in abundance. But the very fact of this abundance has rendered the task of sifting, collation, condensation, and rejection very laborious, since the volume had to be kept within reasonable bounds, and the Authors were not satisfied to take anything upon trust. Everything that is presented therefore is the outcome of actual examination of the material. Notwithstanding this, the Authors realize that some relevant documents may have been overlooked, indeed, much longer time might have been devoted to the task than the Authors have, in the midst of other professional duties, been able to devote to it. Those duties are their excuse, furthermore, for the time that has elapsed since the volume was projected.

The illustrations of the engines, or the greater part of them, are copied from the original Boulton and Watt drawings in the collection at Birmingham. In some cases they leave much to be desired in point of clearness, but it will be understood that we are concerned with drawings made, say, 130 to 150 years ago, drawings that have been subjected to office handling, and in some cases to shop handling as well, for years, and afterwards have not been preserved with any particular degree of care. We must be thankful that such drawings are in existence at all in whatever condition of dilapidation or obscurity. Clearer illustrations could have been produced by preparing tracings by hand from the original drawings, but it was felt that this would have been to abuse most important historical material and that the correct procedure was, first, in making a selection of the drawings to be reproduced, to give the preference to those that would come out most clearly, and then to rely entirely upon photographic reproduction.

The Authors are under deep obligation to Major J. M. Gibson Watt, the present representative of the family, for giving access unreservedly to the documentary material preserved at his seat, Doldowlod, Radnorshire, and for information supplied. Hardly less important have been the facilities granted by Miss Boulton and the Boulton Trustees for the collation and transcription of the documents formerly preserved at their seat, Tew Park, Oxfordshire, and now deposited in the Assay Office, Birmingham. The Free Libraries Committee of the City of Birmingham and their librarian, Mr. Walter Powell, have been assiduous in helping the

Authors by placing at their disposal the documents and drawings in the Boulton
and Watt and the Muirhead Collections and in giving facilities for making the
numerous photographs required for illustration.   Mr. C. O. Becker has placed at
the disposal of the Authors his wide and critical study of the drawings in the
Boulton and Watt Collection.   Mr. W. M. Michell of Redruth has rendered great
assistance in identifying the sites of the Boulton and Watt engines in Cornwall.
Mr. R. B. Askquith Ellis, the Honorary Secretary of the Centenary Committee, has
been kind enough to furnish the material for the account of the Commemoration
proceedings ; it is to be regretted that considerations of space have prohibited a
more lengthy treatment of this event.   Mr. Edward Collins of the Heathfield
Estate office has been helpful in supplying information about Heathfield and bygone
Birmingham, and Mr. Gilbert C. Vyle of Messrs. W. & T. Avery, Limited, who now
occupy the Soho Foundry, has been of great help in reference to this site.   It is
interesting to record that Mr. A. W. Brown of this firm, with whom the Authors
were in touch, was of the fourth generation of a family who had served the successive
proprietors from the days of Boulton and Watt onwards.   Mrs. C. H. Turner of
Rooksnest, Godstone, has been so good as to allow the Beechey portrait of Watt
(frontispiece) to be copied, and Mrs. Scott of Hawkhill, Largs, has given like assistance in regard to the Partridge portrait.   Friends too numerous to mention
individually have responded readily to specific inquiries and such help will be found
acknowledged generally in the foot-notes.   Finally, the Delegates of the Clarendon
Press are thanked for the care bestowed upon the production of the volume, and in
this connexion the Authors wish to acknowledge the help rendered by Miss V. F.
Boyson in the revision of certain sections of the manuscript.

H. W. D.

December 1926.   R. J.

CONTENTS

           b

LIST OF PLATES

The illustrations of engines and engine details are from drawings in the Boulton and Watt Collection, except where otherwise stated.

MAPS

LIST OF ILLUSTRATIONS IN TEXT

WATT CHRONOLOGY

The first column gives James Watt's age at the date of the event recorded in the second and third columns.

	1736	Born at Greenock.
18	1754	Starts life in Glasgow.
19	1755	Goes to London to learn trade of mathematical instrument-maker.
20	1756	Returns to Greenock.
21	1757	Moves to Glasgow in July and opens shop in the College there in December.
23	1759	Removes to shop in the Saltmarket in partnership with John Craig.
27	1763	Repairs Newcomen engine model belonging to Glasgow College, and experiments on properties of steam.
Do.	do.	Removes to shop in the Trongate.
28	1764	Marries his cousin Margaret Miller.
29	1765	Death of Craig and dissolution of partnership.
Do.	do.	Conceives idea of the separate condenser and constructs experimental engines.
31	1767	Takes up work as a land surveyor.
Do.	do.	John Roebuck of Carron becomes interested in Watt's invention.
Do.	do.	Visits London in connexion with Monkland Canal Bill and calls at Soho, Birmingham, on his way back.
32	1768	Meets Matthew Boulton at Soho.
33	1769	Takes out patent for separate condenser.
Do.	do.	Engine at Kinneil completed.
34	1770	Meets John Smeaton professionally.
37	1773	Roebuck in financial difficulties.
Do.	do.	Kinneil engine sent to Birmingham.
Do.	do.	Death of his wife.
38	1774	Boulton takes over Roebuck's share in the invention.
Do.	do.	Removes to Birmingham and begins experimental work on engine.
Do.	do.	Introduced to circle known as the Lunar Society.
39	1775	Patent extended by Parliament for 25 years and partnership for like term with Boulton, then 47 years of age, commenced.
40	1776	Pioneer engines erected at Bloomfield Colliery and New Willey Ironworks.
Do.	do.	Revisits Scotland, and marries a second time.
41	1777	Second engine at Soho, the first to work expansively, erected.
Do.	do.	First visit to Cornwall and first engine erected in that county.
Do.	do.	William Murdock, at age of 23, enters service of Boulton and Watt.
43	1779	Murdock sent to Cornwall.
44	1780	Patents letter-copying process.
Do.	do.	First engine on the Continent erected.
Do.	do.	James Pickard patents application of crank to steam engines.

45 1781 Jabez Hornblower patents compound engine.
Do. do. Patents substitutes for crank.
46 1782 Patents expansive-working, double-acting engine and rotative engine.
Do. do. Suggests theory of the composition of water.
Do. do. John Southern, at age of 24, enters service of Boulton and Watt.
47 1783 First rotative engine outside Soho erected.
48 1784 Patents parallel motion, balanced pitwork, steam carriage, &c.
Do. do. First double-acting engine outside Soho erected.
Do. do. Royal Society, Edinburgh, elects him Fellow.
49 1785 Patents smoke-consuming furnace.   Royal Society, London, elects him Fellow.
50 1786 Visits Paris in company with Boulton.
51 1787 Introduces bleaching by chlorine from France.
Do. do. Batavian Society of Experimental Philosophy, Rotterdam, elects him Member.
52 1788 Applies governor to steam engines.
Do. do. Relaxes from business, success of engine being assured.
54 1790 Buys landed estate at Handsworth and builds Heathfield.
55 1791 'Church and King' riots in Birmingham.   Soho threatened.
57 1793 Litigation commenced against infringers of first patent.
58 1794 Firm of Boulton, Watt & Sons established : James Watt, jun., and Matthew Boulton
          brought in as partners.
59 1795 Soho Foundry erected.
61 1797 Studies medical chemistry.
62 1798 Buys landed estate at Doldowlod, Radnorshire.
Do. do. Murdock, aged 44, leaves Cornwall and takes up duties at Soho Foundry.
63 1799 Validity of first patent finally established.
64 1800 Extended term of first patent expires and original partnership with Boulton terminates.
66 1802 Travels on the Continent.
68 1804 Death of Gregory Watt, aged 27.
70 1806 Glasgow University confers upon him degree of Hon. LL.D.
72 1808 French Academy elects him Correspondent.
Do. do. Founds Watt prize at Glasgow University.
73 1809 Death of Boulton, aged 81.
74 1810 Commences construction of machine for copying statuary.
78 1814 French Academy elects him Foreign Associate.
79 1815 Death of John Southern.
83 1819 Dies at Heathfield.

INTRODUCTION

THAT Watt was the inventor of the steam engine is an idea that seems to prevail among people who have no acquaintance with industrial history, but, perhaps it is hardly necessary to say so here, it is a mistaken idea. Before Watt was born Thomas Newcomen had produced an engine which did good service in mine drainage and in pumping water for other purposes, and it was to an engine of the Newcomen construction that the crank was first applied for the production of rotary motion for driving machinery.

But if the cylinder-and-piston engine did not originate with Watt, his inventions were of paramount importance. In the first place they enabled an engine of a given size to do more work and with less fuel than the Newcomen engine. More than this, they opened out a vista of extended application, impossible with the earlier engine, relying as it did upon the pressure of the atmosphere to produce the working stroke of the engine. Although Watt himself, like his predecessor, used low-pressure steam and a vacuum, his construction admitted of the use of steam at any pressure and of its employment without condensation, and although all the engines with which he himself was practically concerned were beam engines with the cylinder in the vertical position, it admitted of the cylinder being placed in position other than vertical.

To-day we see the cylinder-and-piston steam engine being displaced by the steam turbine, by the electric motor, or by the internal-combustion engine. The importance of Watt's influence remains undiminished. The enormous development of manufacturing industry and means of transport in this and other countries since the last quarter of the eighteenth century is bound up with his inventions, and without them these later forms of motors would not have come into existence.

The debt that mechanical engineering owes to Watt is very great. Newcomen, one imagines, devised his engine with particular reference to the abilities of the workmen and the means and materials available in his time. He called for no exceptional skill nor for materials not readily procurable. Watt, on the other hand, conceived his engine without regard to such limitations; he called for greater and greater efforts and a higher degree of accuracy in the workmanship. Indeed the business of mechanical engineering may be said to have been set on foot by him; before the production of his engine was commenced there were no steam-engine factories; there were founders, smiths, and millwrights, but the engine was built and a good many of the parts made on the spot on which it was to stand—just as a house is built.

In the development of the engine there has been a tendency to ascribe to Watt personally all the credit at the expense of his partner Boulton and of the members of the staff employed by Boulton and Watt, notably Southern and Murdock. We say development advisedly, for there can be no question that the fundamental ideas, the separate condenser, air-pump, closed-top cylinder with steam instead of the atmosphere used to press down the piston, the use of steam alternately on opposite sides of the piston, and the expansive working of steam, are to be attributed to Watt.

B

As he was the first to use a closed-top cylinder, it follows that he was the first to use a stuffing-box on a steam-engine cylinder ; he was also the first to rely upon the piston packing to keep the piston tight—in the earlier engine it had been possible to maintain a layer of water upon the piston to prevent the passage of air. The application of drop valves for the admission and exhaust of steam from the cylinder and of the centrifugal principle to the governing of the speed of engines are clearly due to Watt, so is the simple straight-line motion composed of three members, and possibly also the ' parallel motion ' derived from it, although the evidence on this point is not so conclusive. Further, he seems to have been the first to rely on the dead weight of a foundation to keep the cylinder in place.

Such a list as the foregoing is sufficient to explain the high rank that Watt holds as an inventor. He was undoubtedly a man of strong imaginative power—essentially an inventor—he could produce not one but often a number of solutions of any practical problem submitted to him, not merely in connexion with the steam engine but with other mechanical arts and manufactures. It is well known, for instance, how the heavy labour entailed in copying his correspondence induced him to consider whether this could be avoided, and thus led to his invention of press copying.

Watt was not an engineer by training ; he did not come of a family of engineers and did not possess that instinct for the right thing in mechanical engineering that is found, say, in Murdock and Trevithick. The result was that although he could conceive a number of ways of overcoming some difficulty, he could not, in his inventive days, determine which of these ways would answer the purpose best. We have said in his inventive days, for after all the patents of James Watt do not range over a great period of time. He did undoubtedly become a very competent engineer, but it was by the knowledge and experience gained in the course of his work after he had entered upon the partnership with Boulton, who likewise became a good engineer in the same way. It is perhaps not too much to say that it was fortunate that Watt had no engineering training at the commencement of his work on the steam engine and was thus incapable of appreciating the full force of difficulties that might have daunted a more experienced man.

Watt found the engine in the same or substantially the same form as it had been left by Newcomen, who died in 1729—an atmospheric engine with a great lever or beam pivoted centrally and having suspended from it at one end the pump-rods and at the other the piston which worked up and down in a vertical cylinder, and valve mechanism capable of being worked by hand or automatically at will. The piston was driven from the top to the bottom of the cylinder by the pressure of the atmosphere upon its upper surface, a more or less perfect vacuum being maintained below it. For the return stroke steam was admitted to the cylinder below the piston, not to impel it upwards, but merely to balance the pressure of the atmosphere and so to allow the piston to be returned to the top of the cylinder by balance weights that had been raised upon the down-stroke. Upon the completion of the upward movement of the piston the steam was condensed by a stream of water injected into the cylinder, a vacuum or at any rate a large reduction of pressure resulted, whereupon the pressure of the atmosphere forced down the piston as before. The engine was designed for pumping, and in pumping it found its main application ; still it was used for driving mills and machinery by the aid of the crank or other devices for converting reciprocating motion into rotary.

As we have said, Watt was not an engineer, but he possessed the great qualities of the inventor, reliance on experiment, acute reasoning power, and imagination. When he took up the problem of the steam engine, the Newcomen engine, within its limits, was meeting the requirements of practice fairly well, and it had not been realized that it possessed an inherent defect—its inefficiency as a heat engine. Watt, by experiment, ascertained the volume of steam produced by evaporating a given volume of water, and he found, again by experiment on the Glasgow model, that the volume of steam supplied for each stroke of the engine was greatly in excess of the content of the cylinder. He set himself to account for this excess and to find the cause of the disappearance of steam and, having found it, to provide a remedy. In the half-century or so that had gone by since the introduction of the Newcomen engine no one had considered the matter in this systematic manner. The men concerned in the building and working of the engines had accepted, without question, the principle of operation and, when they did anything at all, contented themselves with improving or varying the details of construction.

Henry Beighton (1686–1743) might have done something more ; we find him commenting upon the large amount of water, scalding hot, discharged from the cylinder at each stroke, and no doubt he was fully aware that there was a loss of steam consequent upon its condensation on entering the cylinder at the beginning of a stroke, but he failed altogether to realize the magnitude of the loss, so much so that in giving figures for the relative volumes of steam and water we find him comparing the content of the cylinder directly with the consumption of water in the boiler and entirely ignoring the possibility of loss by condensation. Another indication that the loss was not realized, or that it was deemed negligible, is found in the reason given by Desaguliers for his preference for cylinders of brass over those of cast iron ; it is that they can be heated and cooled quicker and so allow the engine to make a stroke or two more per minute ; he says nothing as to the waste of steam.

When, towards the end of 1763, Watt took in hand the repair of the Glasgow model he was not altogether ignorant of the steam engine. Four years or so earlier Dr. Robison, then a student at Glasgow, had discussed with him the idea of driving road vehicles by steam and some rough models had been made. A year or so later he was experimenting with steam used without condensation. He was acquainted with what had been published by Desaguliers on the subject. However, as he tells us himself, he set about his work on the model as a mere mechanician. It was when he found that the boiler, although apparently quite large enough in proportion to the size of the cylinder, could not be made to supply sufficient steam to keep the engine going for more than a few strokes, that he set about considering the matter. He started from first principles. It was common knowledge that steam, coming into contact with cold bodies, communicated heat to them and was condensed ; also it was known that *in vacuo* water boiled at temperatures below 100°. On the one hand, then, to avoid loss of steam by condensation as it entered the cylinder at the beginning of the stroke, it was necessary that the cylinder should be as hot as the entering steam. On the other hand, the production of a vacuum demanded that the temperature of the cylinder should be kept below 100°. How to reconcile these opposing conditions ? The result of his cogitations was the question whether it was necessary to perform the two operations in the same vessel. Could not the condensation be effected in a chamber distinct from the cylinder but capable of being put in communication with it ? He at once set about to satisfy himself that the steam would pass from the cylinder to the condenser with sufficient rapidity to allow the engine to work at its usual rate. This matter he soon settled by experiment. When he had arrived at the idea of the separate condenser the conception of the new engine was complete in his mind—the use of a pump to remove the air and water from the condenser, and the closed-top cylinder with steam acting upon the upper surface of the piston, but there is a long interval before we find the conception embodied in an engine at work. The delay was due to a variety of causes. Watt was a poor man and not in a position to pay for building a full-sized engine himself. But even had he been so it is likely that in the absence of some driving-force behind him he would have continued indefinitely producing models of one scheme after another. He did make a variety of models, but it is not very clear that they had any particular bearing on the construction of the engine in great ; some of them were for different kinds of surface condenser, a form of apparatus that he used on his first engine and on no other.

However, in 1767 Dr. Roebuck, a man who was interested to a large extent in manufacturing industry, took up Watt and his invention. He supplied the requisite financial support and put pressure on him to go forward to practical results. Progress was slow. Watt had now taken up surveying and civil engineering work, he had a family to maintain, and felt that he must not cut himself off from a source of income, certain even if small, for something that was speculative, or at any rate held no promise of immediate results. However, by the year 1769 an engine had been set up on Roebuck's property at Kinneil and a patent had been granted to Watt. The engine could not be made to work satisfactorily, and at this point progress was impeded by the fact that Roebuck's financial affairs became embarrassed.

The great impulse came when Dr. Roebuck's interest in the invention was transferred to Matthew Boulton of Birmingham. The engine at Kinneil was taken down and re-erected in Boulton's Soho Manufactory, where before the end of the year 1774 it was running satisfactorily ; in June 1775 an extension of the term of the patent was granted by Act of Parliament, and now, ten years after the conception of the invention, Watt entered into partnership with Boulton and had the opportunity of seeing his ideas carried into practice. Possibly the long delay was not altogether a bad thing ; in the interval, indeed it would seem only just before Watt came to Birmingham, John Wilkinson had made his improved cylinder-boring machine, without the aid of which it is difficult to see that Watt could have got his engines to work satisfactorily.

With Boulton behind him Watt soon got to work ; in 1776 two of the new engines were set going, one in Staffordshire and the other in Shropshire, and then in 1777 the connexion with Cornwall began.

Watt, in spite of bad health, now got through an enormous amount of work—calculations, designs, and correspondence—and in the early years he carried on this work entirely without assistance. There were long periods in which he was away in Cornwall supervising the erection of engines, but even there he carried on his office work.

Up to the year 1783, apart from some work in connexion with the expansive working of steam, it may be said that Watt devoted himself to the single-acting pumping-engine. The designs, at first crude to our present ideas, soon show considerable improvement and by the date mentioned may be considered to have assumed a standard form.

Before the demands upon his time and powers in connexion with the single-acting engines required for the Cornish mines had ceased, he had embarked upon the design of the double-acting engine and of the rotative engine, and by ' rotative ' is to be understood a cylinder-and-piston engine provided with mechanism for converting the reciprocating motion of the piston into the rotary motion of a shaft.

In the double-acting engine, apart from the modification of the engine-beam and of the valve-gear, the important new feature was the parallel motion. In the single-acting engines with the strains acting in one direction only, a chain had formed an adequate and simple connexion between the piston-rod and the beam. With double action, i. e. with the strokes of the engine in both directions operative for the production of power, the reversal of the strains precluded the use of a single chain. A double set of chains could have been made use of, but this would have been both heavy and clumsy. The plan actually adopted was that elegant contrivance, the parallel motion, a combination of levers and links which at the same time formed the connexion between the piston-rod and the beam and served to guide the piston-rod in a straight line. The same end is attained to-day by a pin-joint and a straight guideway for the end of the piston-rod. This simpler plan Watt was precluded from adopting, mainly no doubt by reason of the cost of producing the guide surfaces by hand-labour.

In the matter of the rotative engine Watt's hand had been forced : he had been compelled to take up the subject sooner than he had wished to by the activities of Pickard and Wasbrough who were setting up Newcomen engines with rotative mechanism. The converting mechanism now in use, universally, is the connecting-rod and crank. The story of how Watt felt himself debarred from using this combination and resorted to the epicyclic train known as the sun-and-planet gear is a most interesting one. Other features called into being by the rotative engine were the centrifugal governor and the throttle valve, by the aid of which the speed of the engine was automatically kept uniform, or rather the variation of speed was kept within certain limits. Neither of these contrivances was new in itself, but Watt was the first to apply them to the steam engine and the first to use them in combination for any purpose.

Apart from questions of design and construction, the introduction of the rotative engine involved a departure which has given us our great mechanical unit of measurement—the horse-power. With the pumping-engine the performance could be conveniently and sufficiently expressed in terms of the quantity of water raised to a given height. This method was not applicable to engines for driving mills and machinery, so Watt took to expressing what such engines could do in terms of the number of horses, working at the same time, that they would replace. His object was merely to facilitate office work. He did not foresee that he was setting up a unit that was to come into universal use, indeed for many years it was not known to engineers generally that the term horse-power had a definite value expressed in pounds, feet, and minutes.

In the pumping-engines, single- and double-acting, the object was to move a pump-rod or rods up and down, and the engine-work proper finished at the outer end of the beam. With the rotative engines the engine-work extended to the shaft from which the power of the engine was taken off. Usually the concern of Boulton and Watt terminated at the shaft, but in cases in which it was used directly to perform the required work, as for working forge hammers, or drums for winding coal and ores from mines, they went farther than this, and we find them sending out designs accordingly. The material in the Boulton and Watt Collection is specially interesting for the early history of winding-engines.

It has been stated above that when Watt entered the field there were no steam-engine manufactories, but that the engines were built where they were to stand and with materials brought together from different parts of the country. It was on these lines that Boulton and Watt commenced their operations. They did not supply engines, they did make the valves and their containing boxes, but otherwise they acted as consulting engineers, sent out designs, and saw that the engines were put together in a satisfactory manner, and then, as owners of the patent, they drew royalties on the working of the engines. But with the introduction of the rotative engine a change begins, and before the close of Watt's active connexion with the steam engine in the year 1800 they were supplying complete engines, and engine-building works had been set up at the Soho Manufactory and at Soho Foundry.

Although in Cornwall, where the price of coal was high, the Watt engine very soon drove the Newcomen engine out of the field, the older form continued to be built in other parts of the country throughout the period of Watt's patent and after, so that Boulton and Watt were not the only engineers at work. Then too the success of the new engine and the large sums that the firm was known to be drawing for royalties helped to turn the attention of inventors to the improvement of the old engine, or to the scheming of new forms. Some of them honestly tried to strike out on new lines, but such was the wide ambit of Watt's patent that it was impossible to do anything without coming into conflict with it on one point or another. This patent had been granted in 1769; normally it would have expired in 1783. The extension granted by Parliament in 1775 kept it in force until the year 1800. While, having regard to the first-rate importance of the invention, the monetary reward of the patentees was not excessive, it seems pretty clear that the extension was too great and that it hindered the development of the steam engine in this country. Boulton and Watt, from the first, had refused to grant licences to other engineers to work under the patent ; the patent blocked the way of other inventors, and Watt himself had come to the conclusion that there was nothing to be gained by trying new schemes.

As to Watt's personal character, we are left with the impression that as a young man he was shy, modest, and unassuming, and it is certain that he had a circle of friends at Glasgow. Later on in life, no doubt in a great measure owing to ill health, he became less amiable, and his correspondence reveals him sometimes in an unpleasant light. Possibly the Cornishmen should share the blame for this ; they were a difficult set of men to deal with, and Watt was by nature quite unfitted for contest with his fellow men, to direct a body of workers, or indeed to carry on any large undertaking. These deficiencies on his side were made good by his partner ; it is impossible to speak too highly of the part that Boulton played in the partnership and in carrying Watt's ideas into effect. It is pretty certain that, working alone, Watt would not have succeeded in getting his engine adopted in practice, or at least not with any monetary reward. Watt was a student, a philosopher, a man of ideas ; he had no business capacity, and he shrank from public life of any kind. In a small circle of friends, free from considerations of money, of bargaining, and of work, he could be perfectly happy, and we find him after his retirement from business life, mellowed by age and success, an ' alert, kindly benevolent old man '.

JAMES WATT THE MAN

EARLY DAYS, 1736-56

SO much has been written [1] on James Watt that little that is new can be said about him. The purposes of this volume would not be achieved, however, were the salient facts of his life not set down ; this will be done succinctly, as far as possible in Watt's own words or in those of his contemporaries.

Little is known of his ancestors except that they belonged to Aberdeenshire.[2] His great-grandfather held, it is said, a piece of land in that county, and was killed (he probably fought as a Covenanter) in one of those broils of the middle of the seventeenth century which left Aberdeenshire almost ' manless, moneyless, horseless and armless'. The Covenanter left a son, Thomas Watt, who was born between 1639 and 1642. Nothing is definitely known of his early years.[3] He settled in Cartsdyke (called also Cartsburn or Crawfordsdyke), a small burgh (it had forty-four houses with outside stairs and forty-four houses with none) then in Easter Greenock. It was a small seaport with a coasting and fishing trade, and of more importance than Greenock, although the latter had been created a burgh of barony by charter of Charles I in 1635, while Cartsdyke was not raised to that dignity until 1669. The Act of Union of 1707 had a fostering effect on trade. A harbour was built in Greenock in 1706 but Cartsdyke was enterprising too and built in 1716 a vessel of sixty tons burden.[4] She sailed in 1718—the first ship sent from the Clyde to Virginia—and so began the tobacco trade which gave to the Glasgow importers the name of ' tobacco lords '.

Thomas Watt was a ' teacher of the mathematicks' and of navigation in Cartsdyke and is mentioned as early as 1683. He rose to be Bailie of the Barony, an office of some importance, and was also an elder of Greenock parish. He married Margaret Shearer, by whom he had six children, of whom only two sons, (1) John and (2) James, reached maturity.[5]

(1) John settled in Glasgow as a mathematician and surveyor, ' teaching arithmetick, book-keeping, navigation and the other parts of the mathematicks.' He was well liked by merchants and shipmasters, for, in 1720, they asked that an annual salary of £5 might be settled on him, ' he being a person well qualified and deserving encouragement.' [6] He seems to have resembled his nephew James in the

---

[1] His three principal biographers are Williamson, Muirhead, and Smiles (see *Bibliography*).

[2] The surname Watt and its variants, Wat, Watts, and Wattie, are very common there.

[3] A Thomas Watt appears in a list of bejans (freshmen) of 1668 in the Marischal College, Aberdeen, but there is no subsequent entry of his name (Anderson, P. J.: *Records of Marischal College and University*, II. 235). Our Thomas would be 26 or else 29 years of age at the time, but that, such was the tradition of the period,

would merely indicate lack of early opportunity and determination to remedy it. In the mathematics course at the College, navigation was definitely taught and with reason for Aberdeen had been for some centuries the chief port in Scotland.

[4] Marwick (Sir J. D.) : *The River Clyde and the Clyde Burghs*, 1909, p. 82, note.

[5] See Genealogical Table, facing p. 10.

[6] Renwick (Robert) : *Extracts from the Records of the Burgh of Glasgow, 1718-38*, 1909, p. 76.

C

care he gave his work, for, more than once,[1] magistrates and town council suggest that he should be paid an extra fee for the ' great trouble pains and care ' he bestowed on his maps and plans. In 1733 he received fifteen guineas for drawing two plans of Port Glasgow.[2] These plans may have been required in connexion with the works at the quay, which had then been in progress for some years. Before his death in 1737 he drew (1734) a survey of the river Clyde which is of permanent value. It was subsequently edited by his nephew John, and sold, as will be seen, by John's brother—James, the engineer. This map is very rare, only three copies being known to exist.[3]

(2) James was born in Cartsdyke in 1698 and settled in Greenock as a ship-wright, ship-chandler, builder, shipowner, and merchant. He subsequently became a bailie, or junior magistrate of the town, and treasurer, refusing the office of chief magistrate; like his father, he was an elder of the Kirk. His name occurs in 1731 as tenant of a house at the Midquay Head of Greenock. A few years later he bought this tenement and a piece of land, leaving the property, in 1774, to his son James. The house faced the sea, from which it was separated only by the breadth of an ordinary road—afterwards known as the High Street—and, later, as Dalrymple Street. The house was the last but one at the eastern termination of the south side of Dalrymple Street, and the entrance to it was in William Street. The old house (the site of which is marked by a tablet) was afterwards rebuilt as the *Greenock* tavern, and is known to-day as the *James Watt* tavern.

It was to this house that James Watt brought his young wife—Agnes Muirhead (or Muirheid), whom he married in 1729, a woman of great sense, judgement, and intellect. Her talents being confined as was then thought proper to the home she was, as one would expect, a notable housewife, and a true helpmate to her husband. She was descended from the ' stout Laird of Muirhead ' of Sir Walter Scott's ballad,[4] and she has been described by one who knew her as ' a braw, braw woman—none now to be seen like her '. Her brother John was associated with her husband in business, and to other members of her family her son James afterwards owed much.

Of the five children—four sons and one daughter—born to her only the two youngest reached manhood: James, who was born Jan. 19, 1736; and John—who was to die at sea before he was twenty-four[5]—born three years later.

James was christened the 25th of January, as is shown by the entry[6] in the Greenock register; and he might well have said of himself, as did Robert Burns, ' a blast of Janwar win' blew hansel in on' me. Always a delicate child, he was very much with his mother, and received from her his first lessons. His boyhood was lit here and there by some sparkle of excitement as when during the ' forty-five' it was rumoured that Prince Charlie had landed at Greenock and was lying hidden there, and a house to house search for him was made in vain. Jamie was sent to M'Adam's school ' held at the Rue-end ' ; later he went to the Grammar School in Wee-kirk Street, where he learnt Latin and some Greek from the master Robert Arrol, ' a learned and a virtuous person '. But he was thought slow at his lessons, and his schoolfellows—healthy urchins running wild—found him dull and laughed

---

[1] *Ut supra*, pp. 285, 375.
[2] Ibid., p. 388. For his nephew James's work on Port Glasgow harbour, see p. 32.
[3] A facsimile of one, with notes in James Watt's handwriting, is printed in George Williamson:

Memorials of the Lineage . . . of James Watt, 1856.
[4] *Minstrelsy of the Scottish Border*, 1803, I. 283.
[5] He died in 1762, while on a voyage to America in one of his father's ships.
[6] See Fig. 1.

THOMAS WATT = Margaret Shearer
"Professor of Mathematicks" m.1679
of Crawfordsdyke co.Renfrew. 1642-1734   1656-1735

Margaret 1682-1683
Thomas 1684-1687
Cathren 1687
Doritie 1688-1706 unmarried

John = (1) Janet.Todd d.1731
Land Surveyor = (2) Christian Tennant d.1735
1694-1737

JAMES = Agnes Muirhead m.1729
merchant in Greenock 1698-1782   1701-1753

Robert Margaret Thomas John Charles
all died in infancy

JAMES
engineer
of Heathfield co.Staffs.
1736-1819
= (1) Margaret Miller m.1764 d.1773
= (2) Ann Macgregor m.1776-1750-1832

John 1739-1763 drowned at sea no issue
Jean d.1771 = John Cochrane
Elizabeth d.1812 = John Wilson

Margaret d.an infant
Agnes 1770-72
Warwick -8- ed
A son stillborn 1773
Gregory 1777-1804 unmarried
Janet 1779-94 unmarried

Jean d1799
Thomas m. no issue
Margaret
Ann
Agnes
Elizabeth = James Birnie

John no issue
Elizabeth Cochrane
Helen Grierson = George Holden of Rio de Janeiro

JAMES WATT 1831-1891 assumed additional surname of Watt by Royal Licence 1856 = Emma Henrietta Hoey m.1874
James 1830-31

Margaret Elizabeth = Henry Marsh M.D.

Ethel Margaret Watt = Hansard Jockney and had issue

Agnes = James Gibson M.D. Surgeon 13th Light Dragoons d.1835

Agnes =:Chilley Pine 4th Dragoon Guards Killed at Balaclava 1854
b.1827

:(1) Thomas Patterson
=(2) Alexander Blackie
Had issue

JAMES MILLER = Marjorie Adela Ricardo m.1911
of Doldowlod
Major South Wales Borderers
b.1875

Geraldine Emma = Algernon Cankrien Thompson 1903

Evelyn Margaret =(1) Philip Prideaux Budge R.F.A. Killed in France 1918 =(2) John Necby H.L.I. no issue

Dorothy Ellen d.1919 = George Badcock A.S.C. no issue

=Gertrude Shortlands
issue

Ricardo Gerald Ronald

at him, for he was quiet and thoughtful. Even thus early, it is probable that he was afflicted with the long-continuing, violent headaches, with consequent lassitude and depression, which darkened his early and middle life. It was with the study of mathematics that his intelligence leaped into evidence. He was then between thirteen and fourteen, and his teacher was John Marr, 'mathematician in Greenock', or ' English schoolmaster' as he is described in certain records of the town.

When he was not at his lessons—and indeed after he left school—James busied himself in his father's workshops. Many people came in and out, such as boat-swains for their stores, to whom the shipwright would give an extra packet of sail-needles and twine, because ' I once lost a ship for want of such articles on board '.[1] The shop itself was full of marvels for any boy. There were the carpenters' tools and benches, parts of the rigging and fittings of ships in many stages of repair—

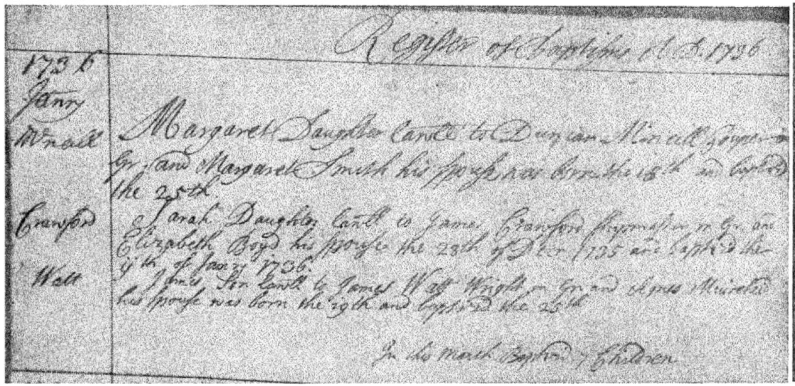

Fig. 1. Entry in Greenock Parish Register of Baptisms, 1736, now in the General Register House, Edinburgh

figure-heads half carved, sails, ropes, blocks, tackle, rudders, compasses (for his father ' touched' compass needles) ; also, perhaps to him most alluring of all, there were quadrants, telescopes, and other optical instruments with which to be-come familiar. All this handling and shaping of intricate forms of wood and metal developed the love of handiwork, the delicacy of touch which remained with him to the end of his long and arduous life. He had his own bench and tools and forge, and made models of whatever pleased him—' Jamie has gotten a fortune at his finger's ends' the workmen would say. These models included such diverse things as a crane (his father made and erected, for the use of ' the Virginia tobacco ships', the first crane seen or used in Greenock), a barrel organ, and a punch-ladle hammered out of a silver coin.

The Watt family were not then living in the house where James was born, but in one partly built by his father—on the north side of what was then the High Street, which was bordered on both sides with stone houses—the back of which looked on the sea, and had a jetty reaching out into deep water. Many of the

---

[1] Williamson (George) : *Memorials of the Lineage, Early Life . . . of James Watt,* 1856, 155.

houses on the river side had ' closes or quays falling a considerable way into the sea, so that vessels of fifty to sixty tons burthen can discharge at those back closes'.[1] On the jetty Jamie, it is said, would stand fishing, or looking for hours over the beautiful Firth of Clyde towards the northern shore. To the mind of one such as he who never read a book or talked to any one ' without gaining information, instruction or amusement', the mere handling of a telescope would lead to its use ; and on the hill behind Greenock church, beyond the elms and beeches of the Old Mansion House, he would watch the stars. His holidays were often spent with his uncle Muirhead at Glasgow ; here he seems to have picked up something of chemistry and anatomy. He once told his son that had he been able to endure the sight of suffering he would have been a surgeon ; and, as will be seen, he became in later life associated with Dr. Thomas Beddoes in an attempt to cure consumption.

Reading everything which came his way, dreaming, thinking, busy with his hands at times and at others seemingly indolent, his boyhood slipped away. It seems odd that he was not apprenticed to a trade ; doubtless at fourteen, the usual age for apprenticeship, he was too delicate. His mother died in 1753. His father seems to have intended James to carry on his own business, but having had many losses (including the wreck of one of his ships) it was decided that James should go to Glasgow with the idea of learning the trade of mathematical instrument-maker. So he packed up his tools and leathern apron, with his few silk stockings, cut velvet waistcoats, and ruffled shirts, and in June 1754 went to his mother's people in Glasgow, where he remained until the following May.

From lonely, untutored meditation, alternating with hours of handiwork, he passed into the orbit of the literary thought and scholarship of Glasgow. It was a cathedral and university city of several thousand inhabitants but it had not yet become an industrial centre, and it is not surprising that Watt found no one who could teach him to make mathematical instruments (for the ' optician' under whom he worked was not much more than a simple mechanic). He came into contact with men who profoundly influenced his after life. Chief of these were his kinsman, George Muirhead ; John Anderson and Robert Dick, all members—as were the brothers Foulis, Joseph Black, and Gilbert Hamilton—of the Literary Society of Glasgow founded in 1752.[2] Professor Muirhead had recently been transferred from the chair of Oriental Languages to that of Humanity, which he held until his death in 1773. He was then, with Professor James Moor, at work on the famous Greek Homer which was afterwards (1756–8) issued by the Foulis brothers, and is the finest monument of their press. Watt must have become familiar with the talk of both scholars and printers, and with the dark irregular buildings of the old College (near to still older St. Mungo), with the carved gateway[3] opening on the High Street. Within the Old College, in a turreted building looking on to the inner quadrangle, might be found Dr. Robert Dick, one of Watt's best friends, who was afterwards described by Joseph Black as one of the most sensible and manly fellows he ever knew. Dick had succeeded his father as Professor of Natural Philosophy in 1751. Natural philosophy was an elastic term, including physics, experimental philosophy, and physical astronomy. In 1754 a number of astronomical instru-

[1] Williamson (George) : *Memorials of the Lineage, Early Life...of James Watt*, 1856, 136.
[2] Duncan (W. J.) : *Notices and Documents illustrative of the Literary History of Glasgow*, 1831, pp. 132–3.

[3] Incorporated in the new buildings at Gilmorehill. The site of the old buildings is occupied by College station of the London & North Eastern Railway.

1A. THOMAS WATT, GRANDFATHER OF THE ENGINEER

From the oil-painting, artist unknown, in the possession of the Watt family

1B. MARGARET SHEARER, GRANDMOTHER OF THE ENGINEER

From the oil-painting, artist unknown, in the possession of the Watt family

ments had been acquired for Dr. Dick's class, and in June that year [1] the University meeting approved of a plan for raising subscriptions to build an observatory in which to house them. The plans for the observatory hung fire, but the instruments remained, and Watt may well have helped Professor Dick to arrange them. Dick noted Watt's enthusiasm, his knowledge and delicate handling of the objects. He saw that Watt could learn nothing of his craft in Glasgow and, being long past apprenticeship age, advised him to go to London, and gave him an introduction to a fellow Scotsman who had settled there. This was James Short,[2] the well-known optician, one of the few men who could have helped Watt at that time.

Watt therefore took leave of his friends, including John Anderson who had become Professor of Oriental Languages, and returned in May 1755 to his father in Greenock to discuss his future. It was decided that he should go to London. John Marr, possibly a son of his tutor in mathematics, and a relative by marriage, was then on the point of journeying to London to join, as naval instructor, the *Hampton Court*, a sixty-four-gun ship refitting in the Thames. On June 7, 1755, therefore, in Marr's company, with his father's blessing and two guineas in his pocket Watt set forth. They travelled on horseback [3] by way of Coldstream and Newcastle, where they joined the great north road, and reached London in twelve days, Watt's chest having been sent by his father to Leith ' to be shypt for London to ye care of Captain William Watson, at the Hermitage, London'. [4]

Arrived in London Watt had necessarily to find some one who could and who would teach him his craft. There were many things against him. To the guilds he was a ' foreigner ' ; he was in his twentieth year yet had never served an apprenticeship, so could not rank as journeyman ; moreover his desire was not to be bound for any length of time but to learn as much in as short a time as possible. One of his first acts, naturally, was to take Dr. Dick's letter to Mr. Short, whose workshop was in Surrey Street, Strand.[5] Meanwhile he did a little work for a watch-maker known to Marr. ' I have not yet got a master', he writes to his father early in July, ' they all make some objection or other. . . .'

Short, however, soon introduced him to John Morgan, mathematical instrument-maker of Finch Lane, Cornhill. ' If it had not been for Mr. Short,' wrote Watt some weeks later,[6] ' I could not have got a man in London that would have undertaken to teach me, as I now find there are not above five or six that could have taught me all I wanted.'

Morgan undertook to give Watt instruction for one year in return for twenty guineas and his pupil's labour. The terms were agreed to, and Watt settled down to work. He wrote now and again to his father. He described his master [6] as being ' of as good a character, both for accuracy in his business and good morals, as any in his way in London. Though he works chiefly in the brass

[1] Coutts (James) : *A History of the University of Glasgow*, 1909, p. 229.

[2] James Short, 1710–68. His most celebrated instrument was a Gregorian of eighteen inches aperture, completed for the King of Spain in 1752, for £1,200. He was a good general scholar as well as a mathematician and optician.

[3] The first stage coach between Glasgow and London was started in 1758, when the Newcastle wagon ran from the *Swan with two Necks*, Lad Lane, London, to the Gallowgate, Glasgow

(*Glasgow Past and Present*, 1884, III. 422–3). Lad Lane has now been swallowed up into Gresham Street. Readers of *Vanity Fair* will remember that Becky Sharp's bandboxes were flung into the gutter of the ' Necks '.

[4] Smiles (S.) : *Lives of Boulton and Watt. Principally from the original Soho MSS.*, 1865, p. 101.

[5] *Gents. Mag.*, 1768, p. 303.

[6] Muirhead (J. P.) : *The Life of James Watt*, 1858, pp. 36–8.

way, yet he can teach me most branches of the business, such as rules, scales, quadrants &c.'

But that year in London left its mark upon Watt. He had never before lived among and worked for strangers. He had never before given all his time to mechanical work, unrefreshed by the least recreation. At night he dropped exhausted on his bed, his hand shaking ' from ten hours hard work '. ' We work to nine o'clock every night,' he writes, ' except Saturdays.' [1] Moreover he was very poor. His father sent him money and he earned a little now and then by rising very early or sitting up late. But after the payment of the twenty guineas premium by his father he was anxious to cost the latter as little as possible, lodging under the roof of his master, so we are led to believe, but not receiving board, he lived on eight shillings a week. He could not live on less, he said, without pinching his belly. But he made steady progress. By the 5th of August he was at work on Hadley quadrants ; by the 23rd he had done a Hadley's quadrant better than his fellow worker, an apprentice of two years' standing. In October he had begun to make rules, ' there being only one man who could make them perfectly well, and he having taken to other work '. In November he was busy with azimuth compasses ; by December he ' could work tolerably well '.

He hardly stirred abroad for fear of press gangs—or worse, kidnappers. Impressed for the navy one might return ; kidnapped for the plantations men disappeared and were never heard of again. In March 1756 he wrote : ' they now press anybody they can get, landsmen as well as seamen, except it be in the liberties of the City, where they are obliged to carry them before my Lord Mayor first, and unless one be either a 'prentice or a creditable tradesman, there is scarce any getting off again. And if I was carried before my Lord Mayor, I durst not avow that I wrought in the City, it being against their laws for any unfreeman to work, even as a journeyman, within the Liberties.' [2]

The confinement, the draughty workshop, the poor food, the lack of congenial intercourse, the daily shifts and frets pressed heavily on his health and spirits. With the spring came a longing for the free winds and blent sweetness of sea and moorland. But he held on ; he must complete his year. In April he wrote proudly : ' I think I shall be able to get my bread anywhere, as I am now able to work as well as most journeymen, though I am not so quick as many.' When June came he was able to ' make a brass sector with a French joint, which is reckoned as nice a piece of framing work as is in the trade'. So July came round again. But the winter's racking cough still distressed him, with ' a gnawing pain in his back ' and ' weariness all over his body ' ; so he made ready to leave London. With money his father sent him and some few poor savings of his own he bought a kit of tools, with some ' absolute necessary' materials for others which he knew he must make himself ; together with a copy of Edward Stone's translation of Bion's [3] *Construction and Use of Mathematical Instruments.* Then he took leave of his master.[4] With the craftsman's pride strong within him, Jamie Watt came home again.

[1] Muirhead (J. P.) : *The Life of James Watt,* 1858, pp. 36–8.
[2] Muirhead (J. P.) : *Mech. Inv.* I. lxxviii.
[3] Bion (Nicholas), 1652–1733 : *Traité de la* construction et des principaux usages des instrumens de mathématique, 1723.
[4] John Morgan died two years later, so never came to know of his pupil's success.

II B. AGNES MUIRHEAD, MOTHER OF THE ENGINEER

From the oil-painting, artist unknown, in the possession of the Watt family

II A. JAMES WATT, FATHER OF THE ENGINEER

From the oil-painting, artist unknown, in the possession of the Watt family

II

GLASGOW DAYS, 1756–64

WATT recuperated in Greenock until October, then went to Glasgow. The city had then but four main streets, leading out from the Cross : High Street, Saltmarket Street, Gallowgate, and Trongate ; there was one little bit of flagged pavement at the Gallowgate known as the Plain Stanes, and here the ' tobacco lords' strutted up and down. Argyll Street was still often called by its old name of Wester Gate ; King Street—made across cornfields thirty years before —was still to the older people ' the new street'.

He went back to the College—eager to see Dr. Dick. There was a stir in the department of Natural Philosophy over the expected arrival of some fine astronomical instruments from Jamaica. The previous January it had been announced that Alexander Macfarlane, a rich merchant who had established an observatory at Port Royal, had bequeathed all the valuable instruments which it contained to the University where he had been educated. They arrived [1] while Watt was in Glasgow. He helped to unpack them, and, as several had ' suffered by the sea air ' Professor Moor and Dr. Dick desired ' him to stay some time in town to clean them and put them in the best order for preserving them. . . .' [2] He was given quarters in the College, near the Natural Philosophy class, a room with two windows looking on to the inner quadrangle, and here he settled down for the next six weeks or so. It was work he loved, and he was never solitary. Professors gathered round the Macfarlane telescopes and quadrants and young Watt with his lofty forehead and deep-set eyes (shadowed already by an indefinable look of suffering) and skilful, delicate hands. Here would come Dr. Dick ; Dr. Moor ; Dr. Adam Smith, afterwards of *Wealth of Nations* fame ; Dr. Simson, Professor of Geometry. Or there might stroll in, whistling softly (he played the flute), a tall, slender, gentle-mannered, irreproachably dressed young man of twenty-eight, with ivory-white face and large, dark, tranquil eyes—the newly appointed Professor of Anatomy and Chemistry (afterwards of Medicine), Joseph Black.[3] Even then he was meditating on his doctrine of ' latent heat ', which was to form the basis of modern thermal science.

Treading on the heels of the professors, or looking over their heads (for he was very tall) there came often to Watt's room an eager-faced boy of seventeen, impetuous of speech, with handsome head. He was anxious to examine the instruments, anxious to make friends with Watt ; he regarded Joseph Black with a touch of hero-worship—his whistling, he wrote long afterwards, ' thrilled me to the heart '.

---

[1] Captain Wylie, of the *Caesar*, was paid by the University £8 for their freight, and £2 as a gratuity for taking care of them on the voyage. Coutts (James) : *A History of the University of Glasgow*, 1909, p. 229.
[2] Minute of a University meeting held Oct. 26, 1756, quoted by Muirhead in his *Life of Watt*, p. 42.

[3] ' The wildest boy respected Black,' wrote Lord Cockburn of Black as he knew him afterwards at Edinburgh. ' No lad could be irreverent towards a man so pale, so gentle, so elegant and so illustrious.' (*Memorials of his Time*, 1856, p. 51.) In 1789 he was chosen as one of the six *Associés étrangers* of the Royal Academy of Sciences in Paris.

This was John Robison, later described as ' one of the greatest mathematical philosophers of his age'. Of all those who came to Watt's workroom he had the most chequered career. A graduate at seventeen,[1] tutor and midshipman at twenty ; with Wolfe [2] when he rowed down to the Anse du Foulon in the darkness of that September night before the capture of Quebec ; he was in the ship which brought the hero's body home ; a few years later he was sitting in his house at Cronstadt, drinking tea and amicably chatting in their own tongue with Russian professors ; then back as lecturer at a Scottish University again. Never once did his deep and loyal affection for Black or Watt fail. A notable threefold friendship was born in 1756 in the little college workroom—between Joseph Black, James Watt, John Robison. Who dares measure now the relative depth of their attachment ? Looking on the surface only can one say that Black gave, Robison gave—both, with open hands, whole-heartedly—and Watt received.

In December Watt was paid £5 for his work on Macfarlane's instruments, and he then seems to have returned to his father at Greenock. A *Waste Book* [3] dated ' Greenock, January 3 1757 ' is headed : *An Inventory of the Goods, Money, Debts, &c. Belonging to me James Watt Junr. also what I owe to others*. His total capital is returned at £21 4s. 2d. This suggests that he started in business in Greenock, but there are practically no transactions recorded. Perhaps he concluded that he must turn his eyes elsewhere and we find this item on August 2nd : ' Expenses removing to Glasgow 7s. 8d. ' He may have gone to his relatives to give him time to look round for a shop or he may have returned to his quarters within the University, which was to be his home for the next six years. His first friend, Dr. Dick, had died in May 1757. Dick was succeeded as Professor of Natural Philosophy by John Anderson,[4] who is known to have given Watt the run of his library and to have employed him in repairing the instruments (astronomical and other) used in connexion with the department of Natural Philosophy. Watt also helped Dr. Black with his experiments. ' I ', wrote Black afterwards, ' had many opportunities to know that he was as remarkable for the goodness of his heart, and the candour and simplicity of his mind, as for the acuteness of his genius and understanding. I therefore contracted with him an intimate friendship, which has continued and increased ever since that time.' [5] The tribute honours the writer as much as the man of whom it was written.

With such friends among the professors it is not strange to find that Watt was

---

[1] He entered Glasgow University in 1750 at the age of 11, matriculating and graduating (as M.A. in April, 1756) under the name of John Robertson, his true name ; but he always afterwards called himself Robison.

[2] Robison is responsible for the well-known anecdote of Wolfe and Gray's *Elegy*. See Robert Wright : *The Life of Major-General James Wolfe*, 1864, p. 577 and note, where Robison's name is given as Robison. He and Wolfe may have met in Glasgow, since Wolfe was attending classes at the University in 1753 or 1754.

[3] Muirhead Papers, in the Birmingham Reference Library.

[4] He is but little heard of in connexion with Watt. He held his post for over thirty years, his career being stormy and combative. He

was one of the first professors to teach French in the University. Inventing a cannon which was refused by the English Government, he went to Paris in 1791 and offered it to the National Convention, who placed a model of it in their hall. He was a capable teacher and allowed artisans to attend his lectures in their working clothes. He endowed Anderson's College (David Livingstone was one of its students) which, since 1886, has been merged in what is now the Royal Technical College, Glasgow.

[5] From a MS. in Dr. Black's handwriting written in 1796 in connexion with the actions against Bull, Hornblower, &c., for infringement of Watt's patent. Printed in Muirhead : *Life of Watt*, pp. 58-9.

III. THE COLLEGE OF GLASGOW IN 1093

From the engraving in Slezer's Theatrum Scotiæ

allowed to open a shop in the University itself [1]—we may fix the date from a heading in the *Waste Book* already alluded to : ' College Glasgow, Dec. 6th 1757 '. The shop was on the ground floor of an old house—' with a sort of arcade in front, supported on pillars ' [2] —forming part of the College buildings ; and it opened on to the High Street. Moreover he was privileged to call himself ' mathematical instrument maker to the University'. Even after he rented a shop in the town James Watt retained his shop in the College.[3]

There were, at that time, two reasons which made him glad to be allowed to establish himself within the precincts of the University, which was a little world in itself : his poverty ; [4] and the fact that he was not a burgess of the Burgh, and had therefore no legal right to set up as a craftsman within the Regality. The only contemporary information that we possess on this latter point is a remark in Dr. Black's MS. referred to above : ' Mr. Watt came to settle in Glasgow as a maker of mathematical instruments ; but being molested by some of the corporations, who considered him as an intruder on their privileges, the University protected him by giving him a shop within their precincts, and by conferring on him the title of Mathematical-instrument-maker to the University.' The ' corporations' were the fourteen incorporated trades ; ' intruder' was the word popularly used to describe a person who attempted to start a business without acquiring the legal qualifications of burgess-ship. That there was molestation is clear, but of what it consisted there is nothing to show ; yet Black's few words have been added to—here a little, and there a little—by successive biographers of Watt until a load of blame has been laid on the shoulders of the burgesses and trades—especially the Incorporation of Hammermen—of Glasgow, who have been accused of persecuting Watt. An intruder could be dealt with by the Dean of Guild, or by the Provost and Magistrates on the information of a burgess or of a guild, but there is no trace in the records of these bodies, of the Hammermen, or of any other Incorporation, of proceedings against Watt.

Mr. Harry Lumsden, in a brilliant article in *The Glasgow Herald* [5] of Dec. 26, 1911, traces the ' persecution' legend in detail through twenty books between 1829 and 1911, and declares the accusation to be without foundation, in fact we have here a literary ' snowball' of no mean dimensions. In any case, two years later Watt did actually set up a shop in the city—in Saltmarket Street.

Meanwhile in his College shop Watt was not only showing instruments for sale, but making and mending flutes, guitars, harps, and barrel organs. He made a small

---

[1] There were precedents for such a course. The printers to the University had been allowed to set up workshops in the College : Donald Govan, printer of the first Glasgow news-sheet— *The Glasgow Courant* in 1715 ; the brothers Foulis ; Alexander Wilson, the typefounder ; and others.

[2] Smiles : *Lives of Boulton and Watt*, p. 107, quoting Professor Fleming.

[3] Coutts (James) : *A History of the University of Glasgow*, 1909, p. 266.

[4] More than fifty years later Watt, on a visit to Glasgow, was asked if a certain shop, in the window of which the questioner remembered seeing optical and other instruments for sale, was his. Watt smiled and said ' Na, na, lad ;

it was not mine. I was not so rich as to have a shop of that kind.' (*Trans. of Glasgow Arch. Soc.* 1859, Ser. I, I. 3.)

[5] Reprinted in Lumsden and Aitken : *History of the Hammermen of Glasgow*, 1912. Appendix XII, pp. 394-404. Mr. Lumsden assures us that the molestation must have been ' confined probably to an occasional gentle reminder that he [Watt] had failed to take out a burgess ticket ' and further, that ' mathematical instrument making was never a pendicle of the hammermen trade'. Although to which of the fourteen incorporated trades a mathematical instrument-maker—and Joseph Moxon refers to such a craft in his *Mechanick Dyalling*, 1668—would belong, if not to the hammermen, it is difficult to say.

wind-organ [1] for Dr. Black, and also an ingeniously contrived machine for drawing in perspective. He did not care for music, neither did he know one musical note from another, but he studied at some length the mathematical theory of music, and to this knowledge was added his wonderful mechanical skill.

After about a year he wrote to his father (Sept. 15, 1758) : ' Unless it be the Hadley's instruments there is little to be got by it, as at most jobs I am obliged to do the most of them myself ; and as it is impossible for one person to be expert at everything, they often cost me more than they should do.'

If he was not then making money he had many friends. Writing of those early days Robison describes Watt gaily as ' a philosopher as young as myself and always ready to instruct me. I had the vanity to think myself a proficient in my favourite study and was rather mortified at finding Mr. Watt so much my superior. . . . I lounged much about him, and, I doubt not, was frequently teasing him.' [2]

The two philosophers, of twenty-three and twenty respectively, discussed many subjects, including steam engines which, suggested the younger, might be applied ' to the moving of wheel-carriages and to other purposes'. But Robison was not able then to pursue the subject of steam carriages for he went off to London, and in February 1759 sailed for Quebec, in charge of Admiral Knowles's midshipman son, in H.M.S. *Neptune*.[3]

In August 1759, Watt also went to London,[4] no doubt to get orders for his mathematical instruments. Back again in Glasgow we find him advertising for sale his uncle's map of the Clyde, which had been completed by his (Watt's) father and brother :

' Just published and to be sold by James Watt, at his shop in the College of Glasgow, price 2s. 6d.

' A large sheet map of the River Clyde, from Glasgow to Portincross, from an actual survey.

' To which is added a Draught of part of the North Channel with the Frith of Clyde according to the best authorities.' [5]

An entry in the *Waste Book* (already referred to) for November 4th shows that ten copies were sold on that day.

The same year he entered into partnership with John Craig (from whom he had already borrowed money, as is shown by entries in the *Waste Book*) and they set up a shop in Saltmarket Street, nearly opposite St. Andrew's Street. Among the Doldowlod Papers is a MS. entitled : *An Inventory of Tools, Goods, &c. belonging to us James Watt and John Craig each one half. Taken Oct. 7th. at Glasgow.* Of the tools the most costly is a ' *great turning lathe* ' £3 ; the total value of the tools being £22 17s. 4d. The stock includes Hadley quadrants, Davis hard wood quadrants, Gunter scales, compasses, pencils, microscopes, temple frames, burning glasses, magic lantern glasses, &c. The stock is valued at £69 1s. 11½d. ' Cash in hand £108 ' brings the total up to £200.

There is also a *Day Book* [6] in which are entered the ready-money sales from Oct. 20, 1759, until April 1765, in which year Craig died. The goods sold on the first

[1] One of Watt's organs is still in existence. See Thomson (Rev. James, M.A.) : *History of St. Andrew's Parish Church, Glasgow*, 1905, p. 25.
[2] Muirhead : *Life of Watt*, p. 61.
[3] During the voyage Edward Knowles was promoted to the rank of lieutenant on board the *Royal William*. Robison went with him, and was rated as midshipman.

[4] An entry in his *Waste Book* runs :
Expenses of journey to London £4 6s. 3d.
Expenses there   .   .   . £5 6s. 10½d.
Do. returning and keeping the
    mare there   .   .   . £4 18s. 6½d.
[5] *Glasgow Courant*, Oct. 22, 1759.
[6] Doldowlod Papers.

day were : a fish-skin case, 2s. 6d. ; a case of instruments, 6s. ; a pair of bow compasses, 4s.—not such a bad start.

No Hadley quadrant was sold until Jan. 1, 1760, when two were sold at 27s. each (the price was raised later). Repairs to musical instruments were also undertaken ; in the following April occur the entries : ' mending a flute 1s. 6d.; mending a bagpipe 4s. ; mending a flute 10d.'

Between May and September are entries showing that Watt borrowed £60 from Craig. On October 4th are entries of wages paid to journeymen for piece and day work. There is also the entry : ' paid James Watt's first year's salary £35'. The same entry occurs in the two following years.

Meanwhile, in his College workshop Watt was working for the University ; a sum of five guineas for repairs to instruments was ordered to be paid him June 26, 1760.[1]

Looking on himself as definitely settled in Glasgow, he made an offer in 1761 to lease a garden from the University for nineteen years ; ' but the only minute on the subject ends with the tantalizing statement that the proposal was to be considered at the next meeting.' [2] Another subject—possibly connected with the above—on which the lack of information is even more tantalizing is his courtship at that time of Margaret Miller. That she lived in Glasgow ; that she was his cousin ; that he had for her an ' early and constant attachment ' [3] is all that we are told. He must have been very fully occupied in 1761, for although in that year Dr. Black made public his discovery of the doctrine of ' latent heat ',[4] Watt does not seem to have known of this until later. He took up the sale of toys, i. e. the steel ornaments for which Birmingham was so well known ; the removal to another shop being announced as follows :

' James Watt has removed his shop from the Sautmercat to Mr. Buchanan's land on the Trongate, where he sells all sorts of mathematical and musical instruments, with variety of Toys and other goods.' [5]

By that time he was employing several journeymen and taking apprentices.[6]

On June 10, 1763, Watt was made a master of the Lodge of Free and Accepted Masons of Glasgow.[7]

In 1763 he left his rooms (but not, apparently, the shop) at the College—where he had made another friend in the person of George Jardine [8]—and seems to have gone to Delftfield, where was the pottery in which he had become interested. The Delftfield Pottery Company was established in 1748 on grounds which lay to the west of the city. In those days Jamaica Street was but a by-street, only partly built upon ; at its head was a rope walk.[9] On the south of Anderston Walk, reaching down to the Clyde, lay the lands [10] belonging to the Delftfield Company, en-

[1] Coutts (James) : *The History of the University of Glasgow*, 1909, p. 265.
[2] Coutts (James) : Ibid., p. 266.
[3] Muirhead : *Life of Watt*, p. 50.
[4] In 1756 Black began to meditate on the slowness with which ice melts and water is dissipated in boiling. He ascertained the cause in 1761 and from that time onwards taught the doctrine of latent heat in his lectures.
[5] *The Glasgow Journal*, Dec. 1, 1763.
[6] B. & W. Colln. The indenture, dated Jan 11, 1762, of one of his apprentices, James Dunshie,

' to the art and business of instrument making ', is preserved.
[7] Muirhead Papers.
[8] George Jardine, 1742–1827. Licentiate of the Church of Scotland ; afterwards Professor of Logic in the University of Glasgow from 1774 to 1827. In 1792 and later the two handsomest and most brilliant boys in his class were the friends Thomas Campbell, the poet, and Gregory Watt, James Watt's younger son.
[9] *Glasgow Past and Present*, II. 192.
[10] McArthur : *Plan of the City of Glasgow*, 1778.

III

THE WIDENING CIRCLE, 1764-74

AMONG the apparatus belonging to the Natural Philosophy Class of the College of Glasgow was a model of the 'fire engine' of those days, the invention of Newcomen[1] and used for pumping water from mines or for waterworks. Briefly the arrangement was such that steam was admitted to a cylinder below a piston and there condensed. Consequent upon the vacuum thus formed the piston was forced down by the pressure of the atmosphere and, by a flexible connexion to a lever or beam, made to work a pump. Moreover, the machine was fitted with automatic valve-gear which enabled it to repeat its operations indefinitely. Such was the piece of apparatus submitted by Prof. Anderson to Watt for repair,[2] curiously enough, one hundred years after Newcomen's birth. Prof. Anderson had been authorized on June 25, 1760, by the University to 'lay out a sum not exceeding two pounds sterling to recover the steam engine from Mr. Sisson[3] instrument maker at London'. The model had 'been in such a situation that it did not answer the end for which it was made'. Whatever were its defects, Watt overcame them,[4] and he could not have done so without having made himself thoroughly conversant with the principles of its construction. While experimenting he was struck by the enormous consumption of steam, due obviously to the fact that at every stroke the cylinder and piston had to be heated to the temperature of boiling water and cooled again; this prevented the model from making, with the available boiler capacity, more than a few strokes. He saw readily that this waste of steam was inherent in the construction of the engine and his mind became obsessed with the idea of finding some remedy.

The determination of such constants as the volume of steam at atmospheric pressure that is produced from a given quantity of water, the heat that is required to evaporate it, and the quantity of cold water that is required to condense it, called out and exercised his experimental skill because few data were then available. It was at this juncture that Watt learnt from Dr. Black of the latter's discovery—to which he was led by Jean de Luc's experiments on the atmosphere[5]—of 'latent heat'. Black pointed out that bodies on changing their physical state—e. g. ice turning into water at freezing-point, or water at boiling-point turning into steam—give out or absorb a certain amount of heat without its becoming sensible; this was the heat he called 'latent'. Watt had noticed this phenomenon in the production of steam, and he came to realize that the loss of latent heat was the most serious defect in the Newcomen engine.

[1] Thomas Newcomen (1663–1729). His life and work on the steam engine are dealt with fully in *Trans. Newcomen Soc.* IV. 113.
[2] For his work on the model he was paid £5 11s. in June 1766.
[3] Sisson was a well-known man; his work is praised by Delambre in his *Histoire de l'Astronomie au dix-huitième siècle*, 1827, p. 387.
[4] The model is still preserved at Glasgow, but it is not possible to say what were the alterations made to it by Watt.
[5] See p. 39.

IV. NEWCOMEN MODEL REPAIRED BY WATT

In the possession of Glasgow University

He carried out many experiments (which will be referred to in another chapter) in the cellar-like shop of the house which jutted out into a little court at the north end of the beef market.[1] Dr. Black was always ready to help him with money and with counsel ; and John Robison had returned from his seafaring life [2] all strength and enthusiasm. Step by step he followed Watt in all the latter's trials, ransacking for him the University and other libraries for every book on mechanics that had any bearing on the subject of 'fire' engines; but to Watt alone—of whom Robison once said ' everything became science in his hands'—came the sudden idea of the separate condenser—the result not of inspiration but of months of torturing thought—in May 1765. Long afterwards [3] he told the story to Robert Hart, an engineer in Glasgow :

' It was *in the Green of Glasgow*.[4] I had gone to take a walk on a fine Sabbath afternoon. I had entered the Green by the gate at the foot of Charlotte Street—had passed the old washing house. I was thinking upon the engine . . . and gone as far as the Herd's House, when *the idea came into my mind that as steam was an elastic body it would rush into a vacuum, and if a communication was made between the cylinder and an exhausted vessel, it would rush into it and might be there condensed without cooling the cylinder.* I then saw that I must get quit of the condensed steam and injection water, if I used a jet as in Newcomon's engine. Two ways of doing this occurred to me. First the water might be run off by a descending pipe, if an offlet could be got at the depth of 35 or 36 feet, and any air might be extracted by a small pump ; the second was to make the pump large enough to extract both water and air. . . . *I had not walked further than the Golf-house when the whole thing was arranged in my mind.*'

Thus was thought out the separate condenser—the greatest single improvement ever made in the steam engine. But it is a far cry from theory to practice, as Watt soon found. We must remember that he knew nothing, practically, about atmospheric engines and their construction ; moreover, the few people to whom he felt he could confide a knowledge of his invention were not more competent.

At first Watt lived, like every inventor, in the region of dreams, among engines perfect in design and flawless as his thought. ' I can think of nothing else but this machine,' he wrote, April 29, 1765,[5] to Dr. Lind.[6] The craftsman and scientist in him triumphed, and he was strong in his pride. Robison relates how he went to see Watt after a short absence and found him in his parlour with ' a little tin cistern ' on his knee. Robison was eager to tell Watt some thought of his—' something about steam', when Watt looked up and said briskly : ' You need not *fash* yourself any more about that, man ; I have now made an engine that shall not waste a particle of steam. . . .' ' I put a question to him about the nature of his contrivance,' continues Robison, humbly ; ' he answered me rather drily.'

---

[1] Now Miller's Place.

[2] An account of Robison's seafaring days and adventures will be found in Playfair's edition of Robison's *Works*, 1822, Vol. IV, pp. 123-37.

[3] In 1813 or 1814. The story was related by Mr. Hart to the members of the Glasgow Archaeological Society (*Transactions*, 1859, Ser. I, I. 1). The Editors have found no contemporary record of this event among Watt's papers.

[4] The large open space which lies on the north bank of the Clyde.

[5] Unless otherwise stated, the letters from which phrases are quoted will be found in Muirhead, *Mech. Inv.*, under the date cited ; if the letter should be a long one, the page number will be added.

[6] James Lind (1736-1812), physician and astronomer, one of Watt's earliest friends. He sailed in 1765 as surgeon in an East Indiaman to Madras and China but returned in 1768 in ill health. He went to Iceland with Joseph Banks in 1772 ; later was physician to the royal household, and settled at Windsor. Miss Burney, in her *Diary*, refers to him several times. He was good to Shelley, then at Eton, and is the old hermit in Shelley's *Laon and Cythna*, and Zonoras in the fragment *Prince Athanase*.

That touch of arrogance remained with Watt—however deep his dejection, however broken his strength—to the end. But failure followed on elation when he started out to make a working model. For one thing, he who was used to working with delicate apparatus had now to trust to parts made by indifferent workmen. Watt was no easy person to work for ; he was continually changing his designs in details, aiming always at perfection ; moreover, the master mechanic was not a master of men.

Among those whom Dr. Black's brilliant lectures on chemistry had attracted to Glasgow [1] was John Roebuck, who had been practising as a doctor in Birmingham, and was then a man of forty-seven. He gave all his spare time to chemistry, being helped in his experiments by Samuel Garbett, a merchant of Birmingham. Dr. Roebuck, after introducing a new process for the manufacture of sulphuric acid, had recently turned his attention to the manufacture of iron in Scotland, and had formed a company—in which he was the chief figure—composed of himself, his brothers, Garbett, and the firm of Cadell & Sons. Roebuck chose for the site of the new works a spot on the river Carron, some nine miles south-east of Stirling, not far from where the Carron flows into the Firth of Forth. For water-power there was the Carron, for water-way the Forth, while all about them were ironstone, limestone, and coal. Much of the machinery was contrived by Smeaton, and the first furnace was blown in at Carron Jan. 1, 1760.[2]

All this time Roebuck was in communication with Dr. Black, and often afterwards acknowledged how much he owed to him. When the Carron works soared into prosperity and Roebuck was fast making money, in order to procure more coal he leased from the Duke of Hamilton large coal-mines and salt-works at Borrowstounness (Bo'ness) ; and, about 1764, went to live at Kinneil House, which went with the lease. He then set to work to sink for coal and open up new seams. But his pits became flooded, and the atmospheric engine he used was not powerful enough to unwater them. It was then that Black spoke to him of Watt and Watt's engine.

Black had been lending Watt the money necessary for him to carry out his experiments ; but larger sums were needed.   Besides Watt required the best workmanship for the parts of his engine, and influence to put it, when completed and tested, on the market. Roebuck, a warm-hearted (if warm-tempered), generous man, had money, sympathy with enterprise, was enough of an engineer to appreciate the possibilities of Watt's discovery, and very badly needed powerful pumping plant. From the summer of 1765 a correspondence and friendship began between them. But the aims of the two differed. The elder man wanted an engine as soon as possible—any kind of engine that would cope with the water—the younger thought only of perfecting his invention.

' Press it forward with all speed', writes Roebuck, in a pleasant, friendly letter from Bo'ness, Sept. 13, 1765.

' I must remedy everything before I send it away', replies Watt from Glasgow, October 11. On the 16th : ' I still propose improvements on my piston, with which

---

[1] The group of scientific men then teaching at the universities of Edinburgh and Glasgow were famed throughout Europe, whose every learned society was watching their discoveries.

[2] B. Faujas de Saint-Fond in his *Voyage en Angleterre* . . ., 1797, gives a vivid account of the Carron Ironworks as he saw them in 1784 :

the place covered with cannon, mortars, bombs, carronades (of which the first was cast in 1779), siege-guns, field-pieces ; sugar-boilers for the West Indies—in short, they manufactured everything in iron, down to cast-iron hinges and bolts for doors.

I am confident it will succeed to my utmost expectations.' Again, on the 18th he has ' another contrivance that I think will do better '. In November came bitter disappointment over the workmanship at Carron—a cylinder bored there was useless, ' though the best Carron could make '. Doubtless this was what is referred to in the ledger of the Carron Company under date of Sept. 19, 1765 : [1]

' To 1 pipe bored—2 cwt. 2 qr. 21 lb. at 30s. . . . . £3 16s. 7d.'

In December he has ' thought on something new about the pump of the condenser, & also for the pump in the pit. Thinking on these things is a kind of relief amid my vexations.'

In April 1766 Watt was at Carron on matters connected with the engine ; leaving Glasgow unwillingly and glad to return to it.

In the spring of 1766 Matthew Boulton in Birmingham,[2] Dr. Erasmus Darwin [3] in Lichfield, and Benjamin Franklin in London, were thinking, writing of, and experimenting with, fire-engines. Boulton was discussing his model with Franklin ; Darwin was working on ' a fiery chariot'—a locomotive steam engine—and sending his diagrams to Boulton with the suggestion that they should become partners and take out a joint patent for it, while Dr. William Small,[4] lately come to Birmingham, was becoming the close friend and confidant of them all.

Darwin's projected partnership with Boulton—like his fiery chariot—came to nothing. Roebuck, just before he started the Bo'ness business, also asked Boulton to join him as partner ; but Matthew Boulton, dissatisfied with his cramped premises at Snow Hill, had his hands full with the erection of his famous manufactory in the parish of Handsworth, co. Stafford, about two miles north of Birmingham, on the Wolverhampton road. The site, lately unenclosed warren, he named Soho. The buildings, opened in 1765, consisted of ' four squares with shops, ware-

[1] The following entries under the like heading ' Goods invoiced to James Watt, engineer, in Glasgow ' are also of interest. The Editors are indebted for them to the courtesy of Mr. George Pate, manager for the Carron Company :

			£	s.	d.
1766	21 Mar.	To 1 cylinder . .	0	17	4
	19 May	1 Plate . .	3	7	10
	16 June	1 bored Cylinder.	3	6	8
1767	27 Jan.	Wrot Iron Frame, rolls, &c. .	66	0	3
1770	30 July	Ballast, 214 pieces	175	9	9
	10 Aug.	Ballast, 6 pieces .	5	3	10
	11 Dec.	Sundries · Working Barrels, plug door pipe &c., steam chest and cover, with a branched pipe, &c	83	4	5

With regard to the goods invoiced under date of Jan. 27, 1767, the firm wrote : ' The great loss of time attending to these things has given us much vexation. It was unfortunate we undertook them as smith work is out of our way and we believe we are losers by the transaction. The Brasses we were obliged to leave to you to cast,

that no further time might be lost.'

[2] Matthew Boulton (1728–1809) is the most conspicuous figure next to Watt in the present story. A Life of him still remains to be written. It need only be said that on the death of his father, a silver stamper and piercer, in 1759, Matthew inherited his fortune. He married an heiress, Anne Robinson of Lichfield, in spite of opposition from her friends.

[3] Erasmus Darwin (1731–1802), grandfather of Charles Robert Darwin, was a successful physician, the friend of Rousseau, a scientist, something of a freethinker and radical, a generous friend, and author of the *Botanic Garden*, with its well known verses on steam. The work had, in its day, a remarkable success, and, bad poetry as it is, shows a powerful mind.

[4] William Small (1734–75) was a native of Carmylie, county of Angus. He was Professor of Mathematics and Natural Philosophy in the College of Williamsburg, Virginia, but the climate did not suit him. Returning in 1765, he was introduced to Boulton by a letter from Benjamin Franklin describing Small as ' an ingenious philosopher and a most worthy, honest man ', and was thus led to settle in Birmingham. Muirhead : *Life of Watt*, 242.

houses, etc., for a thousand workmen'.[1] Boulton took as partner John Fothergill, an active business man, and sank his wife's fortune and his own in the buildings and equipment. Singularly strong, serene, generous, and far-sighted, he was anxious to wash away from Birmingham wares the 'Brummagem' taint, and to attain in his works perfect workmanship allied with highest artistry. His was the same spirit that animated his friend Josiah Wedgwood—a potter from the age of nine—whom he helped with advice on business organization when the latter built Etruria (on a spot two miles from Burslem, near to the course of the projected Grand Junction Canal), opened in 1769.

Into this circle Watt had not, as yet, come. In 1766 he lost his best friend from Glasgow when Dr. Black was appointed Professor of Medicine and Chemistry in the University of Edinburgh. Black was on his own recommendation succeeded at Glasgow by his pupil and admirer—and Watt's staunchest friend—John Robison. It was in the summer of this year, it seems, that Watt gave up his shop, took an office in King Street, a little south of Princes Street, and set up as a land surveyor and engineer. We do not know what he did with the instrument-making business ; we have no record of its sale. This step may have been taken under the influence of Dr. Roebuck, for as early as October 1765 Watt made a ' calculation of experiments ' ' on one of Dr. R's common engines . . . to show the great waste of fuel and steam '.

On Dec. 6, 1766, we find that Robert Mackell[2] and James Watt propose to erect for the Carron Company at Carron ' a fire engine of a new construction (according to a plan given in), with a 6 feet cylinder and 2 boilers of 20 feet diam. each with a pump of 52 inches diam. to raise water to their works '.[3] From another letter,[2] dated May 15, 1769, we learn that the above-mentioned Mackell had erected a fire-engine for Lord Kennet. The writer of the letter complains of defects in the engine but does not blame Watt personally ' because I know you trusted McKell with that part of the work'. Watt himself states that at this time he needed ' experience in the construction of large machines', and was ' concerned in making some very indifferent common engines '.

By the end of 1766 he was engaged, with Robert Mackell, in a survey for the Carron Company of a projected canal between the Clyde and the Forth. The report of the two surveyors appeared in 1767.[4] The project was abandoned for a time, owing to the objection of the landlords, but was later carried out, not by the Loch Lomond passage as favoured by Watt and Mackell, but by the direct route. In March 1767 Watt went to London in connexion with the promotion of the necessary Bill. There he made the acquaintance of a Parliamentary Committee but failed to convince them of the desirability of the project. It was perhaps this that caused him to write to his wife (April 5) : ' I think I shall not long to have anything to do with the House of Commons again—I never saw so many wrong-headed people on all sides gathered together. . . . I believe *the Deevil* has possession of them.'

Watt took the opportunity on his outward journey to London to call at Birming-

[1] Swinney's *Directory of Birmingham,* 1774, quoted in Langford (J.A.) : *A Century of Birmingham Life . . . 1741–1841,* 1868, II. 47. Cf. also p. 29.
[2] A surveyor of Falkirk, associated with Watt and Andrew Meikle.
[3] Doldowlod Papers.
[4] *An Account of the Navigable Canal, Proposed to be cut from the River Clyde to the River Carron,* as surveyed by Robert Mackell and James Watt, 1767. The alternative tracks are described and discussed.

ham, bearing, no doubt, introductions from Roebuck. His Journal[1] has the following entry:

> ' 1767 Mar. 16. Called at Birmingham, found Mr. Garbett gone, proceeded to Oxford that night.'

Next day he was in London. Obviously Watt could not have found time to visit Soho, but he might have seen Dr. Small. Very probably he called again on his way back, for he was certainly at Lichfield, in order to call on the great-hearted—if somewhat sharp-tongued and unwieldy—genius Erasmus Darwin, who had turned from his ' fiery chariot' to some other experiment—perhaps the windmill to grind flints, which he was designing for his friend and patient Wedgwood. To Darwin, under a strict promise of secrecy, Watt revealed his invention of the separate condenser.

Some months later Watt received from Dr. Darwin the following letter, dated Lichfield, Aug. 18, 1767 :

> ' Now, my dear new friend, I first hope you are well and less hypochondriacal, and that Mrs. Watt and your child are well. The plan of your steam improvements I have religiously kept secret, but begin myself to see some difficulties in your execution which did not strike me when you were here. I have got another and another new hobby-horse since I saw you. I wish the Lord would send you to pass a week with me, and Mrs. Watt along with you ;—a week, a month, a year ! You promised to send me an instrument to draw landscapes with. If you ever move your place of residence for any long time from Glasgow, pray acquaint me. Adieu, Your friend, E. Darwin.'

Watt also took occasion on this journey to visit the Calder and the Bridgewater Canals, which shows that he was fitting himself for the increasingly important civil engineering work that was coming his way. Had his steam engine never been heard of, there is every indication that Watt would have reached the front rank as a civil engineer, just as Smeaton did.

The winter time permitted Watt to engross himself once more with the steam-engine experiments. He wrote quite gaily to Dr. Lind from Glasgow, Jan. 5, 1768 :

> ' *Nil mihi rescribas, attamen ipse veni.* . . . I am going to be at home, God willing, for some time. I am going to try some things I am persuaded you would like to see (perpetual mobiles, the elixir magicum, and some other trifles of that kind). Seriously, it would give me great pleasure if you could spend a few weeks with me ; I think I could entertain you. What I know about the steam-engine before you went away [i.e. in December 1765] was but a trifle to what I know now.'

By the spring of 1768 Watt had made a model, or rather a small engine, with which Roebuck was so far impressed that he agreed to become a partner with Watt in his invention to the extent of two-thirds. In return he took on himself the debt Watt had incurred to Black and Craig—over £1,000 ; agreed to pay the expenses of the patent which was to be applied for ; while Watt was to attend and conduct the experiments. A period of strenuous work followed, the details of which Watt communicated to Roebuck by letter.

' I intend to have the pleasure of seeing you at Kinneil on Saturday or Friday', wrote Watt to Roebuck (May 24). ' I sincerely wish you joy of the successfull result, & hope it will make you some return for the obligations I ever will remain under to you. . . .'

Roebuck naturally was not interested in details but evidently he was satisfied

---

[1] Muirhead Papers.

that it was time to apply for a patent. Accordingly, he sent Watt to London by coach to secure protection. Watt's Journal[1] shows that he took the oath on the patent on Aug. 9. While in London he received a letter (dated Aug. 12) from Dr. Small, begging him to ' get your patent and come to Birmingham'. He also informs him that a Mr. Edgeworth,[2] a gentleman of fortune, ' young, mechanical and indefatigable . . . (has) taken a resolution of moving land and water carriages by steam '.

Watt left London on the 27th and returned via Lichfield, Newcastle-under-Lyme, and Stoke. At Birmingham, for the first time, he met Matthew Boulton, and stayed a fortnight with him at his ' Hôtel d'Amitié sur Handsworth Heath '—as he called Soho House. Boulton loved to gather people round him, and to his house came English and foreign scientists and celebrities—even Catherine II of Russia once accepted his hospitality. Here Watt renewed his friendship with Dr. Darwin, and was a fellow guest with the soldier-chemist James Keir, who had a glass works near Birmingham. Watt afterwards (Oct. 25, 1768) described him to Lind[3] as ' a mighty chemist before the Lord, and a very agreeable man '. Watt was also the guest of an informal association of scientific men brought into being about 1766 by Boulton, Darwin, Keir, and Small. They met at one another's houses for dinner at 2 o'clock on the Monday nearest to the full moon—hence they came to be known as the Lunar Society[4]—in order to have the benefit of its light in returning home (they generally parted at 8 o'clock). Watt afterwards became a member, as did Edgeworth, Dr. Withering, the botanist and chemist (who analysed the mineral Witherite) ; John Baskerville, the printer ; and the two Galtons, father and son. Priestley was one of the most notable members, but he did not go to Birmingham to live until 1780.

The link between such diverse men was—until his death in 1775—William Small, ' by all who were admitted to his friendship beloved with no common enthusiasm ', says Richard Lovell Edgeworth.[5] Small had a disarming way of going to drink tea with those who spoke against him.[6] He was not above match-making, being successful in finding a wife for Thomas Day, who expected a paragon to ' live in a cottage on love '. When a young man Day carried his devotion to the simple life to the extent of seldom combing his hair, ' though he was remarkably fond of washing in the stream.' Edgeworth in his *Memoirs* gives a charming little sketch of his fellow Lunarians, whose intimacy was ' broken but by death' ; they ' formed altogether such a society as few men have had the good fortune to live with, such an assemblage of friends, as fewer still have had the happiness to possess and keep through life '.

It was an inspiring visit for the retiring Watt, and not the least interesting time was that spent in the manufactory with Boulton. By 1768 the fame of Soho

> Soho ! Where Genius and the Arts preside,
> Europa's wonder and Britannia's pride,

[1] Muirhead Papers.
[2] Richard Lovell Edgeworth (1748–1817), the buoyant, light-hearted Irishman. A charming *camaraderie* existed between him and his daughter Maria. Another daughter, Anna, married Dr. Beddoes, with whom Watt was afterwards associated in an attempt to cure consumption. Edgeworth's closest friend was Thomas Day (1748–89), who originally wrote his *Sandford and Merton* for Maria's *Moral Tales*.

[3] Keir's mother was a Lind. Born in 1735 Keir had known Dr. Darwin since boyhood and was a friend of his son Charles. Keir was the author (1791) of a *Life* of Thomas Day. (See also note to page 51.)
[4] A full account of this Society is given by H. Carrington Bolton : *Scientific Correspondence of Joseph Priestley*, 1892, App. ii.
[5] *Memoirs*, 1844, p. 116.
[6] Krause (Ernst): *Erasmus Darwin*, 1879, p. 37.

—as a local poet [1] describes the manufactory—had spread all over Europe. Boulton and Fothergill had correspondents at Marseilles, Lyons, Montpellier, Paris, and Rouen ; while the trade with Russia was sufficiently important to keep Fothergill in Russia and Sweden for a year.

Boulton, and Small who had become associated with Boulton in his clock-making business, took Watt over the factory. Watt looked about him with seeing eyes, noting the making of

' steel, gilt, and fancy buttons, steel watch chains and sword hilts, Plated wares, ornamental works in Or moulu, Tortoise shell. . . . A mill with a bearer which was employed in laminating metal for the buttons, plated goods, &c. . . . and I was informed that Mr. Boulton was the first Inventor of the inlaid buttons, and the first who had applied a mill to turn the laps . . . Besides . . . I saw an ingenious lap, turned by a hand wheel . . . and a shaking box put in motion by the mill, for scowering button blanks . . . which was also a thought of Mr. Boulton . . ' [2]

As he walked through the workshops—so much finer than anything at Carron—Watt recognized that here was the place where he could build his engine and fulfil his dream.

Boulton and Watt conceived a hearty liking for one another, perhaps from the very dissimilarity of their character. The invention, the importance of which was already fully appreciated by Boulton and Small, was discussed and the prospect of their being admitted to the partnership with Roebuck was mooted. Another matter discussed was the water-supply for driving the factory—never too large and in the summer months all too small. Boulton had been making experiments with a rotary engine to pump the water back again to the water-wheel but decided to wait till Watt had perfected his engine.

Watt arrived in Glasgow on Oct. 11th and as soon as possible thereafter he wrote [3] to Boulton making the position clear :

' When you were so kind as to express a desire to be concerned in my fire engine I was sorry I could not immediately make you an offer. The case is thus—By several unsuccessful projects & expensive experiments I had involved myself in a considerable debt before I had brought the theory of the fire engine to its present state. About three years ago a Gentleman [i. e. John Craig] who was concerned with me dyed. As I had at that time conceived a very clear Idea of my present improvements & had even made some trial of them tho' not satisfactory as has been done since, Doctor Roebuck agreed to take my debt upon him & to lay out whatever more money was necessary either for Experiments or securing the Inventions for which cause I made over to him two thirds of the property of the Inventions ; the debt & expenses are now about £1,200. I have been since that time employed in Constructing several working fire engines on the common principle as well as in trying experiments to verify the theory. As the doctor from his engagements at Bon-ness & other bussiness cannot pay much attention to the executive part of this, the greatest part of it must devolve on me, who am from my natural inactivity, want of health & resolution, incapable of it. It gave me great joy when you seemed to think so favourably of our scheme as to wish to engage in it.'

He added that he had called on Dr. Roebuck, and had suggested that Boulton should be offered one third share in the business. Roebuck had promised to write to Boulton.

Roebuck, naturally, was now keener than ever on the engine, and wrote to Watt (October 30) : ' I shall be very glad to see you and Mrs. Watt soon at Kinneil.

[1] James Bisset, in his *Poetic Survey round Birmingham* . . . (1800).
[2] *Memorandum concerning Mr. Boulton commencing with my First Acquaintance with him*, by James Watt, 1809, Sept. 17, preserved among the Boulton Papers, now at the Assay Office, Birmingham.
[3] Boulton Papers, 1768, Oct. 20.

I want much effectually to try the machine at large. You are letting the most active part of your life insensibly glide away. A day, a moment, ought not to be lost. And you should not suffer your thoughts to be diverted by any other object or even improvement of this, but only the speediest and most effectual manner of executing one of a proper size. . . .' He also pointed out that he did not wish to hasten the patent.

On November 9 Watt replied giving details of the engine which he proposed to erect at Kinneil; on December 12 he informed Boulton that he had ' almost finished a most complete model of my reciprocating engine'.

On December 22 Roebuck wrote to Watt that he was anxious to see him, adding ' write to your friend to take out the patent'.

Ill health and depression were overclouding Watt, and he unburdened himself to Small. Writing from Glasgow, January 28, 1769, in reference to the specification for the patent, he sighs over the ' much contrived & little executed, how much would health & spirits be worth to me ! . . .' In the same letter he refers to the Delftfield business : ' Our pottery is doing tolerably, tho' not as I wish. I am sick of the people I have to do with, tho' not of the business, which I expect will turn out a very good one. I have a fine scheme for doing it all by fire or water mills, but not in this country nor with the present people'. Later he wrote : ' I have had another three days fever . . . this cursed climate and constitution will undo me. . . .'

The memorable patent for ' a new method of lessening the consumption of steam and fuel in fire engines' was granted to Watt, Jan. 5, 1769. In accordance with official procedure the specification had to be enrolled within four months ; it was in relation to this that Small, who was acting as his adviser in drafting it, wrote (February 5) ' Mr. Boulton and I . . . think you should give neither drawings nor descriptions of any particular machinery, but specify in ye clearest manner that you have discovered some principles. . . .' Unfortunately Watt acted on this advice, although he had drawings ready.[1] He patented a principle and not the application of a principle ; this led to trouble later when the validity of the patent was assailed.

Meanwhile, Roebuck had seen Boulton and offered him the privilege of manufacturing Watt's engine in the three Midland counties, Warwick, Stafford, and Derby. In reference to which offer Boulton wrote to Watt (February 7)—two days after the birth of a little James Watt : [2]

' Dear Watt—By this time, I dare say, you have fully concluded that I am a very queer fellow, I having never answered your last friendly letter of the 20th October, nor your last of the 12th December . . . the plan [Roebuck's three-county share plan] proposed to me is so very different from that which I had conceived at the time I talked with you upon the subject that I cannot think it is a proper one for me to meddle with as I do not intend turning engineer. I was excited by two motives to offer you my assistance which were love of you and love of a money-getting ingenious project. I presumed that your engine would require money, very accurate workmanship and extensive correspondence to make it turn out to the best advantage, and that the best means of keeping up the reputation and doing the invention justice would be to keep the executive part out of the hands of the multitude of empirical engineers, who from ignorance, want of experience and want of necessary convenience, would be very liable to produce bad and inaccurate workmanship ; all of which deficiencies would affect the reputation of the invention.

[1] See page 101, and Pl. XIV.
[2] James Watt, born Feb. 5, 1769, died unmarried at Aston Hall, Warwickshire, June 2, 1848, the last of Watt's lineal descendants. Four children were born to Watt by his first wife ; but only two survived infancy : a daughter, who married a Mr. Miller of Glasgow, and James.

To remedy which and produce the most profit, my idea was to settle a manufactory near to my own by the side of our canal where I would erect all the conveniences necessary for the completion of engines, and from which manufactory we would serve all the world with engines of all sizes. By these means and your assistance we could engage and instruct some excellent workmen (with more excellent tools than would be worth any man's while to procure for one single engine) could execute the invention 20 per cent. cheaper than it would be otherwise executed, and with as great a difference of accuracy as there is between the blacksmith and the mathematical instrument maker. It would not be worth my while to make for three counties only, but I find it very well worth my while to make for all the world.

'What led me to drop the hint I did to you was the possessing an idea that you wanted a midwife to ease you of your burthen, and to introduce your brat into the world.' Boulton goes on to say : 'Although there seem to be some objections to our partnership in the engine trade, yet I live in hopes that you or I may hit upon some scheme or other that may associate us in this part of the world, which would render it still more agreeable to me than it is by the acquisition of such a neighbour . . . pray remember, when you come, always to put up at the *Hôtel d'Amitié sur Handsworth Heath*, where you will always find a friendly reception. . . .'

It was a masterly letter ; sympathetic, firm, showing full comprehension of the present difficulties and clearest insight into the future ; and it held open the door for future negotiations. On these Watt set his heart.

Small wrote teasingly from Birmingham : [1]

'A linendraper at London, one Moore, has taken out a patent for moving wheel-carriages by steam. This comes of thy delays. I dare say he has heard of your inventions. . . . Do come to England with all possible speed. At this moment how I could scold you for negligence ! However, if you will come hither soon, I will promise to be very civil and buy a steam-chaise of you and not of Moore. And yet it vexes me abominably to see a man of your superior genius neglect to avail himself properly of his great talents. These short fevers will do you good.'

Watt retorted (28 April) :

'If Linen-Draper Moore does not use my engines to drive his Chaise, he can't drive them by steam—If he does, I will stop them. I suppose by the rapidity of his progress and puffing he is too volatile to be dangerous. . . . I wish Mr. Boulton and you had entered into some negotiation with the Doctor about coming in partners. I am afraid it is now too late for the nearer it approaches to certainty he grows the more tenacious of it.'

Then the fire dies away and depression sets in : he is not so capable as he was four years previously, he sighs: [2]

'I have mett with many disappointments. I must have sunk under the burthen of them if I had not been supported by the friendship of Doctor Roebuck. . . . I have now brought the engine near a conclusion, yett I am not nearer that rest I wish for than I was 4 years ago. . . . Of all things in life, there is nothing more foolish than inventing. . . . You talk to me about coming to England just as if I was an Indian that had nothing to remove but my person. Why the devil do we encumber ourselves with anything else ? I can't see you before July at soonest, unless you come here. If you do I can recommend you to a fine sweet girl. . . .'

In reply Small (Apl. 28, 1769) asked Watt to come to Birmingham, and ' bring this pretty girl with you when you come. . . .'

Not only Roebuck, Boulton, and Small were waiting for Watt's engine, but in 1769 Dr. Darwin was writing to his friend Wedgwood (who was then erecting a flint mill to drive which Darwin was inventing a powerful windmill) with fine generosity and self-forgetfulness : ' I wanted to learn in what forwardness Mr. Watt's Fire-

---

[1] 1769, Apr. 18. Smiles : *Lives of Boulton and Watt*, p. 188.   [2] Muirhead : *Mech. Inv.* I. 53.

Engine was in. . . . I would recommend steam to you if you can wait awhile, as it will on many accounts be preferable I believe for all purposes.' [1]

Watt then was occupied with survey work in connexion with a canal nine miles long from Glasgow to Monkland, ' the encreasing price and the scarcity of coals in the city of Glasgow' inducing ' the magistracy to consider of the most probable means of remedying these evils '.[2]   The University voted £200 towards the scheme, on which Watt was engaged until 1772.   He was also engaged on work in connexion with the deepening of the harbour of Port Glasgow.[3]   ' This I would not have meddled with ', he wrote, ' had I been certain of being able to bring the engine to bear, but I cannot on an uncertainty refuse every piece of business that offers.'   He then describes [4] an improvement he had made upon the surveyors' level ; and about 1770 or 1771 contrived what he called a ' micrometer '—or, as we should say nowadays, a telemeter, or range-finder.   This he used on his surveys of the Crinan and Inverness Canals.   In this micrometer one perpendicular and two horizontal hairs were placed in the focus of the eye-piece of a telescope, through which sights were taken on two disks—one fixed and one movable, each white with a red stripe across it—on a rod at a known distance from the observer.   The telescope was turned until the horizontal hairs covered the red stripes and the perpendicular hair was parallel to the rod.   The upper or movable index was then fixed to the rod, which was marked into divisions of so many chains.   It was then, says Watt,[5] only necessary ' to send an assistant with the rod to any place the distance of which was wanted to be measured, and by signs to make him move the upper index up and down until the two horizontal hairs covered the red stripes on the upper and lower indexes ; the divisions on the rod then showed the distance, which I found could be ascertained to within less than one-hundredth part of the whole distance '.

Later he modified the micrometer by using a sliding object-glass and a scale on the telescope, but he never got beyond making a model of this apparatus.

Watt also invented about the same time another telemeter.   This had a prism of an angle of one or two degrees which was cut in two, and one part was made capable of movement on a centre relative to the other.   The latter part refracted the rays of light more than the former, so that when fixed in the focus of the object-glass of a telescope two images were formed.   An index and divided sector served to measure the refraction.   In this case again Watt only got as far as making a model.

Watt was still very anxious that Boulton should join in partnership with himself and Roebuck.   In a long letter to Small (Sept. 20) relating to his engine, he says : [6]

' I have had some conversation with him [Roebuck] about making Mr. Boulton & you a proper offer, which I expect he will do when with you.   From what I now write you, in which I assure you I have not concealed any circumstances that makes against the engine, you will judge how far it may be to your interest to engage in it on its own acct.   As to the Doctor, he has been to me a most sincere, generous friend, & is a truly worthy man.   As for myself I shall say nothing, but that if you three can agree among yourselves, you may appoint me what share you please, & shall find me willing to do my best . . . or, if this should not succeed, to do any other thing I can to make you all amends, only reserving to myself the liberty of grumbling when I am in an ill humour. . . .'

[1] Fleming (J. Arnold) : *Scottish Pottery*, 1923, p. 271.
[2] *A Scheme for making a navigable canal from the City of Glasgow to the Monkland Coalierys*, by James Watt [1770].
[3] *Report Concerning the Harbour of Port Glas-* gow *made to the Magistrates of Glasgow*, by James Watt, Engineer, and submitted to the Consideration of the Merchants, Aug. 9, 1771.
[4] *Proc. Inst. Mech. Eng.*, 1915, p. 490.
[5] Muirhead : *Mech. Inv.* I. cxxxiii.
[6] Ibid., I. 72.

Four days later he wrote on the same subject to Roebuck :

' When you are at Birmingham I expect you will try what can be done with Mr. Bolton & hope you will be able to conclude some bargain which, even though it should appear a little hard for us, I would wish you to accept. . . . The assistance of Mr. Boulton's and Dr. Small's ingenuity . . in improving and perfecting the engine may be very considerable, & may enable us to gett the better of difficultys that might otherwise damn it. Lastly, consider my uncertain health, my irresolute & inactive disposition, my inability to bargain & struggle for my own with mankind ; all which disqualify me for any great undertaking. . . .'

Small wrote a friendly letter in reply (Oct. 10, 1769) :

' Boulton and I will do anything we can do to have you here, and to forward your success ', but he states that both he and Boulton had but little money to spare just then, having ' just engaged in another scheme '.

Roebuck's offer to Boulton and Small was the option of coming in as partners to the extent of one-third within a period of twelve months, paying him (Roebuck) not less than £1,000. Small wrote (November 30) : ' Mr. Boulton and I have agreed with Dr. Roebuck.' This option, however, was never exercised, although Watt wrote joyfully (December 12) on the receipt of Small's note : ' I received yours & shake hands with you & Mr. Boulton on our connexion which I hope will prove agreable to us all. I have just got the new alterations of the engine ready to carry to Kinneil.' He then gives Small a detailed account of his work on the Monkland Canal and the dry dock of Port Glasgow. He continues this subject in a letter of Jan. 3, 1770, adding ' The *vaguing* about the country, and bodily fatigue have given me health and spirits beyond what I commonly enjoy at this dreary season. . . . Hire yourself to somebody for a ploughman ; it will cure ennui. . . .'

Roebuck was still pressing Watt to go to Kinneil : ' the sooner you come the better,' he wrote on Jan. 17, 1770, ' that the further trial may be executed of the engine. I have much to say to you of Dr. Small. . . . I very much esteem the Doctor. . .'

Watt, apparently, did not go, for on February 7 Roebuck wrote somewhat despairingly to Boulton and Small from Bo'ness : ' a single step has not been advanced towards the engine since my last letter. The fact is Mr. Watt hath been constantly and necessarily engaged in planning the Glasgow canal. But in three days I expect him here, and he has promised not to leave this place till the whole of the alternate engine has been effectually tryed. . . .'

Watt wrote at great length to Small in February and March—sometimes from Glasgow, sometimes from Kinneil—in reference to his engine. One letter contains a reference [1] to his being obliged in May to ' remove my dwelling from one part of Glasgow to another, and as I have any quantity of . . . moveables . . . I cannot leave this business wholly to Mrs. Watt. . . .' He also refers to his engagement by the Commissioners of the Forfeited Estates to make a survey for a proposed canal from Perth to Cupar of Angus. In his Journal [2] for July are some entries showing his friendly relations with Smeaton :

' July 29. Mr Smeaton breakfasted with me. . . . I had a good deal of conversation with him about fire engines ; he is trying experiments on that head. I promised to send him some account of what I had known engines do.

' Aug. 1. Attended General Meeting (of the Canal Proprietors) where my salary was fixed at £200 per annum. Gave them acct. of proficiency of the work & of Mr Smeaton's opinion about it. . . . They agreed to give Mr Smeaton £15 15s. for his trouble.'

---

[1] Muirhead : *Mech. Inv.* I. 95.   [2] Muirhead Papers.

Apparently Watt also discussed his own engine with Smeaton, who ' said a good deal in favour of the contrivance and desired me to push it, but that cannot be at present '.

He could not push it because of his growing employment as a civil engineer. The struggle that was going on between his desire to follow up his invention and the necessity of earning his bread is shown in a letter to Small : [1]

' I had just gott to Kinneil to finish my experiment on the engine, when on a days warning I was called to Glasgow to attend a survey of the river. . . . I had now a choice whether to go on with the expts. on the engine the event of wch was uncertain, or to embrace an honorable & perhaps profitable employment attended with less risque of want of success—to carry into exe- cution a canal projected by myself with much trouble or to leave it to some other person that might not have entered into my views & might have had an interest to expose my errors (for everybody commits them in those cases), Many people here had conceived a much higher idea of my abilities than they merit—they had resolved to encourage a man that lived among them rather than a stranger. If I refused this offer I had little reason to expect such a concurrence of favorable circumstances soon. Besides I have a wife & children & saw myself growing gray without having any settled way of providing for them. There were also other circumstances that moved me not less powerfully to accept the offer which I did tho' at the same time I resolved not to drop the engine but to prosecute it the first time I could spare.'

Alas for good intentions ! we cannot find any mention of experiments on the engine during the next three years. The ' other circumstances ' alluded to were doubtless Roebuck's growing financial difficulties.

Small suggested to Watt that he should send drawings of his rotary engine to Soho, so that a trial engine might be made there. Boulton and Small wanted to use one of these engines, but without his condenser, for the propulsion of canal boats ; the thought being suggested to them by the recent opening of the first part of the Birmingham Navigation Company's canal to Wolverhampton.[2]

In a letter dated September 30 Watt asked Small what method of applying the motive power they intended to adopt, and used the pregnant words : ' Have you ever considered a spiral oar for that purpose or are you for two wheels ? ' He gives a rough sketch which leaves us in no doubt that his idea was that of the screw pro- peller but it must not be thought that this remark had any influence on the applica- tion of the screw to steam navigation that took place more than half a century later. It must be taken rather as an indication of Watt's fertility of intellect. Afterwards he changed his mind about the suitability of spiral oars.

Small replied (Oct. 5, 1770) :

' I have tried models of spiral oars, and have found them all inferior to oars of either of the other forms. . . . I admire your scheme for spirals, but shall soon be in a rage with workmen. . . . Pray tell me what Smeaton said about the common engine, but do not let him pump you. . . . I grow infamously lazy, and think of buying a small annuity and passing the rest of my life in sleep, which I have discovered to be the best of all human things. . . . The French, you know, offer large præmia for time-keepers. Were I idle I should try to win one of these. But physic exhausts my whole faculties, and pays but indifferently. . . . Get into a warm bath, especially of a decoction of rosemary, when your head aches, or when you are dull. . . .'

But the rosemary bath, if tried, did not cure dejection, for Watt wrote to Small, October 20 :

' I really am become monstrously stupid, and can seldom think at all. . . . I greatly doubt

---

[1] Boulton Papers : Watt to Small, Sept. 9, 1770.

[2] The engine was put in hand but the castings proved to be defective and it was never com- pleted.

whether the silent mansions of the grave be not the happiest abodes. I am cured of most of my youthful desires. . . . Smeaton started some objections . . . which I answered. . . .'

Among those who were awaiting the completion of Watt's engine was Josiah Wedgwood, who asked Darwin, with some impatience, when Watt could construct an engine for him. Darwin replied (December 9) : ' Mr. Watt's Fire-Engine, I believe, goes on, but I don't know at what rate.' [1]

Watt's hands were full of other work ; on December 21 he wrote to Small :

' Notwithstanding the desperate weather I am almost constantly at the canal. It costs me many a fit of chagrin ; shows me many of my imperfections. . . . I have lately made a plan and estimate of a bridge over our river Clyde eight miles above this [i. e. Hamilton Bridge] ; it is to be of five arches and 220 feet water-way, founded upon piles on a muddy bottom. So much of self. . . . Your acquaintance, my friend Robison, is just gone to Russia with Admiral Knowles,[2] as his Secretary ; they say this will be worth £400 per annum to him. . . .'

Robison was not long in Russia before he wrote from St. Petersburg (Apl. 22, 1770 O.S.) [3] with official cognizance sounding Watt as to whether he ' would relish the Scheme of coming here in quality of Master Founder of Iron Ordnance to Her Imperial Majesty [Catherine II]. ' I am not as yet able to tell you what may be your encouragement, but I am sure that her Majesty's generosity will make it in every way suitable.' Watt replied : ' The proposal you make me is highly flattering both to my vanity and Ambition.' However, he declined the offer ' from being sensible that I by no means merit your recommendation and from my advancing in the estimation of mankind here as an Engineer'.

Watt was wise in declining, although the offer coming at a time when his employment was anything but congenial must have been tempting.

Later when Robison was at Cronstadt [4] he asked Watt to undertake the erection of an engine to empty the graving dock there, a duty being ineffectively discharged by windmills. Watt declined : ' I think you are fully able to conduct that project, and it will do you credit in the country where you are.' Robison therefore had to communicate to his friends Watt's ' genteel declinature '. Smeaton eventually supplied an atmospheric engine for this work.

Small continued to remind Watt of his engine—had the latter ever wished to forget it—and he had something of consequence to tell :

(Feb. 14, 1771) ' A friend of Boulton and me in Cornwall sent us word four days ago that four or five copper-mines are just going to be abandoned because of the high price of coals, and begs me to apply to them instantly. The York Building Company [5] delay rebuilding their engine . . . waiting for yours. Yesterday application was made to me, by a mining company in Derbyshire, to know when you will be in England about fire-engines, because they must quit their mine if you cannot relieve them. . . . I have perfected my clock with one wheel of nine inches diameter, which is to tell hours, minutes, and seconds, and strike, and repeat, and be made for thirty shillings.'

---

[1] Fleming, *op. cit.*, p. 271.

[2] Admiral Sir Charles Knowles, with whose son Robison went to Quebec in 1759 : Knowles, in 1770, accepted a post as president of the Russian Board of Admiralty. Russia was then at war with Turkey, but Knowles's service was entirely administrative. Robison was with him at St. Petersburg until the summer of 1772, when he was offered another appointment.

[3] B. & W. Colln. Robison to Watt and

draft of the latter's reply, 1771, June 9. On the same date in Watt's Journal (Muirhead Papers), there is a note of this reply.

[4] Apparently in 1773 ; cf. Muirhead : *Mech. Inv.* I. cxcvii.

[5] The York Buildings Company among its manifold activities had a pumping station near the present Adelphi and supplied Piccadilly with water.

It was in 1771 that Watt first met Wedgwood; the friendship then formed lasted until Wedgwood's death. ' Both had gentle natures, philosophic minds, and a modest bearing, largely the outcome of the physical sufferings they were compelled to endure'; [1] both reached through failure upon failure to ultimate success. Never once through Watt's dark days did William Small fail him; always putting aside his own suffering, his own work and hopes to enter into those of his friend. ' Nothing of late years', he wrote (Oct. 19, 1771),[2] ' has vexed me so much as the peculiar circumstances that have retarded your fire-engine . . . you have as much genius and as much integrity, or more than any man I know . . . farewell . . . Your most faithful friend and humble servant. . . .'

But now further misfortune happened to Watt in 1772. Roebuck, who had been losing heavily at Bo'ness where he had sunk his own and his wife's fortune, became insolvent. In that and the succeeding year severe commercial depression, following restriction of credit in consequence of bad harvests and unwise speculation, affected business and caused unemployment. In the same year came the failure of Neale, James, Fordyce, and Downe, a large London firm; this affected the Scottish banks along with other mercantile banking houses. Nearly every private banker in Scotland failed during this period.[3] Boulton was one of Roebuck's creditors.

' Had I conceived you or our friend Mr. B. to be interested people,' wrote Watt to Small in August 1772, ' I should never have entertained an idea of a connexion among us . . . there is no person to whom I have so fully explained my inmost thoughts as I have done to yourself, and I have no fear of ever having cause to repent it. But,' he continues, with quick sensitiveness and a certain proud humility, ' I do fear that in this affair I may have urged you too far and with too little delicacy; and that you have some reason . . . to think more meanly of me than I deserve. . . . Although I am out of pocket a much greater sum upon these experiments than my proportion of the property of the engine, I do not look upon that money as the price of my share, but as money spent upon my education. I thank God that I have now reason to believe that I can never while I have health be at any loss to pay what I owe, and to live at least in a decent manner. More I do not violently desire. I therefore beg, my dear Sir, if I have descended anything in your good opinion, you will allow me to climb up again, if not at once at least by degrees.'

Roebuck no longer could afford to bear the cost of experiments on the engine, and instead of giving help himself needed it. ' It is a matter of great vexation to me ', Watt wrote to Small (Aug. 30, 1772), ' that the Doctor should be out so great a sum upon this affair. . . . What little I can do for him is purchased by denying myself the conveniences of life . . . or by remaining in debt where it galls me to the bone to owe. . . .'

The Monkland Canal was soon to stop for lack of funds. Watt turned again to the thought of perfecting his engine at Soho. He was also reproaching himself for his share in Roebuck's misfortune. ' Nothing ', he wrote to Small (Nov. 7, 1772),[4] ' gave me so much pain as the having involved Dr. Roebuck so deeply in that concern,' and he again asked Small's help in inducing Boulton to take a share so

' that at least half of the property should belong to Mr. Boulton and you. At any rate let us be on such a footing that the experiments may go on, and the matter be concluded. . . . I have very much to tell you I cannot write, and there are many things I want to learn from you. . . . Our canal has not stopped but is likely to do so . . . everything has been turned over upon me, and the necessary clerks grudged to me; I am also indolent and fearfull, terrifyed to make bargains, &

---

[1] Fleming, op. cit., p. 269.
[2] Muirhead: Mech. Inv. II. 18.
[3] Lord (John): Capital and Steam-power,

1750–1800, 1923, pp. 83–4.
[4] Muirhead: Mech. Inv. II. 27. The full text of this long letter occupies pp. 27–31.

I hate to settle accounts. Why therefore shall I continue a slave to hatefull employment while I can otherwise by surveys and consultations, make nearly as much money with half the labor and I really think with double credit ; for a man is always disgraced by taking upon him an employment he is unfitt for. . . . Remember in recommending me to business that what I can promise to perform is ; to make an accurate survey & a faithfull report of anything in the engineer way ; to direct the course of canals, to lay out the ground and to measure the cube yards cut, or to be cut ; to assist in bargaining for the price of work ; to direct how it ought to be executed. . . . But I can on no account have anything to do with workmen, cash, or workmens accts. nor would I chuse to be bound up to one object that I could not occasionally serve such friends as might employ me for smaller matters. . . . I have no great experience and am not enterprising, seldom chusing to attempt things that are both great & new. I am not a man of regularity in Business & have bad health. Take care not to give any body a better opinion ot me than I deserve, it will hurt me in the end. I cannot forgive you writing me so seldom. . . '

FIG. 3. Micrometer attributed to Watt, in the Science Museum,
South Kensington

It is a singularly modest self-valuation of an intellect which was always attempting new things. To rest his mind from the great engine-question he thought out new ' gimcracks '—as Small calls them in a letter of Nov. 16, 1772. Watt replied (November 24) :

' As to gimcracks, I have contrived a new micrometer made by drawing two converging lines upon glass. I believe from trial it will answer. I mentioned a dividing screw : it has a wheel fixed upon it with 150 teeth and only ¼ inch diameter. It is moved any portion of a turn or number of turns by a straight line rack, the teeth of which fit it without shake, & is moved by the hand or foot. It divides distinctly an inch into 400 equal parts.'

A report of a trial of this screw on Mar. 3, 1774, states that ' it divided 9 inches into 20 and did not err the 1/200th of an inch in the whole 9 inches'.

In the Watt Workshop, now in the Science Museum, South Kensington, is preserved a dividing engine that agrees with the description here given ; possibly it is one and the same. In the Museum, too, there is a micrometer which came from Soho and is stated to have been invented by Watt. It embodies some of the features here mentioned, but so compactly that one would say it was a later production. In fact, as engineers will see from the illustration, Fig. 3, it is in nearly all essentials the micrometer of to-day. That Watt did make a micrometer as early as 1772 for

Macfarlane's Observatory is shown by a letter to him from Dr. Irvine,[1] July 2, 1778.[2]

Writing to Small again Watt says : [3]

' I am making a new surveying quadrant by reflection having the uses of a semi-circle as taking angles to 180°, the principle that of Bird's octant in which the objects are only once reflected. In this I am making, the fixed glass stands at 45° to the first radius and by shifting the place of the eye, the head is never in the way. I am going to make another altogether of glass with nonius of the same.'

It is not easy to say what idea was here present to Watt's mind. He is trying to get away from the double reflection inherent in the sextant, but he has to shift the eye at every observation, so that probably he found his idea was only an alteration without being an improvement. Watt refers again to these pursuits as follows : [4]

' My dividing screw can divide an inch into 1000 tolerably equal and distinct parts on glass. I have invented two problems for clearing the observed distance of the moon from a star of the effects of refraction and parallax : one trigonometrical by Mercator's sailing—the other instrumental by a sector having a line of chords on each limb and a moveable portion of a circle of the same radius which if of three feet the problem may be solved to ten seconds. If I have time I will make a model of it and bring it when I come. Moreover, I can solve the same problem according to Dunthorne's method by two lines of natural cosines upon a sliding rule.'

Here again there are not sufficient data for us to be able to say what was in Watt's mind ; one method is by calculation ; the other by an instrument which, however, if as long as he says it is, would be clumsy to use.

Meanwhile Roebuck's distress was increasing. At a meeting of his creditors Watt acted as Boulton's attorney, having been so appointed by Boulton in a generously expressed letter of Mar. 29, 1773. The next important step is a discharge by Watt dated May 17, 1773, to Roebuck for all the sums he may be owing under the partnership ' in consideration of the mutual friendship existing between Dr. Roebuck and myself and because I think the thousand pounds he has paid more than the value of the property of the two-thirds of the invention '. One result of this arrangement was that the Kinneil engine became Watt's property. He therefore took it to pieces and shipped the parts to Soho via London. Roebuck's creditors, it may be remarked, did not value the engine at one farthing.

Roebuck owed Boulton and Fothergill about £1,200, of which it seems that £500 had been advanced by Boulton in the ' faith of being assumed as a partner ' in the engine. Boulton, however, lest he should seem to be taking an unfair advantage over the other creditors, decided to wait and buy the share from the receivers of the estate. The following August Roebuck had to make a composition with his creditors, and it would seem that Boulton got the two-thirds share into his hands ; his partner, Fothergill, would have nothing to do with the matter, and his share of the £1,200 was paid to him by Boulton.

The same month Watt wrote to Small (Aug. 17, 1773) that he had been appointed to survey the wild, rough country between Inverness and Fort William for a projected canal.[5]

[1] William Irvine (1743–87), chemist. He studied medicine under Dr. Black, and succeeded Robison in the chair of Chemistry in Glasgow University.

[2] Muirhead : *Mech. Inv.* II. 108.

[3] Doldowlod Papers, 1773 Jan. 17.

[4] Ibid., 1773 Mar. 7.

[5] Watt's survey for the Fort William and Inverness canal was the last and most remarkable of his civil engineering works. Telford, who afterwards, in 1801, carried the Caledonian Canal through this tract, bore generous testimony to the accuracy of Watt's survey.

' I have been reading De Luc,' [1] he adds ; ' and I have tried a curious experiment to determine the heats at which water boils at every inch of mercury from vacuo to air. De Luc's observations and mine agree ; but his rule is false. I have some thoughts of writing a book, the " Elements of the Theory of Steam-engines ". . . . This book might do me and the scheme good. . . .'

While Watt was in the north he was summoned to Glasgow by tidings of the illness of his wife, who was expecting her fifth child. Of his dreary journey southward a few entries will be found in his *Journal* : [2]

' 1773 Sept. 26th. Sunday about half past ten received letter from Mr. Muirhead advising that Mrs. Watt was dangerously ill & her life despaired of & desiring me to come home with all speed. I immediately set out, with a very sad heart & came to Fort William that night having a heavy rain the whole way.'

On the following day he reached Tyndrum, and on the 28th ' came to Dumbarton about ten at night, where the certainty of my loss bore so strong upon me that I could come no further ', so ordered a chaise for the following day.

' Sept. 29th. About ten o'clock I saw ye chaise arrive and Mr. Hamilton in it ; [3] by his black coat and his countenance I saw I had nothing to hope.'

Watt then learnt that his wife had died on the Friday previous, September 24. He went to Mr. Hamilton's house as he ' feared to come where I had lost my kind welcomer'. He then learnt the details of his bereavement and says : ' in her I lost the comfort of my life, a dear friend and a faithful wife.'

She left two children—the elder only six. At her loss Watt's life seemed at first to break in two. In a fragment without date written to Small, Watt says :

' I know that grief has its period ; but I have much to suffer first. I grieve for myself, not for my friend . . . I am left to mourn. . . . I had a miserable journey home, through the wildest country I ever saw, and the worst conducted roads : an incessant rain kept me for three days as wet as water could make me. I could hardly preserve my journal-book. . . . Adieu, God bless you.'

This was a cruel misfortune—the cheerful helpmate, the mother of two surviving children the elder only six, the careful housewife, was taken from him just when the day of better fortune was about to dawn !

' Come to me as soon as you can,' wrote Small, and later (October 27) begged him not to indulge his grief, but to forget it in work. The following November Watt was absent from Glasgow on

' a survey of some improvement of the Upper Forth. I had many cold fingers and feet, and have much boiled my brains suiting a report to Lord Cathcart's [4] genius, which I do assure you

[1] Luc (Jean André de), 1727–1817, had recently come to England from Geneva. Watt had probably been reading his *Recherches sur les modifications de l'atmosphère*, 1772. He made numerous experiments on the atmosphere and was one of the first to notice that when ice thaws there is a disappearance of heat. On this Dr. Black founded his theory of ' latent heat '. In 1771 he contributed to *Phil. Trans.* the first correct rules ever published for measuring the heights of mountains. His chief discovery was his ' Dry Pile ', or ' Electric Column ', which he published in 1810 in *Nicholson's Journal*. He was appointed reader to Queen Charlotte, and

read *Cecilia* to her, although, laments the author who mentions him several times in her *Diary* —he ' can hardly speak four words of English '.
[2] Muirhead Papers.
[3] Gilbert Hamilton, agent for the Carron Company in Glasgow, afterwards Dean of Guild, and Watt's brother-in-law, his wife being the sister of Watt's second wife ; at Hamilton's death in 1808 Watt said ' I never knew a more friendly or worthy man '. Gilbert Hamilton, the younger, came to Soho.
[4] Charles, ninth Baron Cathcart, 1721–76, for many years Lord High Commissioner in the General Assembly of the Kirk of Scotland. For

is no easy matter. He is, however, a most honest and friendly man, and much my friend.
. . . This ennui of yours is vilely infectious ; I believe, like the plague, it can come by post. . . .
I am not melancholy, but I have lost much of my attachment to the world, even to my own
devices. . . . I long much to see you—to hear your nonsenses and to communicate my own ;
but so many things are in the way, and I am so poor. . . . I am heartsick of this country ; I am
indolent to excess, and, what alarms me most, I grow the longer the stupider. . . . I tremble when
I hear the name of a man I have any transactions to settle with. . . . This last year my whole
gains do not exceed £200. . . . There are two things which occur to me, either to try England, or
endeavour to get some lucrative place abroad ; but I doubt my interest for the latter. . . .' [1]

But the links binding Watt to Glasgow were breaking one by one ; he turned
his back on Scotland, and on May 17, 1774, set out for Birmingham, whither his
engine had preceded him.

three years—1768–71—he was ambassador extraordinary at the court of Russia, and must
have known Robison at St. Petersburg. When
Lord Cathcart went to Russia he took with him
a wonderful dessert service made by Wedgwood.
This was so much admired that the Russian
market was thrown open to Wedgwood's ware.
[1] Muirhead : *Mech. Inv.* II. 68–9.

IV

PARTNERSHIP WITH BOULTON.
STRENUOUS DAYS, 1774–84

WATT arrived in Birmingham on May 31, 1774. It was the second great turning point in his career for, although he did not realize it perhaps, the times were ripe for the reception of his invention. The possibility of applying animal and water power for the needs of ever-growing industry had been explored and their limitations had been realized, so that there was a wide field awaiting any new means of supplying power, especially one that could be located at any desired spot. The immediate need, however, was that of power for mine drainage, particularly in Cornwall. When it became known that Boulton was likely to become partner in a new fire engine, inquiries and applications for it came in from many quarters, particularly from that district. Watt, however, was not in a position to start making engines at once—he had first to resume experimental work on the Kinneil engine, and to this task he addressed himself with vigour.[1] A fragment of a log book in which Watt recorded his results from November 1774 onwards has been preserved.[2]

With the skill now devoted to it the engine worked satisfactorily, as shown by a letter Watt wrote to Roebuck this month, and by another written to his father. The latter contained these modest words:

' The business I am here about has turned out rather successful, that is to say the fire engine I have invented is now going & answers much better than any other that has yet been made, and I expect that the invention will be very beneficial to me.'

There remained, however, one further difficulty : six years out of the fourteen for which the patent had been granted had already run their course. Boulton was shrewd enough to foresee, what in fact proved to be the case, that the patent would expire before he could recoup himself for the outlay that, as he had foreshadowed in 1769,[3] would be required. It was therefore necessary to find out whether the patent could be extended or a new one obtained. Accordingly Watt came up to London in the spring of 1775 and took counsel's opinion on the subject. He wrote to Boulton :[4] ' I have taken advice of several people whom I could trust about the patent. They all agree that an Act would be much better and cheaper, a patent being now £130, the Act, if obtainable, £110.' Application for an Act to extend the patent for a term of twenty-five years was decided upon. Small drew up the petition for the Bill which was presented to the House of Commons on Feb. 23, 1775. In its passage it met with ' violent opposition from many of the most powerful people in the House '. Watt drew up and printed for distribution to members of the

---

[1] Boulton had installed Watt, apparently, in his old quarters at Newhall. The directory of Birmingham, 1777, has this entry : ' James Watts engineer ' at ' 1 Newhall Walk '.

[2] B. & W. Colln. ' Experiments on ye first engine at Soho.' Cf. also p. 108.

[3] See p. 30.

[4] Boulton Papers, 1775 Jan. 31.

House a memorandum in which, after describing the Newcomen engine and pointing out its defects, he showed what were his improvements. He continues :

' The inventor of these new engines is sorry that gentlemen of knowledge and advowed admirers of his intention should oppose the Bill by putting it in the light of a monopoly. He never had any intention of circumscribing or claiming the inventions of others ; and the Bill is now drawn up in such a manner as sufficiently guards those rights and must oblige him to prove his own right to every part of his invention which may at any time be disputed. If the invention be valuable, it has been made so by his industry and at his expense ; he has struggled with bad health and many other inconveniences to bring it to perfection, and all he wishes is to be secured in the profits which he may reasonably expect from it—profits which he cannot obtain without an exertion of his abilities to bring it into practice by which the public must be the greatest gainers and which are limited by the performance of the common engines ; for he cannot expect that any persons will make use of his contrivance unless he can prove to them that savings will take place and that his demand for the privilege of using the invention will amount only to a reasonable part of them. No man will lay aside a known engine, and stop his work to erect one of a new contrivance unless he is certain to be a very great gainer by the exchange ; and if any contrivance shall so far excel others as to enforce the use of it, it is reasonable that the author of such a contrivance should be rewarded.'

On March 9th the Bill was brought in and read for the first time. Watt came up to town again in May, and Boulton was also there, adding his influence. The outcome of it was that the Bill was passed through all the stages and received the Royal Assent on May 22, 1775.[1]

A point of interest is that whereas the original patent covered only England, Wales and the Plantations (i.e. the Colonies), the Act extended it to cover Scotland. It is open to doubt whether the extension granted by Parliament did not err on the side of too great liberality. Certainly in its later years it retarded greatly the development of the engine that many ingenious men were desirous of making, but who were obliged to turn their talents elsewhere.

While Watt was in London on his first visit this year, he received from Boulton [2] the melancholy news of the unexpected death of Dr. Small from ague. This was a great blow to both Boulton and Watt, especially to the latter whose present position was due in no small measure to Small's persistent advocacy and his encouragement of Watt during his periods of greatest despondency. Their letter correspondence breathes an intimacy, a comradeship, and, at times, a happy humour which reveal another and a different Watt.

Boulton in the letter just alluded to says : ' Your going to Russia staggers me. The precariousness of your health, the dangers of so long a journey and my own deprivation of consolation render me a little uncomfortable ; but I wish to assist and advise you for the best, without regard to self.' Apparently this was the outcome of another attempt on the part of Robison to get Watt to go to Russia, in what capacity is not clear, but the salary offered was £1,000 per annum. Evidently Boulton had been speaking highly of Watt, for in a later letter (Mar. 11, 1775) he says : ' I find I love myself so well that I should be sorry to have you go to Russia and I begin to repent sounding your trumpet at the Ambassador's.' When Darwin heard about it he wrote (March 29) : ' Lord how frightened I was when I heard a Russian bear had laid hold of you with his great paw, and was dragging you to Russia. Pray don't go if you can help it. . . . I hope your fire engines will keep you here.'

[1] *Journ. of Ho. of Commons*, xxxv. 142, 168, 185, 280, 313, 387.

[2] Muirhead ; *Mech. Inv.* II. 81. Boulton to Watt [1775 Feb. 25].

They did. The die had been cast and there could be no turning back.

While Watt was in London in January this year, he took occasion, on Boulton's advice, to test the atmospheric engines at the New River Head, York Buildings Waterworks, and Chelsea Waterworks. This practice of testing engines became habitual with Watt, indeed a note-book is preserved[1] tabulating data of tests, many by Watt himself, extending from 1764 to 1779 ; it is hardly too much to say that Watt was the originator of systematic engine testing.

In April Boulton reported the arrival of the cast-iron cylinder to replace the tin one of the Kinneil engine.

This was supplied by John Wilkinson of Bersham, a remarkably forceful character with a passion for cast iron and a believer in it as a constructive material for all purposes.[2] The reason for ordering the cylinder from Wilkinson was because he had made a machine suitable for boring cylinders and other hollow articles. He had previously invented and patented, in January 1774, a machine for boring guns. The improvement in his machine over its predecessors was that it not only bored a cylinder circular at any one diameter but also made it cylindrical throughout its length. The importance of the invention generally can hardly be over-estimated, and it may be said to have been vital to the success of Watt's engine.

The famous partnership between Boulton and Watt commenced on June 1, 1775, for a period coterminous with that of the Act of Parliament. The articles of agreement were of such importance that Watt's letter[3] to Boulton reciting them must be given *in extenso*. It should be mentioned that Watt had gone to Glasgow to bring his two children, left there in charge of relatives, to their new home in Birmingham.

' Glasgow, July 5th, 1775.

' As you may have possibly mislaid my missive to you concerning our contract, I beg just to mention what I remember of the terms.

1. I to assign to you 2/3 of the property of the invention on following conditions :
2. You to pay all expenses of Act, or others incurred before June, 1775, and also the expense of future experiments, which money to be sunk without Interest by you being the consideration you pay for your part, but the experimental machines to be your property.
3. You to advance stock in trade bearing Interest, but having no claim upon me for any part of that further than my Intromissions, but the stock itself to be your security & property.
4. I to draw one-third of the profits so soon as any arise from the business, after paying the workmens wages and goods furnished, but abstract from the stock-in-trade excepting the Interest thereof, which is to be deducted before a balance is struck.
5. I to make drawings, give directions, & make surveys, the company paying travelling expenses to either of us when upon engine business.
6. You to keep the books & balance them once a year.
7. A book to be kept wherin to be marked such transactions as are worthy of record, which, when signed by both to have the force of the contract
8. Neither of us to alienate our share without consent of the other, & if either of us by death or otherwise shall be incapacitated from acting for ourselves, the other of us to be sole manager without contradiction or interference of Heirs, executors, assignees or others but the books to be subject to their inspection & the acting partner of us to be allowed a reasonable commission for extra trouble.
9. The contract to continue in force for 25 years from the first of June, 1775, when the partnership commenced, notwithstanding the contract being of later date.
10. Our Heirs, Executors & Assignees, bound to observance.

[1] Boulton Papers.
[2] Cf. Dickinson (H. W.): *John Wilkinson,* *Ironmaster*, 1914.
[3] Boulton Papers, 1775 July 5.

11. In case of demise of both parties, our Heirs, &c. to succeed in same manner, & if they all please may burn the contract.

If anything be very disagreeable in these terms, you will find me disposed to do everything reasonable for your satisfaction, but wish the deed sent to as conformable to this or any former missive as is agreeable and drawn up by an able lawyer with expedition.

The buttons ordered by Mr. Hamilton much wanted, also a tea kitchen, & some other things, which were ready when I came away.

I believe I shall have no occasion to draw for any money, having got in some of my old scraps which will serve or nearly serve my occasions here.

With comp^{ts} to friends, I ever am,

Yours most sincerely,

JAMES WATT.'

Matthew Boulton, Esqr.
Birmingham.

The book ' wherein to be marked such transactions as are worthy of record', if kept, has not been found by the Editors.

Watt got back to work at the end of August, and Boulton, as was the nature of the man, threw himself heart and soul into the new enterprise. This very fact invested the new engine with value to those, and they were many, who laid store by Boulton's judgement. He had by judicious replies to correspondents on the subject raised expectations which Watt had now to fulfil. So far the latter had only the data supplied by the Kinneil engine to rely upon, and there is not much doubt that on these data he based his first actual engines.

It was not possible, however, to improvise workshops and plant to make engines, so that following the practice of the time we find the firm getting the castings and much of the smith-work from works outside, while the smaller but more important parts were made at Soho. As for Watt, he carried out his drawing and calculating at his own house, where he could be quiet and not liable to interruption.

One of the first engines made was that ordered by John Wilkinson, to blow the blast furnace at his New Willey Ironworks, Broseley, Shropshire. Watt went thither on December 14th to see what was wanted. Great pains were taken by him with the drawings, for much depended on its success with the neighbouring ironmasters. Apparently none of the parts was made at Soho ; everything was done at Wilkinson's works. Watt returned to Broseley in February 1776, to superintend its erection, and as a result the working of the engine was quite satisfactory.

At the same time another engine for pumping at Bloomfield Colliery, near Tipton, was being erected. Boulton seems to have had some interest with the proprietors of this colliery and the engine was undoubtedly meant to serve as a pattern to convince mine adventurers elsewhere, e.g. in Cornwall, of its value.

It is not, however, our intention here to relate in detail the technical difficulties that Watt had with this and succeeding engines, the improvements that he successively introduced, or the methods of remuneration adopted by the firm for their services : these matters are dealt with elsewhere. Boulton was a tower of strength in all directions for, in spite of his having said he would not turn engineer, he was not far behind Watt in fertility of expedient for getting over difficulties. He wrote to Watt while the latter was at Broseley : [1] ' I have fixed my mind upon making from twelve to fifteen reciprocating and fifty rotative engines per annum. I assure you

---

[1] Boulton Papers, 1776 Feb. 24. It should be observed that by ' rotative ' Boulton meant the rotary engine, i. e. the steam wheel. The rotative engine properly so called did not make its appearance till some years later.

V b. ANN MACGREGOR, SECOND WIFE OF JAMES WATT

From the oil-painting, artist unknown, in the possession of the Watt family

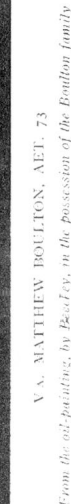

V a. MATTHEW BOULTON, AET. 73

From the oil-painting, by Beechey, in the possession of the Boulton family

that of all the toys which we manufacture at Soho, none shall take the place of fire engines in respect of my attention.'

In June Watt revisited Glasgow for a very important reason, namely, the taking unto himself a second wife. This lady was Ann Macgregor or McGrigor, daughter of a dyer. Watt was now forty years of age, and the marriage was not altogether a love-match. In a letter to Boulton,[1] he says ' I consider it as one of the wisest of my actions.' A rather amusing incident is related in this same letter and we give it in Watt's own words :

' The only disagreeable part of the business that remains to be done is the settlement, & I find that the old gentleman wishes to see the contract of Partnership between you and I, and as that has never been formally executed, I must beg the favour of you to get a legal contract written & signed by yourself, sent by return of post or as soon as may be. Lest he should have called my prudence in question, I have been obliged to allow him to suppose such a deed did exist, but was single, so what you send must pass for a duplicate and another may be actually written which I hope you will not doubt of my readiness to execute as soon as I return, and in fact this deed should have been executed long ago, in common prudence upon both sides, particularly upon yours, who have no legal assignation to the Act of Parliament. I therefore hope that you will excuse the old gentleman's caution. If you do not chuse to send the deed itself, you may have a scroll of it made, without date which you may send as the copy the deed was drawn from.'

Boulton rose nobly to the occasion, and extricated his partner from his predicament with an unblushing white lie in the form of a letter to produce to Mr. Macgregor, containing these words :

' I would without hesitation have sent you the assignment and the article of partnership had it been in my power ; but Mr. Dadley, the lawyer, is suddenly called to London, and it cannot be had before his return ; but if you want to show it to any of your friends, you may give them a copy of the following heads which I have extracted from our mutual missives, and are to the best of my knowledge all that our articles contain.'

Then follow the articles of agreement substantially as already recited.[2] Had Mr. Macgregor known what was going on behind his back he might well have doubted whether Watt was a fit and proper person to whom to entrust his daughter. The astonishing fact disclosed is that the cautious Watt and the business-like Boulton had not—more than a year after the beginning of their partnership—formally executed the indenture, such was the absolute confidence they had in one another.[3] A further interesting fact is mentioned in Watt's letter, and that is, that he had better health than had been his lot for years, so rapidly does congenial occupation react on the nervous system of a man like Watt.

While on this visit to the North, Watt spent some time in collecting old debts and in setting in order the affairs of his father who was now seventy-eight and very infirm. He also got orders for some engines, notably one for Torryburn Colliery in Fifeshire.

On the arrival of the newly wedded pair in Birmingham, it seems probable that

[1] Boulton Papers, 1776 July 3.
[2] p. 43.
[3] In fact the Editors are unable to state positively that the deed was executed, because this important document has never yet been found. In Watt's Journal (Doldowlod Papers) under date 1779 Apr. 20, appears the entry :

'Executed deed of partnership with Mr. Boulton.' What deed this was, unless the one under consideration, it is impossible to say. It cannot refer to the establishment of the firm of James Watt & Co. for letter-copying business, because this was not started till 1780 ; see p. 51.

the bride was not favourably impressed with dingy Newhall Walk and that her dissatisfaction induced Watt to leave it ; for on Mar. 17, 1777, as noted in his *Journal*, the family moved to Regents Place, Harper's Hill.[1] This situation was chosen because it was the part of Birmingham as it then existed most convenient for Soho. Regents Place was swept away about 1872 and the site is now covered with warehouses.

While he was away Boulton wrote to tell Watt about the inquiries and orders for engines that were coming in. He further sent the disagreeable news that one Humphrey Gainsborough, an Independent Minister at Henley-on-Thames, had made the discovery of an engine claimed to be seven times better than the common engine. Boulton wrote [2] half humorously :

' If we had a hundred wheels [i. e. rotary engines] ready made and a hundred small engines like Bow engine, and twenty large ones executed, we could readily dispose of them. Therefore, let us make hay while the sun shines, and gather our barns full before the dark cloud of age lowers upon us, and before any more Tubal Cains, Watts, Dr. Fausts or Gainsboroughs arise with serpents like Moses's that devour all others . . . as to your absence say nothing about it. I will forgive it this time provided you promise me never to marry again.'

The Bow engine was an important one, as it was the firm's first incursion into the London area ; consequently much depended on its success. The materials were dispatched in September, and Watt came up to town in November. He wrote [3] that he ' found that the engine had gone very well'. ' I propose that he [Joseph Harrison] should not leave it for a few days until both his health and that of the engine be confirmed. A relapse of the engine would ruin our reputation here and indeed elsewhere.' [4]

While in town, in 1776, Watt called, at Boulton's request, on representatives of the Shadwell Waterworks Company to examine their engines and make terms for one of the new construction, but did not get very far in the negotiations. Meanwhile Boulton wrote to Watt, ' We have a positive order for an engine for Tingtang Mine, and from what I hear this day from Mr. Glover, we may soon expect other orders from Cornwall.'

Such was the beginning of the long and important connexion of the firm with mine drainage and mining in Cornwall. The expected happened, and the order for Tingtang was followed by one for Wheal [5] Busy, Chacewater. The materials for the Wheal Busy engine were dispatched by the middle of 1777, and those for Tingtang soon after. Watt went down, accompanied by his wife, on August 4th, arriving on the 9th to superintend their erection in person. They went by stage coach or post-chaise via Bristol to Exeter, Lostwithiel, and Truro, spending four nights on the way. On his arrival Watt found the materials for Wheal Busy had been delivered and some progress made in the erection of the engine.

The first impression made upon him by the Cornish folk was that they ' have the most ungracious manners of any people I ever was ammong '.[6] It would appear that the mines were in a bad way, for he says, ' Poldice is grown worse than Wheal virgin was. They have sunk £700 this last month, and £400 pr. month for some

---

[1] See map, facing p. 262.
[2] Boulton Papers, no date. Cf. Muirhead: *Mech. Inv.* II. 102.
[3] Boulton Papers, 1776 Dec. 3.
[4] A relapse did occur the following spring ; see p. 116.

[5] Wheal, or whele, variants of huel, Cornish *hwel*, a work, a mine. Pryce (William): *Mineralogia Cornubiensis*, 1778, p. 323.
[6] Boulton Papers, 1777 Aug. 4, from Chacewater.

months past, they will probably soon give up. North Downs seems to be the next card.' In his next letter from Truro [1] he says :

'I have been at Chacewater and have seen Messrs. Wilson, Dudley and Hornblower,[2] the latter seems a very pleasant sort of old presbyterian. Wheal Bussy is in considerable forwardness, and what iron work has been made there is little inferior to our own if any—all the world is agape to see its performance.'

Watt in his next letter [3] says, of Jonathan Hornblower, ' Hornblower seems a good sort of man and carries himself very fair but is *I hear* an unbelieving Thomas.' Watt gives a great deal of information as to the mines and the men there, and says of the atmospheric engines I ' shall soon try them all & make a table of them, as I am now situated in the middle of them '.

There was another prominent engineer who, like Hornblower, resented the intrusion of Boulton and Watt into Cornwall with their newfangled engine, and

FIG. 4. Watt's house at Harper's Hill, Birmingham, from Smiles's
Boulton and Watt

that was ' Old Bouge ' as Watt calls him, going by the local pronunciation. His real name was John Budge. He spread abroad many false reports about the engine, but afterwards became a convert to the new order of things. In his letter [4] of September 20 Watt says :

' Bouge has been to wait upon me & promises to read his recantation so soon as he is convinced, & never to touch a common engine again. The voice of the country seems to be at present in our favour, and I hope will be much more so when the engine getts on its whole load which will be by tuesday next. So soon as that is done I shall sett out for home.'

When he arrived at Soho he found that a number of orders for engines had been received and demanded his attention, for, so far, he had no one to whom he could entrust any of the technical part of his work.

We must now show how a knowledge of Watt's inventions was spreading on the Continent. In 1777 M. Périer [5] had obtained from Louis XVI an exclusive privilege

[1] Boulton Papers, 1777 Aug. 9.
[2] For the Hornblower family and John Budge; see Chap. XXII.
[3] Boulton Papers, 1777 Aug. 14.
[4] Ibid., 1777 Sept. 20.
[5] Jacques Constantin Périer (1742-1818), a

for raising water from the Seine for the supply of Paris and had formed a company to reap the benefit of the concession.   He came to England for the purpose of getting a common engine for the work and went to John Wilkinson whose brother William had been to Paris to solicit orders for the cast-iron pipes required.   Périer then learnt about Watt's engine and was easily convinced of its superiority.   He tried to induce Wilkinson to supply one but this the latter declined to do because it was his interest, legally and otherwise, to keep on good terms with Boulton and Watt.[1] This incident showed the partners that their position would be strengthened if they could obtain a patent or privilege in France.

Meanwhile the Comte d'Héronville, through M. Magellan as agent, approached the firm for an engine to drain some marshes in the neighbourhood of Dunkirk.   The firm consented only on the condition that the Count would use his influence to obtain for them an exclusive privilege for France.   On the petition of Boulton and Watt an ' Arrêt de Conseil ', dated Apl. 14, 1778, was granted for fifteen years.   Its most important clause runs as follows :

' LE ROI, en son Conseil, a permis et permet aux Srs. Boulton and Watt de faire construire, vendre et débiter dans toute l'étendue du Royaume pendant l'espace de quinze années, exclusivement à tous autres, les nouvelles machines à feu de leur invention après néanmoins que l'essai aura été fait d'icelles, soit dans la Ville de Paris, soit dans les Moeres de Dunkerque, en présence de tels Commissaires que seront nommés par le Conseil et après que la dite machine aura été reconnue supérieure pour l'effect et pour l'économie aux anciennes machines à feu.'

A final clause fixes a fine of 500 livres for infringement, &c.   Nothing came of the Dunkirk scheme, however ; Watt's advice was,[2] ' I would have nothing to do with the Dutchman at the Moere.'

Shortly after the date of their decree the firm was approached by a M. Jary, who styled himself ' Concessionnaire des Mines du Nord en Bretagne' ; these were coalmines, near Nantes, Loire Inférieure.   He procured for Boulton and Watt a new decree granting permission for the requisite trial to be made on this engine instead of on that proposed at Dunkirk or that at Paris.   Following upon this, Jary was furnished with materials, drawings, and instructions for the erection of an engine which, in consequence, was the first of the Watt type in France—or indeed on the continent of Europe.   Special permission was required to export the parts as Great Britain and France were at war.[3]   Perhaps Jary thought the decree was sufficient return for their services but the fact remains that the firm received from him neither payment nor return of any other kind.

Périer now came to England a second time, visited Soho, and concluded a bargain with the firm.[4] ' We have found ourselves obliged to grant him on very moderate terms, because our arrêt not having yet the force of a patent, we durst not risk any opposition.'[5]   On Périer's return to Paris two engines were erected in due course at Chaillot just outside Paris on the banks of the Seine.

The year 1778 was one of extraordinary mental and physical exertion for Watt. There were calls upon his time from all quarters.   He wrote to Boulton [6] ' Tingtang people want me too among others.   I fancy I must be cut in pieces and a portion

well-known mechanician.   He and his brother came five times to England in order to learn all that was known there on steam engines.
[1] Watt to Boulton, 1777 May 2, in Muirhead : *Mech. Inv.* II. 104.
[2] Watt to Boulton, 1778 Aug. 6, in ibid., II. 110.

[3] B. & W. Colln. Lord Dartmouth to Watt, 1779 Apl. 6, and draft reply.
[4] The agreement is in the B. & W. Colln.
[5] Watt to Black, 1779 Jan. 13, in Muirhead : *Mech. Inv.* II. 113.
[6] Boulton Papers, 1778 Feb. 2.

VI. TROUBLES AT CHACEWATER

Page from Watt's Journal, 1778, Doldowlod Papers

sent to every tribe in Israel.' In May he set out with his wife for Cornwall, and arrived at Redruth on the 28th. They rented a house at Plain-an-guary Green, on the outskirts of Redruth. The letters which followed his arrival are full of complaints of mistakes made in castings, misunderstandings, and blunders of work-men, difficulties and cost of carriage of the goods, and the low price of ore. The fact of the matter seems to have been that Watt was overburdened with detail and was attempting more than he could manage. The most serious difficulty, however, was the pecuniary embarrassments in which Boulton had become involved, due to the large amount of capital sunk in his other businesses. Watt[1] made a suggestion:

'You know I am a bad ways & means man, but however, the following thought may merit your consideration. You know Wn. [Wilkinson] has several times hinted a wish to be concerned in this scheme, to wch we have had material objections but rather than founder at sea we had better run on shore.'

Watt was opposed to Boulton's suggestion to borrow from the Truro bankers on the security of the royalties on their engines so far erected because he thought this would become known and damage their credit in Cornwall. Boulton's courage and determination were equal to the emergency, and after long negotiations he raised £7,000 from a Mr. Wise, and £14,000 from London bankers on the security of the royalties to be derived from the engine patents. Watt wrote congratulating him:[2] 'I am happy to hear that you have weathered the storm and have now a fair pros-pect of surmounting all your difficulties.' Farther on he says: 'Both the engines [Chacewater] go on exceeding well and give great satisfaction.' Chacewater large engine which Watt looked upon as 'their capital card' continued to pump steadily and satisfactorily.

In his letter of September 6th[3] Watt says of the movements among the mine adventurers[4] with regard to engines, 'all this depends on ye success of Chace-water, which God protect.'

'The terms' to the miners, he says, 'must be ye cubic foot & fathom, the coals being ascert[nd] by Chacewater & ye present consumption of others in ye Country without any other terms on our part than the delivering them an equally good engine with Chacewater.' In other words the performance of a representative atmospheric engine in Cornwall was to be compared with the Watt engine at Chace-water, and from that comparison the savings in coal would be fixed. The engines at Poldice were considered a good average and they were tested by a committee of representatives from both sides. Their report[5] dated Oct. 30, 1778, gave figures whence it appears that the duty of the engines was 7,047,429 lb. per bushel. The royalty that Boulton and Watt decided on was the value of one-third of the savings. The first engines of which we have spoken were erected without agreements as to this payment, but for subsequent engines agreements were entered into in which the payment of royalty is very clearly defined.[6] It was not long, however, before this arrangement was altered and a royalty, based on inch of piston diameter, sub-stituted.

---

[1] Boulton Papers, 1778 July 8.
[2] Ibid., 1778 Aug. 3.
[3] Ibid., 1778 Sept. 6.
[4] Shareholders or partners in a mining enter-prise. The system of adventuring reached its height in Cornwall during the eighteenth and nineteenth centuries.

[5] See *Phil. Trans.* 1830, 122.
[6] Several such agreements are in the B. & W. Colln. Two have been printed. See Commis-sioners of Patents—Contributions to the History of the Steam Engine, 1872. Agreements for the Ale & Cakes Mine and Werneth near Oldham.

In a letter [1] to Zaccheus Walker, Soho, Watt says : ' I send enclosed a drawing for 63 inch Cyl^r for Poldice mine which please cause Mr. Playfair copy accurately and forward the *Original* to Bersham with all convenient despatch as the Engine is much wanted.' This shows that at last Watt was getting a little assistance in the technical work. The letter goes on : ' If Wm. Murdock is not at home he should be sent for immediately as he understands the patterns and care must be taken to avoid mistakes of which our engine shop has been too guilty.'

The tale of William Murdock's call at Soho in 1777 to ask for work has been told by his son. How impressed was Boulton, whom he interviewed, when he found out that the hat which Murdock was twirling in his hands was a wooden one made ' on a bit lathey' of his own making. That was enough—a mechanic who could do a job like that would never be at a loss—he was engaged the following day. Murdock was not only diligent and conscientious, but he thoroughly mastered the engine, so much so that later on in Cornwall he was everywhere in request.

Just after the date of the above letter to Walker, Boulton came to Cornwall. He managed to get the terms fixed for the use of the engines already erected, entered into agreements for others that were building, and finally succeeded in borrowing £2,000 from Elliot & Praed, bankers, of Truro.

Watt left Cornwall on New Year's Eve, 1778, and arrived in Birmingham on Jan. 4, 1779, only to find himself almost overwhelmed by the press of work. He wrote in May to his partner, relative to two new customers, ' I beg that you will not undertake to do anything for them before Christmas. It is, in fact, impossible at least on my part, I am quite crushed.'

It seems to have been characteristic of Watt that when he was down in the depths like this he was generally on the eve of making some brilliant invention. In this instance, it was the invention of letter-copying. Watt first directed his attention to the invention of a copying machine employing two pen nibs. Entries in Watt's Journal [2] under Jan. 14 and 26, and Feb. 13, 1779, give an account of his trials, but his conclusion was ' unless one can be made a deal lighter, the machine may be given up as it has too much friction '.

In Apl. 27, 1779, appears the first mention of press copying : ' Made fair copy of the answers [i.e. to Bedworth Papers] took impressions of it.' Watt's Journal shows that the experiments continued with patient assiduity during May and June. Then he communicates his invention to Boulton in the following letter : [3]

' I send you enclosed some of Mr. Nobody's draughts with authentic copies of them. You will observe that that which appears full black will not preserve even the colour it has, but will grow browner by keeping. The paler kind are nearly if not quite fully as white as the paper was before the operation, and will stay so.

' I will be much obliged to you to procure me a quire or two of the most evenly & whitest unsized cambric paper and also specimens of the cheaper kinds with their wholesale prices. . . . It is absolutely necessary that there be no size in the paper which [you] may know by touching it with a wet finger.

' The copy will continue to grow blacker than they were before copying and as far as I can judge not in the least defaced.'

On September 18th Watt's Journal has this entry : ' tryed the taking the proofs of writing by a rolling press which answers extremely well without being hard sett.'

With this addition the invention was complete. Press-copying was one of Watt's most important minor inventions. The process is the same as transfer printing,

---

[1] Boulton Papers, 1778 Oct. 2.   [2] Doldowlod Papers.   [3] Boulton Papers, 1779 June 28.

but as the copy is on thin paper it may be read the right way by looking at the back. The invention was of the greatest use to Watt himself, as it saved him hours of unprofitable drudgery—he even copied letters to his wife, so enamoured was he of the method—while in the firm it came to be employed extensively, not only for correspondence, but also for drawings ; indeed in the latter aspect, apart from the fact that on such paper the copy was of course ' reversed' as compared with the original, we may say that the method was the first of those copying systems we see so extensively used in the drawing office of to-day.

The ink was, if anything, more important than the paper, and Watt spent a long time in experimenting to obtain the right ingredients. A letter is in existence [1] addressed to Mr. Keir,[2] whom he pressed into his service in this business, giving his ' thoughts on Ink-making' with instructions for its preparation and for further experiments.

A patent was taken out by Watt for the copying method in February 1780, and on March 20 a fresh business under the style of James Watt & Co., of which the partners were Watt, Boulton, and Keir, was started to exploit it with the commercial public. The invention was at first received with distrust, and even with alarm by the banking fraternity, especially from the feared danger of forgery, but the inherent value of the new process forced it to the front and it remained indispensable for business purposes for a century, i. e. till the advent of the typewriter.

To return to the steam engine, we may be forgiven for quoting the only letter [3] known to us where Watt really lets himself go and gives way to exuberant feeling. It was written on the occasion when a satisfactory agreement with the Bedworth Colliery people had been arrived at. It runs as follows :

' Dear Sir,
            Hallelujah ! Hallelujee !
We have concluded with Hawkesbury—£217 pr. annum—from
Lady Day last £275 5s. for time past £117 on account.
We make them a present of 100 guineas
Peace and good fellowship on earth—
Perrins and Evans to be dismissed—
3 more engines wanted in Cornwall—
Dudley repentant and amendant—
                        Yours rejoicing,
                        James Watt.'

Watt seems to have repented of his letter almost as soon as he had penned it, for he adds at the bottom, ' Please burn this nonsense.'

Mrs. Watt and the children were away at this time on a visit to Glasgow and Watt on September 25th set out for Cornwall, arriving on the 29th at Chacewater. Not many weeks passed before he felt the need for the strong arm of Boulton to support him against the mine adventurers who were clamouring for abatement of their premiums, using as their argument the depressed state of the mining industry. As usual, also, Watt wanted Boulton to be there to execute agreements. Watt did

[1] Ibid., 1782 Jan. 15. See *Proc. Inst. Mech. Eng.* 1915, p. 510.
[2] James Keir (1735–1820), the chemist. In 1778 he gave up his glass business to take charge of the engineering works at Soho while Boulton and Watt were in Cornwall. He was offered a partnership but declined, on account of the position of the firm at that time. He was a friend of Dr. Small, Erasmus Darwin, Boulton, Thomas Day, Watt; on affectionate terms with Watt's younger son, Gregory; and one of the first members of the Lunar Society. (See Amelia Moilliet, *Life of James Keir* [1868].)
[3] Boulton Papers, 1779 June 31.

not wish to depart from the tables based on the relative performances of his own
and the common engine that he had calculated at considerable trouble. He writes:[1]

' It is absolutely necessary that you come here without the smallest delay : for besides the dissatisfaction caused by Trevethick, which some others foment, here is Wheal Virgin fairly set
on its arse this day, and is to go on with our engines, if they can beat us & the Landlords into
low enough terms. They have therefore proposed a meeting to settle an annual fixed sum with
us, to which I have answered that according to the resolutions of B. & W. a certain mode of proceeding was fixed for all & that I cannot act otherwise without your express consent & that
their present request is contrary to my own opinion, but that I shall wait upon them with our
proposals in the common way whenever they please, & that I shall summon you to attend them,
which I accordingly do.'

In his next letter to Boulton in London,[2] alluding to a meeting of the adventurers
at Hallamannin Mine, which he could not attend owing to a violent headache all
night : ' If you had been here and had gone to that meeting with your chearfull
countenance and a good heart perhaps they would not have been so resolute.'
Watt continues : ' If you have time I wish you would call at Nairnes [3] and gett
one of Mountain's [4] Sliding rules. I believe they are about 2 feet long, the longer
the better, but if any of [them] have the double radius of line of squares next one
side of the slides they will answer our operations best ; but bring one at any rate.'
As far as we can learn, Watt was the first to use the slide rule in engine calculations. This extract shows that he did not design a new rule for the purpose as some
have thought, but utilized the most suitable existing pattern.

The upshot of Watt's importunity was that Boulton went to Cornwall, but on
this occasion failed to come to an agreement as to the royalty to be paid. On
February 22, 1780, he and Watt came back to Birmingham together, leaving in
charge Lieutenant Henderson, who was employed by them to travel and superintend the erection of engines. He was not a really competent man, and Watt had
many complaints to make about his deficiencies. We must, however, remember that
technically trained men were not to be found at that time, so Watt was really
asking for the moon when he expected so much from their employees and mechanics.
Boulton took much the saner view and would not turn away any one for incompe-
tence. He knew he might as well empty the office and close the shop.

In the early part of 1780 Boulton was again in London interviewing the bankers,
to whom he now owed upwards of £17,000 advances by way of overdraft. As
security for the money, Boulton had to make further shift with the engine royalties
and give personal bonds for repayment within a given time. Watt's consent was
necessary to the first of these expedients, which he reluctantly gave it, but he would
have nothing to do with the latter. The money owed by Boulton and for which he
was in part responsible was a perpetual nightmare to Watt. His extraordinary
timidity in business matters led him to apprehend ruin and misery to himself and
family whenever any proposition concerning raising money was made.[5] It was not
as if this sort of thing was exceptional—it went on year after year. Every time the
mortgage came into Watt's mind he had a fit of the ' blues '. How Boulton could

---

[1] Boulton Papers, 1779 Oct. 28.
[2] Ibid., 1779 Oct. 30.
[3] Edward Nairne, 1726-1806, an instrument-
maker in Cornhill ; he constructed and patented
—on plans supplied by Priestley—Nairne's
electrical machine, 1782.
[4] John Mountaine, who wrote a booklet :

*A Description of Lines drawn on Gunter's
Scale as improved by Mr. John Robertson.*
London, 1778.
[5] It is difficult in these days of large capitaliza-
tions to realize that at the end of the eighteenth
century loan capital was most scarce, and even
when obtained was only short-term money or bills.

have stood so much of it, even if he were the cause of it, we confess we can hardly understand; it only throws into relief his magnificent patience, cheerfulness, and tenacity of purpose.

In the autumn of 1780 Boulton went down into Cornwall, and the first business he had to deal with was a proposition put forward by the adventurers of the Consolidated Mines that they should pay a fixed sum of £2,500 per annum for their five engines, in lieu of payment on the basis of work done, as preferred by Watt. Reluctantly the latter gave his consent, observing : ' These disputes are so very disagreeable to me that I am very sorry I ever bestowed so great a part of my time and money on the steam engine.' Boulton advised him to cheer up by making a calculation as to ' what all the engines we shall have in eighteen months erected in Cornwall will amount to : you will find it good for low spirits '.

Now that it was clear that the steam-engine enterprise would be a success, another and greater danger presented itself. This was no less than a movement the aim of which was to repeal the Act and upset the patent. It was argued that the abolition of the patent dues would be a great boon to the mining industry and also enable the adventurers to reduce the cost of their product which was the raw material for other industries. The first point may readily be conceded, not so the second : it would simply have enabled the mine-owners to pocket an increased profit. When Boulton informed Watt of this movement, he replied : ' I suspected some such move as this ; and you may depend upon it the mine owners will never be easy while they pay us anything. This is a match of all Cornwall against Boulton & Watt.' In a later letter Watt contended that if the product of brains was to be pooled for the benefit of the community, much more should land and minerals be held in common. He goes on : [1]

' They charge us with establishing a monopoly, but if a monopoly, it is one by means of which their mines are made more productive than ever they were before. Have we not given over to them two-thirds of the advantages derivable from its use in the saving of fuel, and reserved only one-third to ourselves, though even that has been further reduced to meet the pressure of the times ? They say it is inconvenient for the mining interest to be burdened with the payment of engine dues ; just as it is inconvenient for the person who wishes to get at my purse that I should keep my breeches pocket buttoned. It is doubtless also very inconvenient for the man who wishes to get a slice of the squire's land that there should be a law tying it up by an entail. Yet the squire's land has not been so much of his own making as the condensing engine has been of mine. He has only passively inherited his property, while the invention has been the product of my own active labour and of God knows how much anguish of mind and body. . . . Why don't they petition Parliament to take Sir Francis Bassett's mines from him ? He acknowledges that he has derived great profits from using our engines, which is more than we can say of our invention ; for it appears by our books that Cornwall has hitherto eaten up all the profits we have drawn from it, as well as all that we have got from other places, and a good sum of our own money into the bargain. We have no power to compel anybody to erect our engines What then will Parliament say to any man who comes there to complain of a grievance he can avoid ? . . . We are in the state of the old Roman who was found guilty of raising better crops than his neighbours, and was therefore ordered to bring before the assembly of the people his instruments of husbandry, and to tell them of his art. He complied, and when he had done said : " These, O Romans ! are the instruments of our art ; but I cannot bring into the forum the labours, the sweat, the watchings, the anxieties, the cares which produced these crops." So everyone sees the reward which we may yet probably receive from *our* labours, but few consider the price we have paid for that reward, which is by no means a certain annuity but a return of the most precarious sort.'

[1] Boulton Papers, Watt to Boulton, 1780 Oct. 31.

It is pleasant to be able to record that, at the end of 1780, Watt (but not Boulton) had really turned the corner. Up to Dec. 31st the firm had received from Cornwall in premiums nearly £2,600. In a letter to his old friend Dr. Black [1] enclosing a draft for £40, in repayment of a loan made to him years before, and promising to pay the balance at once or when they met, Watt says :

' I am very much ashamed of having been so long your debtor, but I could not avoid it without putting myself to some inconvenience, as our business never defrayed its own current charges until last year, and the product of that was swallowed up in a very large pay[t] several sums we were obliged to make at Christmas. I have therefore always been in debt to the partnership, but am now clear or nearly so.'

Another more graphic reference to the state of affairs occurs in a letter [2] from Watt to Gilbert Hamilton :

' Our general expenses have hitherto been very great so that the business never paid its way before 1780, when I guess it got about £2,000 which was all swallowed up by original sin as more must be.'

Boulton, however, was not in the same happy state ; he was again under the necessity of obtaining further capital and, in the early part of 1781, was in London, leaving Watt behind at Birmingham in a misery of apprehension, so much so that Mrs. Watt wrote to Boulton a long letter about her husband's depression. To try to dissipate Watt's gloom, Boulton wrote : ' I cannot help recommending it to you to pray morn[g] and even[g] after the manner of your countrymen [' Oh, Lord, gie us a guid conceit o'oorsels.'] for you want nothing but a good opinion and confidence in your self and good health. . . . I pray God to give you good spirits and thereby overcome all your infirmities.' [3]

Writing three days later Boulton commented on his difficulties in settling accounts with John Fothergill who was his partner in the ' toymaking' business, and added : [4]

' However, as to you & I, I am sure it is impossible we can disagree on the settleing of our acc[ts] as there is no sum total in any of them that I value so much as I do your esteem & the promotion of your health & happiness. Therefore I will not raise a single objection to anything you shall think just, as I have implicit confidence in your honor.'

This is a fine thing for one man to say of another after seven years' partnership, especially in the matter of money, which is a greater test of character than almost anything else.

Boulton's partnership with Fothergill was finally dissolved at Christmas, which must have been a great relief to Watt.[5]

Although they had now got William Murdock as their right-hand man in Cornwall—where there were twelve engines under construction—one or other of the partners could never be long absent from the county. The depressed state of the mining industry was such that it was not unnatural that the firm should be asked to take part payment for their royalties in the form of shares in the mines. Unwise as they felt such a course to be, as it thereby delayed the time when they could redeem their capital indebtedness and bank overdrafts, they had to yield.

To make their periods of residence there more agreeable, Boulton decided to take a house in Cornwall, and eventually they found what they wanted at Cosgarne in the Gwennap Valley, within a mile or so of the United Mines, yet secluded from the

[1] Doldowlod Papers. 1781 July 23.
[2] Ibid., 1781 Apl. 9.
[3] B. & W. Colln. Boulton to Watt, 1781 Apl. 16.
[4] Ibid. Boulton to Watt, 1781 Apl. 19.
[5] Ibid. Boulton to Watt, 1781 Oct. 30.

VII a. WATT'S HOUSE IN CORNWALL, PLAIN-AN-GUARY
HOUSE, REDRUTH

VII b. WATT'S HOUSE IN CORNWALL, COSGARNE HOUSE

actual mining area. Possession was obtained before June, for at the end of the month Watt went down with his wife and the two younger children and entered into residence. The change evidently did him good, for Mrs. Watt was able to report : ' James's spirits are surprisingly mended since his arrival.' Perhaps this was because of the amenities of the situation, for he speaks of it in a satisfied tone in the letter [1] to Hamilton, already quoted :

' This is a most delightful place, a neat roomy house with sash windows double breadth, the front to the south covered with vines loaded with young grapes. A walled garden with excellent peaches and plums—plenty of currants—two orchards a lawn before the door. The whole surrounded with elm trees. It stands in a valley safe from the north and south winds, but the east & west can get at us . . . the worst thing about the house is the bad roads which lead to it in every direction ; however, they are rideable.'

He still found subjects to depress him in the conduct of Dick Cartwright and others of the Soho men in their work on the engines, and in the mortgage with the bankers.[2]

' When I executed the mortgage,' he wrote, ' my sensations were such as were not to be envied by any man who goes to death in a just cause, nor has time lessened the acuteness of my feelings. . . . I thought I was resigning in one hour the fruits of the labour of my whole life—and that if any accident befell you or me I should have left a wife and children destitute of the means of subsistence, by throwing away the only jewel Fortune has presented me with. . . . These transactions have been such a burden upon my mind that I have become in a manner indifferent to all other things and can take pleasure in nothing until my mind is relieved from them ; and perhaps from so long a disuse of entertaining pleasing ideas never may be capable of receiving them any more.'

Watt descends in this letter to a depth of pessimism that makes him look almost ridiculous. It is difficult to reconcile the feelings about money matters exhibited by Watt at this period with those shown when he was in Glasgow. It is conceivable that the change was due in part to the influence of the second wife, for it is readily understood that a man of his temperament would be easily influenced by domestic worrying. Boulton, although he was gifted with a sense of humour, nevertheless could not help feeling sore with his partner's implied reflection on himself. It would have done real service, we feel, if Boulton had said to Watt's face that which he wrote in a letter to Matthews, his London financial agent : [3]

' When I reflect on his situation in 1772 and my own at that time and compare them with his and mine now, I think I owe him little. . . . I some time ago gave him a security of all my two-thirds after paying off L. V. & W. [i.e. Lowe, Vere & Williams, bankers] from which you may judge how little reason he has to complain. He talks of his wife and children, by the same rule I ought not to neglect mine. His wife's fortune joined to his own did not amount to sixpence : my wife brought me in money and land £28,000. I advanced him all he wanted without security but in return he is not content with an ample security for advancing nothing at all but what he derived from his connexion with me.'

Decidedly our sympathy is with Boulton.

On the very day that Watt was writing the above gloomy letter, Boulton wrote to him : [4]

' The people in London, Manchester and Birmingham are *steam mill mad*. I don't mean to hurry you, but I think that in the course of a month or two we should determine to take out a patent for certain methods of producing rotative motion from . . . the fire engine.'

[1] Doldowlod Papers. Watt to Gilbert Hamilton, 1781 July 23.
[2] Boulton Papers. Watt to Boulton, 1781 June 21.
[3] Ibid., 1781 June 28.
[4] Ibid., 1781 June 21.

This statement was less than the truth, for the demand in the industrial world for such engines was springing up everywhere with mushroom speed. Compared with this field for exploitation, that for the pumping engine was comparatively negligible. The far-seeing Boulton put the matter in a nutshell when he wrote to Watt:

'When Wheal Virgin is at work and all the Cornwall business in good train, we must look out for orders, as all our orders are seemingly at an end, having none now on the tapis. There is no other Cornwall to be found, and the most likely line for increasing the consumption of our engines is the application of them to mills which is certainly an extensive field.'

Accordingly Watt turned his attention seriously to the problem of making his engine capable of producing rotative motion. Nothing could be simpler, we may think to-day, than applying the crank and its concomitant, a flywheel, to an engine suitably modified. The story of Watt's share in it is not quite so simple as it appears; on the contrary it bristles with technical points, and an account of it is therefore reserved for another part of this volume.[1] Forestalled in this application, Watt now busied himself in thinking out mechanisms which would do away with the necessity of using a crank. Eventually, in July 1781, he felt that he was in a position to make an affidavit that he had invented certain substitutes for the crank, and in October a patent was granted. During the interval of four months that elapsed before the enrolment of the complete specification, he worked out no fewer than five different substitutes. Of these the sun-and-planet motion, as it was called, was the only one which was subsequently used in practice by the firm; it remained a distinctive feature of their rotative engines as long as the crank patent lasted. Watt's account of it is as follows:[2]

'I wrote you on the 31st since which I have tried one of my old plans of Rotative engines revived and executed by W. M. [William Murdock] and which merits being included in the specification as a fifth method. It has the singular property of going twice round for each stroke of the engine, and may be made to go oftener round if required without additional machinery.'

It is but fair to state that there is evidence in one of Boulton's letters to show that this mechanism was not merely revived by Murdock, but was independently invented by him. The specification, with drawings prepared by Watt himself—he was dissatisfied with those prepared by Playfair—was duly enrolled in Feb. 23, 1782.

Scarcely had he completed this task when he addressed himself to the task of rendering the engine double-acting, which was the crowning improvement that made it suitable for rotative motion, without the expedient of clumsy weights on the flywheel or beam. This improvement he patented on Mar. 12, 1782, the specification being enrolled in July. In it he included the rack and sector to give positive direction to the piston-rod, and also a rotary engine. From this time onward Watt's engine entered on a career of world-wide utility as a general prime mover.

The steam-engine monopoly had recently been threatened with an attack by powerful interests in the legislature, but the partners had now to face another danger, namely, competition by engineers seeking other means of accomplishing the same ends as did Watt's engine, but without using the separate condenser. We need only mention in the order of date, Jonathan Hornblower, Edward Bull, and William Symington. Although the schemes of these men were ingenious, yet the separate condenser was really present in all. Hornblower's engine, patented in 1781, was remarkable in that it was the first compound engine, i.e. he employed more than

---

[1] See p. 148.   [2] Boulton Papers, 1782 Jan. 3.

one cylinder to carry out the expansion. At the outset no information as to the principle of the engine could be obtained, and Watt was, as he says, ' much vexed' with the affair, and continues, ' *Jabez* does not want abilities, the rest are fools. If they have really found a prize it will ruin us. . . . Bankruptcy might ensue to us both.' [1]

When Watt learnt the nature of Hornblower's invention, he claimed that it was an old idea of his own, yet nevertheless he got alarmed when, with the assistance of some men of substance, the Hornblowers erected an engine at Radstock, near Bristol. However, this engine had to pass through the experimental stage exactly as Watt's had done. Boulton wrote to Watt : [2]

' As to ye General wish of the people being in favour of Horn[r], you mistake it, for I am sure they [i. e. the Hornblowers] are neither loved nor liked, but if the Devil came & said he would save Cornwall our premiums, they w[d] encourage him.'

Boulton sent down James Law, one of the Soho men, to Bristol to find out what progress was being made, but he could only learn that the engine was at a standstill owing to some defect. In a long letter Watt says : [3]

' How ever much I am vexed, I am rouzed and shall prepare myself to meet the worst and not lie down to have my throat cut. I beg you will summon up your resolution and not lose the battle before you fight it.'

The idea of Watt tendering this advice to Boulton is almost ludicrous ! Watt went down in person to Bristol in November, but got no satisfaction, so he inserted in the local paper an advertisement cautioning the public against using the engine as being an infringement of their patent. Watt really did get alarmed about the Hornblowers ; he called them ' Horners ', ' Trumpeters ', ' horned imps of Satan ', and other hard names. However, the Hornblowers were by no means to be despised, and later we shall hear more of them and of Edward Bull. Meanwhile Watt did not allow these distractions quite to obsess him and went on steadily with the rotative engine ; one application was to a tilt hammer, and another to a corn mill. Boulton was so impressed with the latter application that he actually had a mill put up at Soho with the sole purpose of gaining experience in driving it by steam power. As to the tilt hammer, Watt writes : [4]

' I have been at Soho . . . and saw the tilt go admirably from 18 to 24 strokes p[r] minute, and it could have gone much faster, but our men could not work the Iron under it.'

The first rotative engine actually put up [5] was in 1783 for John Wilkinson of Bradley Forge. The conservative Watt, however, seems to have still hankered after pumping-engines, and an inquiry from the Fen district led him to write to Boulton : ' I look upon these Fens as the only trump card we have left in our hand.' Boulton showed a wider and truer vision when he said :

' For my part, I think that mills, though trifles in comparison with Cornish engines, present a field that is boundless and that will be more permanent than these transient mines, and more satisfactory than these inveterate, ungenerous and envious miners and mine lords.'

In August 1782 James Watt's father died at Greenock. He was unable to go there but wrote : ' there is some thing so afflicting in the thought of the final solemn departure of a beloved friend and revered parent that though I have been by his

---

[1] Boulton Papers, 1781 July 16.
[2] B. & W. Colln., 1781 Aug. 14.
[3] Boulton Papers, 1782 Sept. 28.

[4] Ibid., 1782 Nov. 30.
[5] That is, outside Soho.

long illness and declining state prepared for the event, the account of it has given me much pain.'

In the spring of 1783 Watt was again in Cornwall, and wrote long letters telling Boulton of the ins and outs of affairs in the mines and among the adventurers. A letter of February 20 contains the information, ' Murdock has gotten the absolute power over Wheal Virgin engines, . . . Wm. is also put in full power over engines and engine men' at Poldice. This shows what a commanding position Murdock was attaining in Cornwall. In the autumn of this year Boulton's health showed signs of giving way ; he went for a tour in the north of England and Scotland, and returned restored to health.

It is pleasant to reflect how Watt, in his new surroundings was, in spite of his unremitting attention to business, steadily maturing intellectually and expanding socially. There were the visitors to Soho on business whom Watt had to entertain, at any rate when Boulton was away from home, but it was the group of literary and scientific men forming the Lunar Society that we allude to especially. We can picture the kind of atmosphere at the meetings from Darwin's excuse for his absence on one occasion :

' I am sorry the infernal divinities who visit mankind with diseases and are therefore at perpetual warfare with Doctors should have prevented my seeing all your great men at Soho today. Lord ! what inventions, what wit, what rhetoric, metaphysical, mechanical and pyrotechnical, will be on the wing bandied like a shuttlecock from one to another of your troop of philosophers while poor I, I by myself, I imprisoned in a postchaise am joggled and jostled and bump'd and bruised along the King's high road to make war on a stomach ache or a fever.' [1]

Watt took his turn with the rest in having the members to dinner, and we have quite an agenda of one of such meetings :

' I beg that you would impress on your memory the idea, that you promised to dine with sundry men of learning at my house on Monday next. . . . For your encouragement there is a new book to cut up ; and it is to be determined whether or not heat is a compound of phlogiston and empyreal air, and whether a mirror can reflect the heat of the fire. I give you a friendly warning that you may be found wanting which ever opinion you adopt. . . . If you are meek and humble, perhaps you may be told what light is made of, . . .' [2]

Darwin replied that the

. . . ' devil has played me a slippery trick and I fear prevented me from coming to join the holy men at your house. . . . I can tell you some secret in return for yours viz . . . that water is composed of aqueous gas which is displaced from its earth by oil of vitriol . . .'

It was a great deprivation to Darwin and a loss to the Society when he removed to Derby in 1782, for as he says, ' I am here cut off from the milk of science, which flows in such redundant streams from your learned lunatics.'

Before his departure the little circle had been enlarged by the inclusion of Dr. Joseph Priestley who in 1780 became Minister of the New Meeting House, Birmingham. He was no stranger to Boulton and he was brother-in-law of Wilkinson. Priestley was most versatile, had rare perception and tremendous powers of application. His studies ranged over history, theology, and science, on all of which he wrote voluminously. As literary companion to Lord Shelburne he had made a Continental tour in 1773. In 1774 he discovered oxygen (dephlogisticated air), and followed this up by the discovery or isolation of many other gases. Small wonder

[1] Smiles : *Boulton and Watt.* Darwin to Boulton, 1778 Apr. 5.

[2] Muirhead : *Mech. Inv.* II. 123. Watt to Darwin, 1781 Jan. 3.

that his enthusiasm should communicate itself to the Lunarians and that chemistry should become the rage. Boulton wrote :

' Chemistry has been my hobbyhorse but I am prevented from riding it by cursed business except now and then of a Sunday.' [1]

Priestley in 1781 observed the formation of moisture when inflammable air (hydrogen) and dephlogisticated air (oxygen) are mixed in certain proportions in a glass vessel, and fired by the electric spark, but failed to grasp the true explanation of the phenomenon.

The experiment, however, aroused the reasoning powers of Watt ; he deduced that ' water is composed of dephlogisticated air and phlogiston deprived of part of their latent or elementary heat ; that dephlogisticated or pure air is composed of water deprived of its phlogiston, and united to elementary heat and light ; and that the latter are contained in it in a latent state, so as not to be sensible to the thermometer or to the eye ; and if light be only a modification of heat, or a circumstance attending it, or a compound part of the inflammable air, then pure or dephlogisticated air is composed of water deprived of its phlogiston and united to elementary heat'. [2]

Such was the conclusion that Watt, after much discussion with his fellow ' lunatics ', embodied in 1783 in a letter to Dr. Priestley for the purpose of being read to the Royal Society. The letter was delayed, however, for about a year because Watt wanted confirmation from Priestley of some of his facts. Meanwhile Henry Cavendish had been engaged in similar experiments, and his paper was communicated to the Royal Society before the letter of Watt. A heated controversy ensued between the adherents of Watt and Cavendish as to which was entitled to the credit of being the discoverer of the composition of water. The controversy is not one that is worth reopening at this date ; it will be sufficient to quote the views of an authority on the history of chemistry, Sir Edward F. Thorpe. [3] His verdict is :

' To Cavendish belongs the merit of having first supplied the true experimental basis upon which accurate knowledge could alone be founded. Watt, on the other hand, although reasoning from imperfect and indeed altogether erroneous data, was the first, so far as we can prove from documentary evidence, to state distinctly that water is not an element, but is composed, weight for weight, of two other substances, one of which he regarded as phlogiston and the other as dephlogisticated air. It would be a mistake, however, to suppose that Watt taught precisely the same doctrine of the true nature of water that we hold to-day. Nor did Cavendish utter a more certain sound. What we regard to-day as the expression of the truth, we owe to Lavoisier, who stated it with a direction and precision that ultimately swept all doubt and hesitation aside—except to the mind of Priestley, whose "random experiment" gave the first glimmer of the truth.'

' As regards Watt this incident serves to bring out only more clearly . . . the true character of the man. It illustrates the vigour of his intellectual grasp, the keenness of his mental vision. At the same time, it exhibits his love of truth for truth's sake ; his unaffected modesty ; and the sense of humility, that was not the less real because accompanied by a sense of what his inherent rectitude taught was due also to himself.'

It was while engaged on reasoning as to the composition of water that Watt found himself ' much plagued reducing French grains to English ones '. [4] He pointed out the trouble of comparison even with the same standard owing to the ' absurd subdivisions used by all Europe ' ; further he argued against the unneces-

[1] Boulton Papers, 1781 Sept. 6.
[2] Phil. Trans., 1784, 331-3.
[3] Thorpe : History of Chemistry, 1902, p. 121.
[4] Watt to De Luc, 1783 Nov. 23, quoted in Muirhead : Mech. Inv. II. 182 et seq.

sary calculation involved in comparing cubic inches with weights. He proposed a
' decimal subdivision of the pound' into 10,000 ounces, ' decimal tables of specific
gravities will give the weights without calculation. All liquids to be weighed.' He
also discusses an absolute measure of length derived from the seconds pendulum.
His practical nature comes out in his conclusion that in the consequential change of
standards : ' I therefore give the preference to those plans which retain the foot
and ounce '. Watt's proposals were communicated to Laplace [1] but we are on un-
safe ground in assuming, as some have done, that they were the basis of the metric
system introduced in France after the Revolution.

STRENGTH OF MATERIALS

THE temper of Watt's mind to ' try all things' is shown by experiments on the
strength of materials carried out by him in 1783 and 1784. It is true that he was not
the first to carry out such tests, but there is a range and variety about them that
strike one as typical of him. The record [2] is as follows :

' Experiment on the stiffness of iron Augt. 1783.

' A Barr of Swedish Iron 1 inch sq^r and 4 feet long was placed perpendicularly, its ends were
made conical, one was fixed in a hole in a piece of iron below and the other in a hole in a piece of
iron fixed to a lever above. The lever was loaded with weights, and the Barr struck on the side
with a hand hammer. This barr supported 11 cwt. 1. 0. before it began to yeald to the blows and
then bent only 1/8 inch.

' A barr of the same dimensions and 2 feet long, carried 17 cwt. 3.14. It ought to have carried
twice as much as the other but I suppose was not such good Iron.

' A barr of *cast iron 1 inch square* was supported by 2 supports at 2 feet distance from each
other, and in the middle between them were suspended weights.

		cwt.
1st Expt.	Bent about 1/8 Inch. with	4
	Broke	12½
2nd	Bent ¼ inch	10
	Broke	11¾ '

A record follows of other experiments on 1½ in. sq. cast-iron bar ; 1 in. and 1½ in.
sq. wrought-iron bar ; 2 in. square both oak and deal ; 1 in. sq. both oak and deal.
In the case of the 1 in. sq. wrought-iron bar there is a note that after bending the
bar ' servives to former Straightness '.

What was the purpose of the first experiment unless it was to simulate a part of
a machine, we cannot say, nor can any useful deduction be made from the data
since they are incomplete.

In the case of the bending tests we are on firmer ground. The measurement of
the deflexion is too rough to base any calculation upon but the breaking load shows
the iron to have been of excellent quality (i. e. modulus of rupture 22·5 and 21·2 tons
per sq. in. respectively).[3]

We might instance further examples, did space permit, of this almost instinctive
resort to actual experiment which was so characteristic of Watt.[4]

---

[1] Laplace (Marquis Pierre Simon de), 1726–
1806, the French astronomer and mathematician.
[2] B. & W. Colln. : Blotting and Calculation
Book 1782 to 1783, p. 55.
[3] *Loc. cit.*, p. 87. There is another series of
experiment recorded on the ' testing of wood and

iron ' but nothing very satisfactory can be deduced from them ; accordingly they are not
quoted.
[4] Cf. Boulton Papers : Watt to Boulton, 1786
Mar. 20 and 22, quoted at length in *Proc. Inst.
Mech. Eng.*, 1915, p. 517.

In 1784 Watt was occupied with another patent, in which he included the application of the double-acting engine to a tilt hammer to wheel carriages and to other purposes. It is in this patent that his most elegant invention, that of the parallel motion, is included. Its purpose cannot be explained better than in Watt's own words : [1]

' I have started a new hare ! I have got a glimpse of a method of causing a piston rod to move up & down perpendicularly by only fixing to it a piece of iron upon the beam, without chains or perpendicular guides or untowardly friction, arch heads or other pieces of clumsiness ; by which contrivance it answers fully to expectation. . . . I have only tried it in a slight model yet, so cannot build upon it, though I think it a very probable thing to succeed. It is one of the most ingenious simple pieces of mechanism I have ever contrived, but I beg nothing may be said on it till I specify.'

A little over a week later, July 11, Watt writes that he has

' made a very large model of the new substitute for racks & sectors which seems to bid fair to answer. The rod goes up and down quite in a perp^{lar} line without racks, chains or guides. It is a perpend^r motion derived from a combination of motions about centres very simple . . . however, don't pride yourself on it, it is not fairly tried yet, & may have unknown faults.'

As mentioned above, the specification of 1784 included the application of the engine, with or without the condenser, to road carriages. There does not seem to be much doubt that the reason why Watt included it was because Murdock that summer had taken up an old project of his, that of steam locomotion, and had occupied his spare time, it is said, in making a working model which he had tried at Redruth. It was a curious trait in Watt's character that no sooner did any one suggest an application of steam than he at once seized on it, and by calculation and reasoning carried the matter on paper to a stage much farther than the original suggestion. It was so in the case of the steam turbine,[2] and now in the case of the steam carriage he writes to Boulton [3] a letter of inordinate length, amounting almost to a treatise on the subject, finishing with these words :

' You will see that . . . the machine . . . will cost much time to bring it to any tolerable degree of perfection, and that for me to interrupt the career of our business to bestow my attention on it would be imprudent.'

In this his judgement, as proved by the experience of succeeding inventors and engineers, was quite sound. The second part of this letter is taken up with his views as to what their policy should be with regard to Murdock, for as Watt says, ' if he succeeds tolerably, he will naturally attach himself totally to it, and we shall lose the benefit of his knowledge and experience in our Cornish business.'

Although Watt says that he does not want in any way to ' prevent him going on with his scheme ', it is quite clear that Murdock was talked over for the time being. He did not entirely drop the subject, however, for in 1786 he undoubtedly made a model. He was still in Cornwall, but this time it was Boulton who was down there, and Watt writes to him : [4]

' I am extremely sorry that W. M. still busys himself with the St^m Carriage . . . in the meantime I wish W. could be brought to do as we do, to mind the business in hand, & let such as Sym^n [i. e. Symington] & Sad^r [i. e. Sadler] throw away their time and money hunting shadows.'

[1] Boulton Papers : Watt to Boulton, 1784 June 30.
[2] i. e. Kempelen's aeolipile patented 1784. The letter (Boulton Papers, Watt to Boulton, 1784 May 11) is given in *Proc. Inst. Mech. Eng.*, 1915, 514. It will surprise the non-technical reader to learn that a steam turbine was thought of at so early a date.
[3] Boulton Papers, 1784 Aug. 31 in Muirhead : *Mech. Inv.* II. 194.
[4] Boulton Papers, 1786 Sept. 12.

V

PARTNERSHIP WITH BOULTON.
ACHIEVEMENT OF SUCCESS, 1784–1800

IT is pleasant to be able to record that by 1784 the tide in the affairs of Boulton had turned, and that he was rapidly freeing himself from indebtedness. He estimated that their dues from Cornwall, for the year 1784, if all paid, would amount to £12,000 and there was in addition other business of rotative engines, &c., which was mounting up. By the end of the following year, 1785, it may be said that all Boulton's overdrafts and bankers' advances had been paid off. Watt, of course, had been in receipt of his salary of £330 per annum and expenses, and he now had the pleasure of sharing in the profits. The effect of this happy state of affairs on the two men was characteristic. Boulton, sanguine and buoyant as he had been a quarter of a century earlier when he married and received a fortune with his wife, was ready, nay anxious, to embark in new enterprises. Not so Watt, who, looking back on nearly half a century spent in comparative poverty and ever contending with disappointments and ill health, was intent on consolidating his gains and would, we feel sure, have been satisfied to have retired at this juncture. However, there could be no question of that for the business demanded ever-increasing attention ; as Watt said : ' I see plainly that every rotative engine will cost twice the trouble of one for raising water, and will in general pay only half the money.'

Moreover, the web of political affairs at this juncture caught them in its toils. William Pitt, on taking office in 1784, projected taxation which, although necessary to restore financial equilibrium in Great Britain, was conceived in a manner that, in the opinion of the manufacturers of the country, would have the effect of seriously injuring trade. Pitt's proposal was to tax coal and other raw materials to the extent of about one million pounds a year. Boulton considered the policy suicidal, at that time especially, when foreign governments were imposing heavy import duties upon English goods. Associating himself with Wilkinson, Wedgwood, and others of his friends, he agitated till he succeeded in 1785 in establishing a Chamber of Manufacturers in London, to watch over their interests ; Boulton was naturally appointed the Birmingham delegate. His dictum was : ' Let taxes be laid upon luxuries, upon vices, and, if you like, upon property ; tax riches when got and the expenditure of them but not the means of getting them ; but of all things don't open the hen that lays the golden eggs.'

At the same time the Irish Parliament was imposing duties on manufactures imported into Ireland from England. Watt was drawn into the fray, much against his inclination, and is believed to have had a hand in drawing up for circulation : ' Remarks on a Government Paper entitled Iron Trade, England and Ireland.' [1] Watt advocated what we should now call free trade within the United Kingdom. The result of this agitation was that the proposals for taxation by the English and Irish Parliaments were materially modified.

[1] B. & W. Colln., printed, n.d.

To return to Soho : the most important work in hand during 1785 was the engine for the Albion Flour Mills in Southwark. As we have mentioned above, Boulton was convinced that the double-acting rotative engine would eventually be in request for grinding corn, and had conducted experiments on the subject. The Albion Mills project had been started as early as 1783, but it met with considerable ·opposition on the ground that it would compete unfairly with existing wind- and water-mills and throw people out of work.· Boulton pointed out that the logical sequence of this argument would be ' to go back again to the grinding of corn by hand labour'. A sufficient number of gentlemen were got together to form a company, however, but their application in 1784 for a charter of incorporation was refused and the company was constituted in the way then usual by partnership. The mill building was erected by Samuel Wyatt, the architect, the grinding and dressing machinery were entrusted to John Rennie, and the engine, of course, to Watt. The erection of the building and machinery occupied the whole of the ·year, and it was not till 1786 that the mill was ready to start. Boulton was in town for the occasion and experienced the difficulties with which all new operations are attended. But Watt tells him : [1]

' Above all patience must be exercised and things coolly examined and put to rights and care must be taken not to blame innocent parts. Everything must as much as possible be tried separately. Remind those who begin to growl that in new, complicated and difficult things human foresight falls short—that time and money must be given to perfect things and find out their defects.'

The cost of erection of the mill considerably exceeded the estimate, but Watt was too experienced to worry about such a matter ; what did hurt him, however, was the news that the mill was being made a show place. The letter he addressed to Boulton [2] is so much to the point and so characteristic of the man that we quote it at some length :

' It has given me the *utmost* pain to hear of the many persons who have been admitted into the Albion Mill merely as an Object of Curiosity. Were there no other loss than the taking up your time it is a very serious one but there are other essential ones which are too obvious to need to be pointed out, among which are that the disgraceful condition in which it has hitherto been has been more likely to do us hurt than good as engineers and the bad management or want of management in oy[r] [i. e. other] respects must hurt the Credit of the Company—I hear from different quarters enough to convince me that we are looked upon by the serious common sense men as vain and rash *adventurers* and that our talking of what we can do is construed into either a want of ability to perform it or the foolish cry of Roast beef [3]—My natural hatred of ostentation may perhaps make me feel these things too strongly, but surely those who say so think they have some reason for the observations & it cannot happen that the most pointed of them can come to my ears, considering how little company I keep—Among other things I heard some time ago that on a certain day there was to be a Masquerade at the A M, and this from persons no ways connected with us & who had heard it as com[n] Birm[m] talk—and I felt it as a severe reproach Considering that we are much envied at any rate, everything which contributes to render us conspicuous should be avoided, let us be content with *doing*   R. [i. e. Rennie] no doubt has vanity to indulge as well as us but he sh[d] be curbed & the bad consequences pointed out to him, it will ruin him, Dukes & Lords & noble peers will not be his best customers—And let me entreat that the doors of the Mill be strictly shut against all comers without an order signed by three comittee men & that only at a comittee meeting on some fixt day of the week

---

[1] Boulton Papers : Watt to Boulton, 1786 Mar. 10.
[2] Ibid., Watt to Boulton, 1786 Apl. 17.

[3] ' Roast beef '—an announcement of one's success or prosperity (cf. *Notes & Queries*, 12 Ser. XII. 176).

& let that rule be inflexibly adhered to. I know that you have been actuated by good motives in showing the mill, namely the desire of getting quit of part of the property we have in it & the hopes of making interest to get a charter, but I conceive these things will be better attained by making it a mystery to the many & by the external appearance of business.

' I cannot think on corn machines till everything else is successful—In the meantime the corn sh⁴ be frequently turned & the flour attended to, for it will suffer much more than the corn.

I remain, my Dear Sir,
Affectionately Your's
JAMES WATT.'

Although technically successful the mill can hardly be said to have been successful commercially, in spite of the fact that the turnover was very large. There were great hopes that it would be able to get over its difficulties, when, in March 1791, it was burned down.[1] Boulton lost about £6,000 and Watt £3,000.

Towards the end of the year 1786 orders for engines poured in apace at Soho, keeping Watt very busy, not only with the preparation of drawings but also with the men. Writing to Boulton in Cornwall he says : ' I foresee I shall be driven almost mad in finding men for the engines ordered here and coming in.' In another letter he writes : ' I have no doubt that we shall soon so methodize the rotative engines as to get on with them at a great pace. Indeed that is already in some degree the case, but we must have more men and these we can only have by the process of breeding them.'

By this time Watt was in an easier situation with regard to the office work. John Southern had entered the service of the firm in 1781, and being well educated and of a mathematical turn of mind was establishing himself as Watt's right-hand man.[2]

In 1785 Watt turned his attention to consuming the smoke from boiler furnaces. ' I have some hope of being able to get quit of the abominable smoke which attends fire engines,'[3] and on December 11th he writes : ' I have accomplished the fire without smoke.' Meanwhile he had taken out a patent for this—the first smoke-consuming device on record.

In the letter just quoted he says :

' I have been turning some of my idle thoughts lately upon an arithmetical machine ... I intend to make an attempt at making it for though the machine is exceedingly simple, yet I have learnt by experience that in mechanics many things fall out between the cup and the mouth.'

The only evidence of this attempt is a wooden model in the Watt workshop from Heathfield.

There is evidence that the partners about this time meditated the acquisition of patent or exclusive privileges outside the United Kingdom. On May 27, 1786, one Barnabas Deane petitioned the Assembly of the State of Connecticut for the grant of an exclusive right for making and erecting steam engines.[4] The application was refused. The names of Boulton and Watt are not mentioned, but the inventors are described as having obtained an Act of Parliament in Great Britain giving them an exclusive right to the invention for twenty-five years. There is therefore no reasonable doubt as to whom the inventors were.

A patent dated Nov. 8, 1786, for the ' States of Holland and West Friezeland ' for a term of fifteen years seems also to have been obtained,[5] but for what particular invention we have not been able to ascertain.

---

[1] For a fuller account of the undertaking see p. 164.   [2] See p. 285.
[3] Watt to De Luc, 1785 Sept. 10 : Muirhead :
Mech. Inv. II. 204, 208.
[4] Report U.S. Patent Office, 1850, p. 444.
[5] B. & W. Colln.

It was doubtless the pursuit of the same policy that caused the partners to accept an invitation [1] from the French Government to visit Paris in order to confer about an engine to replace the celebrated machine of Marly [2]—now grown very leaky after a service of upwards of half a century—that supplied Versailles with water from the Seine. Both Boulton and Watt went and were ' treated with much respect, civility and attention ', the expenses of their journey being paid. The partners made extended calculations about the machine of Marly but would not commit themselves further than to give a general opinion.

They paid a surprise visit to Périer ; who admitted that it was ' un coup de soufflet diabolique pour lui '.[3] Watt continues : ' He has succeeded, however, in having erected a most magnificent and commodious manufactory for steam engines where he executes all the part most exceedingly well. He is a man of abilities and would be very estimable if he were a little more just or more honest.'

There is extant [4] a petition to the King of France from Boulton and Watt dating probably from this time reciting the events leading up to it and praying that His Majesty will deign to recompense them by confirming the Arrêt by Letters Patent. The partners did not propose to disturb Périer in his privilege nor in any improvements he could make in the pumping engine, but desired to cover in France their subsequent improvements as to the rotative engine.

We gather that not only were Boulton and Watt well ' dined and wined' during their visit but, what was more important, they made the acquaintance of a number of influential men among whom was M. de Calonne, Controller General of Finance to Louis XVI. The negotiations appear to have been conducted through the Abbé de Calonne, brother to the Minister. On their return Watt wrote a diplomatic letter [5] of most inordinate length to the Abbé. In it we can see the hand of Boulton equally with that of Watt, both as regards the breadth of view displayed and the business acumen shown. Commenting on the fiscal reforms that M. de Calonne was making, Watt says:

' That part of our patriots who are not over liberal fear that he will by freeing your commerce and agriculture from its shackles make France too rich and too great for the welfare of Britain and its trade. We argue otherwise, for we say if these measures contribute to increase your manufactures or commerce, that our government will be obliged to lay aside the erroneous part of their system, and free our manufactures from their grievances, and that the richer France becomes the better Customer she can be to Britain, and that at the worst if no change of system is adopted here and our trade becomes ruined by the aggrandisement of yours, that you will have made France so desirable a country that all active men who are not rooted to the soil like so many vegetables, will remove thither, and help to make it still greater leaving our ruined land-holders to pay the national debt, to eat their own corn, and muddle their undertaking with their own ale, without one enlivening drop of burgundy, except what the wool of their sheep can purchase.'

This is pure *Wealth of Nations* argument, and shows that Watt must have imbibed freely the doctrines of Adam Smith while he was under the latter's influence at Glasgow. These negotiations were interrupted primarily by the fall from power of M. de Calonne a month later, and secondarily by the troublous times of the French Revolution which supervened.

[1] The text of the letter, dated Sept. 1786, is given by Muirhead : *Mech. Inv.* II. 212.
[2] This once-famous hydraulic machine was the work, in 1676, of Renquin Swalm, of Liège. It gave its name to the village—north of Versailles—of Marly-la-Machine; see V. Dwelshauvers-Dery: *Quelques Antiquités mécaniques de la Belgique,* 1906.
[3] Watt to Roebuck : Muirhead : *Mech. Inv.* II. 215.
[4] In the B. & W. Colln.
[5] B. & W. Colln., 1787 Feb. 10, given *in extenso* in *Proc. Inst. Mech. Eng.,* 1915, 520.

The visit to France in 1786 widened Watt's horizon and was productive of many friendships. We can only mention that with M. Claude Louis Berthollet, the French physician and chemist with whom he corresponded frequently in after years. Berthollet had discovered in 1785 the method of bleaching by chlorine gas and exhibited the experiment to Watt.

On arriving home, the latter made some experiments which he communicated to his father-in-law, Mr. Macgregor, who carried on the business of a bleacher. Berthollet wrote Watt that he did not wish to make anything out of the invention by patent or otherwise, but Watt expressed a wish ' to procure such reward as I could to a man of merit who had made such an extensively useful discovery in the arts ', and at the end of December 1787, when in Glasgow, tried out the process on the practical scale. In the following February Watt mentions that he has seen in the papers an account of a liquor for whitening linen and has no doubt that it ' is the same that Mr. Bertholet, the inventor, showed to us, and which in consequence of his permission and generous refusal to apply for an exclusive privilege to himself, I have used for more than a 12 month, and have made many expensive experiments upon, and which I have communicated to Mr. McGrigor who is now whitening 1,500 yards of linen by the process '. [1]

Naturally in the absence of any restriction several firms had taken up the process. However, the process was so inconvenient and detrimental to health that it was not till bleaching powder—formed by the action of chlorine on slaked lime—was discovered by Sir Charles Tennant in 1799 that this method superseded the old and tedious one of the bleachfield.

By the end of the year 1787 Watt was not only relieved from any responsibility for Boulton's capital but had a large sum standing to his credit. On December 7 [2] Boulton wrote to Matthews in London :

' As Mr. Watt is now at Mr. Macgregor's in Glasgow, I wish you would write him a line to say that you have transferred £4,000 to his own account, that you have paid for him another £1,000 to the Albion Mill, and that about Christmas you suppose you shall transfer £2,000 more to him, to balance.'

In Boulton's case his only use for money seems to have been to launch out into new enterprises. He had in 1785 been the prime mover in forming a Copper Company to take the ore from the different mines over a long period, with the object of steadying the market. So far as the Editors are aware this is the first example of a trust, a development usually thought to be of American origin. One result of this scheme was that Boulton made it his business to find a market for the copper produced ; in fact he took a large quantity himself. When the commercial panic of 1788 occurred, he found great difficulty in holding on to it. In his embarrassment he applied to Watt who, however, had invested his money and could not help him. Boulton in a letter to Wilson, says : [3]

' Mr. Watt hath lately remitted all his money to Scotland, and I have lately purchased a considerable quantity of copper at the request of Mr. Williams. . . . I shall be in a very few weeks in great want of money, and it is now impossible to borrow in London or this neighbourhood as all confidence is fled.'

[1] Boulton Papers : Watt to Boulton, 1788 Feb. 25. The only present Berthollet would accept from the English manufacturers who used his process was ' un morceau de toile blanchi par son procédé '. Cuvier (G. L. C. F. D. de) : *Éloge historique de M. le comte Berthollet*, 1824, p. 14.
[2] Boulton Papers, 1787 Dec. 7.
[3] Ibid., 1788 May 4.

One gathers from this that Watt had been asked to advance money and had declined. Such conduct on the part of Watt towards his partner, and such a partner as Boulton, seems rather shabby. One would have thought that Watt would have stripped himself to have helped Boulton, and we cannot help regarding this as a blot on Watt's conduct.

Boulton took up the coinage of money, and medallic work generally, and to it he devoted most of his latter years. An account of this work [1] written by Watt in a MS. memoir of his friend throws into great prominence the work of this truly remarkable man. It is not the place here, however, to go further into the matter, engrossing though the story is.

While Boulton does not seem to have diminished his interests and enterprises, Watt was beginning for the first time to enjoy life ; in August 1789 we actually hear of him taking a holiday ! He and his wife went to see the beauties of Worcester, Malvern, Hereford, and Chepstow—with consequent improvement in health and spirits. In the following summer he visited London and had the honour of an interview with the King at Windsor, of which he gave a succinct account to Boulton. Watt had been presented to His Majesty on the occasion of his visit to Whitbread's Brewery in 1787.

How easy Watt's financial situation had become will appear from a statement, ' Money received from Cornwall to Dec. 1791', showing that a sum of nearly £76,000 had been received for premiums from 1780 to date.[2]

The momentous times of the French Revolution did not pass without reverberations in this country generally and in Birmingham in particular. Events were followed with the keenest interest by the members of the Lunar Society, more particularly by Dr. Priestley who for a time forgot his scientific pursuits in order to plunge into political controversy with Edmund Burke, a bitter opponent of the Revolution. So vehement were Priestley's attacks that he was acclaimed as a brother by the Société des Amis de la Constitution de Bordeaux. Trouble arose out of a public dinner which was held on July 14, 1791, in the principal inn—the *Hen and Chickens*—in Birmingham to celebrate the second anniversary of the Revolution. A crowd collected and proceeded to smash the inn windows. Although Priestley was not present at the dinner, the mob went on to smash the Meeting House where he ministered, and as night approached made for his house at Fairhill, which in a few hours was gutted of all furniture, apparatus, books, and MSS.—the latter the results of twenty years' study—and the house burnt to the ground. For three days more the mob continued its violence till the military arrived. Both Boulton and Watt were apprehensive that attacks would be made on their houses and on the Manufactory at Soho. The flint-lock muskets that Boulton provided to defend their premises are still to be seen among the objects preserved in the Boulton and Watt Collection at the Birmingham Reference Library. The storm blew over, more, in Watt's opinion, because their houses were at a distance from the quarter where the rioters lived than from any other reason. Watt described the affair in a vivid letter to De Luc.[3]

Priestley escaped and went to London whence he emigrated to America. There his friends saw to his welfare, and supplied him with apparatus for which he was profoundly grateful, but nothing could supply the advantages of the Lunar Society.[4]

[1] Boulton Papers.   [3] Muirhead : *Mech. Inv.* II. 241.
[2] B. & W. Colln.   [4] Ibid. 275.

An intimate glimpse into the life of the Watt household at Harper's Hill in 1790 is afforded by the reminiscences of Mary Anne Galton,[1] who was a playmate of the children.

' Mr. and Mrs. Watt were amongst our most intimate friends, and constantly formed a part of our social circle. They then lived, not in the handsome mansion and domain they afterwards occupied, but in a very moderate house in the suburbs of Birmingham, at Harper's Hill. In this house we were frequent visitors, and there Mr. and Mrs. Watt resided with a very simple establishment of two maids and a man-servant, all brought up under their own eye, and trained by Mrs. Watt in the thrifty and far-seeing habits of the most enlightened Scotch housewifery ; besides which they had two little pug-dogs, which were likewise taught by Mrs. Watt never to cross the unsullied flags of the hall without wiping their feet on the mats, placed at every door of entrance.

' They had one son and one daughter. The son, about thirteen, named Gregory, was a youth of very precocious talents ; the daughter was a pretty-looking girl, of no very decided character ; but I must describe the house and family more particularly.

' Mr. Watt, deeply absorbed in his philosophical pursuits, was simple in all his habits. He had not the domestic cheerfulness of Dr. Priestley, but he was ever ready to give information, even to the most ignorant ; and often do I remember his calling me to sit upon his knee, whilst he explained the different principles of the hurdy-gurdy or monochord, the harp, and the piano, or the construction of a simple whistle or Pan's-pipe or of an organ ; but he never failed to tell me, that the hurdy-gurdy was the most venerable in point of antiquity, being no other than an adaptation of the celebrated monochord of Pythagoras. When I recollect Mr. Watt's philosophic mind, and calm truth and loving-kindness, I have often thought that Miss Edgeworth, in her story of Harry and Lucy, had, in the character of Harry, depicted what she conceived the childhood of Mr. Watt might have been.

' The mental fatigue of Mr. Watt at this period was often so great, that I have heard he required from nine to eleven hours' sleep to recruit his powers, and his evenings were uniformly spent in some light amusing reading. Mrs. Watt was exactly the needful help to her scientific husband, to whom she was wholly devoted, and whose fame she considered her crowning glory.'

Mr. and Mrs. Watt had already begun to think about removing from Harper's Hill to a more agreeable situation, and in 1790 Watt began the erection of a roomy comfortable mansion, aptly named Heathfield from its situation on Handsworth Heath, to the designs of Samuel Wyatt the architect. A garret over the kitchen was retained by him for use as a workshop, access to it being obtained by the back stairs, while in the yard below he had a forge. He took the opportunity afforded by the enclosure of the Heath in 1791 to acquire some of the land and continued to add to the property as opportunity offered till he had about forty acres in his possession. He proceeded to lay out the grounds, wall in a kitchen garden and erect hot houses, build lodges and plant timber, till the place became a scene of quiet beauty ; and such it remained until last year, despite the fact that it had been encroached upon from all sides by bricks and mortar. We learn with regret that the house is being pulled down as the estate has been sold for building purposes.

The early part of the year 1792 was much taken up with litigation against Hornblower and Maberley. In a letter to John Southern [2] Watt says : ' as they have brought Mr. Giddy, the High Sheriff of Cornwall, an Oxford boy to prove by *fluxions* the superiority of their engine, perhaps we shall be obliged to call upon you to come up by Thursday to face his fluxions by common sense.'

[1] *Life of Mary Anne SchimmelPenninck*, by Christina C. Hankin, 1858, p. 285. She was the daughter of Samuel Galton, quaker and Birmingham manufacturer. As she was only twelve years of age at this time and did not record the events till sixty-six years afterwards, too much reliance cannot therefore be placed upon the accuracy of her account.

[2] Southern Papers, 1792 Apl. 21.

This was Davies Giddy who assumed the name of Gilbert, and was afterwards President of the Royal Society. The case, however, was not finished this term.

Watt was again in London in August and went thence, accompanied by Boulton, to Cornwall in September. In a letter to Southern,[1] Watt says :

' The people here are in general as absurd as ever, nor do I believe it possible for an angel to convince them of the inferiority of Hornblower's engine though it is now nearly unable to keep the bottoms dry.'

In another letter (September 22nd) Watt reports tests of engines at Wheal Butson, Poldice, and Tincroft,[2] and remarks :

' No body or few will own conviction though we have circulated accounts of these experiments. But we have reason to believe they have taken considerable effect on men's minds & we hear that Bull has given up one of his orders & that a small engine made by the Horners will not work.'

The following year found Watt in town over the lawsuit with Edward Bull. Watt was again in London for the Michaelmas term, but it was finally found impossible to have the case tried that term. The following term found other delays so that Boulton and Watt were obliged to get injunctions against Bull and others to restrain them in the interim. After going into the minutiae in one letter, Feb. 12, 1794, Watt says : ' All this may be law or perhaps *Justice*, but I shall swear in Chancery if required that it [is] not *mercy*.'

Watt abominated litigation and yet he was involved in still another lawsuit with Hornblower and Maberley. It was not until the Michaelmas term of 1797 that this particular litigation was concluded, and it was 1799 before the validity of the patent was finally established. It was during this period of suspense that the various Cornish mine adventurers, knowing that the patent had only a few years to run, and believing that the partners would not be able to enforce payment, set them at open defiance. Watt wrote to Boulton : [3]

' The rascals seem to have been going on as if the patents were their own. . . . We have tried every lenient means with them in vain and since the fear of God has no effect upon them, we must try what the fear of the devil can do.'

Accordingly legal proceedings were taken against the mine adventurers and eventually after great difficulty most of the arrears were collected.

The younger generation was now taking up the reins of management to the great satisfaction of the two partners, each of whom was represented : Boulton by his only son Matthew Robinson Boulton, and Watt by James Watt, born in 1769, his only surviving son by his first wife. Watt had, it is true, a son Gregory born in 1777 by his second wife, and although eventually he also became a partner, he took very little share in the business. Practically, therefore, each partner had one son to follow him. Both these young men had, as one would expect, intelligence and character ; each one had had a careful education including practical knowledge of foreign languages. There is singularly little reference to young Watt in the correspondence and we conclude that Watt was undemonstrative and averse from the exhibition of feeling or affection even with his own family. Young Watt was sent at the age of fifteen to Mr. Wilkinson's Ironworks at Bersham in Wales, where his father says of him : ' he is to study practical book-keeping, geometry and algebra in his leisure hours ; and three hours in the day he works in a carpenter's shop.'

---

[1] Southern Papers, 1792 Sept. 15.   [2] Tincroft was a Hornblower engine.
[3] Boulton Papers, 1796 Mar. 20.

He remained there a year, after which he was sent to school at Geneva. Returning to England in 1788 a place was found for him by Boulton in the counting house of Messrs. Taylor and Maxwell, fustian manufacturers, printers and dyers, of Manchester. Here he remained for two years. His father's reputation and standing served to introduce him quickly to many people and he soon found the allowance made to him by his father insufficient for his needs. Does he apply to him in his need? Nothing of the kind; he writes to the sympathetic Boulton, the young folks' friend. Young Watt's letter [1] to him is a manly and straightforward one, full of filial respect, yet pathetic withal in that it reveals a certain lack of responsiveness in his father :

' My father probably with a view of preventing me from becoming extravagant has only allowed me £70 for all my expenses this last year and even of this I was obliged to deduct £10 for debts which I had contracted the year before ; the remainder, although I have used all possible economy without descending to meanness, has proved totally inadequate to my necessary wants, for the mere amount of my board lodging and cloaths already exceed that sum, consequently I have not been able to keep myself clear from debts and I have now the disagree-·able prospect of Christmas before me, when my creditors will not fail to send in their bills and I shall be without money to satisfy them. I am at the moment on the best possible footing with my father but were I to inform him of my necessities I do not know what would be the consequences ; not that I suppose the money in itself would be an object to him, but because he would look upon it in the light of encouraging what he would call my extravagance, never having been a young man himself he is unacquainted with the inevitable expences which attend my time of life, when one is obliged to keep good company and does not wish to act totally different from other young men.'

Young Watt did not receive an answer so that on the 22nd December he wrote again to Boulton and his distress of mind may be imagined, arising partly from fear lest his letter had miscarried and partly from his predicament with his creditors. The explanation of the delay in replying was that Boulton had been in London. On Boxing Day, appropriately enough, he sent the much-longed-for draft for £50, the sum for which young Watt had asked, together with sage and fatherly advice, to which the young man promised to conform himself, ' as much as possible being perfectly convinced of the strength of your arguments although I cannot at the same time help seeing the moral impossibility of my living within my present income.'

One of his friends mentioned in the last letter was Thomas Cooper,[2] at that time in a large way of business as a bleacher. He was keenly interested in the political events of the day and inflamed young Watt with the thrilling events then taking place in France. In 1792 Cooper and Watt were in Paris and were deputed by the Constitutional Society of Manchester to present an address to the Société des Amis de la Constitution.[3] Their action was denounced by Burke in the House of Commons on April 30, 1792. Young Watt took part in revolutionary politics, but being denounced by Robespierre as a secret agent of Pitt fled from Paris into Italy, whence

[1] Boulton Papers, 1789 Dec. 4.

[2] Thomas Cooper, M.D. (1759-1840), had a chequered career. He was at Oxford in 1779, and became a barrister in 1787. After the event related he failed in business and went to America. He practised as a lawyer, and became a common law judge of Pennsylvania. He was professor of chemistry in various colleges, 1812-34, and president of one. He compiled manuals of American law and a scientific encyclopaedia. See Malone (Dumas): *Public Life of Thomas Cooper*, 1926.

[3] Discours de MM. Cooper et Watt, députés de la Société Constitutionnelle de Manchester, prononcé à la Société des Amis de la Constitution séante à Paris le 13 Avril 1792 ... avec la réponse du Président [J. L. Carra].

he went to Germany, returning to England in 1794. The role which he had played in France caused him to be received with suspicion on his arrival, and his father feared that his son might be implicated with some of the political societies in this country whose leaders were being severely treated by the Government of the day. However, the apprehension appears to have been unfounded, and young Watt henceforward devoted his undoubted talents for organization to the business of Boulton and Watt, to which he was admitted a partner along with young Boulton.

VIII. JAMES WATT, SON OF THE ENGINEER

From the oil-painting, artist unknown, in the possession of the Watt family

VI

LIFE IN RETIREMENT, 1800–19

FROM 1795 onwards Watt appears to have withdrawn gradually from active participation in the business. When the partnership finally drew to a close in 1800, it was no wrench but a happy and quiet release. A share of the heavy arrears of royalties which had been collected in the preceding year fell to Watt, rendering him really well off, and he indulged in the pleasant task of looking out for a landed estate. As early as 1798 he had fixed upon a property with a farm-house at Doldowlod on the banks of the Wye between Rhayader and Newbridge, Radnorshire. Here he was accustomed to spend some part of the summer, but he was too old to root himself afresh and he returned with zest to his comfortable house at Heathfield, to his mechanical recreations, and to the society of his old friends. Alas! the circle was being steadily narrowed by death, a loss that inflicts on old age one of its saddest penalties. Thomas Day had been killed by a fall from his horse in 1789 ; Wedgwood died in 1795 ; Dr. Withering in 1799, and in the same year died Dr. Black. His loss was severely felt by Watt who wrote to Robison :

· ' I may say that to him I owe in a great measure what I am ; he taught me to reason and experiment in natural philosophy and was a true friend and philosopher whose loss will always be lamented while I live.'

But the loss that was the greatest grief to him was that of the children by his second wife. Jessy, the younger, was an affectionate girl, never very strong, but as she grew up appeared to improve. She died, however, of consumption in 1794 at the age of fifteen. Writing of her death, Watt says :

' Mrs. Watt continues to be much affected whenever anything recalls to her mind the amiable child we have parted with. . . . With me, whom age has rendered incapable of the *passion* of grief the feeling is a deep regret ; and did nature permit my tears would flow as fast as her mother's.'

But it was for his son Gregory that Watt seems to have had very deep affection. He was a boy of undoubted talent, handsome and of a generous disposition. His father sent him to Glasgow College where, among his fellow students, were the poets Campbell and Thomson and Francis, afterwards Lord, Jeffrey. Campbell anticipated success for him at the Bar, and in Parliament. Jeffrey said : ' He is a young man of very eminent capacity, and seems to have all the genius of his father with a great deal of animation and ardour which is all his own.' Watt, however, destined him to take part in the business ; but he never took much part in it as tubercular trouble intervened. Watt took him under advice to the south of England. While in Penzance he resided at the house of the mother of Humphry Davy. This intimacy led to the appointment of the young chemist as demonstrator to the Pneumatic Institution of Watt's friend Dr. Thomas Beddoes of Bristol, a step which had very great influence on Davy's career.

There was an improvement in Gregory's health, but it was temporary, and in 1804 his parents travelled with him to Clifton, Bath, Sidmouth, and Exeter. There he became rapidly worse and died on October 16th at the age of twenty-seven. His remains were deposited in the cathedral, where a tablet marks the spot.

Watt in his grief wrote to Boulton: ' His virtues and merits will be best recorded

L

in the breasts of his friends . . . as soon as we can settle our accounts we shall all return homewards with heavy hearts.'

In a letter to Mr. R. Muirhead, Watt writes: ' We ... cannot help feeling a terrible blank in our family.   When I look at my son's books, his writings and drawings, I always say to myself : Where are the mind that conceived these things, and the hands that executed them?' [1]   Many like references occur showing how deeply Watt was affected.   The books and papers alluded to were placed in a hair trunk and put into the garret workshop where the father's glance might rest on them.

It was the illnesses of his daughter and son that turned Watt's thoughts to remedial medicine by inhalation.[2]   In conjunction with Dr. Beddoes, a famous physician and innovator in medicine, Watt designed an apparatus for the inhalation of oxygen and carbon dioxide for diseases of the respiratory organs.   This apparatus was made by the firm at Soho.   Watt described it with observations in a chapter in one of Beddoes's publications, afterwards published separately as a pamphlet.[3] He aided Beddoes to establish the Pneumatic Institution at Clifton where medical treatment of the nature indicated was carried out, but the results did not realize expectations.

Watt and his wife continued to travel about at intervals.   During the short period of the Peace of Amiens in 1802 they made quite the longest journey they ever undertook—through Belgium, up the Rhine as far as Frankfort, then to Strasbourg and Paris and so back to England.   In 1803, the year of the great scare due to the threat of invasion of England by Napoleon, when the Volunteer Force was first embodied, he was at Glenarback near Dumbarton, the residence of Mr. Robert Hamilton.   In a letter [4] dated from there Watt mentions the volunteer movement, and promises £100 to the fund then being raised at Lloyds and also £10 to the Handsworth parochial fund for the purpose.   In the autumn of 1805 he went again to Scotland via Derby, Sheffield, Doncaster, York, Scarborough, Whitby, Sunderland, Newcastle, Gilsland, Moffat, and Hamilton.   The winter found him in Edinburgh, where Henry Brougham says ' he was a constant attendant at our Friday club, and in all our private circles, and was the life of them all'! ! ' [5]   The club was named from the evening of the week on which the members met to sup.   Besides Henry Brougham just mentioned there were Professor Playfair, Leonard Horner, Francis Jeffrey, and others to keep Watt young by their wit and scholarship, while they in turn were delighted with the extent of his knowledge and with the anecdotes that he could tell, set off by a pawky humour ; in such company he appeared at his best.   We can best describe the effect he made in the words of Sir Walter Scott : [6]

' There were assembled about half a score of our Northern Lights. . . . Amidst the company stood Mr. Watt, the man whose genius discovered the means of multiplying our national resources to a degree perhaps even beyond his own stupendous powers of calculation and combination, bringing the treasures of the abyss to the summit of the earth—giving the feeble arm of man the momentum of an Afrite—commanding manufactures to arise, as the rod of the prophet produced water in the desert—affording the means of dispensing with that time and tide which wait for no man, and of sailing without that wind which defied the commands and threats of Xerxes himself.   This potent commander of the elements—this abridger of time and

---

[1] Watt to Muirhead, 1805 Apl. 8. Muirhead : *Mech. Inv.* II. 292.
[2] Watt to Darwin, 1794 June 30, in Muirhead : *Mech. Inv.* II. 247.   [3] See p. 362.
[4] Boulton Papers : Watt to Boulton, 1803 Aug. 31.
[5] Brougham : *Lives of Men of Letters and*

[6] Scott : *The Monastery*, 1820, 3 vols, I. 61–4. Scott's letter to Captain Clutterbuck.   Scott says he only met Watt once, and if so it must have been the introduction mentioned in a letter to his son (Muirhead : *Mech. Inv.* II. 360) as having occurred in Dec. 1814.

Science in the time of George III, p. 383.

IX. JAMES WATT, AET. 71

From the oil-painting, by Partridge, in the possession of Mrs. Scott, Hawkhill

space—this magician, whose cloudy machinery has produced a change in the world, the effects of which, extraordinary as they are, are perhaps only now beginning to be felt, was not only the most profound man of science, the most successful combiner of powers and calculation of numbers, as adapted to practical purposes, was not only one of the most generally well-informed, but one of the best and kindest of human beings.

' There he stood surrounded by the little band . . . of Northern literati, men not less tenacious, generally speaking, of their own fame and their own opinions, than the national regiments are supposed to be jealous of the high character which they have won upon service. Methinks I yet see and hear what I shall never see or hear again. In his eighty-fifth year,[1] the alert, kind, benevolent old man had his attention alive to every one's question, his information at everyone's command.

' His talents and fancy overflowed on every subject. One gentleman was a deep philologist— he talked with him on the origin of the alphabet as if he had been coeval with Cadmus; another, a celebrated critic—you would have said the old man had studied political economy and belles-lettres all his life,—of science it is unnecessary to speak, it was his own distinguished walk.'

But fresh griefs awaited him in the deaths of other friends: Dr. Darwin died in 1802 ; Dr. Priestley, in exile, so to speak, in Pennsylvania, died in 1803 ; in 1805 Prof. John Robison was released by death from a lingering disease. Watt's opinion of him was : ' He was a man of the clearest head and the most science of anybody I have ever known and his friendship to me ended only with his life after having continued for nearly half a century.' It was his great regard for him which induced Watt, when Robison's collected works were being prepared for the press by Sir Charles Brewster, to add a number of critical notes to the article on ' Steam ' and ' Steam Engines ' which Robison had written for the *Encyclopaedia Britannica* and which were included in the volume. This is what Watt alludes to in a letter [2] to Southern :

' I am at work upon the Proff^ors^ acct of my inventions in the Dictionary which I find it will be a difficult thing to correct leaving any of the Proff^rs^ words.'

Watt succeeded, however, and his annotations and remarks to this account form the only published writing on the steam engine that we have from Watt's pen. Watt certainly made some effort about 1808 to gather material for an account of his contributions to the development of the steam engine. He collected from Roebuck and others all the old letters that he could obtain and annotated them, but there is a suspicion that his remembrance of past history was coloured by his subsequent wider knowledge. We have a remarkable self-revealing letter [3] of this period written to his son evidently relating to Hornblower's account of the history of the steam engine in Gregory's *Mechanics* :

' The great aim of Hornblower's arguments seems to be that the first engines we erected in Cornwall were not so perfect as those we made afterwards, which is perfectly true, the *mechanism* of the engine was not invented all at once, but has been in a course of improvement even unto this day. I was ignorant of many things I know now and possibly of many things which were known to others, neither do I pretend that I ever possessed an eminently inventive genius in *mechanicks*, my forte seems to have been reflection and judgement in combining and applying things invented by myself and others. He finds fault with many of the contrivances described in my patents without considering that it was my business not only to describe the things I preferred but also those by which it could be evaded.'

The truth of this statement about the progressive nature of his invention is brought out remarkably in other chapters, but we can hardly accept Watt's state-

[1] A slight slip—or shall we say literary licence ? In 1814 Watt was only in his seventy-ninth year.
[2] Southern Papers : Watt to Southern, 1813 Dec. 20.
[3] Doldowlod Papers : Watt to James Watt, jun., 1808 Nov. 12.

ment that he was not an ' eminently inventive genius in mechanicks ', although we can admit that he did not possess the true mechanical engineer's ' flair ' for just the right thing among a number of possible alternatives.

Watt had yet another sorrow in his declining years and that was the death of his partner Boulton in 1809. Watt was at Glenarback, near Glasgow, when the news reached him and he wrote [1] to Matthew Robinson Boulton :

' I am very much concerned to have to condole with you on the loss of your worthy Father. I should have wrote to you sooner but found myself unequal to the duty. To offer other consolations than what must occur to your own mind is not in my power, we may lament our own loss but we must consider on the other side his relief from the torturing pains he has so long endured and console ourselves with the remembrance of his virtues and qualifications. Few men have had his abilities & still fewer have exerted them as he has done, and if to them we add his urbanity, his generosity and his affection to his friends we shall make up a character rarely to be equalled. Such was the Friend we have lost and of whose affection we have reason to be proud as you have to be the son of such a Father ! '

Smiles states that one of the fears that haunted Watt as old age crept upon him was that his mental faculties should decay ; to test himself he took up again the study of German which he had allowed to drop. He quickly acquired his old proficiency and indeed surpassed it.

A better proof of his unimpaired mental powers occurred in 1811 when the Glasgow Waterworks Company consulted him with regard to carrying water from the filtering area to the pumping engines on the south side of the Clyde.

' Many plans were designed for this purpose ; that which was adopted was suggested by Mr. Watt ; he proposed the use of cast-iron pipes, fitted where necessary with revolving ball and socket joints, which he then first introduced and of which he sent a wooden model to the company, which model was now presented by them to the Institution of Civil Engineers. These pipes adapted themselves to the form of the bed of the river, and the plan was perfectly successful.' [2]

Watt, as we have seen, took pleasure in the society of men younger than himself and this was some consolation for the loss of so many of his older friends. He found, however, most satisfaction in the pursuit of new inventions, no longer as a business but as a hobby in his old age. He occupied himself from 1804 onwards with machines for copying irregular objects such as busts. No doubt when in France he had seen the *tour à portrait* which was capable of copying in the same or on a reduced scale objects in slight relief—such as a medallion—and this may have given a direction to his labours and stimulated his inventive powers. The principle of the machine was to have a rapidly revolving cutting tool guided by a linkage of bars from a feeler passed over the original object by successive small advances. Watt modified the machine first by mounting the work and the original on axes so that they were capable of a complete turn in step with one another ; in this way the whole surface of a solid could be machined. Secondly the design of the framing connecting the tool and the feeler had to be such that rigidity without shake could be maintained. Several of these frames built up of tubes as well as a complete machine in wood are in the workshop of Heathfield. Probably it was in connexion with the framing that Watt wrote to Southern : [3] ' I return you my best thanks for the Experiments and calculations on the rigidity of tubes that you have been so good as to make for me and which prove satisfactory.'

[1] Boulton Papers : Watt to M. R. Boulton, 1809 Aug. 23.
[2] *Proc. Inst. Civ. Eng.*, 1843, 135. The model is not now at the Institution.
[3] Southern Papers : 1812 Aug. 12.

X A. WATT'S HOUSE AT HEATHFIELD

X B. WATT'S GARRET WORKSHOP

It should be mentioned also that Murdock helped him in the practical construction of the machine. Watt never took out any patent for it although he drafted a specification dated Sept. 21, 1814.[1] He did quite a lot of good work with the machine, as witness the numerous copies of busts that he made for his friends and those preserved in the workshop itself, ' the production ', he writes, ' of a young artist entering upon his eightieth year.' Watt did intend to make a more accurate machine entirely in metal and even prepared some of the drawings, but he did not live to carry out his intention.

The workshop contains quite a wonderful collection of tools and apparatus, some without doubt dating from the Glasgow College days, and we must be grateful to the younger James Watt by whose care its contents were preserved.[2]

It may be said with truth that Watt's career, considered merely from the standpoint of his eminence as an inventor, was in striking contrast with that of other great contemporary inventors—and there were giants in those days—because unlike most of them, Watt reaped a pecuniary reward for his labours and, as is the

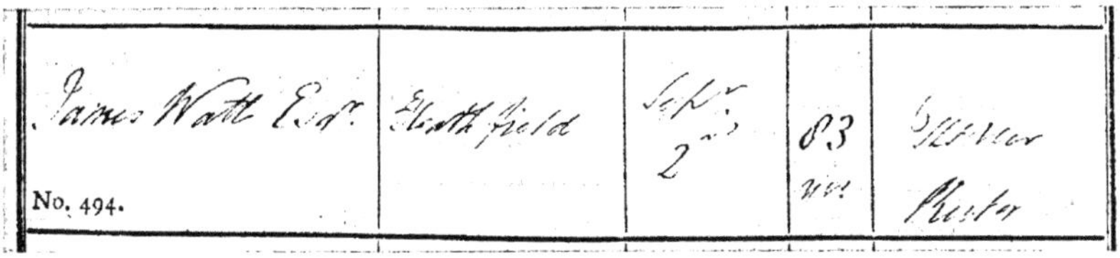

FIG. 5. Entry in Register of Burials, St. Mary's Church, Handsworth

way of the world, honours came to him as his years increased. He was elected a Fellow of the Royal Society of Edinburgh in 1784, and of London on November 24 of the following year, at the same time as Matthew Boulton. He was elected in 1787 a member of the Batavian Society of Rotterdam. In 1806 the University of Glasgow most appropriately honoured itself by conferring on him the degree of Doctor of Laws, *honoris causa*. In return he founded in 1808 the Watt prize in Natural Philosophy and Chemistry. He was made a corresponding member of the French Academy in 1808, and in 1814 had the signal honour of being elected one of its eight foreign associates.

Lord Liverpool, when Prime Minister, offered Watt a baronetcy, but with simple dignity he declined the honour. He was asked to fill the office of High Sheriff of Staffordshire, and later that of Radnorshire, but declined on both occasions because he shrank from public life and felt unfitted for such a duty.

He was honoured in his lifetime by the formation in Greenock in 1813 of the Watt Club. This has been followed by a long succession, named after him, of Colleges, Engineering Laboratories, Chairs, Institutions and Libraries, scholarships, prizes, and medals, too numerous to mention, in different parts of the world. Docks and streets, too, have been named after him.

Watt was above the medium height, not to say tall, of spare figure, and with a decided stoop of the shoulders. His features in repose bore the stamp of deep

[1] Muirhead : *Mech. Inv.* I. ccxlviii.
[2] Owing to the demolition of Heathfield, a new home has had to be found for these relics, and this has been supplied in the Science Museum, South Kensington, appropriately enough close to the Watt Models already preserved there.

contemplation, with marks of the ill health from which he suffered. His eyes were grey, although in some lights they looked blue. His hair early turned white but his complexion was fresh. He spoke deliberately with a pronounced Scots accent in a low but deep—albeit somewhat monotonous—voice. Timmins says [1] that Watt ' was not an early riser and often said that he required ten or eleven hours sleep. He delighted in snuff and was fond of a pipe of tobacco. His habits were simple. His reading included all branches of literature and science. His mind had a sort of intellectual alchemy that enabled him to secure the wheat and reject the chaff of what he read.'

In intercourse with the outer world he was retiring ; he disliked bargaining and wrangling made him ill. With persons he knew and when free from worry his fund of dry humour made him a delightful companion.

It is difficult to say anything as to Watt's religious belief, further than that he was a Deist. When in Glasgow he conformed to the Presbyterian form of worship in which he was brought up. If proof were needed we have the cash memo. book entry : [2]

' 1766. May 14. Pd. seat in Kirk 8s. 0d.'

We have also the remark in a letter to Roebuck [3]

' . . . I shall be glad to see you, but thursday and saturday next are our sacrament days, we would wish you to come before or after them, as we cannot decently try any experiment on them.'

In Birmingham he supported the Church of England. In a letter describing the Church and King riots there in 1791, Watt states in : ' I, among others, was pointed out as a Presbyterian, though I was never in a meeting-house in Birmingham.' In Cornwall, Watt attended chapel on at least one occasion.

FIG. 6. Statue of Watt by Chantrey in the Watt Chapel, St. Mary's Church, Handsworth

The reader will have formed an estimate of Watt's character from the self-revelation given in preceding pages. He was modest to the extreme of self-depreciation. Black wrote to him once : ' Were you to be the first publisher of your discoveries, you would do it in such a cold and modest manner that blockheads would conclude there was nothing in it, and rogues would afterwards, by making a trifling variation, vamp off the greater part of it as their own.' Watt was cautious in the extreme and equally unfitted for high emprise as for high finance. He was unremitting in his application ; his patience was great ; his fertility of intellect was such that he could not avoid scheming improvements on any invention mentioned to him.

Looking back dispassionately over the space of a century, and bearing in mind

[1] Timmins (Sam.) : *Trans. Archaeological Section, Birmingham and Midland Institute,* 1872, 20.

[2] Muirhead Papers.
[3] Doldowlod Papers : Watt to Roebuck, 1769 Mar. 16.

the advances that have been made in the steam engine since Watt's day, we can affirm still that no one has ever made a greater individual contribution to its development, and furthermore that no such contribution has ever been fraught with so great import, for good or ill, to the destinies of the human race.

The end came quite peacefully—he died at Heathfield on Aug. 25, 1819, in the eighty-fourth year of his age. He had been in his usual health the previous month—in fact he had been in London. He was buried beside Boulton in Handsworth Church on September 2nd following. The Watt Chapel was built subsequently.

By his will, which is dated July 7, 1819, he left to his wife Ann £1,400 per annum, with Heathfield for life, and to his son James the residue of the estate. His wife and son were appointed executors. All his books, papers, drawings, tools, &c., went with the residue. There are two codicils by which he left a number of small legacies which, with the exception of £150 to charities, were to old friends, colleagues, and servants. Typical of the man is his request that he might be ' interred in the most private manner without show or parade as soon after my decease as may be proper '. It can hardly be said that this request was faithfully carried out since the expenses of the funeral were over £700. The will was proved by the executors on October 13 for upwards of £60,000

Unlike many other great inventors, Watt has since his death held a high place in the estimation of mankind ; perhaps this is due in some measure to the filial conduct of his son, who honoured his father by establishing or contributing to the cost of memorials to him. We may mention the statues by Chantrey in the Watt Chapel of Heathfield Church, at the Watt Memorial, Greenock, and in Westminster Abbey. The last named was set up by public subscription initiated at a public meeting presided over by the Prime Minister, Lord Liverpool, on June 18, 1824. The inscription upon the plinth is from the pen of Lord Brougham. Lofty as is its language, it has yet the signal merit that every word rings true.

NOT TO PERPETUATE A NAME
WHICH MUST ENDURE WHILE THE PEACEFUL ARTS FLOURISH
BUT TO SHOW
THAT MANKIND HAVE LEARNED TO HONOUR THOSE
WHO BEST DESERVE THEIR GRATITUDE
THE KING
HIS MINISTERS AND MANY OF THE NOBLES
AND COMMONERS OF THE REALM
RAISED THIS MONUMENT TO

JAMES WATT

WHO DIRECTING THE FORCE OF AN ORIGINAL GENIUS
EARLY EXERCISED IN PHILOSOPHIC RESEARCH
TO THE IMPROVEMENT OF
THE STEAM ENGINE
ENLARGED THE RESOURCES OF HIS COUNTRY
INCREASED THE POWER OF MAN
AND ROSE TO AN EMINENT PLACE
AMONG THE MOST ILLUSTRIOUS FOLLOWERS OF SCIENCE
AND THE REAL BENEFACTORS OF THE WORLD

BORN AT GREENOCK 1736

DIED AT HEATHFIELD IN STAFFORDSHIRE 1819

Brougham wrote : ' It has ever been reckoned by me one of the chief honours of my life, that I was called upon to pen the inscription upon the noble monument thus nobly reared.' The inscription is perhaps the finest one of its kind in the English language. Dean Stanley said of it : ' It is not unworthy of the omnigenous knowledge of him who wrote it or of the powerful intellect and vast discovery which it is intended to describe.'

Many have been the tributes to James Watt as a man ; that of Wordsworth [1] deserves a place from its great insight :

' I look upon him considering both the magnitude and the universality of his genius, as perhaps the most extraordinary man that this country ever produced ; he never sought display, but was content to work in that quietness and humility, both of spirit and outward circumstances in which alone all that is truly great and good was ever done.'

We conclude in Watt's own words the spirit of which has been ever present with us in this work : ' Preserve the dignity of a philosopher and historian ; relate the facts and leave posterity to judge. If I merit it some of my countrymen may say : " *Hoc a Scoto factum fuit.*" ' [2]

---

[1] In conversation in 1840 with J. P. Muirhead, the author of the *Life of Watt* (p. 368). Wordsworth was in Paris in 1792 with young Watt.

[2] Watt to Darwin, 1789 Nov. 24. Muirhead : *Mech. Inv.* II. 236. Darwin had asked for material for his projected *Botanic Garden*. The result is to be seen in Canto I, l. 253.

XI. JAMES WATT, AET. 77

From the oil-painting, by Lawrence, in the possession of the Boulton family

VII

WATT PORTRAITURE

IT is a matter of satisfaction that a number of original and contemporary portraits of the great inventor and engineer are in existence—satisfaction tempered by the fact that none of the portraits, much as we should have liked it, is of his earlier years. Doubtless this may be explained by the reluctance that Watt displayed to have his portrait painted ; this diffidence is given expression to in the letter quoted below.

The portraits that really matter can be counted on the fingers ; they are :

    I. Portrait in oils by Charles Frederick von Breda, R.A., painted in 1793, i.e. when Watt was in his 57th year.

    II. Portrait in oils by Sir William Beechey, R.A., painted in 1801, when Watt was 65.

    III. Portrait in pastel by L. de Longastre, drawn about 1805, when Watt was 69.

    IV. Portrait in oils by John Partridge, painted about 1806, when Watt was 70.

    V. Portrait in oils by Sir Thomas Lawrence, P.R.A., painted in 1813, when Watt was 77.

    VI. Portrait in oils by Sir Henry Raeburn, R.S.A., painted in 1815, when Watt was 79.

    VII. Bust in marble by Sir Francis L. Chantrey, R.A., cut in 1815, four years before Watt's death.

Practically all subsequent portraits, statues, &c., have been based on these originals. There are other contemporary portraits, but they are of less importance than those above mentioned. They will be found described below with the others in chronological sequence.

Two or three genre pictures have been painted representing Watt as a boy watching the steam issuing from a kettle, or as a young mechanic repairing the Newcomen model at Glasgow College ; it is perhaps unnecessary to state that none of these has any value as portraiture. These pictures have been reproduced often as cheap lithographs ; the best one of the kind is perhaps that of 'Watt and the Steam Engine', painted by James E. Lander, engraved in mezzotint by James Scott, and published in November 1860.

I. *The Breda Portrait*

No earlier portrait of Watt is known to have been painted than that by von Breda. It would appear that this was commissioned by John Rennie, the well-known civil engineer, or at his instance, as will be inferred from the following letter:[1]

' I have often wished to be possessed of a portrait of yourself & Mr. Watt, but I have had no opportunity for some time passed of obtaining so much of your time as to have them done. In hopes of your managing matters so as to grant me that favor next time you come to London, I have taken the liberty of engaging Mr. Brown (who painted Mr. Smeaton's Portrait) to paint them whenever you will have the goodness to sit. Mr. B. lives in Cavendish Square.'

[1] B. & W. Colln. Rennie to Boulton, 1792 June 7.

    M

To this Watt replied to Rennie, Birmingham, June 9, 1792 : [1]

' I wish you to come this way when you set out on your journey and we can talk about the portrait to which I am rather averse as I think it an honour I do not merit and that my countenance cannot be worth procuring.'

We must suppose that when the opportunity presented itself Mr. Brown was not available, and that the commission was passed on to von Breda.

Watt is represented to below the knees, seated, face nearly in profile to the left. The mien is thoughtful and studious, there is a pronounced stoop, and the whole effect is that of a man of delicate, almost of invalid, constitution.

This portrait, or a replica, was presented to the nation in 1865 by Matthew Piers Watt Boulton, grandson of Watt's partner, and is now in the National Portrait Gallery. Size $49\frac{1}{2} \times 39\frac{1}{4}$ in.

It was engraved in mezzotint by S. W. Reynolds, and published in December 1796, plate size 14 × 10 in., with a similar companion portrait of Boulton. Since then it has been reproduced fairly frequently, e.g. a copy in oils is in the Institution of Civil Engineers.

II. *The Beechey Portrait*

The portrait by Beechey shows Watt half-length, quarter turned to the left. The expression is thoughtful : there is again a stoop of the shoulders, but the physique is more robust. This is in accordance with what we know to be the fact, for when Watt retired finally in 1800, throwing off the cares of business and the anxieties consequent on financial affairs, he seemed to take a new lease of life. The portrait was exhibited in the Royal Academy Exhibition in 1802, and Telford wrote to Watt under date May 3rd : ' I this day paid my respects to you in the Exhibition Room. I think Beechey has succeeded admirably well.' Watt himself seems to have shared this view, for in a letter to Dr. Patrick Wilson, dated May 22, 1809, he says : ' There is no good portrait of me except that painted by Sir William Beechey, still in his possession, and a copy of it by himself in Mr. Tuffen's collection which is more like than the original I having sate again for it. Neither of them is esteemed very like by my son and others.' Muirhead states that : [2] ' On this point his son afterwards rather changed his opinion ; and, indeed, came to prefer the portrait in question to any of the other paintings of his father.' The original portrait hung for many years in the dining-room at Aston Hall, Birmingham, and is now in the possession of the present representative of the family, Major J. W. Gibson-Watt, at Doldowlod, Radnorshire.

The copy above mentioned is believed by Muirhead to be the same portrait as that presented by James Watt, junior, to his friend, Charles Hampden Turner, of Rook's Nest, Godstone, Surrey, and still in the possession of his descendants. It is reproduced as the frontispiece to the present volume.

Copies of the Beechey painting are numerous, e. g. one is in the possession of the Royal Society of Edinburgh. Another portrait, also attributed to Beechey, is in the possession of the Corporation of Birmingham ; its history is, however, not known, and it must be admitted that it differs considerably in features from the other two.

The preference of James Watt, junior, for the Beechey portrait is perhaps the reason why it has been so often reproduced. The earliest and best known of these reproductions is the stipple engraving by Charles Picart from a drawing by William Evans. This was published in March 1809, plate size 15 × $12\frac{1}{4}$ in. Writing

---

[1] Letter in possession of Rennie's descendants.　　　　　[2] Muirhead : *Life of Watt*, 517.

of these engravings on Oct. 29, 1816, Watt says : ' They are not like and the picture from which they were taken was done when I was much younger than I am now. They are, however, the best which have been done.' The Beechey portrait appeared also in Cadell's *British Gallery of Contemporary Portraits*, 1822, and in Knight's *Gallery of Portraits*, 1833.

A miniature by Peter Rouw was in the Royal Academy Exhibition of 1803. This is doubtless the flesh-coloured wax relief, signed ' P. Rouw London 1803', in the style peculiar to that artist, in the possession of the Birmingham Corporation.

A painting in enamel by Henry Bone, R.A., the well-known artist in that medium, faithfully taken from the Beechey portrait, was in the Royal Academy Exhibition of 1804, size 8 × 6½ in. It is in the possession of the Watt family at Doldowlod.

III. *The de Longastre Portrait*

L. de Longastre, who drew the pastel portrait already mentioned, was an *émigré* officer of Louis XVI's Guards, who was introduced to Matthew Boulton in 1805 by the Abbé de Calonne. Watt is shown half-length, facing to right ; the face is full, and the impression is that of a man of bodily vigour. Size of paper, 27 × 22 in. This portrait was considered by Muirhead[1] to be ' a striking portrait and exact likeness'. There was also a duplicate executed for Matthew Boulton, and the drawings are now in the possession of the Boulton and of the Muirhead families respectively. The latter portrait has been reproduced in photogravure by J. L. Caw, in *Scottish Portraits* (portfolio IV, Pl. XC, p. 74, 1903, plate size 7 × 6½ in.), who there gives a critique of the known Watt portraits, deciding in favour of this as the best.

IV. *The Partridge Portrait*

The portrait in oils by Partridge dates from about 1806. It is based on the Beechey portrait, but the artist had additional sittings. It is a half-length, half turned to the left. Muirhead calls it ' a good and pleasing likeness'. James Watt, junior, presented it to Dr. Barr, the medical attendant of his father in his last illness. Subsequently it passed through many hands, notably those of Robert Napier of Shandon, and it is now in the possession of Mrs. Scott of Hawkhill, Largs. It was reproduced in lithograph in Williamson's *Memorials of James Watt*,[2] and other lithographic copies of less merit have appeared. It has been chosen for reproduction in the present volume, facing p. 74.

Peter Turnerelli (b. 1774, d. 1839) exhibited a bust of Watt at the Royal Academy Exhibition of 1808 (No. 865). This is probably the plaster bust marked ' P. Turnerelli fecit 1807 ' in the Watt Room from Heathfield, height 23 in., width 13½ in. The features are much lined and the expression is that of pain or ill health ; that the bust was not carried out in marble need hardly be wondered at. Therefore no copies are known.

In 1809 John Henning, P.R.S.A., of Edinburgh, executed for Lord Jeffrey a black chalk drawing (on paper 20 × 17 in.) as well as a small copy for another gentleman. The bust is shown in profile to the right, the forehead is prominent ; while a vigorous piece of work, the general effect is heavy, and the face lacks sensitiveness. The portrait was shown in 1810 at the Exhibition of the Associated Society of Artists in Edinburgh. The widow of Lord Jeffrey gave the drawing to some cousins of Watt, one of whom presented it to the Scottish National Portrait

[1] Muirhead : *Life of Watt*, 518.   [2] Facing p. 120.

Gallery. The other drawing was reproduced in Williamson's *Memorials* and is now at the Watt Institution, Greenock. Henning also executed from the drawing in the same year two or three medallions in enamel, $3\frac{1}{2} \times 2\frac{1}{2}$ in.

A sepia drawing of Watt, by George Dawe, R.A., small head in profile to right, looking downwards, size $5\frac{3}{4} \times 5$ in., is in the Scottish National Portrait Gallery. It is not dated, but may be about this period.

V. *The Lawrence Portrait*

The large portrait in oils painted by Sir Thomas Lawrence in 1813 shows Watt nearly full face to the left, seated, to below the knees. Size 54×43 in. There is again the stoop of the shoulders, the thoughtful expression, the strong features, and a hint of delicacy of constitution. The forehead is somewhat retreating.

Muirhead[1] states that ' it seems not to have found due favour in the sight of either Mr. Watt or his son '. In the opinion of Muirhead, however, it is ' unrivalled among all the portraits of Watt '. Muirhead further states that the artist considered the head *the finest he had ever painted* ; hence the reason for its having been chosen for reproduction in the present volume (facing p. 80). The portrait was bequeathed by James Watt, junior, to Matthew P. W. Boulton to be preserved as an heirloom, and it is in the possession of the present representatives of the family. It was engraved in mezzotint by Charles Turner and published July 1815, plate size $22\frac{1}{8} \times 16$ in. It has since been re-engraved and printed without date or letters. It is uncommon in either of these states and in the former it is valuable.

VI. *The Raeburn Portrait* ·

As neither Watt nor his son felt perfectly satisfied with any of the portraits previously painted, he sat to Sir Henry Raeburn when in Edinburgh in November 1815. Watt says:[2] ' Raeburn has painted a head of me which, though it has not come up to my ideas of my own face, is more conformable to them than any of the others, and by my friends is said to be a good likeness.' Later he says : ' My Edinburgh picture is come home and is thought like, only it frowns too much.' Muirhead's opinion of it was that it was ' a finely coloured and forcible picture '. In his works published in 1854 and 1859 he states that it was then at Heathfield, but its present whereabouts is not known. A copy was made by the artist for John Rennie, with Watt's full approval, but this also is not to be found. A portrait supposed to be of Watt and attributed to Raeburn was sold at Christie's on Jan. 27, 1912, but experts were not agreed as to the attribution.

VII. *The Chantrey Bust*

The bust in marble by Sir Francis Chantrey was shown in the Royal Academy Exhibition of 1815. It faces to the left and has the characteristic stoop ; the features are strongly marked in repose and the expression is peculiarly benevolent. The bust is in the possession of the Watt family at Doldowlod. Size 18 in. wide, 27 in. high. The artist's original model is at the Ashmolean Museum, Oxford. Muirhead mentions seven replicas in all ; one of these is in the possession of the Boulton family at Tew Park ; another belongs to the Turner family and is at Rook's Nest, Godstone. Three copies, scarcely inferior to the original, were cut, with the consent of the artist's executors, by James Heffernan, Chantrey's assistant. One of these, dated 1843, was presented in that year by James Watt, junior, to

---

[1] Muirhead : *Life of Watt*, 518.   [2] Ibid., 519.

the Royal Society of London, of which Watt was a Fellow; another was presented similarly to the Institut de France, of which he was a Foreign Associate.

Muirhead[1] mentions ' reduced copies of Chantrey's bust which Mr. Cheverton has, with great ingenuity, executed in ivory ', and states that they were produced by a copying machine analogous to that invented by Watt himself. Watt had always hoped to have been able to do this himself, as he says in a letter to Thomas Thomson, May 26, 1818 : ' I do not think myself of importance enough to fill up so much of my friends' houses as the original bust does.' The Editors have not seen any of these ivories.

In 1818 John Jackson, R.A., made a pencil drawing of Watt from sittings, and also a sketch from the Chantrey bust. The former was shown in the Academy Exhibition of 1819. It is a front view with the face finished in water-colour, size $8\frac{1}{2} \times 6\frac{1}{2}$ in. ; although a clever and forcible drawing, Watt considered the expression ' peevish '. The drawing was in the possession of James Watt, junior, at Doldowlod, but has been lost sight of.

A portrait in oils was painted about 1818 by John Graham-Gilbert, R.S.A., partly from the Chantrey bust and partly from the Lawrence portrait. It shows the bust quarter-turned to the left. The portrait is somewhat stern in expression, but is nevertheless a fine one. The artist presented it to the Hunterian Museum in Glasgow University and it now hangs in the Senate Room there. When Mrs. Watt saw it in the Museum she ' thought it so like that she immediately ordered one precisely the same for herself ' ; this replica is now at Doldowlod. The Glasgow University portrait has been reproduced, notably in the *Scottish Historical Review* for January 1910.

The Chantrey bust has naturally been much used by medallists. It is shown on a medal now in the British Museum. Its description [2] is as follows :

> Bust of Watt l., draped. Below, T. & J. and I. M. In arc, above, JAMES WATT ESQR. LLD. F.R.S. LN. AND ED.
> *Rev.* In wreath of thistle, rose, and shamrock, IN / TESTIMONY / OF / NATIONAL / ESTEEM.
> 55 mm. Bronze. I. M. may be J. Marrian of Birmingham, who exhibited an impression from a die in the Royal Academy Exhibition of 1818.

We now come to an interesting etching by William Nicholson, R.S.A., in 1819. He was a well-known portrait painter in Edinburgh and it is more than probable that he had met Watt, who visited that city on several occasions after his retirement. It is a seated figure to below the knees, facing the spectator. The cheeks are baggy, the eyes bulge, and the expression is coarse, the impression conveyed being that of a man of a strong constitution (plate size, $10\frac{7}{8} \times 8\frac{7}{8}$ in.). To the Editors the portrait is quite unconvincing. It is some satisfaction to know that this etching has not been reproduced, but an oil-painting by Henry Howard, R.A. (see below), bears strong internal evidence of having been based upon it.

A medallic portrait by George Mills, who worked at the Soho Mint, was in the Royal Academy in 1821, and is now in the British Museum. Its description is :

> Bust of Watt r., draped. Below, MILLS F ; around, JAMES WATT F.R.S. DIED MDCCCXIX.
> *Rev.* Steam engine. Below, WATTS STEAM ENGINE / MDCCLXXXVII (i.e. the double-acting sun-and-planet rotative engine).
> 45 mm. Silver and bronze. By George Mills (1792–1824).

[1] Muirhead : *Life of Watt*, 522.
[2] For this and succeeding descriptions, the Editors are indebted to the courtesy of Mr. G. F. Hill, Keeper of the Department of Coins and Medals, British Museum.

We have now exhausted all the portraiture which can be called contemporary, but there is still important work to be noticed.

Posthumous Work

First must be mentioned the Chantrey statues, which are really based on first-hand knowledge and should therefore perhaps be placed in the preceding class. Shortly after Watt's death his son, who exhibited marked filial respect for his memory, commissioned Chantrey to execute a statue for his father's tomb. The result was the fine life-size figure seated in a chair holding a sheet of paper and dividers, exhibited in the Royal Academy in 1824 and now in the Watt Chapel in Handsworth Parish Church (reproduced on p. 78). A replica was executed by the sculptor and was presented by James Watt, junior, to the University of Glasgow. Another replica was subscribed for by the inhabitants of Greenock and placed in the Watt Monument in 1838. The latter statue was drawn on stone by A. Haenisch, $10\frac{3}{4} \times 8$ in., and lithographed in colour by McFarlane.

When the national movement to erect a monument to Watt took shape at a public meeting held in 1824, funds were subscribed and Chantrey was again commissioned. He cut a statue closely resembling the Handsworth one but greater than life-size. Space for this was found in Westminster Abbey, in Henry VII's Chapel, with the well-known epitaph composed by Brougham.[1] The position, however, is much too cramped for the spectator to get the proper effect. Another of colossal size in bronze is in George Square, Glasgow.

In 1826 a series of medals of great men were published by Parker. Watt is in company with Scott, Canning, Flaxman, and others. The description of the particular medal exhibited in the Royal Academy in 1826 and now in the British Museum is:

Head of Watt l.; behind, l. WATT; on truncation, A. J. STOTHARD; below, F. L. CHANTREY R.A. D. Border of conventional lilies.
Rev. Muse with pen, leaning on column and holding scroll inscribed TO GREAT MEN; on basis PUBD BY S. PARKER LONDON; below, MDCCCXXVI—T. STOTHARD R.A. D. A. J. STOTHARD. F.
62·5 mm. Bronze. By Alfred Joseph Stothard (1793–1864).

W. Bain showed medals at the Royal Academy Exhibition of 1825 and of 1828 and a head at that of 1846. Of these three the Editors have found no trace. A medal by an unknown artist, based on the Chantrey statue and showing the Watt arms (i.e. subsequent to 1826), is in the British Museum. Its description is:

Statue of Watt seated in chair, three-quarters l., holding sheet of paper and dividers; above, in arc, JAMES WATT; below, *Born 19 January 1736 / Died 25 August 1819.*
Rev. Shield of the Watt arms (i.e. a winged caduceus and a club in saltire) surrounded by a scroll inscribed with the Watt motto INGENIO ET LABORE. Around, HIS OBSEQUIES AT HANDSWORTH STAFFORDSHIRE—SEPT. 2. 1819.
45·5 mm. Bronze.

An engraving of the Chantrey bust in mezzotint was executed by S. W. Reynolds and published in June 1825 (plate size $12\frac{1}{2} \times 9$ in.), but this is not so pleasing as a drawing of the bust made by Edward Finden and engraved by him with great success in stipple with Watt's autograph. Muirhead states that it is 'of all the engravings of his father, that by which the late Mr. Watt wished that his image should be conveyed to posterity'. Hence the unusual step taken by Muirhead

---

[1] For transcript, see p. 79.

in introducing it as the frontispiece to all four of his biographical works: the translation of Arago's *Eloge*, 1839; the *Correspondence on the Composition of Water*, 1846; *Mechanical Inventions*, 1854; and the *Life*, 1859. The Finden drawing was re-engraved in stipple by W. C. Edwards for Weale's Architectural Library (no date).

A statue in marble, and therefore presumably a commission, was cut by William Scoular and shown at the Academy in 1840.

A medal of this period, or perhaps earlier, is in the British Museum. Its description is:

> Bust of Watt l., wearing coat. Around, JAMES WATT. Below, GALLE F.
> No reverse.
> 56·5 mm.   Proof in pewter bronzed.   By André Gallé (1761–1844), who also made a medal of Boulton.

The Royal Cornwall Polytechnic Society instituted a Watt medal in 1833. It appears to have been based on the drawing by Dawe (see *ante*). Its description is:

> Bust of Watt l.; behind, JAMES WATT; below, W. WYON A.R.A. / MINT.
> *Rev.*   Within a wreath of palm and laurel, ROYAL / CORNWALL / POLYTECHNIC / SOCIETY / INSTITUTED / 1833 /   FIRST CLASS.
> 44·5 mm.   Silver and bronze.   By William Wyon (1795–1851).

An obverse die copied from the preceding was in the Academy in 1843 and is now in the British Museum. Its description is:

> Bust of Watt l.; behind, WATT; below, LEONARD. WYON. F. / AETAT. 16 / 1843.
> No reverse.
> 43·5 mm.   Pewter proof of obverse die.   By Leonard Charles Wyon (1826–91).

A medallion of Watt, cast in 1845 by Pierre Jean David d'Angers (1788–1856), is in existence, but the Editors have not seen a specimen.

The Art Union of London, wishing to commemorate Chantrey, in 1846 issued a medal, on the reverse of which the statue was introduced, as it was deemed to be his masterpiece. This is the description:

> Head of Chantrey l.   Around, CHANTREY SCULPTOR ET ARTIUM FAUTOR.   Below, W. WYON: R.A., FEC.
> *Rev.*   Watt Statue, on base of chair, WATT; underneath, FRANCISCI CHANTREY OPUS. Behind, W. WYON. R.A.; FEC. 1846.
> 54 mm.   Bronze by William Wyon (1795–1851).

In 1852, to complete the Edinburgh Memorial to Watt (see p. 368), it was decided to erect a statue to Watt in Adam Square. The Duke of Buccleuch presented a block of freestone from his Granton quarries and the statue, based on that of Chantrey, was executed by Peter Slater. The unveiling took place on May 12, 1854. The statue is now in front of Heriot-Watt College.

It had long been felt in Manchester that the district in general and the engineering industries in particular owed a great debt to the genius of Watt, with the result that in December 1855 a meeting promoted by Sir William Fairbairn and other prominent citizens was held, the outcome of which was a Watt Memorial Committee, who invited public subscriptions. A statue in bronze, really an enlarged copy of that by Chantrey above mentioned, was commissioned from William Theed. On June 26, 1857, this was formally unveiled in Piccadilly, where it still stands, and was presented to the Corporation.

The Institution of Civil Engineers decided in 1855 to institute a Watt medal 'as a means of rewarding strictly mechanical papers'. Again the Chantrey bust was used. Thus :

Bust of Watt r., draped. Behind, JAMES WATT / 1736–1819 ; below, JOSEPH S. WYON S.
Rev. Sun-and-planet rotative engine. Below, STEAM ENGINE / AS CONSTRUCTED BY / JAMES WATT.
47·5 mm. By Joseph Shepherd Wyon (1836–73).

This was presented in silver and in recent years has been given in standard gold. The same artist had a similar medal in the Academy in 1858.

A portrait seated to below the knees, turned to the left, with Glasgow College in the background, was painted by John Blake Macdonald (1829–1901) in 1858, but is not of great value ; this was engraved in mezzotint by John le Conte in May 1860 and published in Edinburgh. Size of plate, 20 × 14½ in.

About this date a small portrait in oils was painted by Henry Howard, R.A., bust only, facing spectator, size 7½ × 5¾ in. This quite looks as if it were based on the Nicholson etching mentioned above. The portrait is now at the National Portrait Gallery, having been presented to the nation in 1882 by Sir Theodore Martin, K.C.B.

A bust of Watt in Parian ware, based on the Chantrey bust, was one of a series prepared by Josiah Wedgwood & Sons. The particulars on back are :

'WATT   JOSIAH WEDGWOOD & SON   PUBLISHED AUGUST 24 1859   E. W. WYON F.'
Height including plinth, 15 in.

The citizens of Birmingham had long been sensible of the reproach of not having in their midst any memorial of Watt, and in 1867 Alexander Munro was commissioned to produce a statue. The result is the well-known life-size statue in bronze. Watt is represented meditatively resting his hand on a cylinder of one of his steam engines and the whole composition is very effective. The statue was unveiled on Oct. 2, 1868, and is at present in Ratcliff Place—a wholly inadequate site. The artist's preliminary studies for the statue are to be found in the City Art Gallery together with a bust and statuette executed at the same time.

The undernoted medal would appear to have been of a commemorative character ; it is of inferior design :

Busts of Matthew Boulton and James Watt, r., jugate. Below, J. MOORE F. 1871
Rev. JAMES WATT & CO / LATE / BOULTON & WATT / ENGINEERS / LONDON & SOHO / BIRMINGHAM
38 mm. Bronze. By J. Moore.

A bust in white marble for the Wallace Monument, Stirling, was commissioned by Mr. Andrew Stewart of Jordanhill from the sculptor, D. W. Stevenson, R.S.A., and was unveiled there on Sept. 11, 1888. The bust is based on that of Chantrey and upon the Beechey portrait.

A statue in marble was executed in 1889 by Henry C. Fehr, and was in the Academy Summer Exhibition of the same year. This is now outside the Watt Memorial School at Greenock. A duplicate was presented by Richard Wainwright in 1898 to the City of Leeds, and is now in the City Square there.

A copy of the Chantrey bust was executed by the same artist in 1896 for the new building of the Institution of Civil Engineers.

The McGill University, Montreal, award a Watt medal for Applied Sciences. A proof in lead is at the British Museum and its description is :

Head of Watt r. Behind, JAMES WATT ; below, ALLAN WYON ; around, PRESENTED AT MCGILL UNIVERSITY, MONTREAL—PRIZE FOR APPLIED SCIENCES.
No reverse.
45 mm. By Allan Wyon (1843–1907).

James Watt would, we are sure, have scouted the idea of becoming armigerous, but as arms are displayed on some of the monuments mentioned above and else-where, it is necessary to mention them and to state that it was his son who took out the grant. In his petition to the Heralds' College in 1826, he recited that he was ' desirous to use his utmost endeavours to preserve in his own family a lasting recol-lection of that fame so justly acquired by his said late Father in the invention and application of various improvements in Natural Philosophy and Mechanics which has already received such unprecedented public celebration '.

On Aug. 3, 1826, the following arms were granted to be borne and used by himself and his descendants and by the other descendants of his Father, ' the Arms also to be placed on a monument or otherwise to the memory of his said late Father ' :

Armorial bearings : ' Barry of six Or and Azure, over all a club in bend sinister proper, sur-mounted by a caduceus saltireways, also proper.'

Crest : ' Upon a fer-de-moline fesseways Or, an elephant statant proper, charged on the body with a cross moline gold.'

Motto : Ingenio et labore.

FIG. 7. ARMS OF WATT, from the original grant 1826, preserved in the Muirhead Papers, Birmingham Reference Library.

WATT AND THE STEAM ENGINE

VIII

THE EXPERIMENTAL STAGE OF THE ENGINE, 1765–74

A FTER two years of speculation and of experimental work on the properties of steam, undertaken by Watt as the outcome of his repair of the Newcomen or atmospheric engine model belonging to Glasgow College, the brilliant idea of condensing the steam in a vessel separate from the cylinder flashed into his brain in the spring of 1765, whilst he was walking one Sunday afternoon on Glasgow Green.

In Watt's own account of what took place, penned, however, thirty years after the event[1] and therefore probably not wholly accurate, he says that ' in a few days he had a model at work, with an inverted cylinder, which answered his expectations. . . . Very simple cocks were employed as regulators or steam valves, and his air pump and condenser were of tin plate. This cylinder, however, was good and of brass 2 inches diameter and a foot long ; the cocks were turned by hand instead of being wrought by the engine.' With this he resolved the first of his doubts, viz. whether the steam would condense quickly enough to form the sudden vacuum required if a working engine were to be based on the invention. The first documentary notice of experiments made to prove the invention, which can only have been with this model, occurs in a letter to Dr. James Lind of Edinburgh.[2]

' I have now almost a certainty of the facturum of the fire engine, having determined the following particulars : the quantity of steam produced ; the ultimatum of the lever engine ; the quantity of steam destroyed by the cold of its cylinder ; the quantity destroyed in mine ; and if there is not some devil in the hedge, mine ought to raise water to 44 feet with the same quantity of steam that theirs does to 32 (supposing my cylinder as thick as theirs) which I think I can demonstrate. . . . In short, I can think of nothing else but this machine.'

The experiments apparently went on for four months, for the earliest definite mention of the model described by Watt above is in a letter to Dr. John Roebuck of Kinneil, to whom Watt had been introduced by Black and for whom he appears to have been carrying out experimental work on the decomposition of salt by lime.

The letter[3] is as follows :

'. I have tryed my new engine with good success for tho' I have not been able to gett it perfectly air tight from its bad materials, yett imediately on turning the exhausting cock the piston when not loaded ascended as quick as the blow of a hammer & as quick when loaded with 18 lbs (being 7 lbs on the inch) as it would have done if it had had an injection as usual. The moment the steam cock was opened the piston descended with rapidity snifting all the while tho' the steam was very weak. On the faith of this I have sett about a larger & more

---

[1] The authors have not discovered any contemporary account. The earliest of several versions of the affair is entitled ' A Plain Story ' and was prepared by Watt in 1796, as an answer to the objections that were raised to the Patent

Specification in the litigation that took place in that year : see Muirhead : *Life of Watt*, 83.
[2] Doldowlod Papers. Watt to Lind, 1765 Apl. 29. Cf. p. 23.
[3] Ibid. Watt to Roebuck, 1765 Aug. 23.

perfect modell having now little doubt of its performing to satisfaction. I think I can now solve the reason why the big cylinder does not take so great a charge as the small but this I refer till meeting.'

On this letter is this annotation[1] or endorsement in Watt's handwriting: ' This seems to relate to the first model with a brass cylinder 1¾ inch diar, and about 10 inch stroke.' It is obvious that his memory was not clear as to details and that he had deduced the size[2] of the cylinder from the figures given in the letter—on the assumption that the piston was directly loaded.

In a further letter to Dr. Lind, September 4th, Watt says: ' I have tryed my small model of my perfect engine which hitherto answers expectation and gives great, I may say greatest, hopes of success (for certainty could not be called hope), in greater model now far advanced.' This confirms the statement to Roebuck that he had started making a second and larger model.

From a letter to Dr. Roebuck of September 9th it appears that the latter had by that time seen the first model. Watt reports further experiments and says, ' I am going on with the model of the machine as fast as possible' ; this must refer to the second model. He remarks also : ' that in proportion as the sensible heat of steam increases its latent diminishes so that in the steam engine working with pressures above 15 lb must be more advantagious than below it, for not only the latent heat is diminished, but the steam is considerably expanded by the sensible heat which is easily added.'

The ' larger & more perfect modell' that Watt speaks of in the letter of August 23rd would now appear to have been made, for on October 11th he says:[3] ' I have made a tryal of my machine. It has not entirely answered my expectations tho it has no faults but what I think I can cure. The principal one & I believe the only one was the untightness of the piston which I think I have found a remedy for.' This is the first mention of what was a continual source of trouble, from this time onwards, viz. piston packing. Apparently packing of the piston-rod was satisfactory from the first, for we hear nothing about it.

Writing to Dr. Lind, on the following day, Watt says:[4] ' I have been making tryal of my machine but have not got the piston steam-tight yett but I hope I shall accomplish it. My error was in applying the (⎍) piston to it ; it being more proper for pumps where the piston is drawn against the pressure than for engines where it flies from it.' He goes on to describe a kind of hat packing made of cloth, which is more fully described in the next letter to Roebuck. He concludes : ' I am at present quite barren on every other article, my whole thought being bent on this machine, so I can write you nothing else.'

Watt's letter[5] to Roebuck is as follows :

' On repeated trials of my machine, I have had better success ; it readily works with 10½ lbs on the inch & sometimes I made it lift 14 lbs. I still propose Improvements on my piston with which I am confident it will succeed to my utmost expectations. This is my present piston. At a are two Collars of varnished cloth, b is the old part of the piston which was made

[1] Succeeding letters to Roebuck preserved among the Doldowlod Papers are similarly annotated and in some cases are dated 1808. About this time Watt was meditating the preparation of an account of his invention and had collected for the purpose a certain amount of documentary matter of which this correspondence with Roebuck is part.
[2] To be exact the diameter works out at 1·8 in.
[3] Doldowlod Papers. Watt to Roebuck, 1765 Oct. 11.
[4] Ibid. Watt to Lind, 1765 Oct. 12.
[5] Ibid. Watt to Roebuck, 1765 Oct. 16.

for Belidor's piston [1] and now remains naked. I propose adding another collar at *c* and another somewhere on *b* with which additions I hope it will be perfectly tight, as you will easily see that the addition of a collar increases the tightness vastly. As to the steam condensed, it is very little ; my little boiler fills the cylinder in less than $\frac{1}{2}$ a second after it has been exhausted. This is the way in wh. I tryed it *a* being the cyl, *b* a leaver fastened down by one end at *c*, *d* is a weight, wh. by being moved backwards & forwards determines the pressure. Now in these circumstances, the weight being in a situation where the engine cannot lift it & a vacuum produced in *a*, it is plain on opening the steam cock the steam will rush into *a* & fill it ; to do which as I said it took less than $\frac{1}{2}$ a second which is known by its opening the snifting valve so soon as it is in equilibrio with the air. So soon as the proposed alterations are made, I shall forward it to you.'

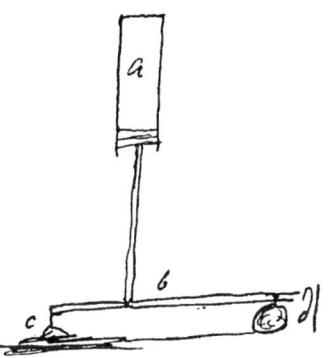

FIG. 8. Watt's piston, 1765, Dol-dowlod Papers

Watt's annotation on this letter reads : ' The cylinder of this engine was 5 or 6 inches dia' with a 2 feet stroke : the inner cylinder was made of copper, not bored but hammered, & not very true. It was inclosed in a wooden steam case & placed inverted, the piston working through a hole in the bottom of the steam case. This was the second of my new engines.'

The figures before us make it clear that the cylinder could not have been ' 5 or 6 inches dia' ', and we can only suppose that Watt has put on this letter an annotation intended for some other document. With a pressure of $10\frac{1}{2}$ lb. on the sq. in., to lift a weight of 14 lb. directly only requires a cylinder $1\frac{1}{3}$ in. diam. While it is probable that Watt did make a model of 5 or 6 in. diam. by 2 ft. stroke in the manner described, it is unlikely that he did so at such an early date, if only because his shop accommodation was so limited.

Now there is a model in existence that, although not definitely known to be of this date, appears to tally remarkably well with the one described in the letter. This model [2] bears the marks of hasty improvisation such as we should expect to find ; it is, moreover, quite elementary, all of which suggests that it is an early construction (see Fig. 11). The cylinder is of sheet iron and the condenser partly of that material, and partly of block tin with soldered joints (see Fig. 12). The cylinder is inverted, single acting, 1·4 in. diam. by about $7\frac{1}{2}$ in. stroke, fitted with a leather-packed piston. The piston-rod passes through a stuffing-box packed with hemp and terminates in a hook. The cylinder is surrounded by a steam case with an opening to connect up with a boiler and with a small hole to drain it. At present the steam or exhaust port is soldered over, whether by accident or design, so that the model is unworkable. The construction is such, however, that the exhaust could have communicated, as shown in

FIG. 9. Watt's experimental cylinder, 1765, Doldowlod Papers

[1] This piston, which is that alluded to in the previous letter to Dr. Lind, was packed with a leather ring held down by a keeper or junk ring resting on the body of the piston ; see Belidor : *Architecture Hydraulique*, VI. 315 & Pl. 3.

[2] In the Watt Collection at the Science Museum, South Kensington. This collection was presented to the nation in 1876 by Gilbert Hamilton, of Soho Foundry, in the name of James Watt & Co.

the drawing, with the pipe condenser which has a snifting valve and a pump barrel. In this position, if the condenser were filled with water and air expelled from the cylinder by steam (which would be known by its appearance at the snifting valve), a rapid stroke of the pump would lay bare an expanse of cold metal which would condense the steam, consequently the piston would rise smartly and lift a weight attached to the hook.

If we admit that Watt's memory had played him false about the material of which the cylinder of the model was composed, we are entitled to say that we have at South Kensington the original model with which he demonstrated the soundness of his great invention of the separate condenser. We are still left with the difficulty that the letter of Oct. 16th appears to relate to the ' larger and more perfect model '. We are in a further difficulty as to whether Watt meant the 14 lb. as the direct lift of the piston or as the weight on the arm of the lever, but we assume the former.

In this letter we find again that Watt is up against the difficulty of keeping his piston tight. His leather rings and multiple hat packing were expedients to get over the difficulty.

Watt, in his next letter to Roebuck,[1] announces his discovery of pasteboard packing.

' I have the pleasure of informing you that I have discovered a substance that will make the piston extremely air tight if oiled and at the same time have little friction itt is pasteboard. I have tryed it, & it was almost perfectly tight with little friction, and does not appear to wear fast.'

Watt rang the changes on this material for packing during many subsequent years.

In a subsequent letter [2] he mentions his scheme for a new form of condenser, consisting of thin cells :

' My old White Iron Man [i.e. tinsmith] is dead, which makes me at a little loss about the condenser but expect to find some other tolerable hand to do it. I have altered the plan of it from small pipes to thin interstices between plates, thus

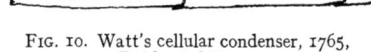

FIG. 10. Watt's cellular condenser, 1765,
Doldowlod Papers

Then follows a calculation as to cooling surface as compared with the tubular condenser: ' I have also thought on something new about the pump of the condenser & also for the pump in the pit. Thinking on these things is a kind of relief amid my vexations.'

Watt's endorsement on this, made about 1808, was : ' Scheme of new condenser for the Iron Cylinder called the plate condenser. It came to Soho and many experiments were tryed with it, but it came unsoldered, and the drum condenser was substituted.'

Muirhead has given us [3] an illustration inscribed in Watt's handwriting : ' Drawing of an inverted Engine with a plate condenser, intended to have been erected at Kinneil in 1765 or 66.' ' The condenser made for this engine was after-

[1] Doldowlod Papers. Watt to Roebuck, 1765 Nov. 18.
[2] Ibid. Watt to Roebuck, 1765 Dec. 5.
[3] Muirhead : *Mech. Inv.* III, Pl. 1, facing p. 1.

Muirhead does not refer to it in the text. The Editors have not been able to trace the original drawing.

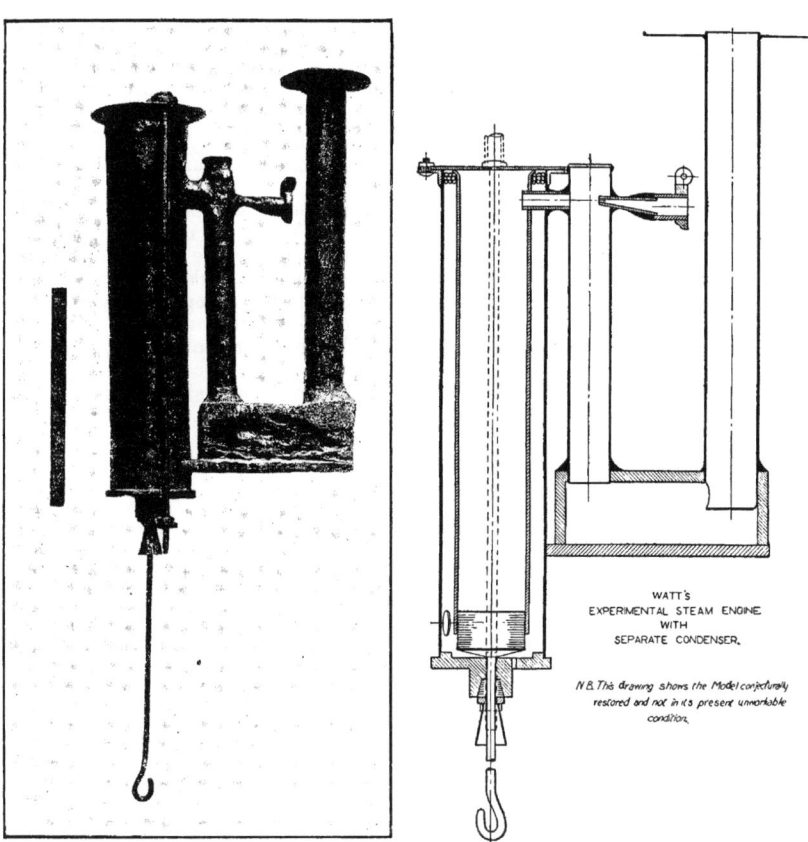

WATT'S
EXPERIMENTAL STEAM ENGINE
WITH
SEPARATE CONDENSER.

N.B. This drawing shows the Model conjecturally
restored and not in its present unworkable
condition.

FIG. 11. Experimental model of the separate
condenser engine, in the Science Museum, South
Kensington

FIG. 12. Section of the experimental model
of the separate condenser engine conjecturally
restored

wards used in ye 18 inch Engine at Soho.' The cylinder shown is about 18 in. diam. by 5 ft. long.

Muirhead has given us also another drawing (Pl. XIII) entitled ' Upright engine with plate condenser 1765 '.

It appears therefore that at this early date Watt had not only designed an engine but that some parts of it were actually made; there is nothing to show that any real progress was made with it.

The value of the drawings, however, lies in the information they give as to the original trend of Watt's mind and as to the course of the development of the invention. We see that Watt has retained scarcely one feature of existing practice even where this was practicable. Referring particularly to Pl. XII, we notice that the old combination of beam and cylinder is replaced by a direct-acting inverted cylinder. A combined steam and equilibrium valve moved by hand is on the top of the cylinder. This has the appearance of being a three-way cock but is more probably the sector-shaped plate valve in a box [1] afterwards used. The condenser is of the surface type already described and situated in a cistern of cold water. The piston is shown directly lifting a weight. The design is far from being a practical one, however, for there is no means shown for working the air pump or the valve ; clearly therefore it was only experimental.

The mention before December 1765 of a condenser formed of small pipes is interesting, because in the Watt Collection at South Kensington [2] there is a circular surface condenser made of 140 tubes, $\frac{1}{8}$ in. diam., soldered into tube plates with an air pump $\frac{7}{8}$ in. diam. The relative positions of the condenser and pump are such that the rising of the water column removes the air before discharging the water. The workmanship is good and there is no evidence of makeshift. It is difficult to imagine that it was made at this date as it looks like an engineer's rather than an instrument-maker's job.

The next letter [3] that we have shows that Watt had divagated in quite another direction : ' I have thought on a simpler circular steam engine than what I mentioned to you and which I expect will be practicable.' This was the steam wheel or rotary engine. Watt appears to have been greatly taken with this scheme, for we hear of nothing else for the next few months, as thus : ' I have set about making the circular engine and I expect to have it done in about a week ' ; [4] and ' The circular machine has taken a great deal more work than I imagined. It is not yett quite finished but near it.' [5] The letters of Dec. 5, 1765, and of Feb. 19, 1766, are of great interest as revealing the true inventor, ever scheming something fresh—indeed Watt would have made more definite progress had he not been quite so prolific in ideas.

For the next eighteen months we know nothing as to the progress of the engine. It will be remembered that it was in the summer of 1766 that Watt started in practice as a surveyor and engineer. Evidence has been adduced to show [6] that he was enlarging his practical knowledge of engine construction. His Journal shows that he was repeatedly at Carron or Bo'ness between March and October of this year, and no doubt he made use of these occasions to keep in touch with

---

[1] See p. 107.
[2] Acquired at the same time as the model alluded to above, see p. 95.
[3] Doldowlod Papers. Watt to Roebuck,

1766 Feb. 19.
[4] Ibid. Watt to Roebuck, 1766 Feb. 28.
[5] Ibid. Watt to Roebuck, 1766 May 14.
[6] See p. 26.

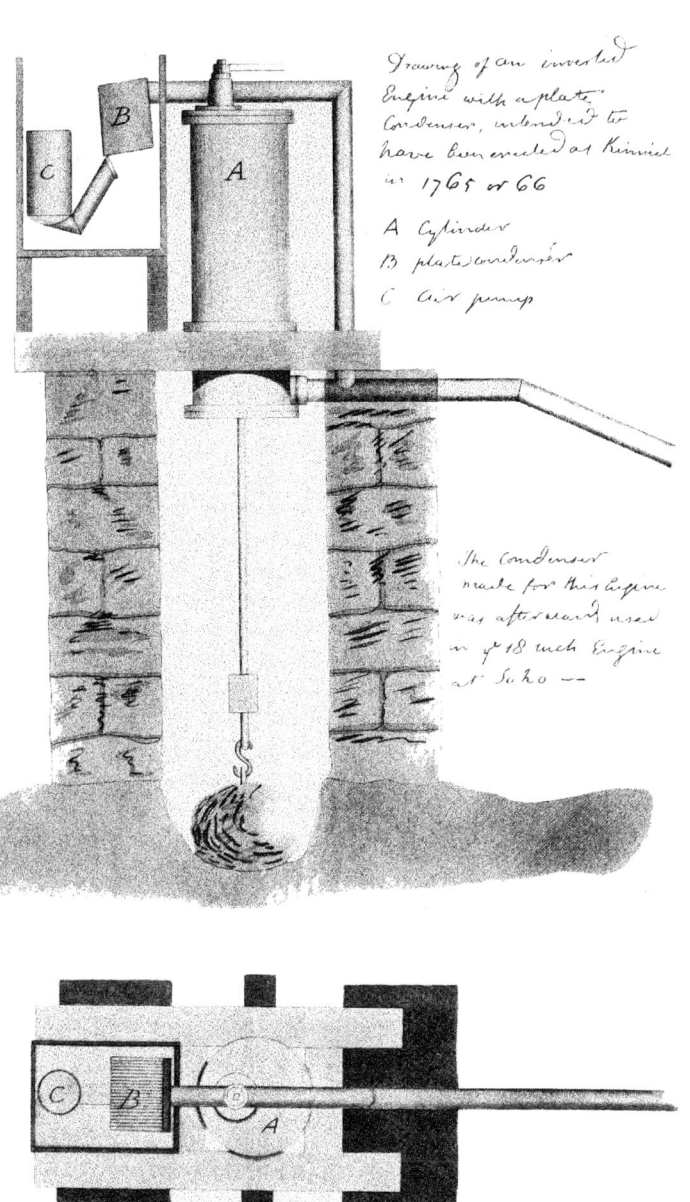

Drawing of an inverted Engine with a plate Condenser, intended to have been erected at Kinneil in 1765 or 66

A Cylinder
B plate condenser
C air pump

The Condenser made for this Engine was afterwards used in ye 18 inch Engine at Soho —

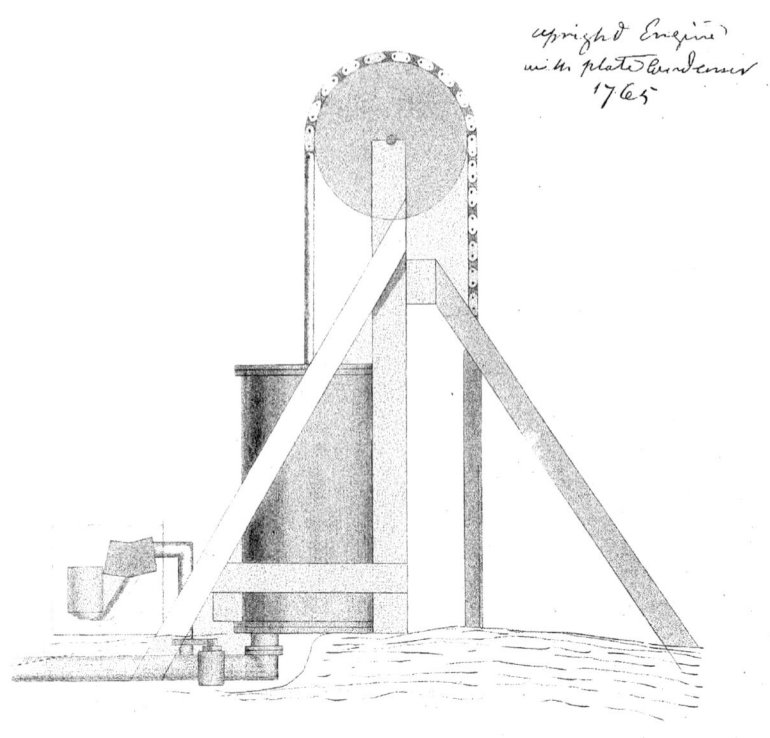

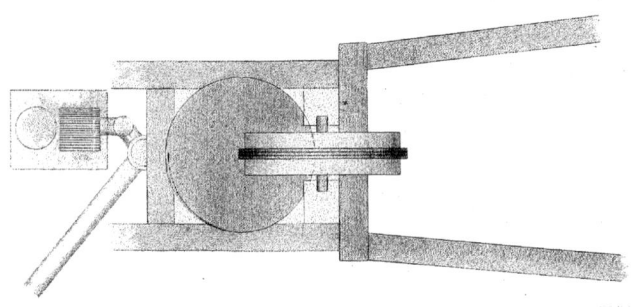

*upright Engine
with plate condenser
1765*

XIII. UPRIGHT ENGINE WITH PLATE CONDENSER

From the drawing in Muirhead) Mech. Inv. III, Pl. 1

Roebuck.   Then again, too, for the first five months of 1767 Watt was fully engaged surveying a canal between the Forth and the Clyde, and in attendance on the Bill in Parliament.   Obviously for this period experimental work was out of the question and as further civil engineering work came his way he had perforce to relax in the pursuit of the invention.   It is clear, however, that progress had been made, for he says to Dr. Lind[1] in January 1768 : 'What I knew about the steam engine before you went away[2] was but a trifle to what I know now.'

At the beginning of 1768 Watt must have had a slack time in civil engineering work, for he took up the engine experiments again.   Writing to Roebuck,[3] he says : ' I would have wrote you sooner but have not yett gott the little wheel compleated . . . meanwhile the other model is going on.'   This shows that he had not relinquished the rotary engine.   The ' other model' must be that referred to in the next letter,[4] which is as follows :

' I have been close working at the Engine since I wrote you, but have not got it perfectly tight yett tho it is much better.   The air seems to come in thro some imperceptible pores in the metal.   I am now getting an Apparatus ready for setting it wholly in steam as before. Excepting this leaking I see nothing else to hinder its performance as the vacuum is formed very suddenly & is best at the begginning when it would be worst if no air got in.   The piston is very tight & seems to last well—I would write you oftener but my health is but indifferent and I have had no good news lately tho I have every day expected to have them to tell you. To-morrow the cylinder will be finished in the new way & also made to act by steam above the piston.'

This letter is endorsed by Watt about 1808 : ' Expt on a large model of the engine 7 or 8 inches dia[r].'   The 'apparatus for setting the engine wholly in steam ' may refer to the steam case.   Four days later Watt reports[5] an untoward accident with this model :

' The mercury in the gage pipe rose to 24 inches & will I hope rise higher when some things are altered but we mett with a sad missfortune : the mercury by an unforeseen accident found its way into the cylinder & has played the devil with the solder.'

This, however, was soon put right :[6]

' I got the damage done by the mercury pretty well repaired & have tryed the engine several times since, as the mercurial gage pipe cannot be used with safety, I estimate the vacuum by weights applyed to raise the piston of the condenser.   This method is not quite so exact, but answers tolerably well.   The condensing pump is 2 inches diameter.   A weight of 47 pounds is in equilibrio with the pressure on it when the condenser has no communication with the cylinder ;   when that communication is open the vacuum is formed directly & the condenser supports a weight of 42 lbs which is coming as near the whole pressure as we can reasonably expect.'   [i.e. 13¼ lb. per sq. in.]

Watt continues the account in a letter[7] two days later : ' The piston of the cylinder that we feared most is extremely tight & may easily be made tighter still ;   it is made of pasteboard baked with lintseed oil and put on like the leather of a pump box.'   He proposes further ' making two exhausting barrels [i.e. air-

[1] Doldowlod Papers.   Watt to Lind, 1768 Jan. 5.
[2] Dr. Lind had sailed for Madras and China in December 1765, and was away till the end of 1767 : see p. 23.
[3] Doldowlod Papers.   Watt to Roebuck, 1768 Jan. 28.
[4] Ibid.   Watt to Roebuck, 1768 Apl. 15.
[5] Ibid.   Watt to Roebuck, 1768 Apl. 19.
[6] Ibid.   Watt to Roebuck, 1768 Apl. 27.
[7] Ibid.   Watt to Roebuck, 1768 Apl. 29.

pumps] instead of one & giving them only ½ the stroke each', and states his reasons. He continues :

'Now I think two of the most important facts are settled the perfection of the vacuum & the suddenty of the condensation. There still remains to determine the quantity of steam used but before I do this I propose making the above Improvement & making the engine go of itself that the strokes may be counted & the water measured that is boiled away.'

This shows that the provision of self-acting valve gear had not yet, viz. in 1768, been made, that the model was wrought by hand, and that the cylinder was still inverted.

A further report is contained in a letter of May 10th : [1]

'I have got the two new exhausting cylinders cast, bored & partly turned also the new condensers made & expect to have it going again by the end of the week. The two Cylinders stand side by side the pipes at bottom communicating with two Condensers which by the pipe c communicate with the large Cylinder. There is a valve at e that prevents the air & water in the condenser from going into the big cylinder at the return of the piston of the exhauster but obliges them to go out at the top of the condenser thro the pipe e having a valve at bottom to prevent their return.'

FIG. 13. Watt's piston, 1768,
Doldowlod Papers

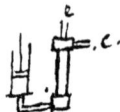 

FIG. 14. Watt's air-pumps, 1768,
Doldowlod Papers

He concludes : 'The vacuum is good & sudden, the consumption undoubtedly the least possible.'

A report of a further trial follows in a letter [2] of May 24th :

'I this day had another tryal of the engine with the double condenser ; the vacuum was as before 14 lbs. on the inch & more readily formed tho this new apparatus is not perfectly air tight. . . . I found the engine could easily make 20 strokes pr minute & snift properly when the steam was middling strong.'

Watt then gives calculations of the quantity of water used to form steam, incidentally stating that the cylinder was 7½ in. diam. by 12 in. stroke. His conclusion is, 'I speak within bounds when I say that every cubic foot of the contents of the cylinder will require only one cubic inch of water to be evaporated. . . . I sincerely wish you joy of this successfull result & hope it will make you some return for the obligations I ever will remain under to you.'

It is obvious that Roebuck was now impressed with the value of the invention, and deemed that the next thing to be done was to protect it by patent. Watt was sent off to London on this business, arriving there in August ; and put the matter in train. He got back home on October 12th, and on the 16th rode over to Kinneil to report to Roebuck. It is clear that they now came to the decision to erect an actual engine, hence the following letter from Watt : [3]

'On considering the engine to be erected with you I think the best place will be to erect a small house in the Glen behind Kinneil the burn will afford plenty of cold water and we will be more free from specⁿ [i.e. speculation] than we can be about Bo=ness. I propose to suspend the Lever so as to take a six foot stroke in the pump for a 4 foot in the cylʳ ; a pump

[1] Doldowlod Papers. Watt to Roebuck, 1768 May 10.   [2] Ibid. Watt to Roebuck, 1768 May 24.   [3] Ibid. Watt to Roebuck, 1768 Nov. 9.

of 18 inches square and 15 foot high will answer then, or a smaller for a greater height. I intend coming next week & concerting & giving the proper directions for chains &c. I have wrote the specification 2 or 3 times over, but am not yett satisfyed with it tho it is better than it was.'

The important point to notice here is that Watt had now dropped his original scheme of an inverted direct-acting engine in spite of the fact that his reasoning had brought him to the conclusion that the beam was not essential to his engine. Perhaps he realized that he had set himself too big a task and was constrained, as the line of least resistance, to adapt his invention to the existing beam engine. It was this very striving of Watt after an entire rebirth of the engine, as well as his fertility of expedient, that caused the experimental stage to be so protracted. It is curious to reflect that with a succeeding generation of engineers the beam type of engine became almost an article of faith.

On the same day Roebuck was writing to Watt [1]:

' I am not so very desirous of hastening the patent. I am afraid we shall be obliged to specify before it is our Interest. . . . You must bring your various attempts at the Spec^n with you. . . . The whole plan of the larger engine must be fixed on when you are here.'

Watt's note-book shows that he was at Bo'ness on November 20th, and doubtless the engine and the specification were exhaustively discussed, for Roebuck changed his mind and wrote, Dec. 22, 1768 :

' I should be sorry to risk the property of the engine ; wherefore by the first post write to your friend to take out the patent. I can spare the money without inconveniency.'

Accordingly instructions were given and the patent was taken out on Jan. 5, 1769. On the 28th Watt writes to Small asking his advice about the wording of the specification which had to be enrolled and as to the drawings for it, and says further : [2]

' I have been trying experiments on the reciprocating engine and have made some alterations for the better and some for the worse which latter must return to their former form. I have improved the condenser by reducing what was before a number of small pipes to one single pipe surrounded at a little distance by another and making cold water run between them. This both answers the end better and will considerably lessen the expense in great. I have misimproved the cylinder by making the bottom of cast iron which being obliged to be thicker than the brass or copper ones used before, does not communicate heat fast enough to evaporate the water left in the cylinder the first stroke without being considerably cooled by it and consequently condensing as much the next. I have also done wrong in removing the outer cylinder which kept the inner one always surrounded by steam, ready to supply any loss of heat that might happen by evaporation from the inside surface.'

Both Small and Boulton advised Watt not to file any drawings in spite of the fact that he had prepared one. Fortunately this has been preserved and is reproduced here. A beam engine is shown with a cylinder of length equal to three diameters and having the upper surface of the piston in communication with the under surface and with the boiler by a pipe. The passage of steam, however, is under the control of a hand-operated regulator in the form of a sector with two ports in it which place the underside of the piston alternately in communication with the upper side and with the condenser. The latter is nothing more than a pipe with another concentric to it outside. The air-pump is a simple bucket worked from a horse-head on the beam. Both condenser and air-pump are in a cistern of cold water. As represented, there is no means for operating the regulator

---

[1] Ibid. Roebuck to Watt, 1768 Nov. 9.
[2] Boulton Papers. Watt to Small, 1769 Jan. 28, quoted in Muirhead : *Mech. Inv.* I. 36.

automatically.   On the drawing is shown also Watt's steam wheel or rotary engine. This consists of an annulus with three flaps hinged on the inner surface, so that they can turn through a right angle and form an abutment for the steam introduced and exhausted at the side by regulators.   The passages communicate with the boiler and condenser respectively through the hollow axis.   The other abutment for the steam is formed of mercury or Newton's metal, whose weight keeps it at the lowest point while the wheel revolves under the steam pressure.   No means for operating the regulators is shown.

On February 10th Watt wrote : [1]

' I have now sett it [i. e. the model] up in a most convenient place where the experiments can be made to your satisfaction.   It 's doing twice as well as the common is, I think, absolutely certain.'

Writing again the following day, he says :

' I thought it was necessary to try if this external cylinder could be wanted [i. e. done without] by introducing steam between the bottom of the cylinder & a false bottom below it.   I also made this bottom cast iron that it might in every respect resemble a large one.   In the first place I found that the sides of the cylinder being exposed to the open air lost a great deal of heat & that the cast iron top condensed much steam which lay on the piston and was troublesome.   I found in the next place that the heat could not sufficiently suddenly pervade the cast iron bottom & that there was a quantity of steam condensed by it.   These things will be remedied by making the top & external cylinder of wood & making the bottom of the internal of copper which transmits heat very fast.   Both the coppersmith & cooper are at work on it & it will be done with all expedition.'

This copper bottom to the cylinder was adopted in actual practice, as we shall see later.   When we reflect on the exhaustive character of these experiments, we can hardly wonder that in after days Watt spoke as if he had explored every possible combination of which the steam engine was capable.

On February 22nd Watt writes to Small : [2]

' I made an imperfect trial to-day of an alteration in the condenser with which I am much pleased, it is this : in the piston of the condenser there are valves as in a common pump, it is 2 inches diameter, stroke 6 inches, contains 18 cubic inches water.   The pipe b of the condenser is surrounded with an outer pipe cold water runs between.   The pipe b and box c contain above 9 cubic inches water ; at c is a sliding valve which is opened and kept open for a little space when the pump a is at the lowest.   The mouth of the pump being above water, the piston when it was at top threw out almost all the water contained above it ; when it is at the bottom, the water in the cistern runs in at the valve c and fills the pump up to the level of the water in the cistern and by that means puts out the water heated in the last stroke; when the piston is raised the valve is shut and the steam, or what of it is not condensed, pushes the water before it into the pump and endeavours to follow it into the pump but is condensed with a crack.   It is not possible if the water be cold that any of it can survive this operation and the machine is simple and works easy.'

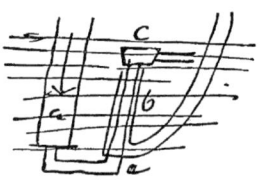

FIG. 15. Watt's displacement air-pump, 1769, Boulton Papers

There is no evidence that Watt used this kind of air-pump, but the reader will recognize that it embodies the principle of the displacement, or Edwards's air-pump, considered quite a recent invention.   We have got to remember that all along Watt's idea was to clear the condenser by flooding it at each stroke.

[1] Doldowlod Papers.   Watt to Roebuck, 1769 Feb. 10.   [2] Boulton Papers.   Watt to Small, 1769 Feb. 22.

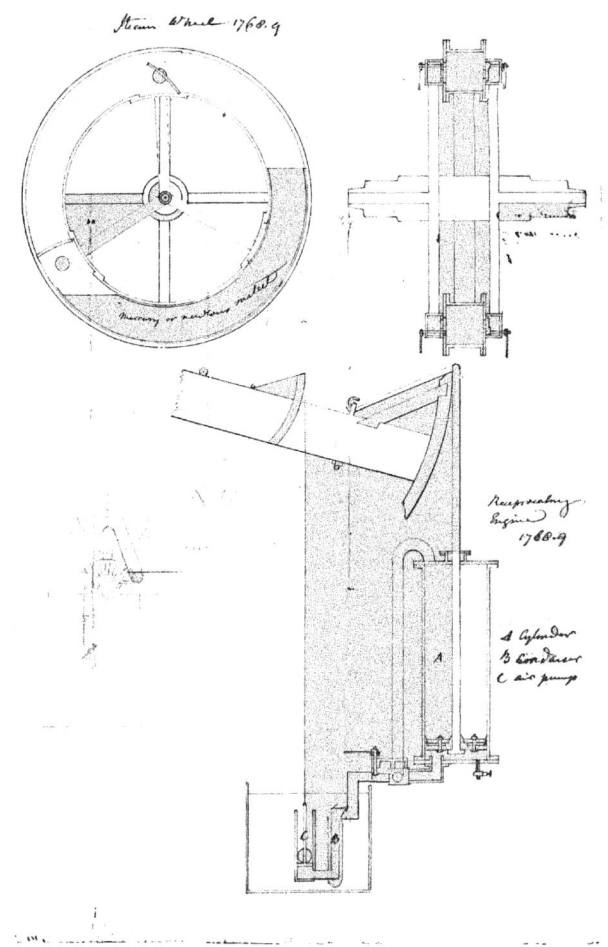

Steam Wheel 1768.9

Mercury or new force rachets

Reciprocating
Engine
1768.9

A Cylinder
B Condenser
C air pump

XIV. DRAWING PREPARED FOR THE PATENT SPECIFICATION, 1769

By Watt. Doldowlod Papers

On March 14th Watt writes : [1]

' I this day made trial of the power of the engine & found that after it was clear of air it readily lifted 620 lbs. & I believe would have lifted more, but had none at hand. The whole pressure on the piston is about £740 [*sic*] this was with a steam not able to support one or at most 2 inches of mercury.'

This letter is endorsed in 1808 : ' Exp^t with cast iron model, seems to have been 8 inch dia^r.' Clearly Watt was not at all certain, and the diameter is given on the assumption that the vacuum was 14 lb. It seems possible that the model under trial was the one with the cylinder 7½ inches diam. already mentioned.

Two days later Watt writes again :

' Since I wrote you I added 80 lbs to the load of the engine making in all 704 lbs neat which it lifted easily. However, the additional weight being a large mortar & inconvenient, I took it off & wrought it a few strokes with the 624 lbs which it raises on half a revolution of the handle of the condenser or the stroke of one pump. This seems to depend wholy on the coldness and quantity of water that passes thro the condenser which I have an easy method of encreasing. I find this cylinder being of cast iron & near ⅛ inch wider at one end than it is at the other, is not so steam tight without oil as the block tin one I had last was. I therefore make use of the oill pump & train oil which answers perfectly well keeping a constant circulation.' [2]

It was an important experiment and Watt felt it to be such, for he wrote an account of it, only in different words, both to Dr. Black and Dr. Small. A point which may be noticed is that Watt was using a handle to operate the model, which shows that the problem of the valve gear had not even yet been tackled.

In April Watt writes to Small :

' I have still [further] improved the condenser. I have rendered its pumps frictionless. I have therefore increased its size. I have reduced it to the last simplicity that is to have no parts but the pumps and a great valve that admits the water also frictionless though air tight. The pumps have an inch of stroke and are five inches diameter, half a cubic foot of cold water can pass through them during one stroke if necessary.'

On April 28th [3] Watt reports a further experiment on the cast-iron model, the gist of which is that ' the regulator condensed steam by lodging water ' and to remedy this he intends to immerse the regulator in steam. On the same day he writes to Small in a similar strain and adds :

' I have begun to erect an engine of an 18 inch cylinder and 5 foot stroke at Kinneil, the Doctor's place. However, it is to be done as privately as we can as we do not propose to puff ; therefore we expect your silence on this head.'

On May 1st [4] Watt writes :

' I send enclosed plan & elevation [5] of the engine. I think the two supporters of the great beam should be 14 inches square at least ; those of the spring beams of the size commonly used when of that length. The two main supporters must be about 14 inches asunder at the top to admit of the Iron straps that secure the axis. You ask what is the principle hindrance

---

[1] Doldowlod Papers. Watt to Roebuck, 1769 Mar. 14.
[2] In a memorandum book entitled ' Journey to London about Kinneil Engine, 1768 ' (Muirhead Papers) there is the following entry, which can only relate to this model ; the title is of later insertion and a mistake may have been made in the date :
' Glasgow model engine, 1768.'
' The little cast iron Cylinder at Glasgow was 8¼ Inch dia^r : its area 53½ square Inches. It lifted readily 620 lbs being 11½ lbs to the Inch. It lifted 740 lbs but more slowly being nearly 14 lbs to the Inch beside friction in Cylinder. It made a vacuum in a 6¾ inch Cylinder, but was rather slow about it in some tryals, this was only equal to 530 lbs.'
[3] Doldowlod Papers. Watt to Roebuck, 1769 Apl. 28.
[4] Ibid. Watt to Roebuck, 1769 May 1.
[5] This drawing is not now with the papers. (Eds.)

in erecting engines ; it is always the smith work. I would wish the boyler bottom to go on first. I will soon be east, & give directions about the manhole & top. Meanwhile I wish you would sett the chain makers to work ; you know we proposed to make it a 3-ply chain (of the best tough iron that breaks with a dusty surface), the midle link $\frac{3}{4}$ inch thick, the two side ones $\frac{1}{2}$ inch each, the breadth about 3 inches. I wish the wood of the frame to be got ready but not to be put up till I come. The pattern of the regulator is at the founders & part of the condenser is made for the little engine (the model I mean). I am busy drawing & planning the several parts. . . .'

Letters in May, July, and August from John Roebuck, junior, to Watt report slow progress with the erection of the engine at Kinneil. From Sept. 20 to Sept. 25, 1769, Watt was occupied in writing an exceedingly long letter to Small, in which he gives an account of the trial which had taken place about a fortnight earlier than the date on which he wrote. He says :

' Well then the trial has not been decisive. . . . The pump is 18$\frac{1}{2}$ inches diameter and 25 feet high ; the cylinder 18 inches wanting $\frac{1}{8}$ or 17$\frac{7}{8}$; the boyler 5$\frac{1}{2}$ feet at the bottom which is the widest part. The cylinder is then loaded with nearly 12 lbs on the inch for the area is 253·8 square inches, the load 46·1 cubic feet or, at 62 lbs per foot, 2858·2 lbs.'

Then follows a long description[1] of the trials and the troubles that were experienced, principally with piston packing. He then says :

' The only conclusion I can draw from this trial is that, supposing we cannot employ oil to keep the piston tight and that we cannot make it better than we had it, it would work easily with 8 lbs on the inch and would not consume above half the steam used by a common engine. Even this I will not positively affirm although I think there is reason to believe it.

' I am now employed in making the following alterations : enlarging the descending valve of the oil pump to three inches square (above nine times its late size) ; making a new piston of cast iron (the last having been wood) to fit the cylinder within a sixteenth all round, the cylinder is to be made exactly round and straight where it stands ; making a new condenser with two pumps to each of which belong eight pipes eighteen inches long three fourths diameter with cold water running about them, the pumps 4$\frac{3}{4}$ diameter and 4 inch stroke.'

The whole letter is of great length and appears to have been written for the purpose of putting before Small and Boulton the exact position with regard to the engine, so that when Dr. Roebuck should call at Soho, as he was intending to do to submit a partnership offer, they should be in a position to judge ' how far it may be your interest to engage on it [i. e. the engine] on its own account '.

In a letter to Small on October 21st Watt gives further details of the cylinder. It ' was a full quarter inch thick and 18 inch diameter. At the worst place the long diameter exceeded the short about $\frac{3}{8}$; it was made of the best tin not hammered.' He had thought of making it true by ' hammering it with a mallet on a piece of hard wood fitted to it '.

On Feb. 7, 1770, Roebuck writes to Boulton to tell him ' that a single step has not been advanced towards the engine since my last letter. The fact is Mr. Watt hath been constantly and necessarily engaged in planning the Glasgow Canal. But in three days I expect him here, and he has promised not to leave this place till the whole of the alternate [i. e. reciprocating] engine has been effectually tryed.'

Watt did arrive as expected, for Roebuck writes to Boulton, on February 10th : ' Watt and his men are at Kinneil about to attempt to rectify the defects of the cylinder, which he is confident he shall accomplish.' Watt was obliged to leave before they were ready, but on March 9th the trial actually took place, but ' had not the hoped success ', as reported in a letter to Small written from Kinneil the

---

[1] Given at length in Muirhead : *Mech. Inv.* I. 70.

day following. The cylinder and piston were satisfactory and Watt got six strokes a minute. But when he tried his pipe condenser with it he could hardly get the engine to move at all. One result, however, was satisfactory: 'the consumption of steam is as moderate as expected.'

Among the Doldowlod Papers relating to Roebuck is an undated and unsigned memorandum, 'To John Gardiner.' Internal evidence points to its being the enclosure alluded to by John Roebuck in acknowledging a letter, on March 7th, from Watt to his father. The memorandum runs as follows:

' As soon as the condensers are made and put in their place they must be surrounded by a cistern as in the draught—

a. a. are the pumps, *b. b.* the condensers, *c.* the steam pipe, & the black line is the little cistern.

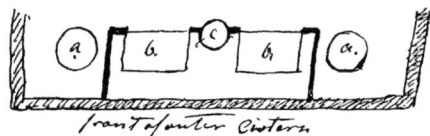

Fig. 16. Cistern for condenser, Kinneil engine, 1770, Doldowlod Papers

' The little cistern may be about 2 inches lower than the great one for the waste water to run over when it cannot pass the condensers. When that is done the Loggerhead & working gear may be sett about—when one stands at the plug frame the loggerhead must be on their left hand & the tail that works the regulator must be above the axis and on their right hand that tail must be about a foot long.

' The axis must be 1½ in. square it must be placed as low as possible that the end of the working tail may come in a line with the regulator handle nearly, but before the axis is fixed the plug should be mounted and hung and the beam moved up & down that it may be seen how the plug will answer to the place of the axis.'

The fact that the valve gear was entrusted to a millwright shows that there was nothing in it that differed materially from that used on the common atmospheric engine. There is no mention made of Joseph Hateley who was borrowed by Roebuck from Carron Ironworks for the job, but doubtless he was the one responsible for the valve gear.

On March 16th–20th Watt writes to Small that he has again tried the pipe condenser without great success till he made 'a small hole to spout into the space between the two valves and immediately obtained a condensation about as speedy as with the

Fig. 17. Working-gear of Kinneil engine, 1770, Doldowlod Papers

former condenser. . . . The reason why the condenser was not tried sooner was that I once imagined that there was much air mixed with steam; I therefore intended my condenser to be perfectly exhausted each time by filling it with water. Now this condenser required two large pumps, and consequently was too great a burthen, but I now find little air to be mixed with steam, and I see by calculation that it will make little difference whether the condenser be filled with water or no.' This was an important step. Watt has now realized that it is unnecessary to flood the condenser at every stroke, and thus the load on the air-pump, or negative work, is diminished.

An account of a trial of the Kinneil engine, just about this time, is given in

Watt's Journal (Muirhead Papers), and is reproduced here (Pl. XV) as illustrating how methodical he was.

Watt was now obliged to fulfil engagements he had entered into to survey a canal between Perth and Cupar of Angus, so that we find his next letter, dated April 8th, directed from the former place and addressed to Roebuck. In it he gives further directions about the working gear and about a new condenser. The letter is as follows :

' The plug frame is to gott fitted up as soon as the patterns for the new condenser are finished & sent off. When the plug is finished compleat the cast iron snout is to be opened & a strict examination past upon the regulator & its joints (but first of all that part of the copper pipe that is without the snout is to be new painted that it may dry while the plug is making). The regulator & snout being judged to be tight, the boiler may be heated & every place where it blows steam to be made close by what John Parker calls caulking ; the wooden cylinder top is then to be made tight & cork to be put in the jack head as directed. The condenser pump rods need not be lengthened for it will do no service. When all these things are done the dimensions are to be exactly taken for the new condenser without taking down the old one & it is to be made as far as it can be but the old one is not to be taken down till I return. A whorle of plaintree as large as can be got on is to be fixed on the axis of the condenser crank near the boiler-end and it may be of this form [1] and about 8 inches thick a block must also be got to hang to one of the couples for the weight to go over that drives this & I beg the Lads to be as diligent as they can as I expect to be back before all this is done & shall not be able to stay long.'

Whether these experiments or some further ones that Watt asked Roebuck to make in a letter of April 15th were carried out or not we do not know. Watt was too busy with his civil engineering work, apart from the expense involved, to go on with them. It was four and half years before circumstances permitted him to take up experimental work again. The situation in what we may call this interregnum may be summarized in Watt's own words : [2]

' I pursued my experiments till I found that the expense and loss of time lying wholly upon me through the distress of Dr. Roebuck's situation turned out to be a burthen greater than I could support, and not having conquered all the difficulties that lay in the way of the execution I was obliged for a time to abandon the project. Since that time I have been able to extricate myself from some part of my private debts, but am by no means yet in a situation to be the principal in so considerable an undertaking.'

To recapitulate : of the whole period of nine years from the time of Watt's invention of the separate condenser till his arrival in Birmingham we find that the time spent in active prosecution of the invention was really only a matter of three years in two periods 1765–6 and 1768–70, and even in these periods the work was not uninterrupted. The net result was that the engine underwent a radical change and assumed eventually the general form of the earlier beam engine, but did not attain complete success, even in that form.

[1] Sketch not reproduced.
[2] Doldowlod Papers. Watt to Small, 1772 Aug. 30. Muirhead : *Mech. Inv.* II. 24.

April 7ᵗʰ 1770

Dated April 8ᵗʰ 1770

XV. DESCRIPTION OF KINNEIL ENGINE

From Watt's Journal, 1770, Muirhead Papers

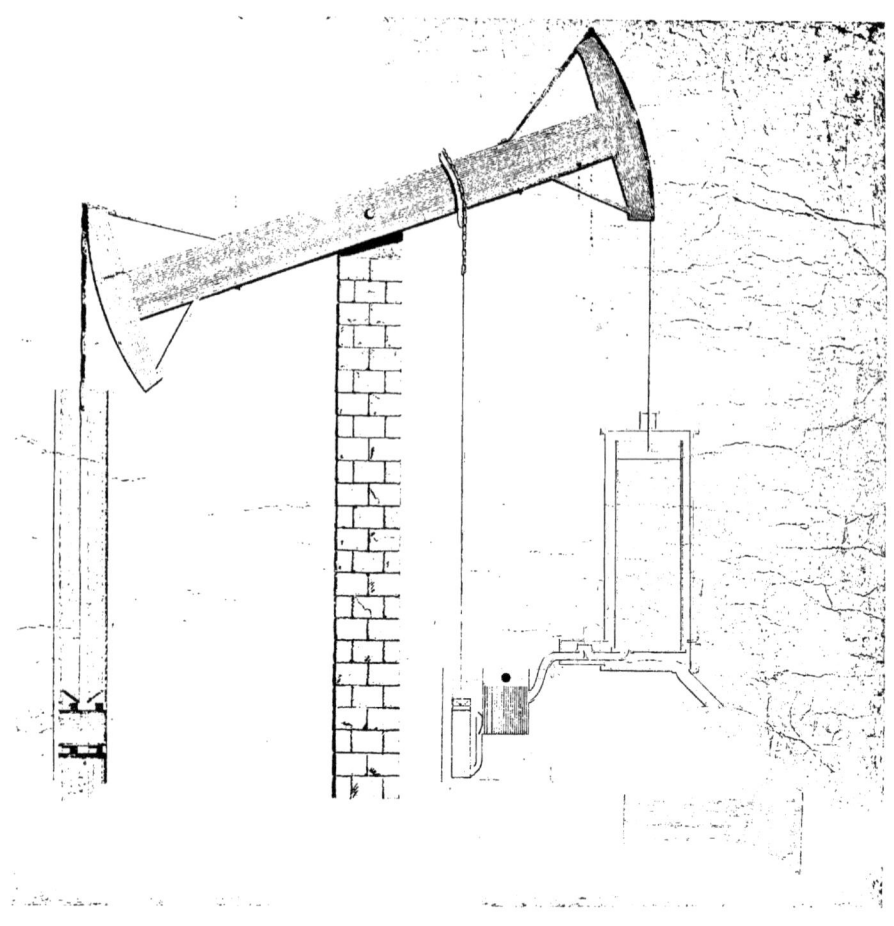

XVI. SINGLE-ACTING ENGINE. DRAWING LAID BEFORE PARLIAMENT, 1775

Drawing by Watt. Doldowlod Papers

IX

THE SINGLE-ACTING PUMPING-ENGINE
1774–7

THE FIRST ENGINE AT SOHO

THE Kinneil engine was taken down and the metal parts forwarded to Birmingham in 1773, and in June of the following year Watt himself was in Birmingham and had begun to re-erect it at Boulton's Soho Manufactory for the purpose of pumping back water for the water wheel employed to drive the grinding-wheels. The engine so re-erected retained the block-tin cylinder; it had an inner cylinder bottom of copper with the branch fixed on it, and the surface condenser was composed of tubes; in other respects we have no definite information as to its construction. However, in connexion with the application to Parliament in 1775 for the extension of the term of his patent, Watt produced to a Committee of the House of Commons two drawings;[1] one of these shows a single-acting engine, and the other a 'double engine', and also a rotary engine. Now there can be little doubt that the drawing of the single-acting engine follows, at any rate in its main lines, the engine that Watt had been setting up at Soho. This drawing, Plate XVI, shows a cylinder enclosed in a steam case and a tubular, or as Watt calls it, a 'pipe condenser', and the air-pump. The steam case is drawn as if of metal, but there is evidence that suggests that that first used at Soho was of wood,[2] and indeed the companion drawing of the 'double engine' shows a similar case, which is described thereon as a 'steam case of wood lined with copper'. The valve construction is of great interest, and is made quite clear by the perspective views, Plate XVII, which appear on the second of the drawings referred to above. Watt retained the oscillating sector-plate of the Newcomen engine, but whereas in that engine the valve was required to control only the admission of steam to the cylinder, he had now to control both the admission and the exhaustion. To do this he contrived what is really a slide valve, a valve with a central cavity working in a steam chest over a face with two ports. The valve moved to one side to uncover one of the ports, and admit steam under the piston, and then to the other side to close that port to the steam chest and to connect it, by the central cavity in the valve, with the port leading to the condenser.

It is clear that the construction is the same as that indicated, less distinctly, in the 1769 drawing, Plate XIV, and we may say with confidence that the Soho engine first went to work with a valve of this sort, and that the working-gear was the same as that in use for Newcomen engines.

Although Watt gave up the use of this valve, it seems to have acted fairly

---

[1] These drawings are at Doldowlod; they are on parchment: that of the single-acting engine was made by Watt and is marked, on the back, 'Draw⁶. that was laid before Parliament, April 1775.' The other drawing bears an inscription to the effect that it was made by one of Mr. Wyatt's draughtsmen from a sketch by Watt.

[2] Doldowlod Papers. Boulton to Watt, 1775 Apl. 24: 'to increase the size of the wooden cylinder, which I have this day put into the hands of our cooper for that purpose.'

well, and certainly was not one of the main sources of difficulty in getting the engine to work. Neither did the condenser and air-pump give much trouble ; indeed it is clear that if Watt had been content to apply his separate condenser to the Newcomen engine, as it then existed, his task would have been a very much simpler and easier one than he actually encountered. In the older engine the piston working in an open-topped cylinder was maintained tight by a layer of water on its upper surface. Watt's plan of applying steam pressure to the upper surface of the piston, instead of that of the atmosphere, at once introduced two practical difficulties : he had to close in the top of the cylinder and to provide a stuffing-box for the piston-rod to work through, and he had to devise some other means for keeping the piston tight. As to the stuffing-box, there was no serious difficulty with it at any time ; but with the piston it was very different, and it took a few years' work to effect even a moderately satisfactory solution of the problem.

The study of the difficulties that had to be surmounted in bringing this, the pioneer, engine into satisfactory working condition is of great interest ; it merits consideration at some length, and fortunately material is available for this purpose. A book in the handwriting of Watt—'Experiments on ye first engine at Soho', in the Boulton and Watt Collection, records a series of trials made with this engine. The particulars registered are, usually, the number of strokes made by the consumption of stated weights of Wednesbury coal, but from time to time notes were taken of the quantity of water evaporated. The steam pressure ranged from 3 to 9 in. of mercury, and the engine, it will be understood, was working against a constant load. In the course of the experiments various alterations were made in the area of the fire grate, and in that of the opening from the boiler flue into the chimney.

For our present purpose we may begin with the entry for Dec. 14, 1774, which tells us that the ' condenser is in good order, and wrought by a beam by the engine'. (In some of the trials the air-pump was worked by hand.) ' Boyler top & steam pipe covered with woolen pasteboard ¼ inch thick.' ' A new piston made of oiled woolen cloth.'

The next entry, Dec. 23rd, says that the ' woolen piston being found wore out & torn by the baked oil sticking to the cylinder, it was taken out and a new wooden one put in . . . the piston was covered with horse muck'. Then, four days later— ' The wood piston being found faulty, a hat piston not oiled was put in ; it was first covered with horse-muck, but it being found that air pervadyd it, it was covered with paper pup mixed with flour paste.' After running for about four hours the piston was examined and ' found smooth and good, but the paper had withdrawn a little from the cylinder, which was cured by screwing down the irons upon it '. The examination made on this day revealed another defect, ' the tin cylinder had been forced in by the steam,' and soon afterwards a cast-iron cylinder was ordered from John Wilkinson.

During the first half of the year 1775 Watt spent a great part of his time in London in connexion with his petition to Parliament ; however, from the time when the cast-iron cylinder was put in, early in May, the story of the engine troubles is continued in a series of letters from Boulton to Watt.[1]

' The engine goes marvelously bad. It made eight strokes per minute ; but upon Joseph's endeavouring to mend it, it stood still. Nor do I at present see sufficient cause for its dullness.

[1] Doldowlod Papers.

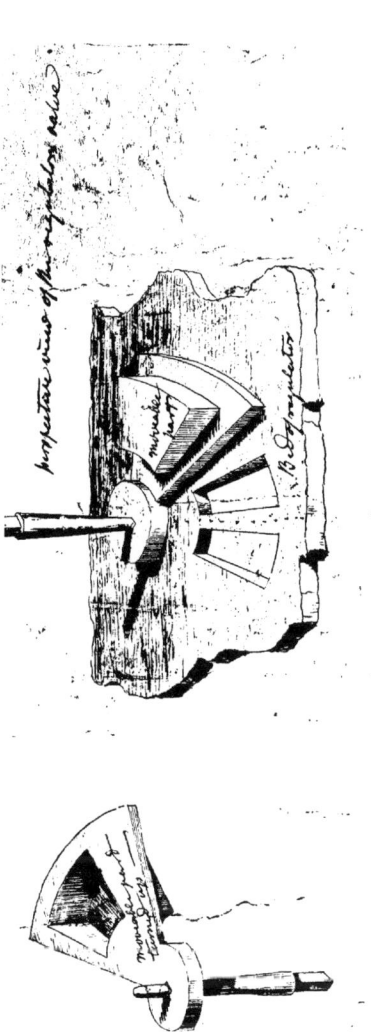

XVII. PERSPECTIVE VIEW OF VALVE. DRAWING LAID BEFORE PARLIAMENT, 1775

Doldowlod Papers

I have a few minutes ago had the top taken out, and find that I can pump down the piston, & although I can hear ye air pass by it into the cylinder, yet the error is not sufficient to account for its bad going. The piston is now taken out, & although ye cylinder is not perfect, yet there doth not appear any gross error. The outside of the piston is hat, filled up with paper chewed. It is 9 inches thick at least, wch. I fear makes much friction. I have ordered a bottle of oyl to be put into ye papier mache, wch. will drain through the hat & lubricate ye sides. It is certain, by another experiment I made, that much steam escapes, as ye water, which passed through ye condenser, continued to be hot. I wish we could learn ye plan of Gainsborrow's piston, as by his own acct. 'tis absolutely perfect for water.'

In the next letter Boulton writes :

' I again attended to the engine, but could not raise nine strokes per minute, even with steam that supported 9 inches of mercury. I therefore this day set myself to examine ye cause. First, I found ye sliding valve very much out of truth, and it admitted air freely ; secondly, the two convex pieces of copper wch. are soddered upon the bottom side of ye cast valve frame were unsoddered in many places and admitted air to pass in quantities. Both of these errors we have rectifyed, & can readily exhaust ye cylinder by hand, & the piston will be I dare say 2 hours in returning ; so that I should suppose that part of the engine perfect enough. The next expt. was to exhaust ye cylinder by hand, and having suddenly opened the valve, I found nevertheless that the beam returned very slowly, which shows there is great friction, either in ye cylinder, piston, or pump, because the pump end of ye beam is about 4 cwt. heavier than ye cylinder end, besides the consideration of the pump end of ye beam being ⅛ longer the cylinder end. I then ordered the cylinder piston rod to be detached from its chain, and suffered to fall freely, which it did, but it descended very slow, even not ¼ so fast as when working, which shows ye friction of that piston is at least equal to 200 lbs that being abt. ye wt. of it, for there is about 90 lbs of lead lying upon ye papier mache.'

Apparently this experiment was carried out on a Sunday, for the letter goes on to say : ' I propose to put in a fire again to-morrow, as Joseph says we have always had the bad luck on Sundays.' Joseph was Joseph Harrison, the leading workman and later the foreman of the engine-yard at Soho. Whether a trial was made on the following Monday is not clear, but what appears to be the next letter in the series is dated ' Sunday, 5 o'clock p.m.', and in it Boulton says :

' I have had a touch with ye engine to-day. 1st. we took the lead wts. off the papier mache, and put on the old light cast metal ones which hath lessened the friction in ye cylinder much. N.B. The papier mache lies within hat, i.e., it is hat that is in contact with ye cylinder, & seems to do very well. I really could not find any fault wth either piston or steam valve. We set to work, & very soon made eleven strokes per minute. The heat of ye water running into ye cistern was 66, and in some parts of ye cistern 68 & the running out water was 80°. The steam raised the mercury to near 9 inches, when it made 11 strokes, but upon loading ye safety valve so as to raise the mercury to 10 & 10½ inches the engine made twelve strokes per minute. Although ye running out water was 80, yet there is no doubt but ye water in the pipes at ye time of condensing is above 100. I presume the present deficiency arises in the condenser. I think it should be larger, and I think some sort of meter should be annexed to it, by which one may see the rate of vacuum, for without an outward & visible sign it is impossible to judge of ye inward and spiritual grace.'

Boulton goes on to give a sketch of a mercury siphon-gauge connected to the side of the condenser and points out that when Watt last tried the engine the temperature of the condensing water was about 40° instead of 66° or 68° as it was in his experiment ; he asks too if he shall try blacklead dust upon the piston, ' as it works well with iron'.

In his next letter Boulton announces that he has worked the engine for the last two days with a mercurial gauge to the condenser. This, he says, indicates 28 in., sometimes 28½, sinking at each stroke about 3 in. ; he seems to have been pleased with the sight of the gauge bobbing up and down. As to the valve, he says : ' I don't think the regulating valve bad but if you have contrived a better I shall be glad, as I am sure it will be a good one.'

In the last letter of this little set Boulton gives particulars of a day's working with the engine, and adds : ' no leaks, the piston and valve good, the piston can't leave the paper behind as it is enclosed in hat, and iron weights lying upon it. I think we must contrive to have all the cold water raised out of the mine pass through the condenser, for I believe the present imperfection arises chiefly from want of more cold.'

We can now return to Watt's Journal of the experiments. He resumes his observations on June 6, 1775, when it appears that the new iron cylinder had been in use for about a month, and had been worked smooth with emery. The piston, he writes, ' was made with paper pulp only in the manner of a common engine piston. It lasted extremely well, but had not sufficient elasticity to apply itself to the cylr.' On June 15th he ' opened the piston lid and found that the paper fitted very well to the cylinder and was wett, applyed more cowdung and put other weights upon it to keep it in place. . . . The additional weights produced a great friction and retarded the going out of the piston.' Accordingly, on the next day, ' the heavy weights taken off and only a shot bag of an inch dia^r. left on and some more cow muck added '. However, the engine was found to be taking too much steam, so it was stopped and again examined, when it was found that ' the valve was hurt partly by the cow muck and partly by the soft metall wh^{ch} had been put upon the piston and had come down. The piston was also bad because the paper had pressed down quite to a sharp edge and was so jammed that it could not play upon the cone and the horsemuck had hardened round it and had furrows in it by which the steam passed. The muck was entirely cleaned out and the piston was laid with paper with a small rope under.'

The rest of the journal is missing. The result of the trials as summarized by Watt was that the engine raised 16,000 cubic ft. of water by the consumption of 1 cwt. of the slack of Wednesbury coal.[1] The hundredweight was 120 lb. The engine had a cylinder of 18 in. diam. and a stroke of 5 ft. It made 14–15 strokes per minute. The pump barrel was 18½ in. diam. and the lift was 24 ft.

By this time, June 1775, Watt was engaged on the designs for two larger engines, the one a blowing-engine for John Wilkinson's blast furnaces at New Willey, near Broseley, and the other a pumping-engine for Bloomfield Colliery, near Tipton. For these large engines, although it seems to have answered fairly well with the Soho engine, Watt decided to throw aside the surface condenser and to adopt injection, in spite of the additional work that he saw would be thrown upon the engine, but before applying it outside he first made trial of a jet condenser on the Soho engine. This was in October 1775, and it answered the purpose very well. Another departure made for the new engines was the substitution of a new form of valve for the sliding valve ; this also was tried on the Soho engine some time in 1775 after June, but the first arrangement was not satisfactory and a new design was got out in January 1776 for ' conical valves and nozzle '.[2] Later on in the year the copper bottom to the cylinder was replaced by one of cast iron, and it would seem that about the beginning of 1777 further changes were made in

---

[1] B. & W. Colln. : Letter Books. Watt to John Bligh, 1775 Oct. 22.

[2] B. & W. Colln. : Drawings, Portfolio 626.

the engine, for in February we read in Watt's Journal,[1] on dates Feb. 6 to Feb. 15 inclusive : ' Engine had been tryed with new pump and had done well' ; 'little engine almost ready for trial'; 'an unsuccessful trial of the little engine, air leaks'; ' all day with little engine, but something was defective about condenser' ; ' little engine in very good going'.

The term ' little engine' requires explanation. A new and larger engine for the Soho Manufactory was now under consideration, the cylinder, a 33-in., was ordered on Feb. 21st and the engine was set to work at the end of July or the beginning of August ; accordingly from this time forward we find the original engine referred to as the ' little engine' or the ' old engine'.

The nozzle design of January 1776 turned out unsatisfactory, and in April 1777 a new nozzle was fitted. Watt seems to have been satisfied with the first trials of this and, referring to certain troubles with the Bloomfield engine, says he hopes it will ' free us of some of these vexations'.[2]

The next thing we learn about the engine is in connexion with forge hammers. ' Wilkinson is going to work in the forge way, & wants an engine to raise a stamp of 15 cwt. 30 or 40 times a minute. I have sett Webb to work to try it with the little engine and a stamp hammer of 60 lb. weight. Many of these battering rams will be wanted if they answer.'[3] The little engine had now been shifted to make way for the new engine and ' now stands below stairs'.[4] No time was lost, and on May 10th Watt writes : ' I saw the Battering Ram, or devil incarnate, go to-day about 60 strokes per minute and work its own regulators.'[5] However, this scheme came to nothing, possibly the machine knocked itself to pieces, and for the next two years this engine seems to have stood still.

In August 1779 we read that ' a customer has appeared for Soho old 18 inch, which I would rather not have had, but lying on hand is a temptation'.[6] However, it seems that the temptation was resisted, and in September we have a drawing of ' Soho old engine with proposed alteration, 1st Sept. 1779'.[7] The alteration appears to be confined to the nozzle. About two years later it is proposed to set the little engine to work two pairs of rolls to roll tin for the wetting-boxes of the copying-apparatus,[8] and in July 1781 the swashplate arrangement for producing rotary motion was applied to it.[9]

BROSELEY AND BLOOMFIELD ENGINES

By the month of June 1775 Watt was engaged on the designs for two large engines, the one a blowing-engine with a 38-in. steam cylinder for John Wilkinson's blast furnaces at New Willey, near Broseley, and the other a pumping-engine with a 50-in. cylinder for Bloomfield Colliery, near Tipton. The blowing-engine was constructed entirely in Wilkinson's own workshops. Watt was at Broseley in February and March 1776 superintending the work of erection, and expressed great pleasure at the excellence of the workmanship—' the cylinder and the fitting

[1] Doldowlod Papers : Watt's Journal.
[2] B. & W. Colln. Watt to Boulton, 1777 Apl. 17.
[3] Boulton Papers. Watt to Boulton, 1777 May 3.
[4] B. & W. Colln. Watt to Boulton, 1777 May 6.
[5] Ibid. Watt to Boulton, 1777 May 10

(Photograph copy).
[6] B. & W. Colln. : Letter Books. Watt to John Wilkinson, 1779 Aug. 25.
[7] B. & W. Colln. : Drawings, Portfolio 626.
[8] Doldowlod Papers. Watt to Boulton, 1781 Apl. 28.
[9] B. & W. Colln. Boulton to Watt, 1781 July 9.

of the piston are beyond my most sanguine hopes. It seems to be truth itself.' [1]
The only drawing of the Broseley and Bloomfield engines that has been found in
the Boulton and Watt Collection is that of the condenser, but a rough outline
sketch of the Broseley engine in the Egerton Collection at the British Museum is
reproduced in Plate XVIII. This drawing is headed : ' Section of the Steam Engine,
on M[r]. Watt's construction, lately erected at New Willey furnace near Broseley,
Shropshire, by M[r]. John Wilkinson—and used for blowing Air into his Smelting
furnace at those works.' The written description upon it in the reproduction is too
minute to be read with ease ; it runs thus :

' A, the Boiler. B, the Steam pipe communicating at all times with the upper part of the
cylinder C, C, C, C, and when the valve *h* is open, with the lower part also, through the tube D.
C,C,C,C, the House cylinder. D, the tube for opening a communication to F through *h*. E,
the piston of the House cylinder. F, the communication pipe, by which a passage is opened
occasionally between the house cylinder & the Condensor Q. I.I, the cistern filled with cold
water & surrounding the Condensor. Q. the Condensor. K, the piston of the condensor with
its valves. L,L, the Beam of the Engine. M, the pivot on which it moves. N,N,N,N, the Blowing
cylinder or Pump. O,O, the piston with its valves. P,P, the pipes through which the air is
forced into the Regulating bellows.

' This section is supposed to be made at the instant the pistons E and K are at their greatest
height, and the piston O at the bottom of the blowing cylinder. The steam passing from
the boiler A, through the pipe B into the upper part of the cylinder C *a*, presses on the piston
E, and a vacuum being formed in the condensor Q, by the rising of its piston K, and a communication opened through F by the opening the valve *g*, the steam which occupied the space
from E to C *b*, rushes into the condensor, and the piston E meeting little or no resistence is
pressed down by the force of the steam to C *b*, when the valve *g* being shut and *h* opened,
a communication is made between the boiler A, through D & *h*, under the piston E, which
being then surrounded on all sides with steam, rises by the weight of the piston & rod O, and
thus a reciprocating motion is continually kept up. The water condensed in the vessel Q is
taken up by the piston K, and discharged at the top of the condensor.'

It will be observed that the cylinder is without a jacket, and that the top is
connected to the bottom by a perpendicular pipe D. The valve admitting steam
to the bottom of the cylinder is shown in the form of a disk, opening by a downward movement, while the exhaust valve is a hinged flap. The condenser is
adjacent to the cylinder, and the exhaust pipe descends directly to the lower end
of the air-pump.

This drawing was made very soon after the engine was erected ; by whom
is not known, and how closely it represents the actual construction we cannot
say, but it is to be noted that the Soho drawing of the condenser (Plate LXXII)
shows a different arrangement ; moreover, it indicates at the lower end of the
cylinder a valve-box, or ' Regulator box', adapted for the oscillating sector-valve,
and not the disk and flap valves here shown.

Another drawing in the Egerton Collection, and by the same hand, headed
' Section of M[r]. Watt's Patent Steam Engine', shows a similar engine operating
a pump. The working-cylinder is in this case enclosed in an outer cylinder,
forming a steam jacket through which the steam passes from the top of the cylinder
to the bottom. The valves are the same (except that the steam valve is shown
opening upwards), also the condenser, but the exhaust pipe descends outside the
condenser cistern and enters it by an elbow at the bottom.

The owners of the Bloomfield Colliery, or some of them, were known to Boulton.
They had contracted with an engineer named Perrins to put up an atmospheric

---

[1] B. & W. Colln. Watt to Boulton, 1776 Feb. 7.

Sketch of the Steam Engine, on Mr. Watts construction, lately erected at New Willey furnace near Broseley, Shropshire, by Mr. John Wilkinson — and used for blowing air into his Smelting furnace at their works.

A. the Boiler
B. the steam pipe, communicating at all times with the upper part of the cylinder CCCC. and when the valve R is open with the lower part also through the cock D.
CCCC. the steam cylinder
D. the cock for opening a communication to F through R
E. the piston of the steam cylinder
F. the communication pipe, by which a vacuum is opened occasionally between the steam cylinder & the condenser L.
LL. the steam fluid with cold water surrounding the condenser
G. the condenser
H. the piston of the condenser with its valves
LM. the Beam of the engine
N. the power in which it moves
NNNN. the Blowing cylinder or Bags
O.O. the piston with its valves
P.P. the pipes through which the air is forced into the Regulating vessel

This Sketch is supposed to be made at the instant the pistons E and K are at their greatest weight. set the piston O at the bottom of the Blowing cylinder — the steam passing from the boiler A through the pipe B into the upper part of the cylinder C presses on the piston E, and a vacuum being formed in the condenser L by the rising of the piston K, and a communication opened through F by the opening the valve g, the steam which surrounded the steam from E to G rises into the condenser, and the piston E meeting little or no resistance is pressed down by the force of the steam to C, when the valve g being shut the h vessel a communication is made between the boiler A through D & R under the piston E, which then being surrounded on all sides with steam rises by the weight of the piston K on its end that a reciprocating motion is continually kept up. The water condensed in the vessel L is taken up by the piston K, and discharged at the top of the condenser.

XVIII. BROSELEY BLOWING-ENGINE, 1776

(*Egerton MSS., British Museum*)

engine, and the work had been commenced when Boulton persuaded them to adopt the new plan. The bargain seems to have been settled at the dinner table ; at any rate when later on it became a question of paying for the engine the Bloomfield people considered that they had a grievance. However, it was agreed that an engine on Watt's system should be erected and that Perrins should do the work except 'the new invented· parts', and in June 1775 Watt wrote to the proprietors : [1] 'If the first cost of ye new engine should exceed that of the old engine wch. it cannot do in any considerable degree, I shall agree that such extra expense shall be deducted out of the saving before I require any profit.' The cylinder that had been made for the atmospheric engine was 72 in. diam., and Watt, after first considering a 46-in. cylinder, finally settled on a 50-in. In ordering this cylinder it is worth remarking that he did not send a sketch, but satisfied himself with a description in writing :

'Messrs Bentley & Co., having agreed to erect one of my steam engines at their Colliery at Bloomfield and the cylinder they have got from you being too large for my way of working their pumps, this comes to desire you to get another cylinder cast as soon as possible of the following dimensions, viz., 9 feet long and 50 inches diameter with a flanch at each end and one upon the body of the cylinder 6 feet from the bottom end, this latter is to have no holes in it and to be quite plain one inch thick and projecting about 3 inches all round the cylr. The bottom flanch is to be pierced with 20 holes for 1¼ inch screws and the top flanch (which serves to screw on a lid) is to be divided into 16 equal parts, 14 of which are to be pierced for 1¼ inch screws and the two spaces to be left blank, at one place. I must beg of you that the cylinder may be very smooth and truely bored and got ready with all convenient dispatch. The drawings for the cylinder bottom, lid, and piston shall be sent to you in a post or two, meanwhile I hope you will order the cylinder to be put in hand. The cylinder I had from you was rough at first but is now got smooth & does pretty well, I may say very well.[2] The thickness of Mr. Bentley's cylinder must not exceed an inch, but you may put the customary number of beads round it. There is no occasion for having so many as are upon the 6 foot one they have got.' [3]

The cylinder was on the ground early in August, and Watt wrote of it as ' by much the best I ever saw of that diaᵗʳ. and I expect will answer well '.[4]

The erection of the engine took some time, and no doubt there were several false starts ; however, it went to work in March 1776, apparently at about the same date as the Broseley engine. The occasion was made one of public ceremony, and now for the first time we have the newspaper men dealing with the Watt engine. *Aris's Birmingham Gazette* for Mar. 11, 1776, has quite an interesting account of the affair :

'On Friday last a Steam Engine constructed upon Mr. Watt's new Principles was set to work at Bloomfield Colliery, near Dudley, in the Presence of its Proprietors, Messrs. Bentley, Danner, Wallin, and Westley ; and a Number of Scientific Gentlemen whose Curiosity was excited to see the first Movements of so singular and so powerful a Machine ; and whose Expectations were fully gratified by the Excellence of its performance. The Workmanship of the Whole did not pass unnoticed, nor unadmired. All the Iron Foundry Parts (which are unparalleled for truth) were executed by Mr. Wilkinson ; the Condensor, with the Valves, Pistons and all the small Work at Soho, by Mr. Harrison and others ; and the Whole was erected by Mr. Perrins, conformable to the Plans and under the Directions of Mr. Watt. From the first Moment of its setting to Work, it made about 14 to 15 Strokes per Minute, and emptied the

---

[1] Ibid. Watt to Bentley & Co., 1775 June 15.
[2] This refers to the cylinder for the Soho engine.
[3] B. & W. Colln. : Letter Books. Watt to John Wilkinson, 1775 June 27.
[4] Ibid. Watt to John Wilkinson, 1775 Aug. 10.

Engine Pit (which is about 90 Feet deep, and stood 57 Feet high in Water) in less than an hour. The Gentlemen then adjourned to Dinner, which was provided in that Neighbourhood, and the Workmen followed their Example. After which, according to custom, a Name was given to the Machine, viz., PARLIAMENT ENGINE, amidst the Acclamations of a number of joyous and ingenious Workmen. This Engine is applied to the working of a Pump 14 Inches and a Half Diameter, which it is capable of doing to the Depth of 300 Feet, or even 360 if wanted, with one fourth of the Fuel that a common Engine would require to produce the same Quantity of Power. The Cylinder is 50 Inches Diameter and the length of the Stroke is 7 feet. —The liberal Spirit shewn by the Proprietors of Bloomfield in ordering this, the first large Engine of the Kind that hath ever been made, and in rejecting a Common one which they had begun to erect, entitle them to the Thanks of the Public ; for by this Example the Doubts of the Inexperienced are dispelled, and the Importance and Usefulness of the Invention is finally decided.—These Engines are not worked by the Pressure of Atmosphere. Their Principles are very different from all others. They were invented by Mr. Watt (late of Glasgow) after many Years Study, and a great Variety of expensive and laborious Experiments ; and are now carried into Execution under his and Mr. Boulton's Directions at Boulton and Fothergill's Manufactory near this Town ; where they have nearly finished four of them, and have established a Fabrick for them upon so extensive a Plan as to render them applicable to almost all Purposes where Mechanical Power is required, whether great or small, or where the Motion wanted is either rotatory or reciprocating.'

Curiously enough, Watt had not, in April, seen the engine at work, at least so he informed Smeaton.

These early engines, although clearly more economical and of greater power for a given cylinder capacity, were sadly defective as machines. A few years after the erection we find Watt writing : ' As to the weakness of things at Bloomfield, we knew no better then, & they might easily have mended many a thing, that they would not & made more fuss about than it was worth. The joints tho' not turned, were good & that plagued joining of the copper bottom to the nozzle pipe which we could not get at & always went wrong.'[1] Twelve months after the start, the engine seems to have been taken apart, and among other things new valves were fitted in the nozzles,[2] but seven or eight months later we read that ' Bloomfield is in bad order, & loaded to nearly 12 pounds per inch '.[3] Then again in March 1778 the engine refused to work and Harrison is sent over to investigate the cause of the trouble. However, a month later we find Watt writing : ' The engines we have erected lately please everybody, and we have altered Bloomfield so as to be as good as any of them.' By this time further experience had been gained and the nozzles, and, no doubt, other parts of the engine, had been replaced by better constructions.

We have no further information as to the early difficulties with the Broseley engine. No doubt Wilkinson met them as best he could, but he was quite familiar with the improvements that were going on, and apparently in August 1779 he had written to Watt about it in disparaging terms, for we have Watt's letter in reply in which he says :

' I allow that it is in many respects a bad one and I should be sorry that it were in anybodies hands but your own tho' I wish you to have the best engines in the world, and that very engine may be made a very good one. I hope you will do me the justice to remember that it was against the grain with me that either Bloomfield or it were undertaken untill the matter had been more fully experimented upon at Soho and upon a larger scale many things might have

---

[1] B. & W. Colln. : Letter Books. Watt to John Wilkinson, 1779 July 16.
[2] Doldowlod Papers : Watt's Journal, 1777

Apl. 14.
[3] B. & W. Colln. : Letter Books. Watt to Wilson, 1777 Nov. 23.

been foreseen and prevented but all could not. Our business has been pushed too hard and to much has lain upon my shoulders and a man less able to support the burden could not easily have been found. The time which should have been spent in experimenting has been lost in erecting. . . .'[1]

It may be that Watt was right in saying that they had embarked on the construction of big engines too hastily—certainly it was a big step from the Soho 18-in. engine to the Bloomfield 50-in.—and that it would have been better to have spent more time on experimental work. On the other hand, Boulton probably felt that Watt, if left to himself, might go on experimenting indefinitely, and that the wisest course was to make a beginning commercially without delay, trusting to gather experience in carrying out the work.

However, as a result of Wilkinson's grumble, certain alterations were made, and a new valve nozzle was fitted to the Broseley engine. Neither of these engines had a steam jacket, both had jet condensers.

BROSELEY INVERTED ENGINE

The blowing-engine, however, was not the only engine erected at the New Willey Works at about this time. There is evidence that an inverted-cylinder engine was put up, but we have no particulars as to its construction. The writer of the paper in the Egerton Collection, already referred to, states that ' Mr. Wilkinson is erecting another fire engine on Mr. Watt's construction, in which the mode of working will be reversed, the piston being driven upwards from the house cylinder. This is to raise water to turn the wheels for boring cannon.'[2] Watt, at the end of his letter from Broseley, dated Feb. 7, 1776, referred to above, says : ' I have got everything my own way, but must make an inverted cylinder, which I believe I can make a good job of.' It is probable that he made a drawing or a sketch of such an engine before he left Broseley, and that Wilkinson at once put it in hand. At any rate, in July 1777, J. Harrison had ' put Mr. Wilkinson's inverted engine to rights'.[3] That there were two engines at New Willey is made clear by a letter from Watt to Wilkinson in reference to the premiums to be paid on the engines he was working. Watt proposed to allow him ' to use the two engines now erected at New Willey, and the one you propose to erect at Wilson House, as well as a devil to be erected where you please' for a purely nominal consideration, ' one shilling yearly when demanded '. As to the engine to be erected at Snedshill, Boulton and Watt were prepared to take the premium in coal.[4]

Many years after, in 1793, when the case was being prepared for the action for infringement against Bull, Watt writes to Southern : ' If you can find any drawing of the inverted engine made for J. Wilkinson, please bring it with you, if not it does not signify, as we have got a witness in town who saw it.' Bull's engine, it will be remembered, was an inverted engine. A few years later than this letter we find James Watt, junior, referring to Wilkinson's ' Topsey Turvey Engine ', and in a list of premiums due from John Wilkinson in 1796 mention

[1] Ibid. Watt to John Wilkinson, 1779 Aug. 27.
[2] British Museum, Egerton MS. 1941.
[3] B. & W. Colln. : Letter Books. Watt to Dudley, 1777 July 1.
[4] Ibid. Watt to John Wilkinson, 1777 July 17.

is made of the ' Topsey Turvey' engine ; this, no doubt, is the same machine, but young Watt speaks of it as ' an old engine converted into one of ours '.

Two other engines were built by Wilkinson for his own use soon after ; one was set up on his property at Wilson House, Lancashire, and the other at Snedshill Colliery. The latter was completed early in 1778. Watt inspected it on April 1st, and notes that it works very well, but had not yet got its load on.[1] Seven months later, however, it was going very ill, and scandalously kept.

BOW ENGINE

ANOTHER engine erected in the year 1776 was made for a distillery at Stratford-le-Bow, London. It had an 18-in. cylinder and the stroke was 3 ft. 2 in. Watt was in London in December 1776, when this engine was started, and wrote to Boulton :

' I went down to Stratford where I found the engine had gone very well. I caused it to be kept going all afternoon & this morning new beat the piston and kept it going till dinner-time about 15 strokes per minute with a steam of one inch or at most 2 inches strong. The longer it went the better it grew. The barometer is still to fix and some other little jobbs.' [2]

We find very little about the engine in the Boulton and Watt correspondence ; so presumably it worked satisfactorily on the whole. In April 1777, however, there seems to have been some little difficulty, and Boulton, from London, informed Watt that he would take off the cylinder lid and examine the piston and asked : ' Shall I venter to drill 4 holes through the top in order to put in grease or suit.' [3]

Then in the following year it was fitted with a new nozzle and working gear,[4] and in 1779 the cylinder was found to be badly worn, and a new cylinder was made for it. As to the cause of this undue wear, Watt says : ' from what we are informed the iron was porous perhaps on the upper side as cast, and at first the devil prompted them to put sand in the piston in which way it wrought a long time, this wore it some wider than it should have been.' [5]

This was the engine that Smeaton inspected with disastrous results. Boulton's account of this visit is :

' Hadley also told me that Mr. Smeaton & Holmes called upon him to request him to attend them to Stratford le Bow, which he did. Smeaton said it was a pretty engine, but it appeared to him to be too complex ; but that might in some degree be owing to his not clearly understanding all ye parts. He gave the engineer money to drink & the consequences of that was that ye next day the engine was almost broke to pieces. Wilbey was very angry, turned away the engineer, and told Hadley the least amends he could make was to put it into order again, wch he did do, but was obliged to put in one new valve.' [6]

It should be explained that ' Wilbey' was one of the partners in the firm (Cook, Adams, Wilbie, and Sager) to whom the engine belonged, and that Hadley was an engineer who had assisted Joseph Harrison in the erection of the engine.

[1] Doldowlod Papers : Watt's Journal. The Wilson House engine was set to work, as it seems, about the same time as the Snedshill engine.

[2] Boulton Papers. Watt to Boulton, 1776 Dec. 3.

[3] B. & W. Colln. Boulton to Watt, 1777 Apl. 17.

[4] Ibid. Boulton to Watt, 1778 June 23.

[5] B. & W. Colln. : Letter Books. Watt to John Turner, 1779 Sept. 9.

[6] Boulton Papers. Boulton to Watt, 1777 Apl. 20.

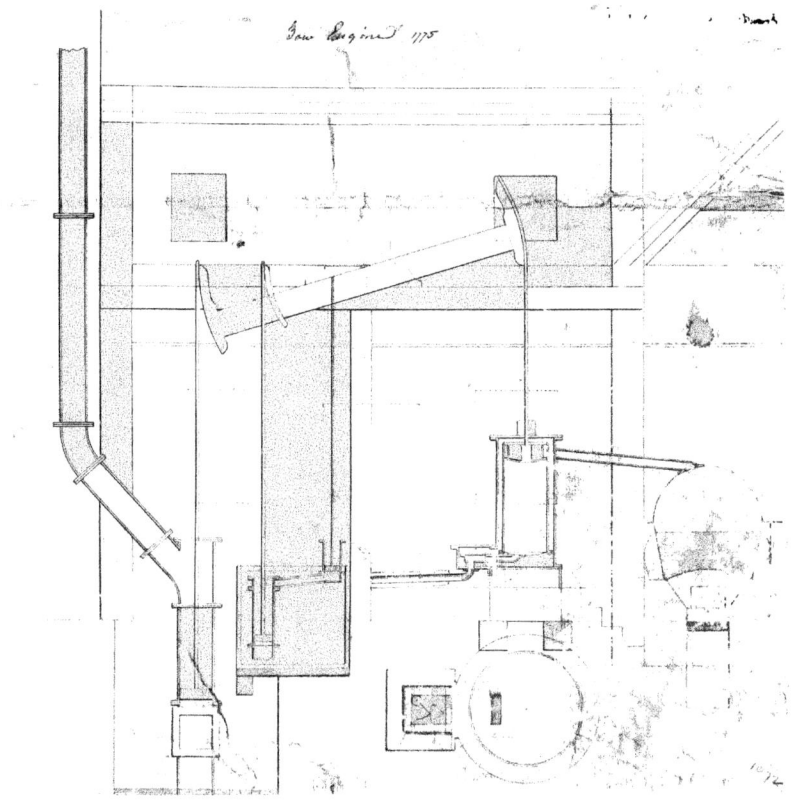

Bow Engine 1775

XIX. SINGLE-ACTING PUMPING-ENGINE, 18-INCH CYLINDER, FOR STRATFORD-LE-BOW

Drawing by Watt, dated 1775

Watt in 1777 stated that this engine worked a pump with a 15-inch barrel and 36 to 38 feet lift, 15 strokes per minute, and that it made 2,000 strokes with 53 lb. of Newcastle coal.[1]

The general drawing of this engine, Plate XIX, dated 1775, shows in the nozzle, which is at the lower end of the cylinder, two of the oscillating sector-valves ; whether these valves were actually used is not known.   The eduction pipe is shown enclosed in a jacket, and the air- and hot-water pumps are worked from the outer end of the beam.

BEDWORTH ENGINE

THE first engine set up in the year 1777 was at Hawkesbury Colliery, Bedworth, near Coventry.   This was the largest engine that the firm had yet been engaged upon ; it had a 58-inch cylinder and a stroke of 8 ft. and was fitted with a steam jacket.   The pump had a working barrel $14\frac{1}{2}$ in. diam., and the lift was 130 yds. Watt went over to supervise the erection at various times in January, February, and March, and the engine was started on March 10th.   Three weeks later it was reported to be going well although not so fast as wished for.   It was, however, an unfortunate engine in more respects than one.   There were many defects to cure, and when the engine had been brought to a fairly satisfactory working condition there was considerable difficulty in getting the owners of the mine to sign an agreement to pay the premium demanded.   At last, after having thoroughly overhauled their old engine and fitted it with Watt's drop valves, a comparative trial was made in March 1779 in the presence of an arbitrator, who awarded Boulton and Watt a premium of £217 per annum, and a few months later a deed was completed for the payment of this sum.   The colliery owners had thought £30 a sufficient payment.   The trial had shown that the new engine was better than the old in the proportion of 411 to 96.[2]

As to the misfortunes of the engine itself, an account in April 1777 is quite a chapter of accidents.   First we are told that the packing in the condenser joints gave way ; this was put right, and then ' the martingale of the lower regulator broke '.   After this was mended the engine was found to be ' in better order than ever, the vacuum at 27 inches and stood almost an hour at 22 after the engine stopt '.   Next, ' the pump rod of the lower lift broke off by the top of the pump, and the rods below it fell down the pumps where they have fixed themselves in such a manner that the capstane rope was broke in attempting to draw them up and there they must in all probability stay untill the old engine getts down the water '.[3]

Various parts of the engine were renewed soon after, and the beam was strengthened by ties and struts ; in July 1778 new valves, nozzles, and working-gear were supplied, and then in 1789, when it was moved to Exhall Colliery, it underwent extensive repairs, and seems to have had a parallel motion applied to it.

This engine had the outer cylinder made in two lengths and the cylinder cover in halves flanged and bolted together.   It had two air-pumps.

---

[1] B. & W. Colln.: Letter Books. Watt to Chapman, 1777 Feb. 22.
[2] Ibid. Watt to Meason, 1779 Mar. 20.

[3] B. & W. Colln. Watt to Boulton, 1777 Apl. 14.

COLEVILE'S ENGINE

On his visit to Scotland in 1776 Watt arranged to put up an engine with a 44-inch cylinder for Peter Colevile at Torryburn, Fifeshire (three miles from Dunfermline). Although this engine was not set to work before January 1778, the drawings, or some of them, had been prepared before the end of 1776, so that in respect of design it takes the precedence of any engine erected in Cornwall. The 'General Section' and 'Ground Plan of Engine, house, Boilers, and Condenser cistern' are here reproduced in Plates XX and XXI.[1] In the section we see the 'Inner Cylinder 44 inch dia^r.' enclosed in the 'Outer Cylinder', which is shown as cast in two lengths and provided with a lid also cast in two halves, as indicated by the flange for bolting the two parts together. The outer cylinder rests upon a stone foundation in which is formed a furnace for supplying heat to the cylinder bottom, and through which, and the 'Cylinder beams' below it, pass the 'screws to hold down cylinder'. To the lower end of the cylinder is secured the 'nozzle' or valve box in which are mounted a valve to admit steam from the jacket to the working cylinder beneath the piston, and a valve to open communication from the cylinder to the condenser. Both valves are here shown as plain disks acting as plugs in the apertures which they close; they open by moving downwards and are actuated by links, and arms on spindles extending outwards through the wall of the nozzle and operated by the 'plugtree'. From the lower side of the nozzle extends the 'condenser steam pipe'; it passes through a water jacket and terminates in a box connecting the bottoms of the two air-pumps. The upper ends of these pumps are connected by another box, on the middle of which stands the 'hot water pump'. The air-pumps and the lower end of the condenser steam pipe are immersed in a cistern of water into which the 'injection pipe' opens. The injection pipe has three limbs, one delivering into the condenser and the others into the upper ends of the air-pumps. A pipe connexion is shown between the lower part of the condenser steam pipe and the upper part of the air-pumps, and from the jacket round the same pipe to the hot-water pump. The engine beam has a 'King post', and is pivoted on the wall of the engine-house adjacent to the pumps, which wall is made of greater thickness than the others. The 'lower floor' of the engine-house is at the level of the bottom of the outer cylinder; another floor is supported on 'beams' just below the level of the cylinder top. From the plan it will be seen that the engine was to have two circular boilers. No stop valve is shown in the 'steam pipe to Boiler', but an explanatory note informs us that 'c' is the 'joint at which ye Boyler Regulators are to be put in'. This view gives us also 'Size of fire Grate 5 feet by 3 feet 6 in', 'Fire door, 2 feet', 'Flues, 18 in wide', and 'chimney 2 feet square'. As to the engine-house, we have at the top of the figure the inscription, 'arch to take in cylinder', and below that 'firing place for fire under the cylinder'. Then on the right-hand side we have 'Enginemans door, Door underneath to go to Boiler', and on the left-hand side 'House door, 3 feet door underneath for 2^d boiler fireplace'. The explanatory note says that 'E' are 'stairs to go down to fires', and 'F' 'stairs to go up'. The remaining letters on this little table are: 'A. Flanch of inner Cylinder, B. Flanch of outer Cylinder, D. Injection Pipe.'

---

[1] It is worth noting that the two views were not drawn to the same scale; the section is on a scale of half an inch to the foot, while the scale of the plan is one-third of an inch to the foot.

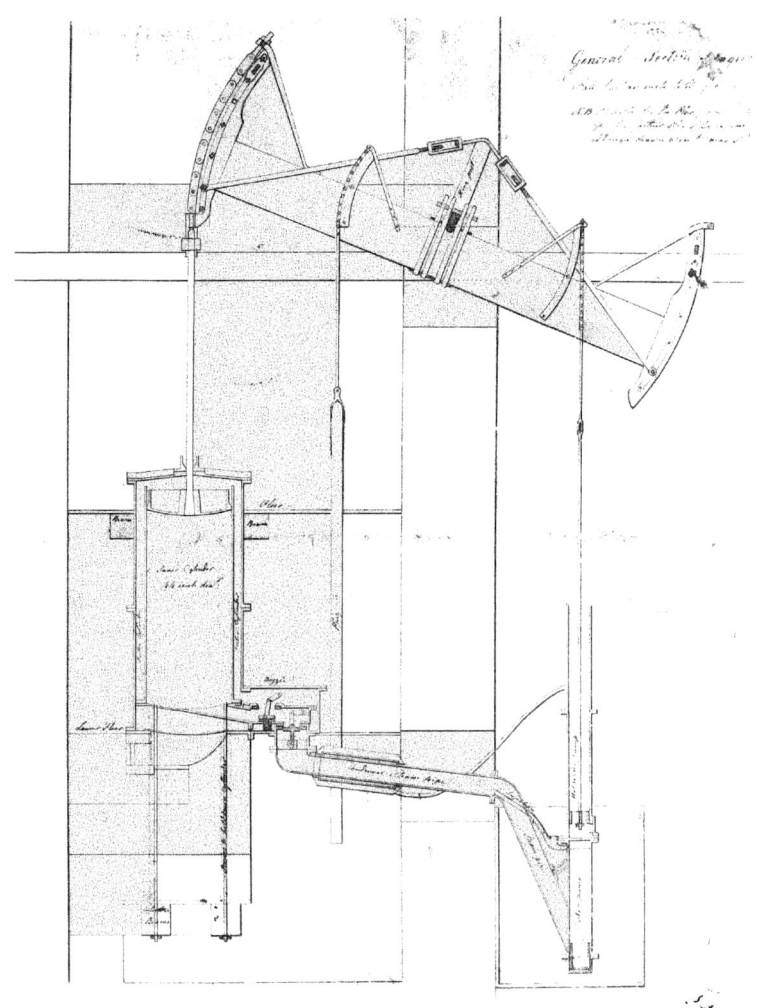

XX. SINGLE-ACTING PUMPING-ENGINE, 44-INCH CYLINDER, FOR PETER
COLEVILE, TORRYBURN, FIFESHIRE

Drawing by Watt, dated Oct. 1776

XXI. PLAN OF COLEVILE'S ENGINE AND BOILERS, 1776

The erection of this engine was carried out by Henderson, with Symington, the father of the better-known William Symington, as his assistant. Colevile became a bankrupt in 1779, and the engine remained idle for a few years, but it was again at work in 1783.

An engine which should be mentioned here, although no particulars of its construction are available, is that made for Sir Harbord Harbord.[1] In 1775 this gentleman made an inquiry for a small engine to supply his house with water, and Watt suggested a cylinder 6½ in. diam.[2] Nothing was done until two years had gone by and then in September and October 1777 it appears from the Manufactory Ledger[3] that the engine was in hand, and that it was put together at Soho.

[1] Sir Harbord Harbord of Gunton, Co. Norfolk, created 1786 Baron Suffield of Suffield, Co. Norfolk.
[2] B. & W. Colln.: Letter Books. Watt to Harbord, 1775 Aug. 4.
[3] B. & W. Colln.: Manufactory Ledger No. 4, 1777–8.

X

EXPANSIVE WORKING

THE SECOND ENGINE AT SOHO

THE idea that economy would result from working steam expansively had suggested itself to Watt as early as the year 1769, when he writes to Dr. Small :

'I mentioned to you a method of still doubling the effect of the steam and that tolerably easy by using the power of steam rushing into a vacuum at present lost, this would do a little more than double the effect but it would too much enlarge the vessels to use it all. It is peculiarly applicable to wheel engines and may supply the want of a condenser where force of steam is only used, for, open one of the steam valves and admit steam untill one fourth of the distance between it and the next valve is filled with steam, shut the valve, the steam will continue to expand and press round the wheel with a diminishing power. . . .'[1]

It must not be thought that it was an absolutely new idea to cut off the steam before the end of the stroke. This was practised with the Newcomen engine, but there it was done with the object of bringing the piston to rest gradually, and so to avoid the striking of the engine beam against the spring beams in the regular working of the engine ; there was no idea of working the steam more economically. As we have seen, Watt at first used a modification of the Newcomen type of valve to control both the admission and exhaust. With this construction, earlier cut-off meant earlier opening of the exhaust, and it would seem that he did not see his way to get over this difficulty. Upon the introduction of separate valves for the admission and exhaust the problem became simple and he was now able to go forward with his scheme of expansive working. It was decided to put down a second engine at Soho, mainly for the purpose of trying the experiment, and in February 1777 a 33-inch cylinder was ordered. Watt was very sanguine of the results to be obtained. For instance, he writes :

'I have made an alteration in the structure of our engines, which renders them better adapted to mines than anything hitherto in use. These new ones have the property that if they were to sett out with a mine only 10 fathoms deep and it were afterwards to be sunk to 50 fathoms deep they would work the pumps as well at the 50 fathoms as at the 10 & yett would burn no more than $\frac{1}{5}$ of the coals at 10 faths. that they did at 50. And if the mine were to be sunk to 100 fathm deep they would still draw the water but would then burn four times the coals to do the same work which they did at 50 fathoms.'[2]

Boulton, too, was an ardent convert to the principle of expansive working. He was anxious to have an engine completed directly, and we find him writing :

'The great thing in ye expansive engine is to prevent the fire men from playing tricks & knocking its own brains out. Suppose a valve was put into the piston & a perpendicular rod rising up from ye bottom of ye cylinder, which rod must open ye valve as soon as the piston is descended so low & if you find you want more or less power you may either raise or fall ye

---

[1] Boulton Papers. Watt to Dr. Small, 1769 May 28.

[2] B. & W. Colln.: Letter Books. Watt to Meason, 1777 Apl. 24.

rod, but that rod must not be in the power of the stoker. I think if that condensing valve was in ye piston & if ye cylinder was within the boyler the devil could not find out our engine by seeing it at work.'[1]

Incidentally this letter is interesting as showing that the plan of mounting the cylinder within the boiler had been discussed at this early date. Boulton's remarks about the 'condensing valve' in the piston require explanation. One of the features of this engine was that the exhaust steam was discharged directly from the upper end of the cylinder to the condenser and that the lower end was permanently open to the condenser, or, as Watt himself described it thirty years later, it was 'an engine in which the piston ascended in vacuo & in which the steam was admitted only during a portion of its descent & acted during the remainder of its stroke by its elasticity'.[2] Several engines, including that for Chelsea Waterworks, were made on this plan, but, as Watt goes on to say, the engines were generally made 'to ascend in steam on both sides of the piston'.

The effect of the valve suggested by Boulton would be to open another passage to the exhaust, and, by adjusting the height of the rod, the valve could be opened at different points in the stroke of the piston.

The engine was built and had been set to work at Soho before Watt departed for Cornwall in August 1777. The performance was not very satisfactory, and its movement was jerky and violent. The cylinder was 33 in. diam., and Boulton came to the conclusion that it was too big for its work; it was, he said, loaded only to about 5·7 lb. per in., whereas if it were loaded to 7½ or 8 lb. 'it would not be so very fierce'.[3] This engine became known as Beelzebub, no doubt from the violence of its action.

Watt was desirous of having trials made to ascertain the consumption of coal, apparently with a view to applying the expansive principle to the engines for Chelsea and Shadwell Waterworks, but the engine, although at work, was in an incomplete state. After a steam jacket had been fitted, and after Watt had returned from Cornwall, Boulton made a two-hours' trial of the engine. He reported that the engine worked very well at 20 strokes per min., but he was not satisfied with the number of strokes per cwt. of coal, and thought it would be possible to improve the performance considerably. He suggested the application of a cataract, the clothing of the steam case and boiler top, and repairs to the valves. At this time the piston had been fitted with the valve referred to above, and alterations had been made in the plug-frame that had much improved the action. Upon the whole, Boulton thought the engine worked as well as he expected it would do, but he still held that it was underloaded.[4]

Watt had now begun to doubt the advantage of expansive working, and we find him writing :

' Query whether Belzebub would not do fully more with the steam & go much more equably if the steam regulator were only opened a very little so as to give the ordinary velocity & be kept always open at least to near the end of the stroke. Our vacuum regulator does not open half an inch high & the steam one inch and the latter is shut when the engine wants a foot of being out.'[5]

---

[1] Ibid. Boulton to Watt, 1777 May 16.
[2] Doldowlod Papers. Watt to James Watt, junior, 1808 Nov. 12.
[3] B. & W. Colln. Boulton to Watt, 1777 Aug. 18.

[4] Ibid. Boulton to Watt, endorsed 'Concerning the performance of Soho engine, 1777.' (Not dated.)
[5] Boulton Papers. Watt to Boulton, 1777 Sept. 20.

The reference is to the Chacewater (Wheal Busy) engine which had just been set to work. No doubt Watt had tried cutting-off and found the action too violent.

A couple of years later it would appear that Watt had lost interest in the expansive engine. He had designed some schemes for equalizing the power of Beelzebub soon after it was set to work, but none of these seems to have been carried out. Presumably the engine continued to work as an expansive engine, for in July 1779 Boulton writes :

' I wish we had the last edition of nozzles to Soho engine, as we might then try some accurate expts. & ascertain ye maximum & Ratio of ye expansive principle, wch. hath not yet been accurately done.' [1]

Indeed, we have now reached a point at which Boulton is the advocate of expansion, while his partner seems to favour throttling. In August 1779 Boulton paid a visit to the Donnington Wood engine, and found that the top valve did not shut until the piston reached the bottom of the cylinder, although the load was not more than 5 lb. per in. ; he directed that the gear should be arranged to shut the valve when the piston had made two-thirds or three-fourths of its stroke.[2]

In spite of the efforts made to improve it, the Soho 33-inch engine continued for some time to be an unsatisfactory machine, and Watt in 1781 agreed with Boulton that it was one of the worst engines they had.[3] It worked a pump, 24 in. diam., with a column of water of 24 ft. The cylinder was 8 ft. 3 in. in length, long enough, says Boulton, to admit of a stroke of 6 ft. 9 in., but in 1781 it was actually making a stroke of 6 ft. 1 in. only.[4] At first sight it might appear that the consequent large clearance space was one reason for the inefficiency of the machine, but probably it had no such effect, as it was the practice to set the piston so that when at the top of its stroke its upper surface stood above the lower edge of the port, and as the lower end of the cylinder was in permanent connexion with the condenser, the dead space there was immaterial.

In its early days this engine was certainly unfortunate. One morning in July 1778 the engine-house was discovered to be on fire, and we read that ' the fire burned with such rapidity that in little more than half an hour the whole of the wooden work & roof of the engine house, at Soho, was burned to the ground floor, save the engine beam is not totally consumed, but is rendered unfit for that service, by being in a great measure consumed. All the soldering and parts of the copper pipes are melted. These are the rough sketches, the particulars of the damage can not be ascertained untill the rubbish &c is cleared. . . . Please to note that in the engine house the cylinder with its joints &c are all supposed safe.' [5] However, the damage was not very serious, for three weeks later Watt, then in Cornwall, is informed that ' Soho engine will be in working order again on Monday or Tuesday next '.[6] He had, in the interval, expressed the view that ' no repair

[1] B. & W. Colln. Boulton to Watt, 1779 July —. It is interesting to note that Boulton had, before this, recognized that there was a limit beyond which expansion could not be carried with economy, and that this limit could be determined only by experiment.

[2] Ibid. Boulton to Watt, 1779 Aug. 12.

[3] Boulton Papers. Watt to Boulton, 1781 Aug. 15.

[4] B. & W. Colln. Boulton to Watt, 1781 Sept. 13. It may be noted that the engine was designed for a stroke of 7 ft. (B. & W. Colln. Portfolio of Drawings, No. 626. Expansive engine for Soho, Feb. 21, 1777.)

[5] Ibid. Boulton & Fothergill to Watt, 1778 July 11.

[6] Ibid. Zach. Walker to Watt, 1778 Aug. 1.

XXII. OLD BESS AT THE SCIENCE MUSEUM, SOUTH KENSINGTON

farther than the roof ought to be gone about at present'.[1] Possibly he thought the opportunity a good one for reconstructing the engine.

This, the second engine put up at Soho and first known as Beelzebub, is identical with the engine known later as ' Old Bess', now one of the treasured relics at the Science Museum, South Kensington ; see Plate XXII.[2] No doubt a good many modifications and renewals have been effected since the engine was first built, but the cylinder dimensions are substantially the same—as we have seen, Beelzebub had a cylinder 33 in. diam., 8 ft. 3 in. long ; the cylinder of the existing engine is slightly larger in diameter, $33\frac{1}{2}$ in., and its length is 8 ft. $2\frac{1}{4}$ in. Moreover, no evidence whatsoever has been discovered of the erection of another engine of this size at Soho.

In 1887 Wm. Henry Darlington, then manager at Soho Foundry, stated that Old Bess was the first engine made on the expansive principle and that it was a pumping-engine to the end ;[3] this agrees with the result we have arrived at, but he goes on to say that the first pumping-engine at Soho they called Beelzebub, that this was burnt down in 1777 and replaced by Old Bess. Mr. Darlington was no doubt giving the Soho tradition, but it has been made clear in the foregoing pages that the first engine at Soho had an 18-inch cylinder, that the second engine was called Beelzebub, was the first made to work expansively, and was only damaged by fire. This engine seems to have remained at work at Soho up to the year 1848. Of its final days there is an account by Joseph Harrison, ' Artificer of the Soho Mint', preserved at the Science Museum, South Kensington. Harrison says that :

' Old Bess (which is a 30 horse-power engine) worked at Soho from about 1771 to 1848, when it was sold by auction with the other parts of the mint machinery, on the occasion of Mr. Boulton giving up the business of coining. It was bought at the auction for £48 by Mr. Lewis, who sold it to Mr. Walker, a metal roller, for £58, and he placed it in front of his works on an island in Derrington Pool, near Birmingham. Mr. Walker at length sold it to Messrs. Branson and Gwythen's Builders, of Birmingham, who subsequently presented it to the Commissioners of Patents for the Patent Museum.' In taking down this engine, after the auction, a ' fireplace was found built in the brickwork beneath the cylinder'.

From a statement by William Buckle it appears that the engine was working satisfactorily up to the time it was taken down in 1848, so that in the course of its working life of seventy-one years it had lived down the evil reputation of its early career and had earned a familiar appellation suggestive of faithful service.

It has been remarked above that about the year 1779 Watt seemed to have lost interest in the expansive working system. A few years later, however, in 1781, Jonathan Hornblower came upon the scene with his compound engine, and Watt again directed his attention to the subject, with the result that we find the specification of his patent of 1782 includes a considerable amount of material dealing with expansion. He explains the action of steam in a single cylinder, by the aid of a diagram showing a curve produced by calculation according to Boyle's law, and then goes on to say that ' the powers which the steam exerts being unequal, and the weight of water to be raised, or other work to be done by the engines, being supposed to resist equally throughout the whole length of the stroke, it is necessary to render the whole acting powers equal by other means'. Quite a number of devices for attaining this end are described ; they are considered under six

[1] Ibid. Watt to Boulton, 1778 July 20.
[2] The Museum possesses also a working model of the engine with the pump and water-wheel.
[3] B. & W. Colln. Wm. Henry Darlington to Col. H. Stuart-Wortley, 1887 Jan. 26.

heads, but in several instances there are modifications so that altogether we have twelve schemes. Some of these may be grouped as linkage arrangements with varying leverages ; others provide for shifting the fulcrum of the beam, for shifting a weight along the beam, or for throwing the centre of gravity of the beam from one side of the fulcrum to the other. In another device Watt uses two auxiliary cylinders, the pistons of which are suspended from opposite ends of the engine beam ; in their ascent and descent these pistons move a body of water from one cylinder to the other, the action being that the down-stroke of the steam piston is assisted by the increasing weight of water on the piston in the auxiliary cylinder adjacent to it. The last arrangement described is a flywheel driven by a pinion and a segment on the arch at the end of the engine beam ; this made several revolutions for each stroke of the engine and moved first in one direction and then in the other. As a modification of this last arrangement, Watt gets in a fly-wheel on a continuously rotating shaft driven by connecting-rod and crank, or other means, from the beam, but he is careful to explain that he intends to cover only the application to pumping-engines.[1] Pickard's patent had already described the application of the connecting-rod and crank for rotative engines.

The specification includes also a ' compound or double engine' arranged for expansive working. The arrangement is somewhat curious. The engines are described as two distinct engines, each complete in itself and having cylinders of the same size. Engine No. 1 takes steam from the boiler, and its perpendicular pipe is connected to the steam pipe of engine No. 2. Engine No. 1 makes a down-stroke with steam above the piston, the equilibrium valve is opened and it makes the up-stroke ; it then makes a pause, the steam valve of engine No. 2 being opened and the steam expanding from below the piston of No. 1 into the space above the piston of No. 2, which then makes its down-stroke. Piston No. 1 is now at the top of its stroke, and piston No. 2 at the bottom: Connexion to the condenser is opened, a vacuum is formed below No. 1 and above No. 2. No. 1 makes its down-stroke under steam, and No. 2 its up-stroke by the weight of the pump-rods. (The lower end of cylinder No. 2 is permanently open to the condenser.) Piston No. 2 makes a pause at the top of its stroke, while No. 1 makes its up-stroke, then it moves down while No. 1 remains stationary at the top of its stroke. In a modified arrangement the lower ends of both cylinders are permanently open to the condenser, and the steam is passed directly from the upper end of the cylinder of No. 1 engine to the upper end of the cylinder of No. 2, each engine making a pause while the other performs its down-stroke, and the up-strokes of both being simultaneous.

It was decided to build an engine on this plan, and in September 1781 Watt writes from Cornwall: ' I send enclosed drawing of a double expansive engine, which I beg may be set about immediately. The drawing contains full instructions of all that concerns the working of it. and as to minutiæ, you must do the best you can, as I cannot at present attend to them.' This drawing is reproduced in Plates XXIII and XXIV ; the instructions written upon it read :

' Drawing of the expansive Cylinder of double engine, Sepr. 3. 1781. Scale inch to the foot. Radius described by the centre of the pin A = 2 feet 7 inches. Radius described by B, 2 feet 6 inches. FA = 13 inches, GB = 24 inches. These dimensions suppose the stroke = 4 feet 4 in.

[1] The suggestion of this arrangement seems to have been due to Boulton : ' If an engine was to work regularly as fast as Shadwell, I should propose a crank & fly.' B. & W. Colln. Boulton to Watt, 1782 Jan. 22.

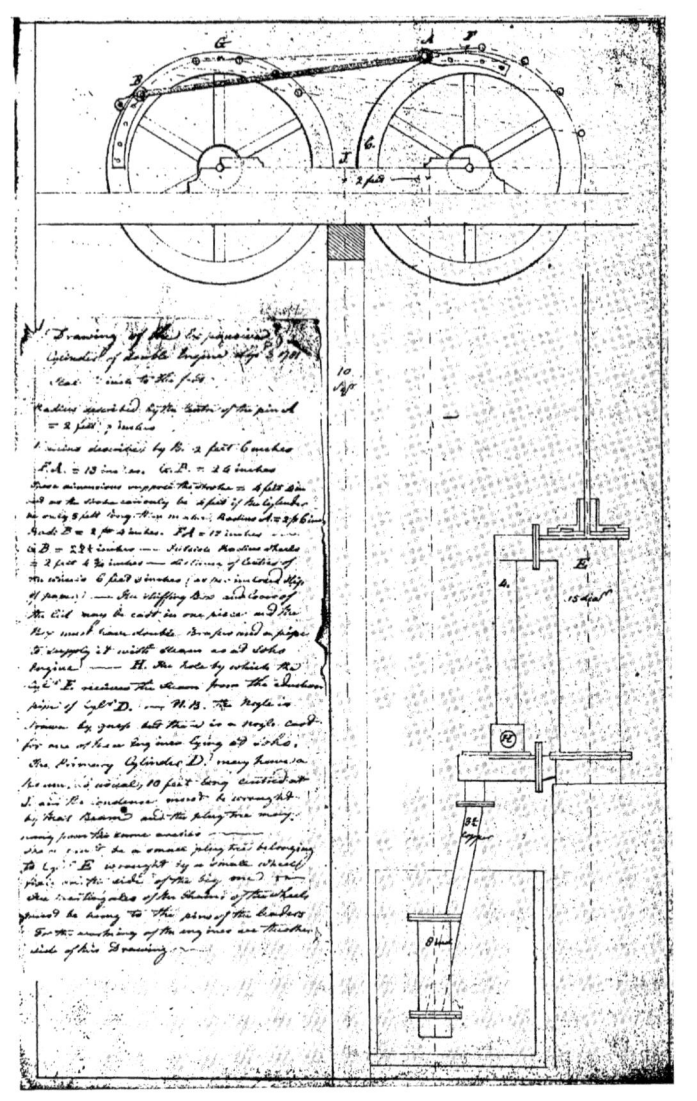

XXIII. 'DOUBLE EXPANSIVE ENGINE', 15-INCH CYLINDERS

Drawing by Watt, dated Sept. 3, 1781

and as the stroke can only be 4 feet, if the cylinder be only 5 feet long, then make radius A = 2 ft. 6 in. ; Rad. B = 2 ft. 4 inches; FA = 12 inches; GB 22½ inches. Outside radius wheels = 2 feet 4¾ inches. Distance of centres of the wheels 6 feet 3 inches (as per inclosed slip of paper). The stuffing box and cover of the lid may be cast in one piece and the Box must have double Brasses and a pipe to supply it with steam, as at Soho engine. H, the hole by which the cyl^r. E receives the steam from the eduction pipe of cyl^r. D. N.B. The nozle is drawn by guess, but there is a nozle cast for one of these engines lying at Soho. The primary cylinder D may have a beam as usual 10 feet long, centred at J, and the condenser must be wrought by that beam and the plug tree may hang from the same arches. There must be a small plug tree belonging to cyl^r. E wrought by a small wheel fixed on the side of the big one. The martingales of the chains of the wheels must be hung to the pins of the leaders. For the working of the engine see the other side of this Drawing.'

On the back of the drawing the working of the engine is thus described :

' The cyl^r. D must have a plug tree and a steam and exhausting regulators as usual, and must work the condenser which need only have one pump (unless that sent you from Bersham have two). The cyl^r. E need have no top regulator nor exhaustion one, only the middle or steam regulator. Both cyl^r. must have steam cases and steam on the lids, and a pipe must be brought from the cross pipe of D to the steam case of E to supply it with steam, and that pipe should be wrapt up to keep it from the air. The cyl^r. E must have a small plug tree and a Y-shaft to work its own regulator. D must be loaded with a column of ▽ [water] = 10 ^{lb} on the inch, and E = 7 lb. ; the pumps ought to be the same height but different dia^{rs}. Suppose the engine to be in motion, when D is filled with steam both pistons are at the top, but E is totally exhausted. Open the exhaustion reg^r of D and the steam will instantly press upon the piston of E and it will make its stroke. When piston E is down its regulator will open, and D will be totally exhausted and make its stroke, open its steam regulator and rise again. Piston E will rise while D is making its stroke. The utmost care must be taken to make E perfectly steam and air tight, which will be effected by making all its joints quite flat by chisseling and filing before they are joined. Perhaps it may be found a better construction to make cyl^r. D with both valves at top, like Chelsea engine and to conduct the steam directly from the under-side of the top nozle of D to the upper side of the top nozle of E, but that requires consideration. PUSH FORWARD.'

It will be observed that the first cylinder is to have a beam as usual, while the second is to have the equalizing-device. One cannot avoid thinking that Watt, possibly owing to pressure of work, had not given this matter the consideration it required, for in view of the fact that the second cylinder was of the same capacity as the first (both cylinders were 15 in. diam. and 4 ft. stroke) this strikes one as particularly a case in which the equalizing-device was not called for. The engine was put in hand at once and Boulton had the valves and connexions arranged in such a manner that it could be worked as a compound engine or as two independent engines. It was intended to apply this engine to turn a shaft to work a hammer (further remarks on the engine considered as a rotative engine will be found in a later chapter) and although the cylinder end of the engine was completed before the end of the year, there was considerable delay about the rotative mechanism and the hammer, and it was not until November 1782 that it was set to work. There is some indication that it was tried as a compound, for we find Watt writing, in reference to one of the trials, of the ' communication being cut off with the sleeping cylinder which it seems, somehow or other, destroyed much steam '. This is the only indication we have of the cylinders being coupled together while at work. Ultimately the cylinders were used for two separate engines.

Neither the double expansive engine nor the equalizing-devices became part of the Soho practice, and these schemes had no bearing on the development of the steam engine.

Indeed, although in 1784 we find Boulton writing in reference to the Wheal Maid whim engine : ' I never saw an engine take so little steam as this in my life & you may be assured that where a fly can be apply'd so as to go 300 or 400 ft per minute, the expansive principle in practice will come up to theory,' [1] and no doubt a number of engines were made to work with a limited degree of expansion ; Watt had realized that there was very little to be gained by it (that is to say, in the temperature range under which he was working).   There is a paper in his handwriting that sets forth the views he had arrived at.[2]   The document is not dated, but probably is of later date than the period we have just been considering ; it is headed : ' Expansive action of steam, or more properly its action by dilation.' After explaining by the aid of a diagram the action of air expanding in a cylinder, Watt goes on : ' This reasoning is strictly true when air is considered as the elastic fluid employed, but is very fallacious if applied to steam, for steam is cooled by dilation, loses part of its bulk & is partly reduced to water, without it can borrow heat from the surrounding bodies ; what effect it can produce by its dilation has not yet been determined by any experiments, though from what has been tryed it is evident that it falls far short of what the above theory promises.'

A fact that makes it clear that expansive action had ceased to be regarded as of any consequence is that when in 1784 a new valve-gear was designed for the double-acting engines the steam valve of one end of the cylinder was linked to the exhaust valve at the other end, whereby early closing of the steam valve was rendered impossible.   This remained the standard practice till 1800 when the eccentric gear devised by Murdock began to be used.

This chapter should not be closed without mentioning that in 1798 a scheme for expansion with automatic regulation by the governor was under consideration at Soho.   A brief description of it will be found in Chapter XV.

[1] B. & W. Colln.   Boulton to Watt, 1784 July 8.
[2] Ibid.   Endorsed ' Explanation of the ex- pansive principle of using steam ', and enclosed in a parcel of papers relating to Jonathan Hornblower.

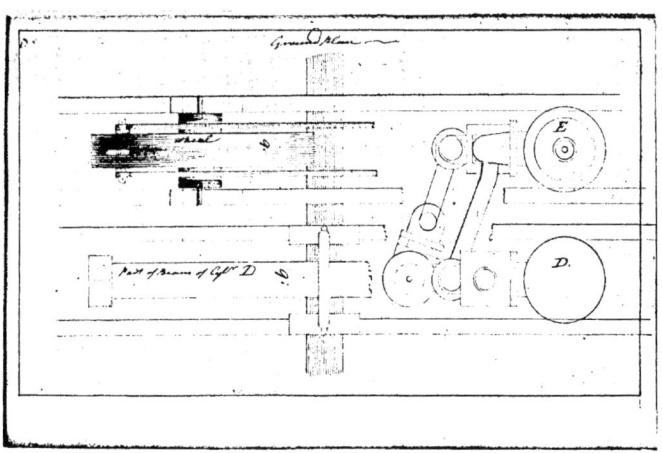

XXIV. PLAN OF 'DOUBLE EXPANSIVE ENGINE'

XI

THE SINGLE-ACTING PUMPING-ENGINE,
1777–83

T HE consideration of expansive working has carried us in advance of the regular line of development of the engine, and we have now to go back to the year 1777, the date of the second Soho engine, and here we may pause to consider some remarks of Watt himself on the construction of the engines as then made. These remarks are taken from his correspondence with Jonathan Hornblower, the elder, then one of the leading engine-builders in Cornwall, who, it should be explained, had ordered an engine for the Tingtang mine; they have therefore the peculiar interest that they are addressed to a man fully conversant with the construction of the Newcomen engine, and may be taken as pointing out the essential differences between Watt's engine and the older machine. Moreover, the constructions described are the results of experience with the working and design of a number of engines.[1]

For a proper understanding of the subject it is necessary to bear in mind that at this period Boulton and Watt were not engine-makers and that engines were not built at Soho. 'How the engines were made' forms part of the subject-matter of a later chapter; here it will be sufficient to state that the only parts of the engines regularly made at Soho were the valves and the valve nozzles, the other parts were made in various parts of the country, London, Cumberland, and North Wales, as occasion demanded.

As to the engine-house, Watt says that a house '15½ feet from water wall to back wall, within, and 14 feet wide' is large enough for a 52-inch cylinder engine, but that 'the boilers cannot be admitted into the house in our way of working, because our cylinders require to be fixed by the bottom to beams placed under a stone platform heavy enough to resist the power of the cylinder'; this is in distinction from the Newcomen engine practice in which, commonly, the cylinder was fixed, between beams, vertically above the boiler.[2]

The engine beams are hung 'all under the gudgeons which is a considerable improvement, as it contributes greatly to equalize the motion of the engine'. For the Tingtang engine the beam is not to be more than 24 ft. long and Watt asks if Hornblower could at any reasonable rate procure an oak tree 30 in. square to make it.

As to the beam gudgeons, we learn that 'the rounded part of our cast iron gudgeons is about 12 inches long and six inches dia'. not to encrease the strength, for that may be done to any degree by the depth of the squared part above the round, which is 9 inches by 6. But to give a greater bearing, for if a great weight

---

[1] Soho 18-inch, Bloomfield, Broseley, and Bow were at work. Bedworth (started in March 1777), Soho expansive (cylinder ordered February 1777), Colevile's (ordered 1776, started January 1778), and others were in hand, or had been designed.

[2] Watt's letters are quoted from the Office Letter Book, B. & W. Colln.

be laid upon a small gudgeon it will infallibly tear the brass it works in. As to the friction, that is lessened by contriving it so that the brass stands always full of oil.'

The description of the plummer blocks for the gudgeons seems to indicate that the use of separate brasses supported in cast-iron blocks was new at this time. ' The plummer blocks we make now of cast iron one foot square and two feet long, and thereby save the greatest part of the brass brasses which are lodged in a hole made to fit them in the plummer blocks. A pair of brass brasses for your engine will not weigh above 1¼ cwt. and by being equally supported will be as strong as if they were 10 times the weight.'

Watt was emphatic as to the necessity for a steam jacket to the cylinders, especially ' when coals are dear '. He says that he had erected two engines without them, ' partly against my will, and where coals are cheap, but there is a great loss by the want of an outside cylinder—and in your engine the difference would not exceed £50 which is no object in such an undertaking'.

In consequence of some difficulty about the floor levels, Watt, in the Tingtang engine, had to take in the steam at the upper end of the outer cylinder instead of at the bottom, as he wished to do.

It is known that in some of the earlier Watt engines a small furnace was arranged in the stone foundation under the cylinder, and the correspondence with Hornblower mentions ' the fireplace under the cylinder '.

With regard to the piston, Watt points out that it is worked entirely free from water and is made tight ' by pieces of lead and oakum soaked with grease of any kind & hard screwed down. It scarcely ever needs repairs after the first fortnight, during which time it must be often looked at.'

In reference to condensation, Watt says ' we use as much injection water as usual, but it comes out colder '.

There is a long discussion as to the beam chains, whether they are to be of cast or of wrought iron ; Watt seems to have favoured cast iron, mainly because the links could be produced of uniform size and shape, but, he says with a touch of humour, ' I cannot insure that cast iron chains will not break provided that force enough be applied for that purpose.'

In one of his letters Watt inquires whether there are any good engine-smiths in Cornwall, ' as we would not chuse to do any smith's work ourselves except what is peculiar to our engines'. To this, Hornblower replies : ' Here are very good engine smiths in Cornwall, & some bad ones (all of them love drinking too much).' It appears that hammered ironwork could be done in Cornwall for 3½d. per pound, which was much less than the Birmingham rate. Watt pointed out that this was an additional reason for having as much of it as possible done locally, but, he says, such work as is done at Soho shall be as well done as the nature of the work admits ; it could not be supplied for less than 6d. a pound, but care would be taken to ' leave to you every thing which a workman would reckon a beneficial job '.

THE WATT ENGINE IN CORNWALL

IN dealing with Watt's work in the period now under consideration it will be convenient again to depart from a strict chronological arrangement, and to deal in the first place with the engines put down in Cornwall.

The Cornish mine adventurers were badly in want of more powerful and more efficient machines for raising the water from their mines, and from the time when they first heard of the new engine they had taken a keen interest in what was being done at Soho. About the middle of 1776 a deputation from Cornwall made its appearance in Birmingham for the purpose of inspecting the Soho and the Bloomfield engines; it was under the leadership of Thomas Ennis of Redruth, or at least he seems to have been the most influential man upon it. The Cornish-men came determined to find out all they could, and after the visitors had departed it was found that a drawing of the engine was missing. Boulton upon this wrote to Ennis in a very outspoken fashion—' we do not keep a school to teach fire-engine making, but profess the making of them ourselves.' The missing drawing was returned very soon, it had been taken by Trevithick (the father of Richard Trevi-thick), as he said, under a misapprehension. The year 1777 saw the first Watt engine in Cornwall at work; this was at Wheal Busy, otherwise Wheal Spirit, Chacewater. The engine for Tingtang mine, near Redruth, had been ordered first, but there was delay in getting the parts to Cornwall and it was not at work until the following year; Boulton and Watt then had ten engines in hand. Several of these engines were for the Cornish mines, and it seems that by the summer of 1780 forty pumping-engines had been set up, twenty of which were at work in Cornwall.

The first order received from Cornwall [1] was for a 52-inch cylinder engine for Tingtang mine, of which Jonathan Hornblower the elder was the engineer, but, as we have said, the first Watt engine actually erected in that county was the Wheal Busy engine, a 30-inch cylinder. Except in respect of dimensions the construction of Tingtang engine followed the designs of Colevile's, an engine which had been ordered earlier but was not set to work until January 1778; but the Wheal Busy engine embodied additional improvements, as to one of which Watt writes: 'Chacewater nozzle is the most complete thing of that kind we have hitherto made & I expect will answer very well.' [2]

The cylinders and other castings for both these engines were ready at Bersham [3] early in May 1777, but when it came to shipping them it was found that the hatches of the vessel were too small to pass the Tingtang cylinder, and so the Wheal Busy goods were dispatched first, much to the annoyance of Watt. The erection of the Wheal Busy engine was confided to Thomas Dudley, a man who had been sent from Cornwall to Birmingham and Bersham to press forward the delivery of the materials, and to receive a course of instruction, but Watt went down to supervise the completion of the erection of the engine. Upon his arrival in Corn-wall in August 1777 he found 'Wheel Bussy in considerable forwardness', and that 'what ironwork had been made there is little inferior to our own, if any'. In the same letter he presses for the dispatch of the Tingtang materials, and says that all the world is agape to see the performance of Wheal Busy.' [4]

The engine was soon set going, and the reports on the performance are very good. 'Wl spirit goes on very well. It has forked the water in the engine shaft.' [5] 'The Spirit goes better and better, working well with $\frac{1}{2}$ inch of steam.' [6]

Like Tingtang this engine had an inner and an outer cylinder, with the valves at the bottom, and the eduction pipe was enclosed in a long cylindrical casing;

[1] Tingtang engine was ordered Nov. 23, 1776.
[2] B. & W. Colln. Watt to Boulton, 1777 May 20.
[3] Bersham Ironworks, near Wrexham. One of John Wilkinson's works.

[4] Boulton Papers. Watt to Boulton, 1777 Aug. 9.
[5] Ibid. Watt to Boulton, 1777 Sept. 13.
[6] B. & W. Colln. Dudley to Watt, 1777 Oct. 13.

the piston rod was forged at Soho ; the story of its production is related on another page.

Wheal Busy engine cost something under £800 ; [1] it was erected for temporary purposes, from the first it was not anticipated that it would be required to work for more than a year.

The correspondence in regard to the Tingtang engine is exceptionally complete, and in so far as it concerns the design of the engine it has been made use of at the opening of this chapter. [2]   It begins with a letter from Jonathan Hornblower to Boulton (dated Chacewater, Oct. 10, 1776) asking the terms on which the engines are supplied.   This letter contains an interesting little personal note :

' I don't know whether you are acquainted with the nature of the Welsh coal made use of here ; it is much more durable and will go much further than ours in Shrop. & Staffordshire (I still call it ours tho' I have been in Cornwall near 30 years), but it requires different management.'

To this letter a reply was sent by Watt giving various particulars of the engine, and of the terms upon which it could be had.   Then a few weeks later Watt again writes :

' We have yet received no conclusive order from your country, if therefore it is agreeable to the proprietors of the Tingtang mine, or any other where you are engaged, to try one of our engines, we will furnish what is peculiar to it, that is the cylinder, piston, piston rod, the regulators, the condenser and its apparatus at our own expense, to be removed by us if after a fair tryal the engine should not do double the work with the same coal that any other engine now does in Cornwall.   If it answers we demand the price of the articles furnished and the third of the savings for 25 years.'

However, the Tingtang adventurers did not take advantage of this offer, but gave a definite order for an engine in a letter by Hornblower dated 23rd November, in which it is mentioned that they already have a boiler ' suitable to a 52-inch cylinder in the common way ' with ' a double flue, (viz.) an external and an internal one '.

The drawings of this engine show Watt's original plan of inner and outer cylinders, with the inner cylinder open at the top, and the valve nozzle is at the lower end of the cylinders.   There are two air-pumps and one hot-water pump worked from the outer end of the beam, the condenser being outside the house.

When sending these drawings to John Wilkinson, Watt wrote :

' As I suppose, I shall have no opportunity of seeing the castings before they go to Cornwall, and the hammered iron work must be fitted by the copy of the drawing now sent, I hope Bersham will be accurate in following dimensions.   I have made *few* alterations from Mr. Coleviles unless in point of dimensions.   The principal one is the making the ends of the beating box for lower valves [in the condenser] sloping which will give the valves less motion in opening and shutting. . . . The nozzles will be executed here. . . . The diameter of inner cylinder to be 52 inches and of outer 64 & the flanches of the inner one to be made to go freely into the outer one.   The inner cylinder lower flanch to be strengthened by brackets as Bedworth was.   The snugs for lifting the inner cylinder by ought not to come over the holes for the holding down screws as at Bedworth they gave us some trouble in getting in these screws which are very long and obliged to go down perpendicularly.   Let each of the snuggs have an 1¼ inch hole in it.   Lett all flanches to be strengthened in the corners unless where expressly forbid.   Mr. Hornblower is come over to cast iron chains of himself, but has the Captain of the mine to convert yet.   The water pump bucket to be the same as Mr. Colevilles, you will please to be

[1] B. & W. Colln. : Letter Books.   Watt to Weston, 1779 June 20.   [2] Some of the letters are in the B. & W. Colln., others at Doldowlod.

particularly carefull in boring piston to dimensions. . . . I say nothing about quality or time, you know we are all upon honour.'

To avoid mistakes Watt furnished Hornblower with ' a list of all the parts of the engine, distinguishing where they are to be made '.

'*Bersham.* Inside and outside cylinders, bottoms, piston, lid and stuffing box. The condenser cast iron work complete.

'*Bradley.*[1] The nozzles and short pipe. The gudgeon and bed, the plummer block and brasses. Cast iron chains for cyl^r end.

'*Soho.* The regulators, brass and ironwork. The F and Y, the plug frame chains, two chains for condenser pumps, gearing the buckets and clacks of condenser, two round piston rods for do., all the screws & burrs for the two cylinders, condenser and boiler steam pipes; screwed ends for holding down screws, the iron work for the great piston rod top, the piston rod itself, 3 two inch dia^r. adjusting screws and burrs for top of piston chains, 3 do. for pump chains, if you chuse them. 4 screwed ends for the stirrup to hang the beam in, 2½ screws, screwed ends to shoot to martingale bolts.

'*Tingtang.* The boiler, firedoors and grates. The pumps and ye appurtenances. The pithead iron work, the poise beam, the pump chains, the screw bolts and arch plates for beam, the catch pins, the two glands which keep down the gudgeon, the stirrups for do., excepting the screwed ends which will be brought ready to shoot to. The martingales. The house and all the wood work.'

The piston-rod was made in London by Coulson ; it was ready in May 1777 and was shipped from London to Falmouth. The Bersham goods arrived in January 1778, but the engine does not seem to have gone to work before the end of July.

In Watt's Journal,[2] under the date July 30th, we read concerning this engine— ' The steam regulator is thrown open with great violence without any effort on ye part of ye Y-shaft wch is owing to ye very light rods ye Beam has to pull it out. . . . The bed of fire is 22 inches deep, 30 inches broad and 7 feet long & ye upper side of ye bed is 22 inches from ye boiler lagging.'

A week later it appears that the engine was ' going prosperously '.

Earlier entries in the Journal show that a valve was fitted in the boiler steampipe, and also that a valve was fixed on the clack door of the hot-water pumps ' to serve in place of those which used to be upon ye lids '.

Following the Wheal Busy 30-inch and the Tingtang 52-inch we have two much larger engines for Wheal Union and Chacewater, both with 63-inch cylinders. Wheal Union near Marazion belonged to Edwards & Co. ; it comprised two mines, Owen Vain (and Up Park) and Tregurtha Downs. The Watt engine was installed at Tregurtha Downs ; the other mine had a 67-inch Newcomen engine. The engine was provisionally ordered in July 1777, and in September, after the starting of Wheal Busy, a definite order was given. Budge was the engineer for these mines and, although he had been sent to Birmingham to see the engines at work and have them explained to him, he declined to undertake the erection ; possibly he had not grasped the underlying principles and had no confidence in the success of the Watt engine. The erection was actually confided to Dudley, who seems to have fallen into disgrace over it through bad work and mistakes. The engine was started in September 1778.

Watt sent the drawing of the cylinder to Bersham in November 1777 and in

[1] Bradley was another ironworks owned by Wilkinson ; it was not far from Birmingham.   [2] Doldowlod Papers.

the letter accompanying it he says : ' The principal dimensions are wrote upon the drawing which is of my own hands and tolerably exact to a scale of half an inch to a foot.' The letter is interesting for the directions on points of construction that it contains. Among others a single air-pump 24 in. diam. is to be used, and instead of a cast-iron outer cylinder one of rolled iron ; actually it seems that cast-iron panels were employed. Another point is that the cylinder is to be bell-mouthed at its upper end ' in the same manner as Soho 33-inch '. This engine had an additional valve at the top to regulate the quantity of steam used when the engine was underloaded. In the course of erection there was a little accident with the working-gear, and Watt altered the position of the tail of the detent catch and decided that the alteration was to be adopted in other engines. .

The Chacewater engine was ordered in November 1777 and Wilson, the manager of the mine, which belonged to Fentons and the Yorkshire Copper Company, requested that the materials might be forwarded at the same time as the Wheal Union materials.[1] Like the Wheal Union engine, it had a 63-inch cylinder with a stroke of 9 ft., and it worked a pump 17 in. diam., 53 fathoms deep, at the rate of 11 strokes per minute with a coal consumption of 128 bushels in 24 hours, ' and moreover puts in motion a very strong connection rod, 25 fathoms long, before it comes to the pump head, which rod and the others which belong to three lifts of pumps weigh about nine tons.' [2]

Watt, in the letter from which we have just quoted, goes on to say : ' We were three months going at the above rate in forking or unwatering the mine ; the whole country declared it impossible, some on account of the known great quantity of water, and others from a belief that the engine could not work the pump to that depth.'

The engine was at work at the beginning of August 1778 ; it made a very good start, but by the middle of the month it was doing very badly as Watt sets out at length in a letter to Boulton :

' Since I wrote you last, Chacewater engine wore a very gloomy aspect, as though going eleven strokes per minute it went sluggishly, it could be forced on neither by steam nor by counterpoising a part of the weight. This was the case Saturday all day. On Sunday the disease increased and when I came to it in ye evening it was taking more injection than ye water pump could discharge, and going extremely heavy though ye vaccuum was 27. I took off ye injection alltogether, it mended a little but still drew full pump, concluded a leak in ye condenser, caused in ye first place examine ye injection valve, found it staked open by a piece of wood, on removing which the quantity of injection was reduced, but no amendment except bringing it to 12 strokes p^r minute at which rate went till Monday, when had a general review beginning at ye nozzle, found the upper and under regulators both leaked some steam, for which tinning their edges was prescribed—my suspicions from ye beginning were very strong that we had ye Bloomfield disease of a bitt in our throat, but as after that accident these copper cones had been ordered to be rivitted on, I thought it could scarcely happen again, and thought perhaps a board, a bunch of oakum, somebody's hat or coat had been left in ye cylinder & had come into ye nozzle and was about to search for it, but first lifted out ye exhaustion regulator, when behold the copper cone, which had neither been rivitted nor well soldered, lay upon ye guard—it was needless to search further so we tinned the regulators and went to work imme-diately 13½ strokes to ye minute and these performed with vigour. We take now exceeding little injection, as in the new construction of water pumps, they do not go to the bottom of ye condenser by 5 or 6 inches, so upon ye first letting ye steam into ye eduction pipe it forces ye remaining injection (from last stroke) into ye air pump and follows it itself up through ye water which it finds there and is principally condensed in ye air pump by ye water which had

[1] Wilson subsequently became the agent for Boulton and Watt in Cornwall.   [2] Muirhead : *Mech. Inv.* Watt to Black, 1778 Dec. 12.

served before. This operation is performed with so much violence that it lifts the air pump bucket at least 2 inches and lets it fall again before the engine begins its stroke. . . . On Wednesday morning at one o'clock Chacewater tye lift pump rods broke again at a joint, which had never been sound. This stopt the engine till one o'clock in ye day when I saw it again at work in good order. During ye stoppage ye water rose six feet in ye mine and was then 20½ fathoms below addit, is probably 2 fathoms lower now. The engine burns 1¾ wey per diem minus one weigh in ye week.

'The Connoissieurs say we shall never fork ye water, or *if* we do we may fork anything as ye water is reckoned the heaviest in ye whole county.'[1]

In his Journal on Aug. 24, 1778, Watt makes an entry : 'injection worked by string tyed to plug wch pulled it open a little before the engine went quite out and shut it immediately when begun to come in' ; and again, on Oct. 29th, he says : 'new laid the piston mostly with ropes plaited with 60 rope yarns interlined with oakum, poured in 30 pounds of tallow'.

This engine was not an entirely new construction ; it was an alteration of the celebrated engine by Smeaton, the 72-inch cylinder of which was retained to form the outer cylinder for Watt's 63-inch cylinder. It was proposed to cast a lid and bottom for the cylinder at North Downs, but in fact a furnace was put up at Chacewater and the lid was cast there ; the casting was not a success, and in June 1778 Watt writes : ' The lid for Chacewater cyl^{dr} has been 3 times cast here, and after all will not do. I must send to Wilkinson for one, but I believe the last cast will make shift till it come.'[2] The new cylinder had an open top as in the engines previously described, and a furnace appears to have been arranged under the outer cylinder. Smeaton's great built-up engine beam was retained, also the manner of fixing the cylinder by cross-beams built into the wall of the house. The boiler was a long boiler with two copper flues.

The next engine to note is Hallamannin, a 40-inch, ordered in February and set going in December 1778. This engine was to the same design as the 36-inch for Wanlockhead, and the same drawings did for the two. We learn that the upper connecting-box was to have its cover cast on, and that the valves in the lid of the air-pump were made round conical valves, ' on account of ye difficulty we have always had of making them tight otherwise'.[3] At starting, although the boiler was bad, the engine went ' entirely well', but within a few days it had lost its vacuum, and the cause could not be discovered. Other faults developed very soon, apparently due to mistakes in erection, and at various times in the following year we find notices of trouble, so in October Harrison was sent to overhaul the engine. The adventurers in the mine sought to make Boulton and Watt responsible for the loss they had sustained by these defects, but Watt at once repudiated any responsibility. This engine had a jacket fitted on to the cylinder, a perpendicular steam pipe, and the valve box below ; the upper flange of the cylinder and the lid were turned. A circular boiler was used.

Following Hallamannin we have Poldice and Wheal Chance, 63-inch cylinder engines of like design. The cylinders, &c., of these engines were sent to Cornwall in the same ship and reached Hayle in September 1779. A printed form of list of materials for the engines was now in use ; on this form was stated in writing any necessary directions for making or erecting and the place at which each article was to be made. For instance for the Wheal Chance engine the cylinder

---

[1] Boulton Papers. Watt to Boulton, 1778 Aug. 13.
[2] Ibid. Watt to Boulton, 1778 June 27.
[3] Doldowlod Papers. Watt's Journal, 1778 Mar. 25.

and condenser were to be made at Bersham, the nozzles at Soho, the working-gear and most of the smith's work at the mine. As to the Poldice engine we learn that it was 9 ft. stroke with a cylinder of 10½ ft. long, that the top nozzle was the same as Wheal Union, but that there were to be considerable alterations in the lower nozzle. There was a furnace under the cylinder and the heating-case or jacket was made in panels. Parts of the condenser were to be made of brass on account of the corrosive nature of the water. The engineer of the mine was Budge, but the erection of the engine seems to have been in charge of Law, one of the Soho men. In connexion with Poldice cylinder we come across one of the few cases in which John Wilkinson's boring-mill did not produce correct results; after the first boring the cylinder ' was found to be oval & had to be re-bored '.

The Poldice mines had a number of other engines on Watt's system. The No. 1 engine became known as Poldice Eastern ; the second engine, also a 63-inch, set to work in October 1782, as Poldice Western. Another 63-inch engine, set to work in 1785, was replaced two years later by two double-acting engines, one a 24-inch and the other a 58-inch.

In connexion with an engine put up at Ale and Cakes (United Mines) in 1780 we have an interesting letter which shows that with increasing experience the erectors were becoming more skilful and certain in their operations ; Boulton writes of this engine that it ' is certainly the most free from steam and air leaks I ever saw and as to neatness it exceeds all former exceedings so much that if you were to hang an engine house with silk damask & gilt the cornice it would not exceed Ale & Cakes engine so much as that exceeds the old Cornish engines '.[1]

Jethro Hornblower had been in charge of the erection. The engine had a new form of cataract as to which Boulton in the same letter says : ' The air cataract is very neat & convenient & may be made a good thing.' The upper nozzle was the same as Ketley engine and the lower nozzle was to a new pattern.

The general drawing of Ale and Cakes engine is reproduced in Plate XXV. Comparing it with the drawings of Colevile's engine, it will be seen that the working cylinder, ' 58 inches dia^r. by 10 feet 10 inches long', is closed at the top, the outer cylinder being dispensed with, but the casing, made in panels, used in place thereof is not shown. There is a valve nozzle at the top as well as at the bottom of the cylinder and these are joined by a perpendicular pipe. The furnace under the cylinder and the jacket round the eduction pipe are no longer used. The beam is built up of a series of logs, and its arms are of unequal length, the stroke of the engine being longer than that of the pumps. On the top floor of the engine-house is seen the winch for raising the piston when desired.

We now come to a very extensive undertaking, the engines for the Consolidated Mines, Wheal Virgin and Wheal Maid. In 1780 Boulton and Watt entered into an agreement in respect of five engines for these mines ; they were all of 9 ft. stroke, but the cylinder diameter varied ; two had 58-inch, one a 56-inch, one a 52-inch, and one a 50-inch cylinder. These engines were all at work in October 1782. At about the middle of 1781 Watt was able to report to his friend Hamilton that these five engines were being erected, and that they were erecting another at Poldice, as well as setting to work another already erected there. This made seven engines within half a mile square, which should have brought in above three thousand pounds per annum. (The premiums actually received from Cornwall for the year ending Dec. 31, 1780, amounted to nearly two thousand

---

[1] B. & W. Colln. Boulton (Plengwarry) to Watt, 1780 Aug. 31.

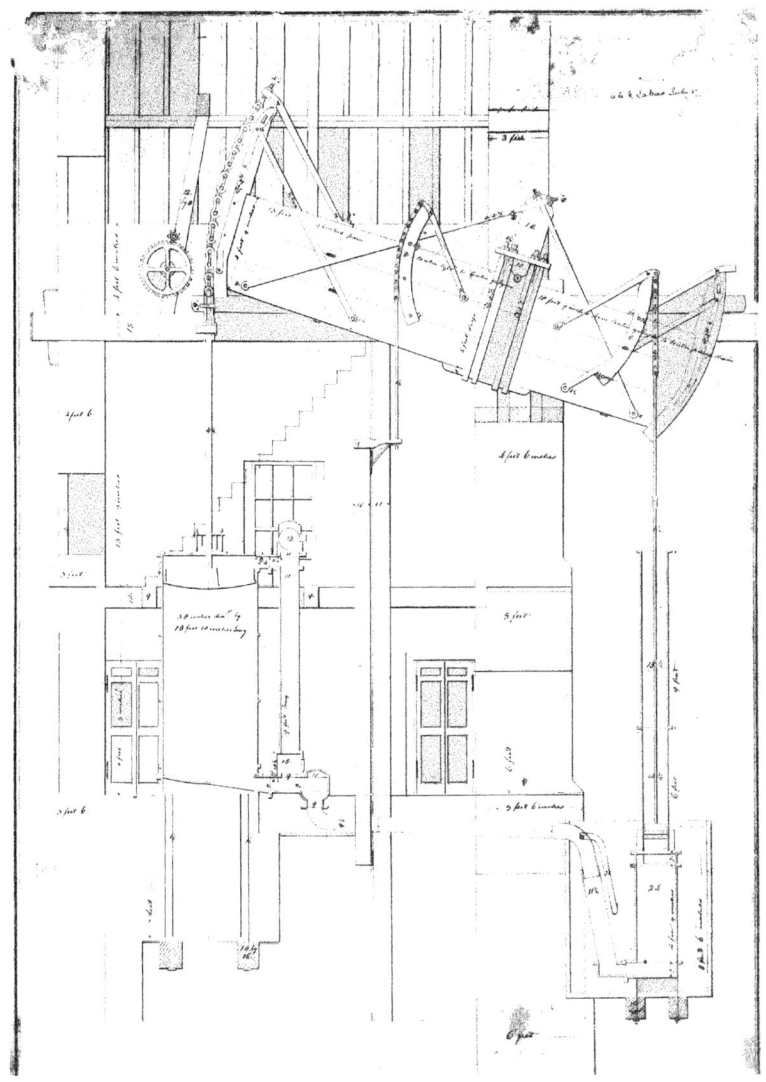

XXV. SINGLE-ACTING PUMPING-ENGINE, 58-INCH CYLINDER,
FOR ALE AND CAKES MINE, CORNWALL

Drawing by Watt, dated July 1779

six hundred pounds.) In addition they were putting up engines at Dolcoath, Pool, Crenver, and Trevaskus, all in Cornwall, and had nine others (say ten) now going here. They had also a few going on in other parts of England, and had been obliged to refuse many orders of lesser value.[1] The engines actually erected in Cornwall at this time numbered eleven ; they were—Wheal Busy, Tingtang, Chacewater, Wheal Union (Tregurtha Downs), Hallamannin, Ale and Cakes, Wheal Chance, Poldice No. 1, Wheal Treasury, Tresavean, and Poldory, so that it would appear that one or two of them were idle.

This period marks the culminating point of the activities of Boulton and Watt in engine-building in Cornwall. Watt in 1783 informs his partner that there is now only one common engine in Cornwall, and that they had erected twenty-one, but that one or two of their engines were not then at work.[2]

However, bad times were setting in. When Boulton went to Cornwall in the following year he found the mining industry in a very poor way, and considered the prospect gloomy in the extreme. He wrote to Watt :

' When I take a survey of the Western mines, they appear as but a barran prospect as well as the adventurers. Wl. Union dead, and its son Castle Addit dead, Hallamanning ye 1st. dead & Hallamanning ye 2nd expiring and cannot outlive 2 mo. Chace ▽ little engine dead, its son Scorrier dead, & its grandson Wl. Virgin St. Hilliry is expected not to live more than another mo. Trevaskus is dying with poverty & so is Wl. Crenver ; all the mines about Roskeer, Wl. Chance & Rosewarn increase the lowness of my spirits when I behold the lifeless Ball and the falling buildings. The fair Wl. Kitty is become a bowling green with a temple erected upon it built by Mr. Harris dedicated to indolence. I verily believe that in the course of the present year there will not be a single engine working west of Redruth unless Dolcoath & Wl. Gons, but I fear ye former. The engines at Pool, Dolcoath, Wl. Gons & Crenver are all in excellent order, vacuum about 27″—but as to Wl. Virgin St. Hillery & Hallamanning abound with dirt, sin and misery.'[3]

As to Hallamannin, Boulton goes on to say that the engine cannot work with less than ten inches of steam and is loaded to eleven pounds on the inch, and that it is in every way so bad that it must give up working, although the mine is good.

The quaint terms used in reference to the Wheal Busy engine, called in this letter ' Chacewater little engine ', mean that it had been moved first to Wheal Chance, Scorrier, and then to Wheal Virgin in St. Hilary.

As to the condition in which the engines were kept, no doubt this depended, in the main, on the mine captain, but it seems clear that the introduction of the Boulton and Watt engines had been accompanied by a general improvement. Watt, soon after his first arrival in Cornwall, writes : ' In general the engines here are clumsy and nasty, the houses crackt & everything dropping with water from their house cisterns.'[4] Three years later we have a very different picture presented in a letter by Boulton : ' You would be pleased & even astonished to see the neatness exactness & quietness of the 2 engines at ye United Mines & at Tresavene. We have at last raised a spirit of emulation amongst the Cornish enginemen, each striving to exceed his neighbour in neatness & good order. I despair of ever seeing any engines kept so well in our country.'[5]

We have now come to a period at which it will be convenient to leave the

[1] Doldowlod Papers. Watt to G. Hamilton, 1781 July 23.
[2] Boulton Papers. Watt to Boulton, 1783 May 18.
[3] B. & W. Colln. Boulton to Watt, 1784 July 8.
[4] Boulton Papers. Watt to Boulton, 1777 Aug. 14.
[5] B. & W. Colln. Boulton to Watt, 1780 Sept. 2.

consideration of the Boulton and Watt engine in Cornwall, and to take up the work that was being done in other parts of the country.

BIRMINGHAM CANAL ENGINES

AN engine for pumping back water at the Smethwick locks on the Birmingham Canal was ordered at the beginning of 1777. It was at work in March 1778, and was said to go 'exceeding well'. In April it was tested by John Smeaton for the Canal Committee, 'much to his satisfaction', indeed Smeaton was so impressed by the performance that he relinquished, in favour of Boulton and Watt, a contract that he had entered into for the construction of a steam engine for the Hull Water-works. The following account of the trial appears in *Aris's Birmingham Gazette* for Apl. 20, 1778.

'The following Letter received last Week by the Committee of the Birmingham Canal Navigation, from their Superintendant of the Locks, affords an irrefragable Proof of the great Utility of a new-invented Steam Engine, lately erected on the said Canal, under the immediate Direction of Mess. Boulton and Watt, the Patentees.

"To the Committee of the Birmingham Canal.

"Smethwick Locks, April 17.

"Gentlemen,—On Wednesday last Mr. Smeaton made an accurate Trial of the Steam Engine erected lately on the Canal at this place, and it appeared that it did not consume more than 64 lb. of Coal an Hour when working at the rate of 11 Strokes a Minute (each Stroke being Five Feet Ten Inches). The Diameter of the working barrel of the Pump is 20 Inches ; and the perpendicular Height of the Column of Water is 26 Feet 10 Inches and a Half, equal to 11-lb. 3-qrs. upon every square Inch of the Piston : The Quantity of Water raised at each Stroke is equal to 12 3 qrs. cubic Feet.

"Mr Smeaton declared, that the best new common Engine, with all his late Improvements (which are very considerable) would have required 194 lb. of Coal to raise an equal Quantity of Water to the same height ; and that a common Engine without those Improvements would consume a still greater Quantity.

"When the Asperities on the different working parts of this Engine are worn off, and the Cylinder is eased and finished, as is intended, I have not a doubt but it will be an Advantage to the Proprietors of 20 per cent. more.

"I am, Gentlemen, your most humble Servant,

"S. BULL." '

Farey, who no doubt obtained his information from Smeaton's Reports, gives the lift as 27 ft., the stroke as 5 ft. 9 in., and the coal consumption as 65 lb. This he works out to be nearly 18 million pounds raised one foot per bushel (84 pounds) of Wednesbury coals. 'The coals used in this trial were broken into small pieces, scarcely any being larger than a hen's egg, and none less than a nutmeg.' [1]

This engine had a 20-inch cylinder, and the 'engine and pumps cost £350, including the putting together'.[2] It is of particular interest in that it is still in working order ; it remained on its original site until 1898, and is now set up at Ocker Hill, Tipton, where it has been worked under steam, notably on the occasion of the James Watt Centenary Commemoration.

At the end of the year 1778 the Canal authorities ordered another engine for Spon Lane Locks ; this was at work in June 1779. It had a 32-inch cylinder, and save in respect of cylinder diameter was a duplicate of a 30-inch engine for Donnington Wood that was being made at the same time. Before the end of 1783 a third engine had been ordered, for Ocker Hill Locks.

[1] Farey : *Steam Engine*, 338.   [2] B. & W. Colln. Watt to Weston, 1779 June 20.

LONDON WATERWORKS ENGINES

BETWEEN August 1778 and May 1779 three pumping-engines were set up for waterworks in the London District, i.e. at Richmond, Shadwell, and Chelsea. Of the Richmond and Shadwell engines there is nothing of particular interest recorded. The Chelsea engine was inspected and studied by John Farey in his younger days, and in his *Treatise on the Steam Engine* he gives a sketch, that he had made in the year 1804, of the cylinder and the arrangement of the valves. The engine is interesting in that it was one of those in which the piston rose in vacuum. The lower end of the cylinder was in permanent connexion with the condenser, and the steam was exhausted from the top of the cylinder direct to the condenser, instead of first passing to the lower end of the cylinder, as in the normal type of single-acting engines. The equilibrium valve in this arrangement became the exhaust valve, and it was placed near the top of the cylinder just below the steam valve. A valve, adjusted by hand, was sometimes placed in the eduction pipe to regulate the flow of steam to the condenser. Farey discusses the pros and cons of the arrangement. The intended advantage of this construction is, he says, that the whole time of the ascent of the piston is allowed for the condensation of the steam, and therefore it might be expected to produce a better vacuum, and a more immediate stroke, than in other constructions, in which the condensation of all the steam in the cylinder must be made after the piston arrived at the top of its stroke and before it can begin to return. He goes on to say that in practice this advantage was not found to be of great importance, and the scheme had the disadvantage that leakage at the stuffing-box and at the joint of the cylinder-cover was greatly increased, and the leakage was now leakage of air into vacuum, instead of being confined to leakage of steam into the atmosphere. Then again, in order that the piston might rise at the same speed as in the ordinary engine, a heavier counterweight was required at the outer end of the beam. The last objection that Farey brings forward is quite interesting; it is that the heat losses in the cylinder are necessarily greater than with the usual way of working, since the steam-jacket was pouring in heat for a greater proportion of the cycle. At the date of Farey's writing (say 1826) the plan had long been disused. The conversion to the ordinary working cycle having been effected by putting in at the top of the eduction pipe a valve operated by the working-gear.

The working-gear of this engine seems to have had some special feature; Boulton on one of his visits to the engine remarked that it answered very well, and wished to have the gear of the Shadwell engine altered to the same arrangement.

Of its earlier days, besides minor accidents, one curious affair is recorded. The links of the chain at the pump end of the beam broke, and the piston came in with such force as to break the cylinder bottom, crack the lower end of the cylinder, and do other mischief. Watt devised a means for closing up the crack and a new bottom was cast in London.

This engine had a 20-inch cylinder, 8 ft. stroke. It seems to have taken the place of one of two atmospheric engines erected in 1741–2. A trial of its performance is summarized by Watt to this effect : Loaded on the average to $8\frac{1}{2}$ lb. to the inch and working at $16\frac{1}{2}$ strokes per minute, 4,103 strokes were made on five bushels of coal weighing 84 lb. to the bushel, ' & very wet, being a rainy day '.

T

Brief mention may suffice for some of the remaining engines of this period. One of the first engines set to work in the year 1778 was for pumping brine at Thirlewood or Lawton Saltworks, Cheshire, belonging to Edward Salmon and partners. The cast-iron parts were made from the patterns of the Bow engine and the erection was commenced early in February 1778 under the supervision of Watt himself, who has left us a full account of the process, which has been utilized in Chapter XX. John Scott of Shrewsbury had two Boulton and Watt engines : the first, for Bog mine, Shropshire, about six miles westward from Church Stretton, was running at the beginning of 1778 ; it was made to the same drawings as Wheal Busy. In 1782 it appears that this engine had been moved to a colliery. The second engine was for Malehurst ; it was on order in the latter half of 1778 ; it had a 24-inch cylinder, but was in all other respects identical with the engine for Hull Waterworks, which had a 22-inch cylinder, and was running in January 1779. In 1778 were put up also engines for Byker Colliery, Ketley Ironworks (Richard Reynolds), and Bradley Ironworks (John Wilkinson).

Before the middle of 1779 an engine had been set up at Wanlockhead mine for Ronald Crauford & Co. (the Countess of Dumfries, Sir Peter Crauford, and Gilbert Meason). This engine had a 36-inch cylinder, but the same drawings served for the Hallamannin 40-inch ; it was ordered in 1777, but it was not until the end of March 1778 that the drawings for the cylinder, &c., were sent to Wilkinson. There was further delay in getting the materials to the mine, and it was in March 1779 that William Murdock was sent to erect the engine. This was Murdock's first independent job. Upon his return south he was sent to finish the erection of the Donnington Wood engine, commenced by Jabez Hornblower.

The first Boulton and Watt engines erected abroad were those for Jary of Nantes, and for Périer for the Paris Waterworks. The materials were sent out in 1779 and 1780. Jary's was the first engine at work outside this island. In 1782 the parts of another engine were sent to France ; they were ordered directly from John Wilkinson, apparently without the knowledge of Boulton and Watt, and the engine was known as Wilkinson's French engine.

The years 1782 and 1783 are represented by engines for Doonane Colliery, near Carlow, Ireland (P. & A. Colclough), for Walker's Ironworks at Rotherham, and another engine for Donnington Wood.

With the year 1783 we take leave of the single-acting engine. Watt's attention was now devoted to the double-acting and rotative engines, and although single-acting pumping-engines continued to be made they do not present any features calling for remark.

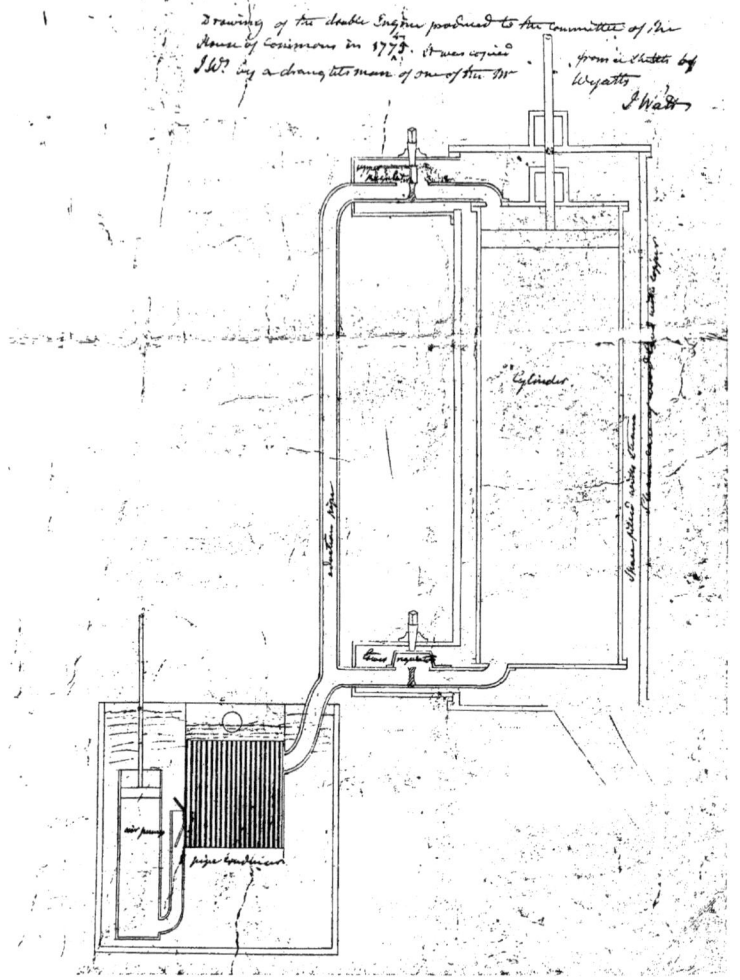

XXVI. DOUBLE-ACTING ENGINE. DRAWING LAID BEFORE PARLIAMENT, 1775

Doldowlod Papers

XII

THE DOUBLE-ACTING ENGINE

THE idea of using steam to press alternately on the opposite sides of a piston, and thus to enable the engine to make a power stroke in both directions, was a great stride in the development of the steam engine, and although the first double-acting engine was not made until 1783, the idea was in Watt's mind as early as 1774 or 1775, and possibly still earlier. It has been stated in a preceding chapter that, in connexion with his application for the extension of the term of his patent, Watt produced before a Committee of the House of Commons two drawings. One of these drawings has a figure, reproduced in Plate XXVI, showing the cylinder and condenser of a double-acting engine. The inscription at the top of this figure reads : ' Drawing of the double engine produced to the Committee of the House of Commons in 1774–5. It was copied from a sketch by J. W. by a draughtsman of one of the Mr. Wyatts. J. Watt.' No doubt this was written at a later date than 1775, but there need be no hesitation in accepting it as a fact that the drawing was exhibited as mentioned, if not in 1774, then in the first half of the year 1775. Upon the drawing are marked the ' cylinder', ' space filled with steam', ' steam case of wood lined with copper', ' upper regulator', ' lower regulator', ' eduction pipe', ' pipe condenser', and ' air pump'. It will be seen that the cylinder has a valve box at the top and another at the bottom, and that the valves, or ' regulators', are of the same construction as those shown in the drawing of the single-acting engine produced on the same occasion, Plate XVI. The top of the working cylinder is closed, and its lid, as well as that of the ' steam case', is provided with a stuffing-box.

The project of a double-acting engine remained in abeyance until 1781, when it was brought forward again by Boulton. Referring to the rotative engine and considering the means by which Pickard's crank patent might be evaded, he says, ' Qr. If an engine was made after a plan & drawing you made some years ago, that is to work down & up, you might by that means turn a crank completely round the circle and then we are secure because a Comn engine can't be made to work up and down.' [1] Then in January 1782 we find Watt writing of such a plan and stating, ' and if a weight equal to half the load be put upon the inner end of the beam, and the engine be made to lift it during the return, by making a vacuum above the piston and using a rack instead of a chain, a cylinder of the present size may work to the same depth by half the steam, and I believe the engine will work very sweetly '.[2]

It will be understood in reference to the loading of the beam that Watt was thinking of the ordinary pumping-engine in which the useful load came on only during the down-stroke of the piston. A little later in the same year Boulton

[1] B. & W. Colln. Boulton to Watt, 1781 Apl. 20.

[2] Muirhead : *Mech. Inv.* Watt to Boulton, 1782 Jan. 16.

suggests the combination of a force and a lift pump,[1] and a few years later he states that the Cornish captains are fond of forcing-pumps.[2]

The double-acting engine was one of the subjects of Watt's patent of 1782 ; and here we may pause to draw attention to the activities of the inventor at this period. Although suffering from bad health and immersed in the business of erecting engines in Cornwall, he was devising schemes for converting reciprocating into rotary motion, elaborating the expansive engine, and then the means of coupling the beam to the piston-rod in the double-acting engine.

In the correspondence the double-acting engine is frequently referred to as the ' double ' engine. The same term was applied to the two-cylinder single-acting rotative engine that was in hand at Soho at the time of the introduction of double action. This has given rise to a certain amount of confusion and misapprehension of the actual facts.

Although he seems to have thought out the idea in relation to a pumping-engine, the first double-acting engines actually made were rotatives ; Watt did not long delay in getting out a design for ' an up and down reciprocator ', and an engine was at work at Soho in March 1783 as to which he says, ' Came home and saw the up and down engine work, which has broke out several teeth of the rack, but works steady.' [3] In the following month Watt informs his friend De Luc that they have succeeded in making an engine act by the force of steam both in the ascent and descent of the piston, and, he adds, ' it acts so powerfully that it has broken all its tackling repeatedly '.[4] To Smeaton he writes, in greater detail, that the engine acts by a rack and sector instead of a chain and has a cylinder 18 in. diam. with 18 in. stroke ; it is to turn a pair of corn-mill stones, $4\frac{1}{2}$ ft. diam., 100 turns per minute. ' The engine is to make 60 strokes (acts up and down) in that time.' He adds ' we are also making a small engine, 1 foot diameter, 1 foot stroke, to work 100 strokes per minute, and to act up and down '.[5] The short length of the strokes of these engines will be noticed. Boulton considered that this was a move in the wrong direction, and while he thought that the short strokes might suit for very small engines and that it was quite proper to try them, he was of opinion that the long strokes would answer best, at any rate when there was no restriction as to expense or to the height of the building.[6] The smaller engine referred to above was put up at Soho for experimental purposes, the larger one was for driving a corn mill, also at Soho.

The first double-acting engine made for a customer was another rotative, for Cotes and Jarratt of Hull, which was completed in 1784. In view of the fact that a later chapter is devoted to ' The Rotative Engine ', it is not necessary to pursue the double-acting rotative engine here beyond dealing with such considerations as arise from the fact that the piston is pressed by steam in both directions, and are common to the rotative and the pumping-engine. The chief of these is the connexion between the piston-rod and the beam.

In the single-acting engine this connexion was always under a tension strain, and a simple chain arrangement answered the purpose very well. In the double-acting engine the reversal of the strains called for some different connexion. The double-chain device was already in use in fire-extinguishing engines. The toothed

[1] B. & W. Colln. Boulton to Watt, 1782 Sept. 30.
[2] Ibid. Boulton to Watt, 1786 Sept. 17.
[3] Doldowlod Papers : Watt's Journal, 1783 Mar. 31.
[4] Muirhead : *Mech. Inv.* Watt to De Luc, 1783 Apl. 26.
[5] Ibid. Watt to Smeaton, 1783 Apl. 27.
[6] B. & W. Colln. Boulton to Watt, 1783 Feb. 11.

sector and rack was an obvious expedient to adopt, and was used in the first engines. Another plan was to connect the piston-rod to the beam by a link and to provide a guide for the end of the piston-rod ; the production of long straight surfaces in metal by hand, however, was a long and expensive process ; moreover, the beam engine did not lend itself very well to the fitting of a guide between the cylinder and the beam (one of the figures in the drawings to Watt's specification of 1784 shows the guide above the beam). Very soon, however, that beautiful contrivance, the parallel motion, was devised. Watt seems to have turned his attention to a linkage arrangement for the first time in June 1784, when he writes to Boulton :

' I have started a new hare. I have got a glimpse of a method of causing a piston rod to move up and down perpendicularly by only fixing it to a piece of iron upon the beam, without chains or perpendicular guides, or untowardly frictions, arch-heads or other pieces of clumsiness ; by which contrivance, if it answers fully to expectation, about five feet in the height of the house may be saved in 8 feet strokes, . . . I have only tried it in a slight model yet, so cannot build upon it, though I think it a very probable thing to succeed, and one of the most ingenious simple pieces of mechanism I have contrived.' [1]

A large model was made soon after and Watt reports that it bids fair to answer and explains that it is a perpendicular motion derived from a combination of motions about centres.

The specification of Watt's patent of 1784 is dated August 24th ; it includes three methods of guiding the piston-rod. The third of these ' consists in forming certain combinations of levers moving upon centres, wherein the deviation from straight lines of the moving end of some of these levers is compensated by similar deviations, but in the opposite directions, of one end of other levers'. One of the devices according to this method is the simple three-bar motion. Shortly after the date of the specification we find Watt describing in clearer terms ' one of the new methods of producing a right-lined motion from a combination of motions round centres. The convexities of the arches described by the ends of the working-beam, and of the regulating radius, lying in contrary directions, there is a certain point in the connecting-lever, which has very little sensible variation from a straight line.' [2] Watt, it will be observed, did not claim that his mechanism produced a perfect straight line.

Whether this was the construction that Watt had in mind in the first instance is not very clear, but there is no doubt that this was the mechanism first applied in practice. The letter from which we have just quoted goes on to state : ' This method we are executing in Mr. Cotes engine.' The drawing of this engine is dated June 1784 and shows the rack-and-sector connexion ; the linkage connexion is, however, roughly sketched in pencil, as may be distinguished in the reproduction, Plate XXXV. This engine was the first made with the three bar motion, and it was put together at Soho. In October Watt was able to report that : ' The new central perpendicular motion answers beyond expectation, and does not make the shadow of a noise ; but for want of some regular work for the engine to do, we have not been able to give it a fair trial.' [3]

This motion was used also for the engine for Stonard and Curtis ; see Plate LXV. The simple three-bar motion worked satisfactorily, but there was a serious practical objection to it—it entailed an extension of the engine framing and of the house

---

[1] Muirhead : *Mech. Inv.* Watt to Boulton, 1784 June 30.

[2] Ibid. Watt to Boulton, 1784 Sept. 11.
[3] Ibid. Watt to Boulton. 1784 Oct. 21.

laterally beyond the cylinder. Watt at once devised means of reducing this evil ; instead of making the radius rod or 'regulating-radius' equal to the radius of the beam and coupling the piston-rod to the centre of the vertical link, he made it shorter ; the rectilinearly moving point was then shifted above the centre of the link, and the piston-rod was joined to it accordingly. This is the construction actually shown in his specification and in the Stonard and Curtis drawing.

But within a few months of the preparation of the specification a far more compact mechanism involving the jointed parallelogram had been arrived at. This was the parallel motion, and we find it shown for the first time on the drawings of Whitbread's engine dated November 1784 ; in the following May we have drawings for 'Parallel Motion for Albion Mill'. The parallel motion is a combination of the three-bar motion and the pantograph, a mechanism that was fairly well known at the time ; it is now a point on the inner vertical link of the parallelogram that has imparted to it the movement in a straight line produced by 'a combination of motions round centres', and the motion of this point is reproduced at the lower joint of the outer vertical link.

We have no particulars as to the birth of this combination and it is not known definitely whether it was due to Watt himself. In a letter to his son James, written in 1808,[1] Watt, after describing the three-bar motion, goes on to say : 'and from this the construction, afterwards called the parallel motion, was derived'. That is to say, he distinguishes between his original contrivance and the parallel motion. Then he proceeds : 'Though I am not over anxious after fame, yet I am more proud of the parallel motion than of any other mechanical invention I have ever made.' But although in this letter Watt distinguishes between his original contrivance and the parallel motion, and then proceeds to credit himself with the invention of the latter, it is possible that we have here some confusion of ideas and that Watt may not have meant to lay claim to the combined mechanism, but rather, and quite rightly, looked upon the three-bar motion as the primary part of the combination and the rest as merely auxiliary ; and it should be observed that while there was a definite and well-known name for the combination, there was no distinguishing term for the simple mechanism. This is well borne out by a comparison of the drawings for the engines of Stonard and Curtis and Albion Mills, Plates LXV and LXVI, both dated May 1785 : the one for the three-bar motion is headed 'Messrs Stonard and Curtis's Motion' while the other is 'Parallel Motion for Albion Mill'. That Watt did later apply the term 'parallel motion' to the original device is clear from the account of the engine that he wrote in 1814.[2] Here, after describing the three-bar motion, he proceeds : 'These were first ideas, but the parallel motion soon was moulded into the form in which it appears in all Boulton and Watt's engines.'

In the foregoing the abandonment of the simple three-bar motion has been ascribed to the desire for a more compact construction, but there was another consideration which may have been of equal or possibly of greater weight—the guiding of the air-pump rod. This, however, seems to apply mainly to the rotative engines ; in the large double-acting pumping-engines the air-pump rod was jointed directly to an auxiliary beam and the parallel motion served only to guide the piston-rod. It will be observed too that in the rotative engines the air-pump rod is not coupled to the vertical link to which the radius rod is jointed,

[1] Muirhead : *Mech. Inv.* III. 89 *n.*   p. 153 (Appendix by Mr. Watt).
[2] Robison : *Steam and Steam Engines*, 1818,

but that a separate parallelogram is provided to work it. This was the practice during the whole period of Watt's connexion with Soho.

The conversion of the single-acting engine to double action called for other changes—the beam had to be strengthened and the valve gear modified. For some years the engines were sent out with mechanism for opening and closing the injection valve ; the erectors appear to have seen, in the case of the rotative engines at any rate, that the periods during which the valve was actually shut were very short, and upon their own initiative disconnected the gearing, so that the injection was always on.

Then there was, in the pumping-engine, the problem of utilizing the power stroke in both directions. We have seen that in 1782 Watt was considering the use of a weight on the beam, and Boulton throwing out the suggestion of a combination of force and lift pumps. A plan, covered by the 1784 patent, in which separate sets of pump-rods are worked from opposite ends of the beam came into most general use. It is shown on the drawing of the Hebburn engine, Plate XXVIII.

The first double-acting pumping-engine was set up at Wheal Towan in Cornwall at the end of 1784 or beginning of 1785. At the same time a double-acting rotative was being put up at Wheal Crane. The drawings for Wheal Towan engine are dated February 1784, and show the rack-and-sector connexion between the piston-rod and the beam ; those for Wheal Crane are dated 1784 and 1785, they show a link connexion and a roller on the piston-rod working against a vertical guide. Wheal Towan engine had an 18-inch cylinder ; Wheal Crane a 14¾-inch cylinder. We have no very definite information in regard to the erection and working of these engines; they are referred to in a letter from Watt to Boulton in September 1784 concerning an engine for Wheal Messa, one of the North Downs group of mines. It had not been decided whether this was to be a single- or a double-acting engine; Boulton wished it to be double-acting, 'with the self-balancing pitwork',[1] but Watt was for allowing the question to stand over until Wheal Towan and Wheal Crane had been set to work. One gathers that these engines worked satisfactorily, for Wheal Messa was built double-acting ; it had a 42-inch cylinder, the condenser and air-pump were within the house, the beam was built with a king-post, and it had a parallel motion composed of wooden rods. The general drawing, dated April 1785, is reproduced in Plate XXVII. This engine was not started until 1786, and before that a still larger engine had been pushed forward and set to work at Wheal Fortune, another mine in the North Downs group.

This had a 45-inch cylinder, a condensing-vessel, and a parallel motion ; it was started in November 1785 ; Boulton, who was in Cornwall at the time, had felt some anxiety about it, and had delayed his departure in order to see it start. ' It is ', he writes, ' a tremendous thing to look at, there are so many large joints and so many combinations of untried things that I am frightened when I behold it.'[2] However, the engine went off very well. A double-acting pumping-engine with a 20-inch cylinder was made for Wheal Mount at about the same time as that for Wheal Messa. It was soon followed by two engines for Poldice (24-inch and 58-inch cylinders, started in April 1787), and in 1788 by one at Wheal Maid (Wheal

[1] B. & W. Colln. Boulton to Watt, 1784 July 22. By ' self-balancing pitwork ' Boulton, no doubt, meant the arrangement included in Watt's patent specification of 1784, in which pumps are worked from both ends of the engine beam.
[2] Ibid. Boulton to Watt, 1785 Oct. 21.

Virgin). This had a 63-inch cylinder, 9 ft. stroke, and was probably the most powerful engine put up by Watt in Cornwall. In 1794 it was stated to be ' the most powerful machine in the world and fully equal to any three of the engines then at work on the mine'. In 1792 this engine was being worked expansively although loaded to ten pounds per inch. In 1793 a new double-acting pumping-engine was ordered for Poldory and another for Huyas.

The double-acting pumping-engine shown in Plate XXVIII is one of those made for the Tyne district—a 63-inch engine for Hebburn Colliery, Durham. It will be seen that the engine has a direct pump-rod from the outer end of the beam, and a ' diagonal pump rod' from the inner end, and that an auxiliary beam, the ' air-pump beam', connected by a link to the inner end of the main beam, is employed to work the air-pump, the hot-water pump, and the plug-rod for the valves. The engine beam is of unequal leverage, and is composed of two logs, 18 in. square, with ' diagonal King posts', 12 in. square. The spring beams, 14 in. square, are supported between the lever wall and the house wall by a cross-beam and by inclined ' spurs' abutting on the lever wall, and are surmounted by ' Springs'. The lever wall is 5 ft. thick at the top and the house wall 3 ft. The direct pump-rod is 12 in. square and the diagonal rod 13 in. by 11 in. The plug-rod is of wood, 10 in. by 6 in., and its guide-posts are 11 in. by 8 in. The rod for working the air-pump is 7 in. square ; the air-pump and the condenser are both 36 in. diam. The engine cylinder is, as stated above, 63 in. diam., and its piston-rod is 5 in. diam. This drawing is dated Jan. 20, 1798, so that this engine may be regarded as representing the practice at the end of Watt's career at Soho ; not indeed that there is any substantial change as compared with the earlier engines made for Cornwall, for instance, the Wheal Maid 63-inch, started in 1788, resembles it very closely. With the drawings of the Hebburn engine were sent out directions of the procedure to be followed in setting it to work. It should be explained that it was designed for expansive working and had the valves operated by two plug-rods :

' The steam handles are shut by one and the exhaustion handles by the other'; and till the engine man is got expert in his business, a man may stand to each plug and take the command of their handles in setting on without incommoding each other. The top and bottom steam valves and lower exhaustion being open, the engine blows and by shutting the lower steam valve for a time a degree of vacuum is formed in the condenser which is supplied by air from the underside of the piston which air is blown out of it by again opening the lower steam valve ; and so by alternately shutting and opening the valve (the others being constantly open) the vacuum will be such that the engine will commence its stroke, but will make only a short one, probably. The top steam and lower exhaustion being shut, and then by disengaging the other valves, they opening, the engine will return to its former situation, called out of house. Here it may be noted that the preponderancy of the rods should be just in favour of the outer ones, that the engine may have no disposition from the gravity of the parts to come in—and it may likewise be noted that in setting on the slider which disengages the lower steam and top exhaustion, should be moved so as not to disengage [i.e. to open the valves], because it may return the stroke sooner than the engineman can pay proper attention to the handles, especially if only one man is at the plugs. . . . The upper steam and lower exhaustion valves are then reopened, and the operation repeated. It will be prudent to prevent the steam valves from opening fully in setting on, by way of checking the velocity of the engine and preventing bangs.' [1]

It remains now to mention an engine of a special type erected at Halebeagle mine (North Downs) in 1796, the only pumping-engine without a main beam for

[1] B. & W. Colln. Drawings Portfolio, No. 575.

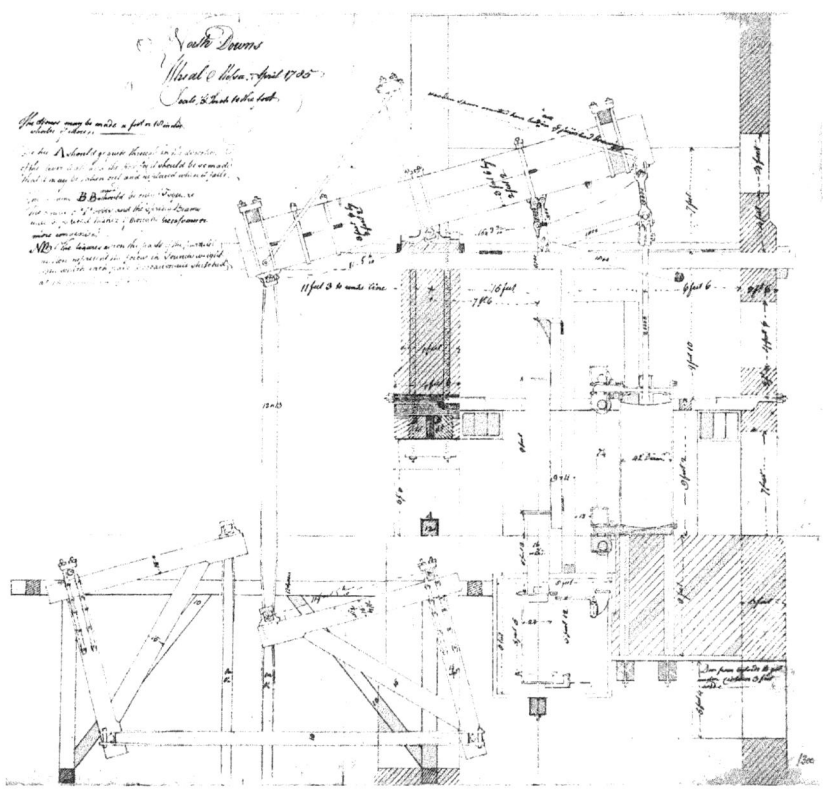

XXVII. DOUBLE-ACTING PUMPING-ENGINE, 42-INCH CYLINDER,
FOR WHEAL MESSA, NORTH DOWNS, CORNWALL

Drawing dated April 1785

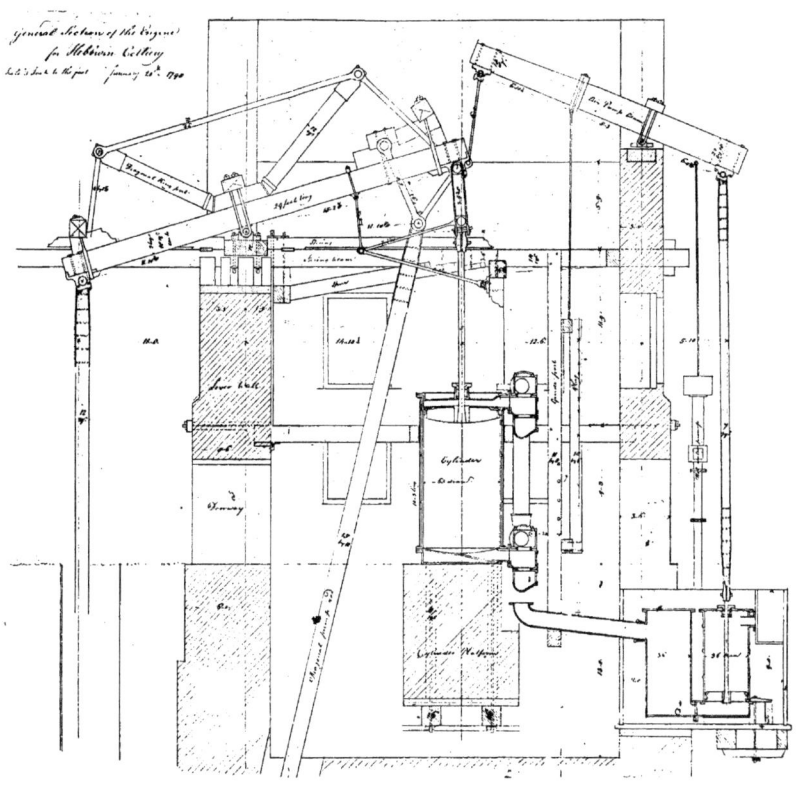

XXVIII. DOUBLE-ACTING PUMPING-ENGINE, 63-INCH CYLINDER,
FOR HEBBURN COLLIERY, DURHAM

Drawing dated Jan. 1798

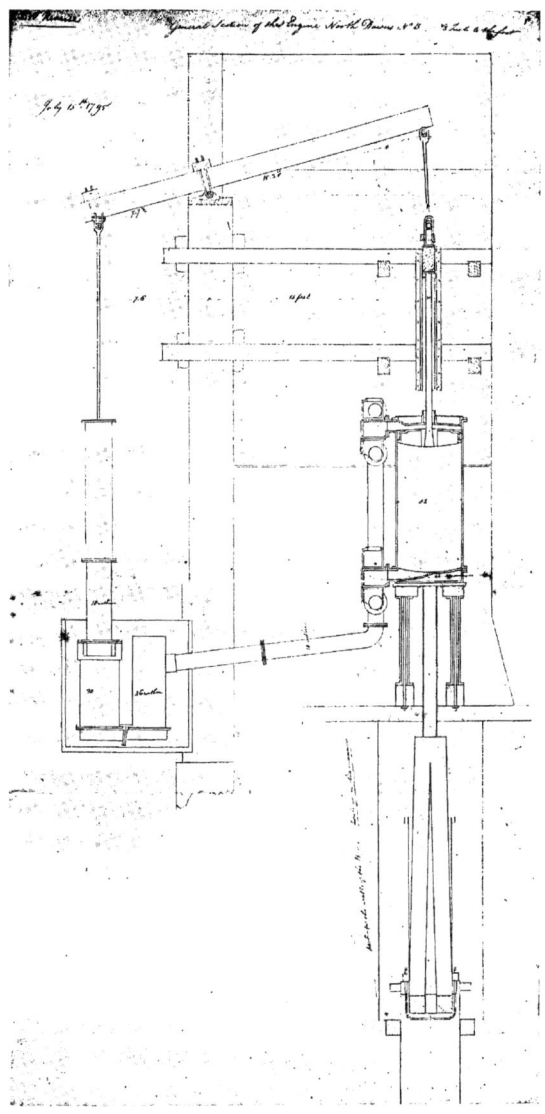

XXIX. DOUBLE-ACTING PUMPING-ENGINE, 52-INCH CYLINDER,
FOR HALEBEAGLE MINE, NORTH DOWNS, CORNWALL

Drawing dated July 15, 1795

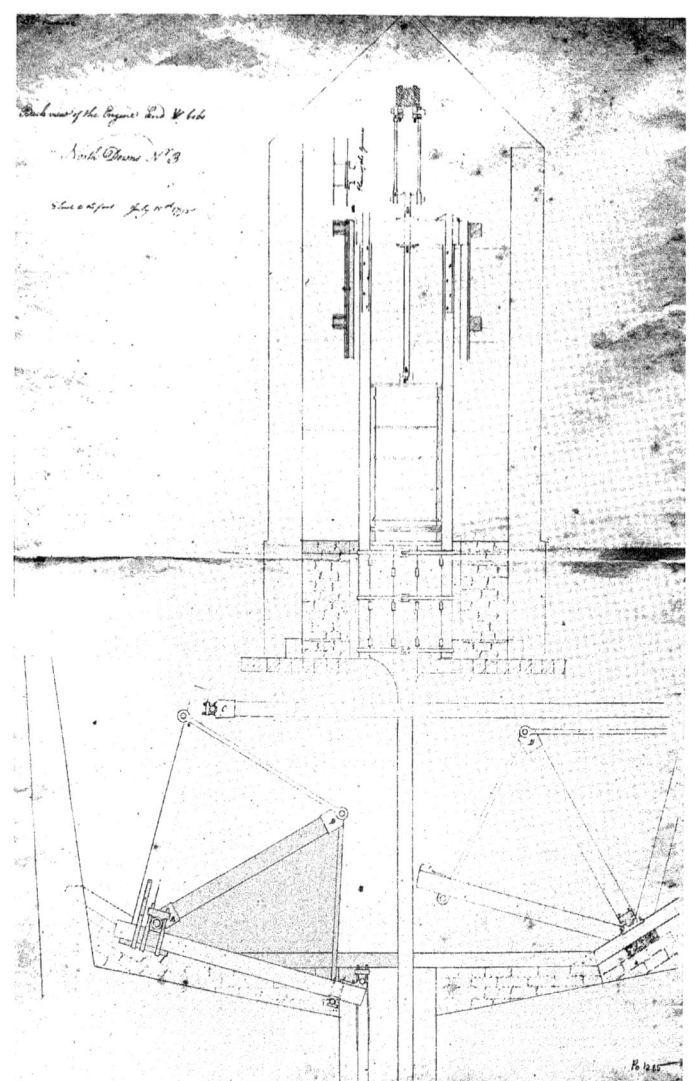

XXX. HALEBEAGLE ENGINE, END VIEW

which Boulton and Watt were responsible. It is not quite clear whether it was on account of particular local conditions, or whether it was to suit the wishes of the mine captain, but the cylinder is placed directly over the shaft. Bull had been putting up engines with the cylinder so disposed, but in his engines the piston-rod came out at the bottom of the cylinder and was directly coupled to the pump-rod, whereas in the Halebeagle engine the piston-rod comes out at the top and its upper end is secured in a cross-head that works in vertical guides and carries a pair of side-rods which work the pumps by means of V-bobs. The air-pump is worked by a beam from the top end of the piston-rod. The design, Plates XXIX, XXX, was based on a sketch sent in by Murdock, who at that time appears to have been the engineer of the North Downs mines. In the letter accompanying his sketch Murdock says that the captain of North Downs insists upon 8 ft. stroke in the pumps, ' which sets aside all our schemes in the old plan. Under you have a sketch for your inspection which I think will answer their purpose ; but the stroke in the cylinder must be the same as in the pumps, for the over reach of T. bobs will be great with a 9 feet.' [1] In a subsequent letter Murdock asks Watt what he thinks of the plan, and says that he is sorry to have had to trouble him with new schemes.[2] This engine was made 8 ft. stroke ; the cylinder was 52 in. diameter.

[1] Ibid. Murdock to Southern, 1795 June 7.   [2] Ibid. Murdock to Watt, 1795 June 14.

XIII

ROTARY MOTION IN THE STEAM ENGINE

THE ROTARY ENGINE

T HE idea of applying the power of steam for the production of rotary motion directly, instead of through a water wheel, to work mills and machinery was in the mind of Watt, as it was in that of Boulton, at quite an early date. But the plan both had in view was the rotary engine, or ' steam wheel '.

The application of his main invention, the separate condenser, to rotary engines is comprised in the Specification of Watt's patent of 1769, and the drawing intended for the Specification, but not filed, Plate XIV, shows the construction he had in view, viz. an annular chamber, carried on an horizontal shaft, fitted with valves, and charged at its lowest part with mercury or Newton's metal. Boulton was desirous of having one or more for his own use, and Watt offered him his advice and such assistance as he could render in constructing them, as well as a licence on very easy terms. Nothing was done at this time, but in 1774 a steam wheel was made and set to work at Soho. When the Bill for the extension of Watt's patent was before Parliament, Boulton, Samuel Garbett, and Joseph Harrison gave evidence before the Committee of the House of Commons. Boulton had tried one of Mr. Watt's steam-wheel engines, Garbett and Harrison had seen it at work, and Harrison added that the wheel was 6 ft. diam. and weighed upwards of a ton.[1]

The drawings laid before the Committee by Watt include sketches which, there can be no doubt, represent the construction of the rotary engine or steam wheel actually made. These sketches are reproduced in Plate XXXI. In the middle is an inscription in the handwriting of Watt, ' Steam Wheel, This drawing laid before the Committee of the House of Commons in 1774–5.' The construction, substantially the same as that shown in the earlier drawing, comprises a hollow annular chamber mounted on an horizontal shaft, but, as stated on the sketch, ' The arms of the wheel are not shown to prevent confusion.' The chamber is divided into three sectors by valves opening one way and each sector is. connected by a radial inclined pipe to a passage in the shaft ; the passages extend along the shaft and form ports at one of its ends which bears against a stationary nozzle or valve, shown separately in the ' view of nozzle ' ; the steam and exhaustion pipes are connected to the nozzle. The lower part of the chamber is charged with mercury, and steam from the boiler is introduced between the end of the mercury column and a valve. The radial pipes, as described on the drawing, are ' pipes which convey the steam from the boiler and into the condenser according to position '.

About the middle of 1775 Boulton informs a correspondent that a steam wheel

---

[1] Doldowlod Papers. ' Copy minutes on recommitment of Mr. Watt's Engine Bill.'

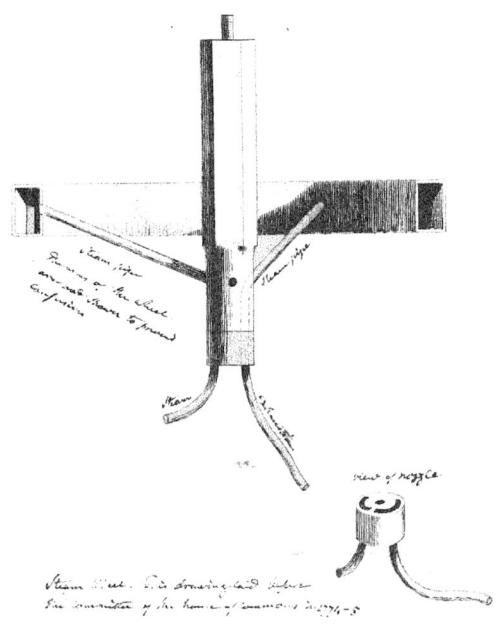

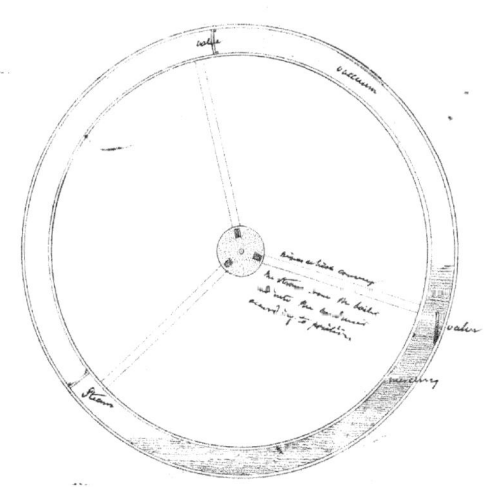

XXXI. STEAM WHEEL. DRAWING LAID BEFORE PARLIAMENT, 1775

Doldowlod Papers

had been made and another is in hand.   His letter is in reply to an inquiry for a steam wheel for turning a silk mill, and he writes :

' We have two sorts of steam engines, viz., the reciprocating engine which is particularly applicable to the raising of water to mills, or out of mines, of which kind we have one in constant work in Soho. . . . Secondly our steam wheel or rotative engine [1] is particularly applicable for all the purposes which water wheels and horse mills are apply'd, and it is of that species of engine that your friend wants, but we are less prepared to speak positively of prices of the rotative engine than we are of the reciprocating engine, having only executed one to satisfy ourselves of the practicability and utility of it, but have now another in hand which will enable us to be more precise in every point relative thereto.   I can only say in general terms that it may be apply'd conveniently and advantageously to the working of a silk mill, that the price will be in proportion to its power & that its power has no limits but that of price, so that the steam wheel may be made equal in power to 2 horses, 4, or 8, or any number & its diameter about 9 and in no case will exceed 11 feet.' [2]

Replying to another correspondent a few months later, Boulton writes :

' If you will suspend your horse mill for two months, I shall be able not only to answer you with precision but shall also be able to show you all you want, vizt., a steam wheel applyed to the turning of laps and grinding stones for the polishing of steel work.   The steam wheel is 9 feet diam^r and turns vertically, consequently the axis lies horizontally and is about 6 feet long ; there must be a little room for the boiler and the condenser. . . . This wheel is capable of working without any cold water to condense, but in that case it will require double the quantity of fuel.' [3]

Soon after this, early in 1776, Henderson, who subsequently entered the service of Boulton and Watt, comes upon the scene.   He had invented a rotary engine and submitted his scheme to the firm.   Boulton seems to have thought it was not so good as their own, and persuaded Henderson to give it up ; he informed him that the steam wheel they worked last year was equal to the power of two horses, but that the last they executed is equal to that of three horses, and as to fuel consumption, he says—' one hundredweight of the slack or small coals from Wednesbury coalpits applyed to our wheel engine, will raise 30,000 cubic feet of water 30 feet high, but ought to raise 36,000 & I doubt not when our workmen are a little more expert in execution that we shall raise the latter quantity.' [4]

We gather from these letters that a small engine had been worked, and that a larger one with a nine-foot wheel was being made, but although Boulton in his letter to Henderson refers to this as ' executed ' it is doubtful whether it was ever completed ; in 1777 the wheel engine was still lying in the yard at Soho and in 1779 patterns for parts of the steam wheel were sent to Wilkinson's foundry at Bradley.   In view of the pressure of work in connexion with pumping-engines, Watt himself could not have given very much time to the rotary engine ; but in 1779 we find him, in reply to a correspondent who had submitted a scheme for a steam wheel, writing :

' The first steam wheel I attempted was on that identical principle, and believe part of the model is now at Soho, but was altered, before any trial of it was made, into a reciprocating circular engine which made a continued rotative motion by wheels & rack work which also was never tryed though I thought it much more practicable than the other and is possessed

---

[1] Boulton frequently applies the term ' rotative ' to what we now term ' rotary ' engine.
[2] B. & W. Colln. Boulton to Samuel Roe, 1775 June 29.

[3] B. & W. Colln. : Letter Books. Boulton to John Collet, 1775 Oct. 18.
[4] Ibid. : Letter Books. Boulton to Henderson, 1776 Feb. 6.

of several advantages over other constructions, better plans at least in my opinion superseded both of them.' [1]

By this time, however, a Newcomen engine fitted with mechanism for converting the reciprocating motion into rotary had been set to work in a mill in Birmingham and no further work seems to have been done at Soho on Watt's rotary engines. Boulton, probably, was reluctant to give up the steam wheel; he had been very optimistic in 1776 when he informed Watt that if they had a hundred wheel engines ready he could dispose of them, so when a few years later one of the workmen—Cameron—came to him with a scheme, he allowed him to proceed to make an engine, or to attempt it, at the expense of the firm. Boulton did this without reference to his partner. Now it seems that Cameron had previously spoken of his scheme to Watt who had sought to discourage him from proceeding with it, pointing out that in so far as it had a separate condenser, and a piston pressed upon by steam, it would infringe his patent ; and saying, in regard to the special features of the engine, that the same idea had been present in his mind some years before and that he had made a model of it. Although it was found later on, Watt was unable at the moment to produce this model for inspection, and Cameron remained unconvinced ; one result was that in the specification of his patent of 1782 Watt included two forms of rotary engine, or rather a semi-rotary, and a rotary engine. The first had a radial vane or piston oscillating in a circular chamber between two fixed radial abutments. In the other the radial piston made complete revolutions and a radial valve or abutment was pivoted at the circumference of the chamber in such a manner that it could fold back upon the wall of the chamber to allow the piston to pass. The first form, no doubt, was what Cameron was working at, and it seems to be the same as the 'reciprocating circular engine' mentioned by Watt in the letter of 1779 referred to above.

Cameron was not able to make his engine work satisfactorily, and his attempts to remedy its defects had to be stopped by the firm. Nor was anything done with Watt's schemes, although early in 1783 Boulton was anxious for him to get working drawings made and sent to the founder.

THE APPLICATION OF THE CRANK TO THE STEAM ENGINE

HAVING disposed of Watt's rotary engine schemes, we can now turn to the production of rotary motion in the reciprocating engine. The idea was not by any means a new one, in fact Watt himself had inspected, about 1768, such an arrangement at Hartley Colliery, Seaton Delaval, Northumberland, used for winding coal and constructed by a man named Oxley.[2] Farey remarks of this engine that it went sluggishly and irregularly, that the mechanism was frequently deranged, and that, finally, the engine was converted into a pumping-engine to raise water for a water-wheel.[3]

It will not be necessary here to enter into a consideration of the various projects that had been put forward by different inventors, but the circumstances connected with one of them must be dealt with on account of their interest in connexion with the introduction of the crank.

[1] B. & W. Colln. : Letter Books. Watt to W. Chapman, 1779 May 23. In this letter Watt suggests a friction brake dynamometer for trying the power of the engine : 'this will be most conveniently done by friction on a round axis with a weight & steelyard.'
[2] Muirhead : *Mech. Inv.* III. 37 *n.*
[3] Farey : *Steam Engine*, 408.

It has been stated above that a Newcomen engine fitted with a mechanism for producing rotary motion was set up to drive a mill in Birmingham. This was in the year 1779 ; the mill was situated on Snow Hill and belonged to James Pickard. The engine had been built by Matthew Wasbrough of Bristol, and the reciprocating motion of the beam was converted into a rotary motion by a pawl-and-ratchet arrangement, the invention of Wasbrough and patented in 1779. The engine seems to have been at work in April 1779, when Watt writes to Chapman :

'Mr. Wasbrough has, I hear, got his engine set to work and assumed two new partners in this place, he says it works a merveille. I have not seen or heard any proper description of it ; he wants engines from us, but we must be better assured of the success of his scheme first. A method of applying a rotative motion to a reciprocating engine has lately occurred to me which will have the advantage of always being acted upon by an equal force, i.e., the uninterrupted power of gravity and will make no set at the return of the engine as I hear Matthew's must do. . . .'[1]

The correspondence in the Boulton and Watt Collection makes it clear that Wasbrough recognized the merits of Watt's invention and was anxious to have Boulton and Watt engines, or a licence to make them, in order that he might apply his own invention thereto. It was only because the Soho firm did not wish this, or were tardy in completing arrangements with him that he had to fall back upon the older form of engine. It appears too that Wasbrough had applied his invention to an engine for his own works at Bristol in February or March 1779, that is to say, before the starting of the Birmingham engine.[2]

The exact position as between Pickard, the mill-owner, and Wasbrough, the engineer, is obscure. Pickard himself invented a mechanism on similar lines to that of Wasbrough; possibly this was after the erection of the Snow Hill engine had been commenced, or even after the engine had been tried at work. At any rate on Apl. 28, 1779, he obtained a patent in Scotland 'for a mill or machine for turning, boring, milling grain, all kinds of milling, or any operation *quod mola facere potest rotativa motione*'. There is no corresponding patent for England and Wales. It appears that Wasbrough and Pickard had agreed to pool their inventions, and Wasbrough already had his patent for England and Wales ; it is dated Mar. 10, 1779. Samson Freeth of Birmingham seems to have put forward capital for the completing of these inventions, and William Chapman of Newcastle-upon-Tyne took some steps towards securing a licence for the counties of Northumberland, Durham, and York.[3] Chapman had one of Boulton and Watt's pumping-engines at Byker Colliery ; he carried on a correspondence with Watt for some time, and it may be that his withdrawal from a connexion with Wasbrough was due to Watt's views on the subject.

The story has grown up, with much picturesque detail, that Watt invented the application of the crank for the production of rotary motion in the steam engine, and that this invention was communicated through the treachery of one of the Soho workmen to a rival engineer, who patented it and thus by fraud deprived Watt for a number of years of the fruits of his ingenuity. In the following

---

[1] B. & W. Colln. Watt to W. Chapman, 1779 Apl. —.
[2] Ibid. Matthew Wasbrough to Boulton & Watt, 1778 July 25 ; 1779 Jan. 30 ; Apl. 3.

[3] See MSS. of R. B. Prosser in the Patent Office Library : 'Abstract of a deed (unexecuted) in the possession of Mr. Wasbrough, Solicitor, of Bristol, grandson of the Inventor.'

pages we shall see that Watt never claimed this invention, and in fact that he did not consider the mere application of the crank to be an invention, or at least a patentable invention, and that what he claimed to have invented, and alleged to have been stolen from him, was the combination of the crank and a particular arrangement of revolving counterweights, an arrangement which, if it was ever applied in practice, was soon abandoned as useless. The evidence certainly seems to confirm Watt's contention that this combination was stolen from him and a patent obtained for it by fraud.

Possibly the story took its birth from the fact that Boulton and Watt refrained from contesting this patent and from making use of the crank until it had run its term. Soho tradition would then take the form that it was the crank itself, that is to say, the single crank used with a flywheel, that Watt had invented, and writers on the subject in dealing with such of the Boulton and Watt correspondence as has been published would approach the matter with this idea in their minds. It must be admitted too that some of Watt's letters, written towards the close of his life, are liable to be misunderstood without a full knowledge of the facts. In what follows contemporary evidence is relied upon as far as possible.

The time was ripe for the introduction of an engine for driving mills; Boulton had said in 1776 that if he had a hundred rotary engines he could dispose of them. The Snow Hill engine soon began to draw a good deal of attention, and there can be no doubt that it was a subject of discussion between Boulton and Watt before the departure of the latter for Cornwall in the autumn of 1779. In September of that year Boulton was getting anxious about the matter, and by his direction Keir writes to Watt :

' Mr. Boulton wishes much that when you have leisure you would turn your thoughts to perfect your scheme of producing a circular motion from the reciprocating fire engine. He finds that many people who have no knowledge of the subject have taken it into their heads that Wasborough's improvement will put your engine out of fashion & although so absurd a notion cannot have any effect with those whose interest leads them to make any enquiry, yet he thinks that it may tend in some degree to affect the general Credit. He therefore would be glad to have a like machine put to the Soho engine.' [1]

Wasbrough's invention, however, was not turning out quite a success in practice ; various repairs and alterations had to be made, and then, towards the end of the year 1780, it was replaced altogether by a connecting-rod and crank. This important event is chronicled in a letter from Watt to Boulton (then in Cornwall) in November :

' Matthew Wasbrough has got a (single) crank to Snow Hill engine and it does very well, better than the rick rack. . . . I think you should call on Matthew as you return and let him know that we will dispute his having an exclusive right to these cranks.' [2]

Boulton soon after this advised Watt to instruct their London solicitor to watch whether any patent was being applied for by Wasbrough or Pickard that they ought to oppose, and he added : ' I think the double cylinder and crank better than any of them & if one were erected it wd crush these quacks & it is a very desirable thing to do.' [3]

Pickard had already obtained a patent, dated Aug. 23, 1780, the specification

[1] B. & W. Colln. J. Keir to Watt, 1779 Sept. 26.
[2] Boulton Papers. Watt to Boulton, 1780 Nov. 19. The word ' single ' before crank is an interlineation in the letter.
[3] B. & W. Colln. Boulton to Watt, 1780 Nov. 26.

of which was enrolled on Dec. 9th of the same year. This document is very brief, and the drawing too is a mere diagram, Fig. 18, but it makes it clear that he had a spear, or connecting-rod, proceeding from the engine beam to a crank at the end of a shaft. There is no reference to the use of a flywheel, but provision is made for carrying the crank over the dead-centres by a weight revolving with a shaft geared to run at twice the speed of the crank-shaft. The weight is applied to the shaft in such a position that it ' will always dissend when the crank and the spear are parallel or nearly so, and the said weight will ascend when they are at, or nearly at, right angles with each other '.

This patent started a long controversy, acrimonious on the part of Watt, who seems to have been furious at the grant, and in April of the following year, 1781, we find him writing : ' I know the contrivance is my own, and has been stolen from

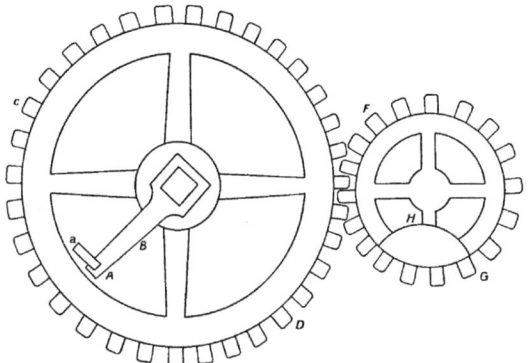

FIG. 18.   Drawing to Pickard's Specification

me by the most infamous means, and to add to the provocation, a patent surreptitiously obtained for it.' [1]   A week later, again in a letter to Boulton, he says : ' being thoroughly convinced in my own mind that they invented no part of it. They got the hint for the crank from Ned Ruston, who had it from a lathe, & the weight they got from us via Cartwright.' [2]   Precisely what Watt meant, if he meant anything in particular, in stating that the patent had been obtained surreptitiously, is not known.   In seeking a patent Pickard presumably followed the established course ; it was not incumbent upon him to inform Boulton and Watt or any other firm that he was doing so.

The allegation that the invention had been stolen from Watt, however, demands examination in the light of the information now available.

A document in the Boulton and Watt Collection, in the form of a statement apparently drawn up with a view to obtaining counsel's opinion, gives an account of the whole affair.   It is obviously inaccurate and in some respects inconsistent ; moreover, it is to be remembered that it is not strictly contemporaneous with the events, and that it presents the case from the Soho point of view.   According to

---

[1] Boulton Papers.   Watt to Boulton, 1781   Apl. 21.   [2] Doldowlod Papers : Letter Book, 1781 Apl. 28.

this statement, Watt had made a drawing and given a description of his invention of the application of the crank to reciprocating engines in the summer of 1779, and very soon after a working-model of it had been made at Soho and another at Newcastle-upon-Tyne. Then, some time in the spring of 1780, Richard Cartwright, one of the Soho workmen, ' who had often seen and worked their modell for producing a rotative motion by means of a crank', met at an ale-house one Sam Evans, the engine-man who had charge of the Snow Hill engine. Evans spoke of the troubles with his engine and of the defective working of the ratchet mechanism, whereupon Cartwright explained and made a sketch of the Soho model. Soon after this, Pickard took out a patent for the aforesaid invention and put in a crank at the Snow Hill engine. ' Saml Evans hath told many of his acquaintances that he was the inventor of the crank machine, and not his master Pickard,' and Cartwright has confessed that he disclosed the invention to Evans, and has given in a statement in writing of the conversation he had with him. The following is Cartwright's statement :

' I mett with Samuel Evens by accident at a publick house and hee described the motion of the engine at Snoe Hill to mee and told mee that their engine was often out of repare by making yuse of that motion that they worked with and I said I thought I cold make a mashine that wold work that engin withought those loose teeth and hee said howe and I said by a motion that I have seen of Mr. Bolton and Watts and I discribed the same motion and hee said howe was it to do it. I said by a crank with a weell wich turns a roller well twice round for the hothers one wich well carries a waight to counter balance the crank and he said my masters have been at tooe much expence already that I daresay they will not put themselves to no mor expence about it ; these words was past between Samuell Evens and mee and to the best of my nolige was in the month of May 1780.

Richard Cartwright.'

Endorsed in the handwriting of James Watt : ' Cartwright's confession of what he told Samuel Evans about Crank engine 1780.'

The document contains a description of the invention drawn up with reference to a drawing ; the drawing is missing, but the description agrees with that in Pickard's specification in that it has a revolving weight on a shaft geared to run at twice the speed of the crank-shaft ; but it has another revolving weight on the crank-shaft itself, which ' it happened by accident that the weight c had fallen off the wheel A when R. C. [Cartwright] inspected the modell & he not being aware of the use of it described the machine to Messrs Wasbrough & Pickard's engineer without that weight c & they have left it out in the specification of their pretended invention '.

Now this is an allegation, not that the idea of using a crank, but that the revolving weight, or the combination of crank and revolving weight covered by Pickard's specification, had been stolen from Watt. On the other side it was said that the invention was the joint production of Wasbrough, Pickard, and Stead, and that it had been made at Bristol. Stead on his own part stated that he had himself thought of applying the crank fourteen years before, and had then made the machine in part.[1]

Indeed it amounts to the same thing as the statements in Watt's letters quoted above. Watt when he first heard of the application of the crank does not consider himself aggrieved, but after he has seen Pickard's specification, he says that the invention, covered by the specification, has been stolen from him, and then later

---

[1] B. & W. Colln.   Boulton to Watt, 1781 Apl. 23.

is still more explicit : ' They got the hint of the crank from Ned Ruston, who had it from a lathe, & the weight they got from us via Cartwright.'

Upon hearing of the erection of the Snow Hill engine it was a natural thing for Watt to turn his attention to the problem of the production of rotary motion in the reciprocating engine ; equally natural was it that he should consider whether the crank could not be adapted for that purpose, and we have definite evidence that he did so turn his attention and that he did consider the application of the crank in the course of the year 1779, in which the Snow Hill engine was erected.

In his Journal under the date Sept. 15, 1779, we read :

' Mr. Chapman arrived . . . mentioned to him means of producing circular motion by means of an axis parallel to the lever and by teeth on the sides of the arches acting on two ratchet wheels, which told him he might take a patent for and enjoy in the counties of Northumberland and Durham, we to have it in the others.' [1]

The scheme was to have a shaft lying parallel to and at the level of the engine beam, toothed segments on the arches at the ends of the beam engaging with wheels coupled to the shaft by ratchet clutches. This was probably Watt's first scheme ; one can imagine from the fact of his having offered it to Chapman that he had a better plan in mind, and it is likely that he did at this time show him the model of the double-cylinder crank engine. At any rate an entry in the Journal on December 5th explicitly mentions ' cranks ' and ' crank-engine ' :

' Wrote to Mr. Chapman in ans^r to his of 3rd Nov.—that the only thing he had added to my crank engine was the Substituting a wheel in place of the lever, and placing the pin wch led the crank at the same radius with that of the crank, which I conceived to be no improvement, and had formerly been considered by myself and disapproved—that it was not true I had tryed experiments on it & had not brought it to answer, that I had desired H^n. [Henderson] to tell him of it to prevent him from wasting time & money on other schemes which were not so good, and not calling for assistance—mentioned that I could make a rotative motion from a fire engine by cranks which would have the power exceedingly nearly equal in all its parts without the help of a flywheel—would turn either forward or backward at pleasure, had no wheels and pinions, ratchets or catches, would make the best use of the steam & would use steam proportional to the power required.'

This correspondence, it must be understood, is of date about a year before the crank was applied to Pickard's engine.

But why, it may be asked, if Watt had invented and had made a model of the contrivance described in Pickard's specification as early as the middle of 1779, had he taken no further steps, either by taking out a patent or by applying it in practice, for instance, on the Soho engine ? The firm recognized this weakness in their case and the statement above referred to informs us, in regard to the failure to apply the invention in practice, that

' the only reason why Boulton & Watt did not execute this application of the crank to the aforesaid purposes was owing intirely to their being so much employed in re-erecting all the old fire engines in Cornwall upon their new principle . . . but it was their intention to have given the rotative engine to the publick & have carried it into execution & apply'd it to most kinds of mills, as soon as they had finished the Cornish engines, which will be in the present year.'

There can be no question that the Cornish business was keeping Watt pretty busy, but did he really think that any one of his schemes was likely to be a satisfactory

[1] Doldowlod Papers : Watt's Journal.

solution of the problem in practice ? It is perfectly clear that at this time, although Boulton thought otherwise, his own opinion was that the rotative engines were not worth spending his time upon in view of the extra trouble and work entailed in comparison with pumping-engines.

Whatever he may have thought later, it is doubtful whether at first Watt considered the idea of applying the crank and connecting-rod to the steam engine to be a patentable invention. The mechanism was in common use for the conversion of rotary into reciprocating motion and it had been used for conversion in the opposite sense. It was one of the contrivances that an inventor would naturally consider in the first instance. The difficulty was to grasp the fact that it could be applied with success to the engine in the only form in which it was then known, the single-acting pumping-engine with its ponderous masses moving slowly, and more or less uniformly up and down, sometimes making a stroke of full length, and sometimes a much shorter one. We find this doubt set forth in a paper presented by John Stewart to the Royal Society in 1777, just a few years before the problem was actually solved. Stewart had invented a mechanism for converting the reciprocating motion of a steam engine into rotary motion, involving a pair of chains and ratchet gear, and in this paper he remarks that the use of a crank for obtaining rotary motion is a plan that naturally occurs in theory, but, he goes on to say, ' in practice he thought it would be impossible from the nature of the motion of the engine, which depends on the force of steam and cannot be ascertained in its length, therefore on the first variation the machine would be either broken in pieces or turned back.' [1]

Stewart was acquainted with the action of a flywheel and indeed proposed to use one with his contrivance, but it had not dawned upon his mind that a flywheel and crank could be applied with success to control and limit the movement of the reciprocating parts. It is not surprising then to find that inventors endeavoured to solve the problem by means which, while producing uniform rotary motion in one direction, would allow the reciprocating parts to retain the motions seen in existing engines, as by the use of some form of ratchet mechanism. Wasbrough's invention was of this nature. In the then existing state of knowledge it would appear that the application of a crank and connecting-rod did not present itself as a satisfactory solution of the problem, and one cannot avoid thinking that this applied as much to Watt as to other inventors. The tone of the letter in which he informs Boulton of the application of the crank to the Snow Hill engine is one of pleased interest, he expresses no surprise or objection, but merely says he will dispute their exclusive right to the crank. In other words, he was content for Wasbrough and Pickard to try the crank in actual work and establish its practicability, and even to use it afterwards, so long as he was not precluded from using it himself. But even after he knew that it was working successfully Watt was not at once convinced that it was the most satisfactory solution of the problem ; in April 1781 we find him writing : ' I know from experiment that the other contrivance, which you saw me try, performs at least as well, and has in fact many advantages over the crank.' [2]

Indeed, later on in life, Watt candidly admitted that he had not appreciated the effect of the flywheel, and gave Wasbrough the credit of having been the first to apply it to the steam engine.[3]

[1] Farey : *Steam Engine*, 408.   Apl. 21.
[2] Boulton Papers. Watt to Boulton, 1781   [3] Muirhead : *Mech. Inv.* III. 37 *n.*

The Snow Hill engine had a flywheel, and when Wasbrough's ratchet gear was replaced by the crank, then, as Farey says, the engine answered so much better than anything which had been tried before that the same principle has been followed ever since.

The merit of Pickard and Wasbrough lies, not in conceiving the idea of using a crank, but in that they had the courage to apply it to the engine and to take the risk of a break-down which was what the prevailing ideas pointed to. Once a crank had been fixed to a shaft provided with a sufficient flywheel, the problem was solved. Whether the high-speed revolving weight of Pickard's specification was applied is doubtful. Farey says that a simple crank was substituted for the ratchet work, and that ' it does not appear that Mr. Pickard's patent was acted upon, although engines with a simple crank and flywheel began to be used very soon after '.

Again one judges from the letter already referred to (p. 150) that Watt had seen the Snow Hill engine at work, and had it been fitted with the revolving weight he would certainly have noticed it and commented upon the fact. As we have seen, it was when he found that a patent had been obtained, the specification of which covered the construction he claimed to have invented, that he began to object.

The exact position in regard to Pickard's patent is not at all clear. The specification, as we have seen, makes no mention of a flywheel, but describes the combination of a crank with a revolving weight on a shaft running at twice the speed of the crank-shaft. It would seem, however, that Pickard and Wasbrough considered, and that it was generally assumed, that the patent covered broadly the use of a crank in a steam engine. This was never tested in a court of law, and Watt, although at first he thought that a patent for the broad idea could not be sustained and did not hesitate to introduce several crank devices into the specification of his patent of 1781, later on thought otherwise and strongly objected to the use of any crank arrangement on the Soho engines, on the ground that it would lay them open to an action at law. At the outset Boulton and Watt talked of having Pickard's patent revoked, but whether they really meant to proceed to extremities in this direction or were merely holding out the threat to bring pressure to bear on the Snow Hill people is not clear. Various attempts were made to come to terms. Pickard and his partners were perfectly willing to give Boulton and Watt the free use of their crank patent in return for the right of using Watt's invention, and apparently they were willing to pay for this. Watt, however, would be satisfied with nothing short of an unconditional assignment of half the patent rights ; as this would be tantamount to an admission that they had obtained the invention by fraud, it is not surprising that the Snow Hill people did not agree. The negotiations seem to have gone on for some time but came to nothing in the end. The crank was applied to quite a number of atmospheric engines, but whether Pickard received any royalties in respect of them is not known.[1] Wasbrough died in 1781

---

[1] ' Sam Evans has eloped from the Snow Hill Company and is erecting a rotative engine with a crank for a gun maker at Liverpool.' (Boulton Papers. Watt to Boulton, 1783 Oct. 18.)

' Sam Evans is chased away with the utmost disgrace by his employers at Liverpool. It is said that his mill was a Gentlemanly mill, it would go when it had nothing to do, but refused to do any work ; they say it has nearly ruined the owner.' (Ibid. Watt to Boulton, 1784 Aug. 17.)

' Banks & Onions have applied a crank to their engine without asking consent of the Snow Hill people.' (Ibid. Watt to Boulton, 1784 Oct. 14.)

and with him, no doubt, the combination lost its mainspring. Pickard lived for some years after the lapse of his patent.

There has been some doubt as to whether his mill was a flour-mill, a rolling-mill, or a mill for grinding metals. A note discovered in Watt's ' Blotting and Calculation Book, 1782 & 1783 ',[1] sets this question at rest ; the mill, at any rate at the period we have been considering, was for grinding metals, but it included a set of rolls for flatting metals.

It seems that later on it was converted into a flour-mill, for in the Birmingham Directory of 1801 James Pickard appears as a ' miller, malster and coal merchant (steam mill) Snow Hill'. In the Directory of 1777 and in his patent specification of 1780 his calling is stated to be that of a button-maker.

From an account given in a local newspaper, it appears that the mill stood at the corner of Snow Hill and Water Street, and that the engine, much altered and repaired no doubt, was in position and in good working order up to about the year 1879, when it was sold and broken up.[2]

WATT'S SCHEMES TO AVOID PICKARD'S PATENT

Up to the date of Pickard's patent Watt had devised at least two schemes involving cranks for converting reciprocating into rotary motion ; the one that which was covered in part by Pickard's specification, and the other a two-cylinder engine with cranks at an angle and a counterweight on the shaft. This, no doubt, is the plan referred to in Watt's letter to Chapman on Dec. 5, 1779, in which he mentioned that he could produce a rotary motion by a crank arrangement ' which would have the power exceedingly nearly equal in all its parts without the help of a flywheel, would turn either forwards or backwards at pleasure, and had no wheels and pinions, ratchets or catches '. It is clearly the same as that referred to by Boulton in his letter of Nov. 26, 1780 : ' I think the double cylinder and crank better than any of them.'

Watt now set about devising other plans for circumventing Pickard's patent ; by the autumn of 1781 he had contrived a number of devices for producing rotary motion and on October 25 a patent was granted him for ' Certain new methods of applying the vibrating or reciprocating motion of steam or fire engines, to produce a continued rotative or circular motion round an axis or centre, and thereby to give motion to the wheels of mills or other machines'. The specification of this patent, enrolled Feb. 23, 1782, comprises five different methods for converting the reciprocating motion of the beam into rotary motion. The first is a swash-plate device, or ' ecliptic ' ; this was made and applied to the Soho engine, and Boulton seems to have been charmed with its working. However, as is perhaps not surprising, it did not come into extended use. Plate XXXII shows a model of

---

[1] B. & W. Colln. This is a manuscript book of foolscap size. The note (on p. 29) is headed ' Snow Hill engine, from Mr. Southern, senior.' and reads : ' Cyl^r 30 inch diameter. Beam 22 ft. long. Crank 3 feet 7 inch—goes from 12 to 20 strokes per minute (at the time he saw it about 16) ; turns a grindstone 5 ft. 6 dia^r wch when new weighed 3 tons, 4 cwt.—thickness about 22 inches, for grinding Gun barrels— 22 laps for grinding buckles, chapes, &c.,—45

spindles for brushing and polishing, one large lap for dressing rolls—6 pairs of rollers for flatting metal (vizt. 2 p^r 5 inch, 2 p^r 7 inch, and 2 of 15 inches in length)—Every strap or brush pays 2/6 per week ; burns 22 to 24 cwt of Bloomfield coal in 12 hours, including smith's shop, lifts condensing water 13 feet.'
[2] *Birmingham Weekly Mercury*, 1886 Oct. 23, A Chapter of Local History.

XXXII. MODEL OF SWASHPLATE MECHANISM AT THE SCIENCE MUSEUM, SOUTH KENSINGTON

this mechanism applied to work a winding-drum, included in the Watt Collection at the Science Museum, South Kensington. The second method consists of an eccentric wheel, ' applying the power of the engine to pull, push or press a friction wheel or wheels against the external or internal circumference of a circular, oval, or double-spiral wheel'. Next we have a disk crank weighted on the crank-pin side to equalize the driving force (the engines so far were single-acting). The fourth arrangement is the combination of two engines with cranks at an angle referred to above. The fifth is the well-known ' Sun-and-planet gear '. The first we hear of the sun-and-planet gear is in Watt's letter of Jan. 3, 1782 :

' I wrote to you on the 31st, since which I have tried a model of one of my old plans of rotative engines revived and executed by W. M. [Murdock] and which merits being included in the specification as a fifth method, for which purpose I shall send a drawing and description next post. It has the singular property of going twice round for each stroke of the engine and may be made to go oftener round if required without additional machinery.' [1]

Boulton at once made a model and was delighted with it. This contrivance was got into the specification at the last moment, and, as is well known, it was a characteristic feature of the Boulton and Watt engines for many years. The sun-and-planet gear will be described more fully in a later chapter.

Of the crank arrangements, the one that is of real importance is the arrangement of two cranks set at an angle of 120 degrees on the same shaft and worked by two cylinders. In July 1781 John Wilkinson was anxious to have a forge worked by a steam engine and Boulton wanted to supply him with a two-cylinder engine with cranks so arranged. He tells Watt :

' I have got a very pretty modell which is either 2 forges, or, when I please is 1 rolling mill & I am persuaded it will answer in great . . . Wilkinson has seen it & nobody else, not even the workman who made it. It is ye double crank worked with 2 engines. I think 2 thirty inch cylinders will work 2 such forges. You have so much more important & full employ abt. ye mines that I don't think you shall be troubled with projects but if you will consent for me and Wilkinson to hammer out a pair of forges you shall have non of the dishonor, plague, trouble, dispute, or expence & you shall have all the profit. I think if we make expts. upon anybody they should be made upon Wilkinson, as he is doubly interested in the expts. Perhaps the first forge may not be the best possible, but I am sure of making a good one. Nothing shall be done without previously acquainting you, nor then, if you object. You can neither loose reputation nor money & you are sure to gain by Wilkinson's experience. He is so very hot upon it that I don't think it possible to appease him without erecting ye forge. It must be done or quarrell.' [2]

A letter of a week later gives a sketch of the arrangement of engines and hammers and states that it is proposed to erect for Wilkinson at Willey two engines with cylinders 27 in. diam., with 7 ft. strokes, to work a 6 cwt. hammer and a 2 cwt. hammer 120 blows per minute, the whole framing of the forge to be of cast iron and the anvil block to be 10 tons weight. Wilkinson, it appears, preferred the tilting to the lifting hammer.[3]

Watt, however, objected to the use of the cranks; he was afraid of an action for infringement. Boulton attempts to reassure him : ' There is no danger of we having a law suit about cranks because we shall neither make nor vend. J. W.

---

[1] Boulton Papers. Watt to Boulton, 1782 July 10.
Jan. 3.
[2] B. & W. Colln. Boulton to Watt, 1781
[3] Ibid. Boulton to Watt, 1781 July 17.

will take ye whole of the contest off our hands wch. is one reason why I prefer advising him to make the double crank. You refine too nicely to suppose they can claim any right to any part of it & I think it an opportunity that should not be lost.' [1]

A day or two later he returns to the charge saying :

' It is impossible we can have any law suit about ye double engine forge. We give a licence for ye engine but 'tis Wilkinson that makes & uses the forge & he alone is willing to support any law play that may arise, but instead of a law suit I perceive it is the only means of bringing about a reconciliation with Snow Hill engine & if you will not include that in your patent I shall be obliged to take out a patent for it, presuming you will not prosecute me for so doing.' [2]

However, Watt's views prevailed and, as we shall see, Wilkinson's forge engine was not made with cranks.

[1] B. & W. Colln. Boulton to Watt, 1781   [2] Ibid.   Boulton to Watt, 1781 July 26. July 24.

XIV

THE ROTATIVE ENGINE

THE patent for Watt's invention of ' Certain new methods of applying the vibrating or reciprocating motion of Steam or fire engines to produce a continued rotative or circular motion round an axis or centre, and thereby to give motion to the wheels of mills or other machines' was issued in 1781, and Watt signed the specification on Feb. 13, 1782. Of the five constructions described in the specification, two are unmistakably cranks; but although he had included them in his specification, Watt, in view of Pickard's patent, would not take the risk of applying them in practice. There remained the swashplate, the eccentric, which after all was but a crank in disguise, and the sun-and-planet gear. It is worthy of remark that, at the outset, Watt was undecided as to the relative merits of these three mechanisms. The first actually applied in practice was the swashplate, and in his first design for a large engine he employed the eccentric; then in the second practical application he made use of the sun-and-planet gear. The first trials with this gear were not entirely satisfactory and he then inclined to the idea that the eccentric would work better. However, the defects in the gear were remedied until a more or less satisfactory result was obtained, and the eccentric was never tried on the large scale, and it is doubtful whether the swashplate was ever applied to an engine built for a customer.

The first sun-and-planet engine was set up at Soho towards the end of the year 1782. By this time the call for rotative engines had become very insistent. In June 1781 we have Boulton saying that ' the people in London, Manchester, and Birmingham are steam mill mad ', and Watt, in September 1782, exclaiming ' Surely the devil of rotations is afoot'. Watt was now hard at work on rotative engines, dealing with schemes and with inquiries from people who wanted engines. The ' Blotting and Calculation Book 1782 & 1783' shows him making experiments on the friction of the engine, calculating the size of flywheels, and working out the power required to drive corn-mills, cotton-mills, and mills for rasping and grinding logwood. In October 1782 we find him complaining to Boulton that the rotatives have taken up all his time and attention for months so that he can scarcely say that he has done anything which can be called business. Watt's idea was that even if they succeeded in contriving a satisfactory engine to drive mills, there was no prospect of any financial return commensurate with the trouble and work involved. In reference to an inquiry for an engine for Richard Reynolds of Ketley, he writes to his partner : ' by any computation I have yet made of the mill for Reynolds, I cannot make it come to more than 20*l* per annum.' It will be understood that the sum mentioned is the royalty that could be charged on the engine.

The first rotative engine erected outside the Soho establishment was that put up for John Wilkinson at Bradley to work a hammer ; it was in operation at the end of March 1783. Watt was now engaged on the double-acting engine and over a year elapsed before the next rotative was set going, this was a winding-engine in Cornwall ; but now the sun-and-planet engine had commenced its career in earnest and several other engines were put up in 1784, including the double-acting engine for Cotes and Jarratt, the first engine with the straight-line linkage. Some of these first engines were single-acting, but within a few years the design had become standardized to a large extent and we have the double-acting engine with parallel motion, sun-and-planet gear, valves worked, as in the pumping-engine, from the beam, wooden beams and connecting-rods, and, in the smaller engines, wooden frames.

Then in 1788 we have the governor making its appearance, and in 1791 the crank beginning to replace the sun-and planet gear. The use of iron for beams does not begin until 1800 and the valve gear continues to be worked from the beam until about the same period. With the exception of three small engines, made just at the end of our period, all the engines are beam engines.

After this rapid glance over the field we may now retrace our steps and go over the ground in greater detail.

In May 1777, just after the 33-inch engine had been set to work at Soho, the original 18-inch engine was arranged to work a hammer, how, we do not know, for this was before the conversion of reciprocating into rotary motion had become a practical issue. The specification of Watt's patent of 1784 shows a direct connexion by means of a link between the engine beam. and a helve ; it is possible that this was the plan adopted in 1777.

This application aroused a good deal of interest at the moment, but the project was soon dropped, possibly the mechanism knocked itself to pieces, and the engine was lying idle for some time. Then in April 1781 it is proposed to apply it to roll tin for the wetting boxes of the letter-copying apparatus, and by July it had been fitted with the swashplate or ' ecliptic ' rotary motion. It does not seem that the engine was ever applied to the purpose mentioned, and we hear nothing further of this particular application of the swashplate mechanism.

The next engine we have to consider is the two-cylinder compound engine of which Watt sent a drawing from Cornwall in September 1781 (see Chapter X). This drawing (Plates XXIII and XXIV) does not show the rotative mechanism, but it was the intention from the first to set the engine to work a hammer, and in October we find Boulton referring to it as the ' double devil '. The engine was, in part, erected in November, the rotative mechanism had not then been fitted ; for in the following January we have Watt making the suggestion that the two engines should have different mechanisms in order to see which lost the least power. The engine was in fact composed of two separate engines the cylinders of which could be connected to work compound, so that it would have been quite easy to provide each with different mechanisms ; whether this was actually done does not appear. However, we hear nothing more about this engine for some months, then early in November 1782 the castings for a hammer are ordered and in a few weeks the hammer was at work. From the correspondence between Boulton and Watt it is quite clear that one cylinder, or one engine, had coupled to the outer end of its beam a pump-rod and a connecting-rod, and that the connecting-rod by means of a sun-and-planet gear imparted rotary motion to a shaft carrying the cam-ring for working the hammer. At first Watt was not quite satisfied with the sun-and-planet gear, he found that the revolving wheel had a tremulous motion that he did not like ; he proposed to try to remedy this by strengthening the swan-neck part of the connecting-rod, but he found that this gear did not admit of being made so strong

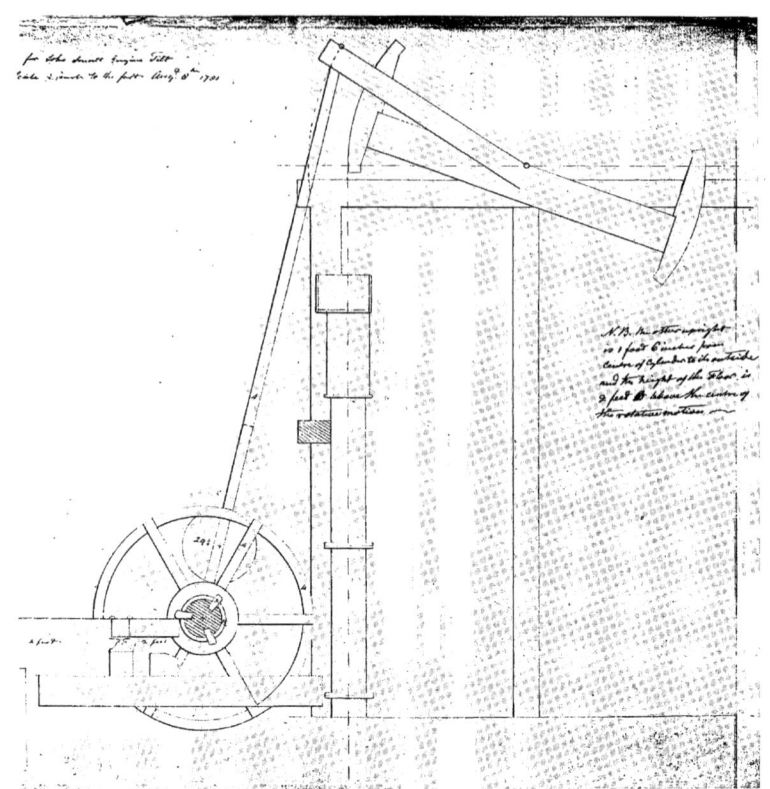

XXXIII. 'SOHO SMALL ENGINE TILT'

as the outside eccentric motion which he then thought would be the best motion for forges and for great powers. However, he seems to have met the difficulty, for, as we shall see, the Bradley forge engine was made with sun-and-planet gear.

It was found necessary to apply a catch to the flywheel to prevent the engine from setting off the wrong way at starting ; this catch was lifted out of action when the engine was at work and thrown in again before it was stopped. Watt also applied some improvements in the working-gear for the top valve to facilitate control by hand.

As we have seen, a rotative motion had been fitted to the 18-inch pumping-engine, but this was the first engine built at Soho as a rotative (it was a pumping-engine as well) and the first to have the sun-and-planet gear. It was a single-acting engine with a cylinder 15 in. diam. and 4 ft. stroke. As to the second cylinder or engine, we have very little information, but it seems that at first it was coupled to the other to work compound, for we have Watt writing, in reference to the steam consumption at one of the trials, that the boiler supplied plenty of steam ' which we attribute to the communication being cut off with the sleeping cylinder which it seems, somehow or other, destroyed much steam '.[1] The second cylinder was of the same dimensions as the first.

The hammer seems to have been put up for experimental purposes, and it was not continued in use at Soho, neither was the double-cylinder engine continued in use as such. It was converted into two separate engines which were sold, the one to Cotes and Jarratt and the other to Goodwin.

The drawing, reproduced from the Boulton and Watt Collection in Plate XXXIII and entitled ' Drawing for Soho Small Engine Tilt ', shows the outer end of a beam working a pump and imparting motion to a rotative cam-shaft by a connecting-rod and what appears to be a sun-and-planet motion. This is very probably a drawing of the engine we have been discussing, but there is a difficulty as to this—it is dated Aug. 8, 1781, which is a date anterior to any we have in reference to the engine and to the invention of the sun-and-planet gear. It is possible that the draughtsman made a mistake in his date and that the year should have been given as 1782 ; the drawing would then fit in very well with the facts as we know them from the letters. (Another drawing, dated August 1782 and headed ' First drawing of small reciprocating engine', shows sun-and-planet gear of the internal type.) Some support to the idea that the drawing does represent the actual construction is afforded by the terms of the note thereon, in Watt's handwriting : ' N.B. the other upright is 1 foot 6 inches from centre of cylinder to its outside and the height of the Floor is 2 feet above the centre of the rotative motion.'

In October 1783 Watt made an experiment with the forge engine to determine the power required to work the rotative motion and shaft when running idle, but with the pump in operation. With the rotative mechanism disconnected he found the coal consumption to be 0·235 cwt. per hour for 20 strokes per minute ; with the mechanism connected it was 0·28 cwt. In both cases the boiler evaporated 11 cubic ft. per cwt.

The next design we have to consider is that for Wilkinson's forge engine. Watt, then in Cornwall, sent a set of drawings to Bersham in February 1782. The engine was single-acting with a cylinder 42 in. diam. for a stroke of 6 ft. ; the condenser and air-pump were adjacent to the cylinder, and the beam had a heavy cast-iron weight at its outer end. The device shown for converting the reciprocating motion

[1] This and other letters referring to this engine are printed in Muirhead : *Mech. Inv.* II. 163 et seq.

of the beam into rotary motion was the 'outside eccentric', an eccentric wheel on the shaft embraced by a frame suspended from the beam and fitted with anti-friction rollers. In the letter accompanying the drawings Watt gives full instructions about the construction of this mechanism and explains that—'we have several other circular motions, but this is preferred because it goes upon the middle of a shaft and most of the others must be at the end of a shaft.' [1] Whether any steps were taken to carry out this scheme is not known. The sun-and-planet gear had been invented about a month before the date of the drawings and in the following July drawings of this gear were prepared for an engine to be erected at Wilkinson's Bradley forge, and there can be no question that this was the gear with which that engine went to work in the spring of the following year, 1783, an event recorded by Watt in his Journal under the date March 25th when he 'called at Bradley and saw the rotative motion go but not the hammer'.[2] The drawing reproduced in Plate XXXIV no doubt represents the construction of the engine and the hammer as finally adopted.

This was the first sun-and-planet and the first rotative engine put up outside Soho.[3] Of the alternative plans covered by Watt's specification the only one applied in practice was the 'ecliptic' or swashplate arrangement which, as stated above, was tried on the Soho engine. The drawings, dated February 1782, of a coal-winding engine for Wilkinson at Bradley show this mechanism, but whether it was actually applied in this instance is not clear.

Very soon after the Bradley forge engine had been started Watt had the first double-acting rotative engine at work at Soho. In order to show that the engine could be applied to work corn-mills, Boulton had one put up at Soho, and this engine was applied to drive it ; it had a cylinder 18 in. diam. and a short stroke. There is some doubt as to the exact length of the stroke ; Watt in one letter gives it as 18 in., in another as 24, while the manuscript book of 'Engines made at Soho'[4] puts it at 22½ in. The connexion between the piston-rod and the beam was by rack and sector, and the sun-and-planet gear was of the internal type, i.e. the sun-wheel was an internal wheel, a construction which is shown in the drawings to the specification of Watt's 1781 patent ; this gear was replaced in August 1785 by 'an outside rotative motion', the normal form ; at the same time the nozzles were fitted with inverted valves. The engine was tried at the end of March 1783, but the mill was not completed until September, when Watt was able to report that the engine goes better than ever, but burns too much coal. In the manuscript book just referred to, it is stated that when the corn-mill was taken down the engine was applied to work lathes and drills at Soho Manufactory, and that it was broken up in 1803.

At the time the corn-mill engine was being completed work had been begun on a small double-acting engine with a 12-inch cylinder, 12 in. stroke. This engine is said to have been put up for experimental purposes, and the drawings show two features which are very remarkable for this date—a trunk piston [5] and

[1] Doldowlod Papers. Watt to John Turner, 1782 Feb. 16.
[2] Ibid. : Watt's Journal.
[3] Smiles (*Boulton and Watt*, 1865, 327) states that the first rotary (meaning rotative) engine was made for Mr. Reynolds of Ketley, towards the end of 1782, and was used to drive a corn-mill. There is no doubt that Watt did make

calculations to determine the amount of royalty that could be asked for such an engine, but the present writers have failed to discover any evidence that the engine was actually made, at any rate at this period.
[4] B. & W. Colln.
[5] The trunk piston is shown in the drawings to the specification of Watt's patent of 1784.

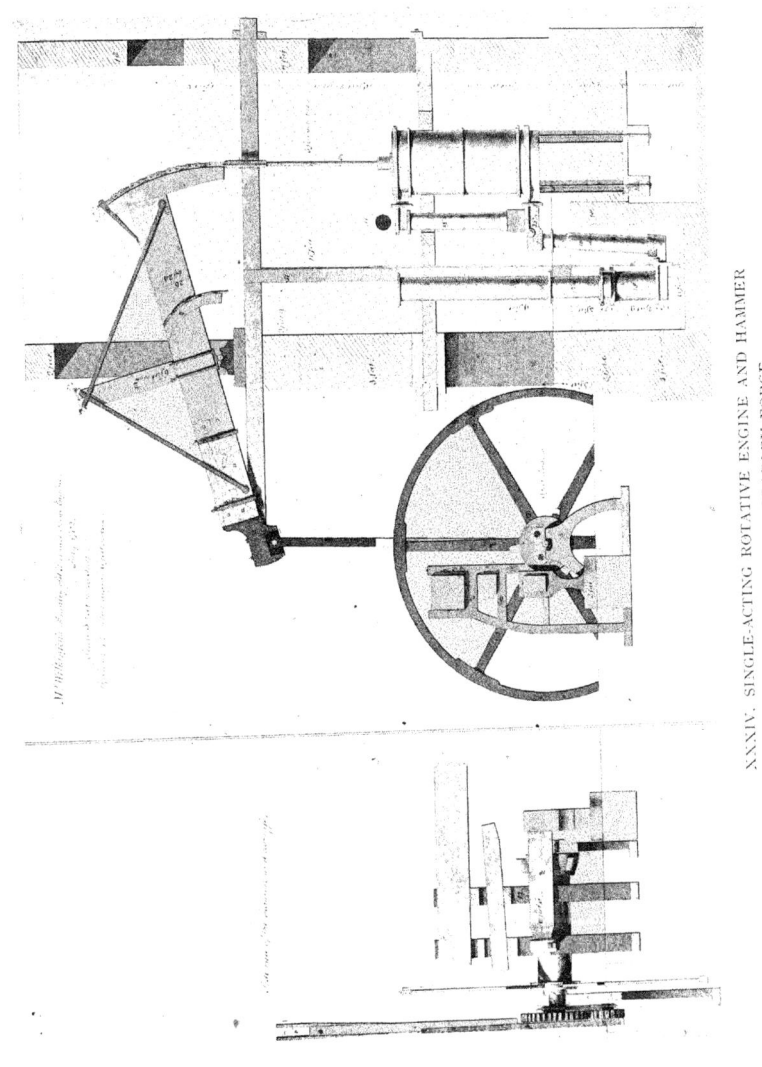

XXXIV. SINGLE-ACTING ROTATIVE ENGINE AND HAMMER
FOR WILKINSON'S BRADLEY FORGE

Drawing dated July 1782

a slide valve. The construction of these parts will be dealt with later on ; there is no reason for thinking they were not incorporated in the engine, but no record has been found of the results obtained with them ; indeed there is little or no information of any description available about this engine, but it is said that in 1802 it was applied as a compressed-air engine to work the lathe in the pattern shop at Soho Foundry. Another engine of the same size seems to have been made at about the same time for Boulton's manufactory, to work cutting-out presses ; this is said to have been sold in 1793 or 1794 to Thos. Fenton & Co., for a works at Morristown, near Swansea.[1]

By about the middle of 1784 a single-acting engine with sun-and-planet gear had been put up for winding ore in Cornwall. In reference to this engine, Boulton writes in June that ' The small rotative engine at W¹ Maid will be set to work in a day or two ',[2] and then, early in July, he says that the whim engine ' went very roughly and particularly when the planet passed over the top of the sun, but by adding 700 lbs of lead to the connecting rod it seems to go better '.[3] Murdock was the erector, and in the first of the letters mentioned above Boulton says ' he hath varied from your drawings, for he hath hung the spear upon the *top* of the beam, i.e., ye centre of working of the top end of the spear or connecting rod is upon the top of the beam, whilst the centre of motion of the beam itself is under the beam. This he hath done in order that the revolving wheel shall always stand right to begin its stroke.' It will be understood that the beam was of wood. After a few months' work it was found that the connecting-rod had become twisted and that the planet-wheel was not in the plane of the sun-wheel, which, writes Boulton, ' induces me to think iron rods wᵈ be best for single engines '.[4] In an earlier letter, that mentioning the lead weight added to the connecting-rod, Boulton had suggested cast-iron connecting-rods, and wished Watt to order one for Goodwyn's engine. He thought it ' much better to have the weight in the rod than upon the rod '. However, some years elapsed before his suggestion was adopted.

Watt was not yet satisfied with the design of the double-acting engine, and this, as well as the engine for Goodwyn referred to in Boulton's letter, was made single-acting. Goodwyn was a London brewer ; his engine was ordered in April 1784, the drawings were prepared in June, and the engine was at work in August. Like Wilkinson's coal-winding engine, it had a cylinder the same size as the Soho double-cylinder engine, 15 in. diam., 4 ft. stroke ; indeed there is reason for thinking that one of the cylinders of that engine was used for Goodwyn—there is in the Birmingham collection a sheet marked ' Full size drawings of the new working gear of the 15 Inch Forge engine. Nov. 1783.', which is marked in red ink ' Messrs Goodwyn's & Co.' The engine had a chain connexion from the piston-rod to the beam, and a wooden connecting-rod. It was the first of Boulton and Watt's rotative engines set up in London. Four years after its erection the brew-house was destroyed by fire, and a new engine was ordered ; this was made double-acting.

In the same year, 1784, another London brewer, Whitbread, ordered an engine with a 24-inch cylinder, 6 ft. stroke. This again was single-acting ; it was made double-acting in 1795, but seems to have been fitted with a parallel motion before this, at least there is a drawing, dated November 1784, in which a parallel motion is shown. This engine is still in existence and is now preserved in Australia in the Technological Museum at Sydney.

---

[1] B. & W. Colln. ' Engines made at Soho.'
[2] Ibid. Boulton to Watt, 1784 June 23.
[3] Ibid. Boulton to Watt, 1784 July 8.
[4] Ibid. Boulton to Watt, 1784 Sept. 25.

Another single-acting engine put up in 1784 was that for Horsehay Forge. This was started at the end of November or beginning of December ;[1] it had a cylinder 26 in. diam. and a stroke of 6 ft.

The first engine with the straight-line linkage motion was that for the oil mill of Cotes and Jarratt at Hull. It was a double-acting, sun-and-planet engine with a cylinder 15 in. diam. and 4 ft. stroke, and in fact was formed from one-half of the Soho double-cylinder engine referred to above. The engine was ordered in 1783 and the general drawing here reproduced in Plate XXXV is dated June 1784 ; it is marked ' This is the one executed ' and contains a note in the handwriting of Watt : ' The wood framing of the engine all made at Soho.' In fact, exceptionally to the usual practice at that period, the engine was completely put together and run under steam at Soho. It will be seen that the connexion between the piston-rod and the beam is by rack and sector. It was about this time, however, that Watt contrived his straight-line linkage—the three-bar motion, a radius rod coupled to the beam by a link ; he decided to apply the new mechanism to this engine and it is sketched in pencil on the drawing. In October Watt was able to inform his partner that : ' The new central perpendicular motion answers beyond expectation, and does not make the shadow of a noise, but for want of some regular work for the engine to do, we have not been able to give it a fair trial. All we could do was to burthen the engine by means of a break, and to apply a pair of rolls, with which we have rolled some copper and red hot iron.'[2]

The next few years saw the erection of a considerable number of rotative engines. Among these may be mentioned the engines for the starch manufactory of Stonard and Curtis (18-inch), for Walker of Chester (20-inch), for Robinson of Papplewick near Nottingham (18-inch), for Dobbs (18-inch), for Felix Calvert (24-inch), for Harris of Nottingham (15-inch), for Howard and Houghton's Oil Mill, Sculcoates (17-inch), for Liptrap of Mile End (17-inch), and the Albion Mills engine. The engine for Stonard and Curtis had a straight-line motion like that of Cotes and Jarratt, consisting of a radius rod and a single link. The same device seems to have been used for the engines for Reynolds's forge, while Howard and Houghton's engine had a rack and sector. The other engines had parallel motions ; all had sun-and-planet gear.

The Albion Mills engines demand notice at some length. The project of these large steam flour-mills excited a good deal of discussion at the time, and the engines and machinery were inspected by a great many people.

ALBION MILLS

THE project for a large steam flour-mill in London was put forward in 1782 or 1783 ; from an early period Boulton and Watt were interested in it, and in October 1783 we find Watt writing that he had nearly finished the drawings for Blackfriars.[3] The mill was erected on the Surrey side of the Thames near Blackfriars Bridge.

It was not the first flour-mill driven by steam power—Wasbrough had previously erected an atmospheric engine for driving such a mill at Bristol—but it was beyond

---

[1] B. & W. Colln.   W. Reynolds to Boulton & Watt, 1784 Nov. 29.
[2] Muirhead : *Mech. Inv.* Watt to Boulton, 1784 Oct. 21.

[3] The drawings from which the first engine for this mill were actually constructed are dated 1784 and 1785.

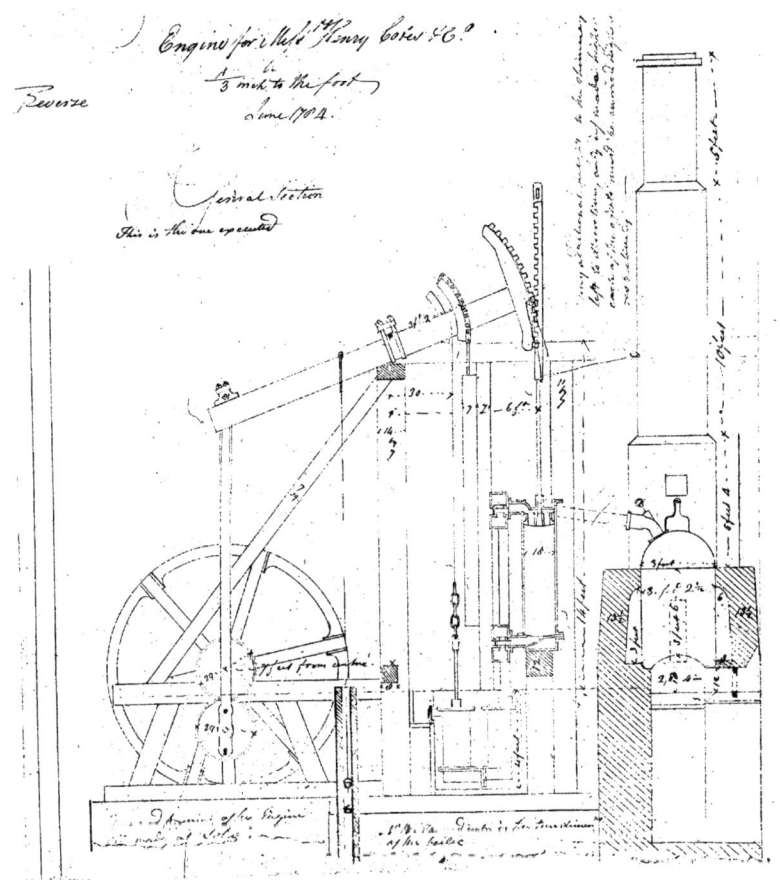

XXXV. DOUBLE-ACTING ROTATIVE ENGINE, 15-INCH CYLINDER,
FOR COTES AND JARRATT

Drawing dated June 1784

doubt the biggest and best-equipped mill of the period. By the middle of 1790 the sale of flour was very considerable, for the time, but alongside this big output there was bad organization and management, and when, in 1791, the mill was burnt down, the blow was so severely felt that the company never recovered.

Unfortunately there are no good general drawings of the engines as first erected, and it is clear that such drawings as have been published represent, at any rate, in so far as concerns the first engine, the construction at a later late, after alterations and additions had been made. However, the engines differed in no essential features from the engines mentioned above, but they were more powerful, having cylinders 34 in. diam. for 8 ft. stroke. The drawings made in 1784 for the first engine show the rack-and-sector connexion between the piston-rod and the beam, but the engine was actually made with a parallel motion. The mill work was distinguished by the extensive employment of cast iron and the mill is said to have been the first in which cast-iron wheels and axles were used throughout.

The erection of the engine and mill work was under the charge of John Rennie ; he was engaged for this purpose by Boulton and Watt and gave up his business in Scotland in August 1784, but as the building was not sufficiently advanced he was detained in Birmingham until the following November. He was paid a guinea a week, but in 1788 we find him applying to the Albion Mills Company for a payment for his services over and above the weekly allowance.

It was originally intended to put down three engines, each with its set of millstones, dressing-machines, and other accessories, but only two engines were erected. The first engine was successfully started on Feb. 15, 1786 ;[1] as, however, the mill was not quite ready, it was not until some weeks later that it was worked under a load. On March 9th Rennie writes that the mill has been at work all day with four pairs of stones on wheat and that it worked well on the first trial which was made before a crowd of spectators, including Sir J. Banks.[2] Under load the engine developed faults, particularly in the sun-and-planet gear, and Boulton says : ' the gudgeons of the sun wheel are loose in the iron shafts from bad fitting ', and ' the iron segments of circles which separate ye 2 rows of teeth in the sun wheels were badly made of bad iron & are obliged to be remade as they are broke to pieces '. The working-gear also was defective, that of the steam valves ' doth not open them quick & even requires ye hand to assist sometimes or a weight '.[3] The upper nozzle too was cracked.[4] The sun-and-planet gear had been made by Wilkinson, and Boulton says later that the sun-wheel gudgeons have been a very troublesome and expensive job and it would have been better to have given him £40 or £50 rather than have had them so badly fitted.

Boulton was in London at this time, and the affairs of the Albion Mills were a source of considerable anxiety to him ; he lived within the walls of the mill from ten in the morning till ten at night. However, before the end of the month, March, he had the satisfaction of seeing the engine go ' by far more sweetly than before. wch I attribute, in a great degree, to the segments in the sun wheels being cottered firm in their places, & turned both on their sides & faces '.[5] The mill was again got to work but within a few days there was another stoppage due to engine trouble, ' a loud slaping was heard in ye cylinder wch obliged us to stop & we now find what we expected, viz., the Wilkinsonian round pin wch fastens ye piston to ye rod

[1] B. & W. Colln. John Rennie to Watt, 1786 Feb. 15.
[2] Ibid. John Rennie to Watt, 1786 Mar. 9.
[3] Ibid. Boulton to Watt, 1786 Mar. 15.
[4] Ibid. Boulton to Watt, 1786 Mar. 22.
[5] Ibid. Boulton to Watt, 1786 Mar. 27.

is again bent & loose . . . they are about to cut an oblong cotter hole through the rod'.[1] This fastening had given trouble during some of the trial runs of the engine ; the cotter was placed above the surface of the piston and was in addition to the round pin.[2] The next difficulty was with the piston-rod stuffing-box and this seems to have been cured by introducing a ring of steam between two rings of packing— 'a ring of steam round ye stuffing box will be made & thereby that air leak will be prevented'.[3]

By the end of April repairs had been made and the mill was again at work; a trial of the engine performance was then made. In ten hours 48 bushels of coal were consumed and 527 bushels of wheat ground ; the quantity of meal dressed was more than this, and at the same time two sack tackles were at work. The engine made 17·9 strokes per minute, and the boiler evaporated $11\frac{1}{2}$ cubic feet of water per bushel of coal. Boulton sums up by saying that one bushel of coal grinds eleven bushels of wheat, besides driving the other machinery, and that one cubic foot of water evaporated will grind and dress one bushel of wheat. However, he was not satisfied with the boiler ; he thought the heating surface insufficient and that there was too much heat going up the chimney, and he declared that 'I shall not be satisfyd untill the heat is as much absorbed by the boiler that one may hold ones hand in the chymney'.[4] A few months later the sun-wheel gudgeons were again loose, this was indeed a constant source of trouble ; in addition the wooden connecting-rod had broken.[5] During his attendance at the Albion Mills Boulton did not allow the engine to monopolize his attention. He interested himself in the milling process as well, and we find him writing : ' It would be a great improvement to this manufacture if we had some cheap & conven^t means of preparing all our wheat for keeping in the bins. 1st, by warming and drying a very little ; 2nd by cleaning it with rotative brushes, & 3rdly by winnowing or blowing dry air through it and blowing light stuff away.' [6]

Then he wished to have a comparative trial of different coals before buying a large quantity of any particular description ; for this purpose ten chaldrons of four or five sorts had been procured and Boulton asks that ' our water gage with all ye apertures belonging to it' may be sent up. Harrison, he says, will be able to state the quantity of water that runs through each aperture per hour.[7]

Another point about the mill comes out in one of Rennie's letters to Watt : ' The method of disengaging the dressing machines separately is of great use, but this is also their ruin, for in stopping them they don't observe when the machine is empty, but stop her full & in this state strike her on again, by which some of them have been torn in pieces.' [8]

By the beginning of 1789 the second engine with its set of millstones had been laid down, and Rennie reports that when grinding wheat the engine will work six pairs of stones, when grinding sharps eight pairs, and when grinding part wheat and part sharps seven pairs.[9] There were twenty pairs of stones in the mill, ten to each engine.

As noted above, the output of the mill was by this time very large for the period.

[1] B. & W. Colln. Boulton to Watt, 1786 Apl. 4.
[2] Ibid. Boulton to Watt, 1786 Mar. 1.
[3] Ibid. Boulton to Watt, 1786 Apl. 15.
[4] Ibid. Boulton to Watt, 1786 May 1.
[5] Ibid. John Rennie to Watt, 1786 July 4.
[6] Ibid. Boulton to Watt, 1786 Apl. 15.
[7] Ibid. Boulton to Watt, 1786 Mar. 22.
[8] Ibid. John Rennie to Watt, 1789 Jan. 27.
[9] Ibid. John Rennie to Watt, 1789 Feb. 27.

The sales of flour in a week in June 1790 amounted to £6,800,[1] but Boulton was not at all satisfied with the state of affairs in respect of finance and organization.[2]

However, his concern on this account was terminated soon after, for the mill was destroyed by fire on Mar. 2, 1791. There were strong suspicions of incendiarism and Boulton thought that the company should offer a reward of £500 upon conviction of the offender ; and he considered, the undertaking being ' a national object ', that the origin of the fire should be investigated by the Government. Rennie and Wyatt, the manager of the mill, on the other hand, thought the fire was caused by accident. Writing to Watt, Rennie says : ' This morning at six o'clock a fire began at the Albion Mills, which in the space of two hours totally consumed the same. There is nothing left standing beside the walls. Various have been the accounts of this accident, the most probable, & that which Mr. Wyatt agrees to, is that it arose from the large corn machine, in front of the kiln, wanting grease.'[3] A few days later Rennie states explicitly that he does not think the mill was wilfully set on fire.[4] In the same letter he says that ' the rejoicings of the mob were very great '. It is beyond question that there was a strong public feeling against the mill, probably fomented by rival millers and the mealmen. The company was accused of monopolizing the flour trade and putting up the price of flour. As a matter of fact the price of flour in London had gone up pretty steadily from 30s. a sack in 1786, the year in which the mill was started, to 42s. in 1790. The rise, however, was consequent upon a rise in the price of wheat, and soon after the fire the company issued a printed statement showing that for the twelve years before the mill was started the average difference between the price of wheat per quarter and the price of flour per sack was 5s. 0¾d. ; whereas for the five years during which the mill was at work the average difference was but 3s. 4d.

The mill was not rebuilt, and it was some time before the affairs of the company were wound up. As late as the year 1800 the erection of a new engine and mill was under discussion. In the same year one of the partners owning a twenty-fifth share wished to withdraw from the concern and offered £400 or even £500 to be released from his liabilities. Boulton and Watt, who held two-fifteenths of the undertaking, objected to the acceptance of this offer ; they said that they were prepared to go out themselves on the same terms.

We have seen in the case of the Albion Mills first engine that a good deal of trouble arose from defective workmanship. Some years later, in 1793, we find a letter from the owners of a cotton mill at Nottingham with a long series of complaints not only of bad work, but of bad material as well. The wheels of the sun-and-planet motion were so bad that they had been obliged to replace them by new ones ; the connecting-rod top had broken ; the coupling-link was badly forged and had broken in three places ; the brasses of the coupling-link and the connecting-rod gudgeons had broken and had had to be replaced ; the fastenings for the gland of the piston-rod stuffing-box and the gland of the air-pump stuffing-box had broken ; the top and bottom exhaust spindles were made of bad iron, and they had broken ; the racks and sectors for the valves were badly fitted and very much out of order ; the spring box at the top of the exhaust rod was in a very bad condition. This list seems appalling enough, but the writers wind it up

---

[1] Ibid. Boulton to Watt, 1790 June 26.
[2] Ibid. Boulton to Watt, 1791 Jan. 10.
[3] Ibid. Rennie to Watt, 1791 Mar. 2.
[4] Ibid. Rennie to Watt, 1791 Mar. 5.

with the words[1] 'and many other matters', and all this concerns an engine that had been started twelve months before!

One of the minor difficulties that had to be surmounted was getting the engine to start running in the right direction. It is mentioned in connexion with the helve engine at Soho in 1782. Dayus, the erector of Robinson's engine at Papplewick, writing in 1786, reports that the engine started very well, but 'is very liable to go the contrary way in setting on . . . which is very hurtful to the machinery of the mill. But we intend to erect a catch for the wheel so that it can't go back.'[2] A catch or pawl to engage notches on a rotating part seems to have been the usual expedient to meet the difficulty, but apparently it was not the practice to apply it in the first instance, at any rate at this period. In the following year we have Boulton writing : 'Something must be done to prevent Goodwin's engine from turning backwards, as it hath done mischief twice by so doing & has again this morning broke the fulling stocks by turning ye wrong way.'[3]

There is no evidence that anything in the nature of a reversing-gear was made in Watt's time, but it is clear that engines were run alternately in opposite directions. Watt, writing in 1795, said that the first whim engine made for Cornwall 'changed the motion by stopping the engine & setting the fly agoing the contrary way'.[4] Boulton, in 1792, referring to the engine of a lead-rolling mill where a difficulty was found in keeping up a supply of steam, writes that he is 'persuaded that if the rolls had been made to return by the millwork instead of stopping ye engine & reversing its motion, the steam might be kept up'.[5]

The second Albion Mills engine may be said to mark the attainment of a certain degree of standardization in practice in respect of the rotative engine, the salient features of which are wooden frames, beams and connecting-rods, the sun-and-planet gear, and the parallel motion ; the valves are worked from the beam and are opened by weights, except for a period from 1790 to 1798 when the practice was to use flat springs. At about the same time as this engine was set going, the centrifugal governor was being tried on the Lap engine at Soho Manufactory, Plate XLI. This was the first steam engine so controlled ; the device seems to have succeeded at once, and soon became known. In 1790 we find Rennie asking for four or five of them to be fitted to engines in London.

Some of the drawings in the Birmingham Collection, however, do represent exceptional constructions, as, for example, the portable winding-engine shown in Plate XXXVI. The drawing is headed 'General Section of the Engine for Winding Coals for Messrs Reynolds & Co, May 1788'. The engine is mounted on a four-wheeled truck, the boiler is a fixed structure, and the winding-drum is mounted on beams that are laid in place on the engine frame after this has been brought into the desired position. The wheel base of the truck is 10 ft. long and one axle is directly under the cylinder and the other directly under the flywheel shaft. The engine beam is pivoted at the top of a simple Λ-frame. The cylinder is 10 in. diam. and the piston-rod is shown linked directly to the beam without any guiding arrangement, but it seems that the engine was to have guides in the form of two upright pieces of wood fastened on the cylinder. The plug-rod is jointed directly to the beam and serves to work the air-pump. The flywheel is composed of a spider

[1] B. & W. Colln. Henry Green & Co. to Boulton & Watt, 1793 Aug. 5.
[2] Ibid. Dayus to Watt, 1786 Feb. 21.
[3] Ibid. Boulton to Watt, 1787 June 20.

[4] B. & W. Colln. : Letter Books. Watt to Jeffries, 1795 Aug. 23.
[5] B. & W. Colln. Boulton to Southern, 1792 Oct. 17.

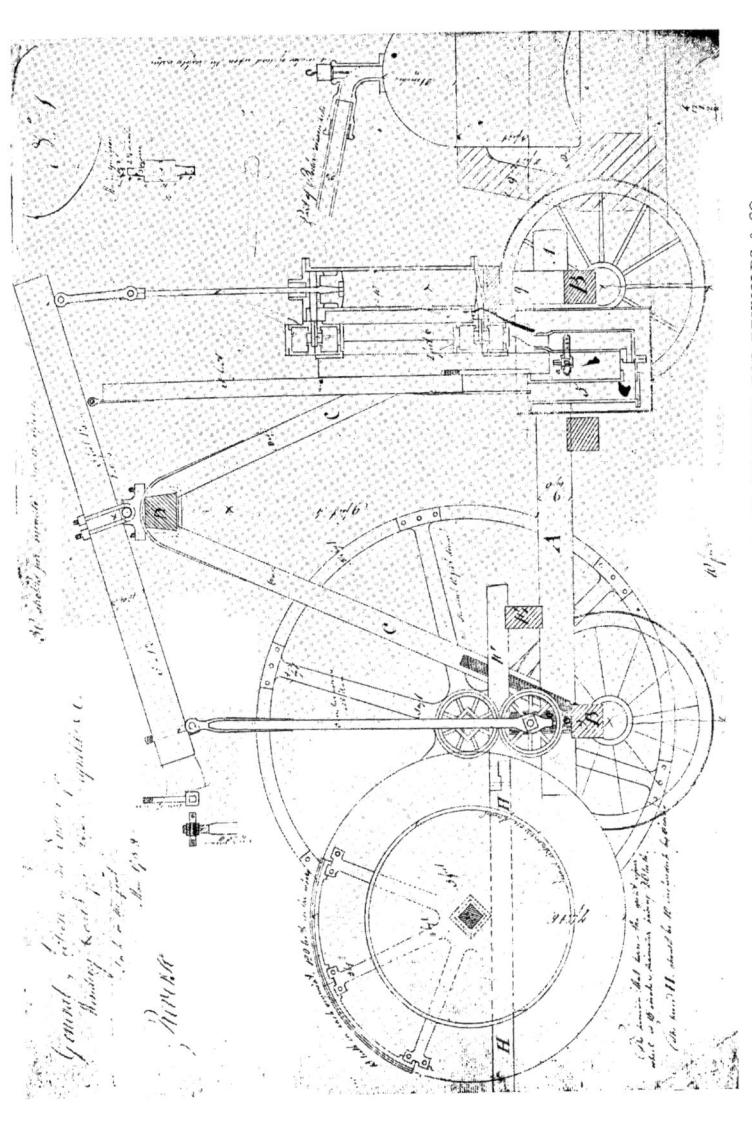

XXXVI. PORTABLE ROTATIVE ENGINE, 10-INCH CYLINDER, FOR REYNOLDS & CO.

Drawing dated May 1788

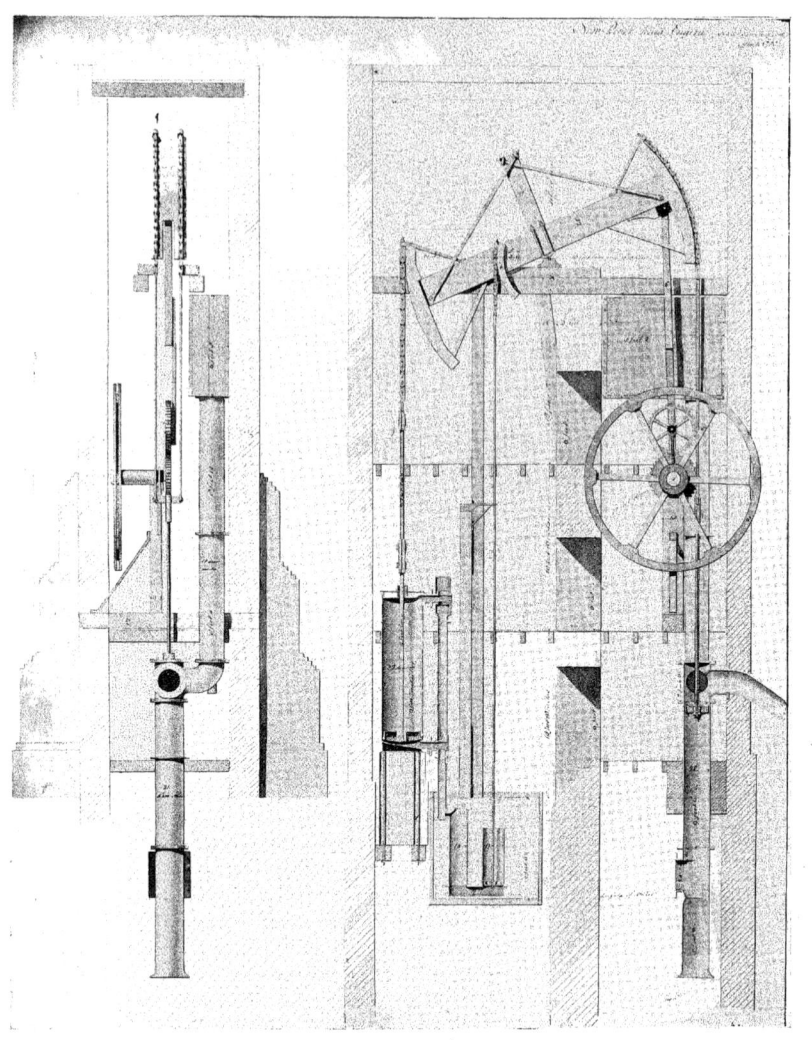

XXXVII. ROTATIVE PUMPING-ENGINE, 32-INCH CYLINDER.
FOR NEW RIVER HEAD

Drawing dated March 1783

and segments; it is 10 ft. diam. with a rim $3\frac{1}{2}$ in. by 1 in. The connecting-rod is of cast iron 3 in. square; at its upper end it is embraced by a wrought-iron strap in which is an eye for the joint pin. The winding-barrel, 5 ft. diam., is geared to the flywheel shaft in the ratio of 1 to 6, and the engine speed was 30 strokes per minute. In the written instructions sent with the drawings we find the earliest reference to a throttle valve; here it is to be worked by hand and is thus described: ' It consists of a ring of cast iron 3 inches thick having a spindle like one of the valve spindles and two thin sheets of copper rivetted together and to the square part of the spindle and wood between them to fill up.' There was to be no injection gear, ' as the injection is to play constantly while eng^n works,' no blowing-pipe and valve, and no top valve for the air-pump. The valves on the cylinder were to be worked by links instead of racks.[1]

The engine shown in Fig. 19, drawing dated June 22, 1790, is another small engine with a cylinder $10\frac{1}{2}$ in. diam. and a frame like that of the engine just described. It appears that a 3 h.p. engine with a cylinder $10\frac{1}{2}$ in. diam. and 3 ft. stroke was supplied to the Birmingham Navigation Company in 1788–9; it was returned in 1791 and then sold to the Dublin Canal. In 1793–4 an engine called ' Birmingham Canal Boat Engine' was made with a cylinder 23 in. diam. and 2 ft. stroke; this was afterwards sold to the Grand Junction Canal.

In the specification of his patent of 1782 Watt describes and illustrates the application of a rotating shaft with a flywheel to a pumping-engine. This feature is found in the drawing, dated March 1783, of ' New River-head Engine', Plate XXXVII. The shaft is rotated by sun-and-planet gear and a connecting-rod jointed to the beam just inside the arch that works the pump. The cylinder of the engine is 32 in. diam., and its lower end is permanently open to the condenser. A similar engine was put up at about the same time for the Chelsea Waterworks at Pimlico wharf. Farey, who inspected this engine in 1803–4, found it in constant work, but with the rotative mechanism disconnected; the mechanism, according to his account, was the crank and connecting-rod; whether this was the form put in in the first instance we cannot say, but it seems unlikely.

The term of Pickard's patent expired in 1794, but Boulton and Watt seem to have been applying the crank to a limited extent a few years before that date. The drawings, dated September 1791, of an engine for Wright and Jesson show a cast-iron crank; the drawings of the years 1792 and 1793 include but two crank engines, 1794 has none, but the years 1795 and 1796 have two each. The sun-and-planet gear continued to be made as late as 1802.

One of the early crank engines is shown in Plate XXXVIII. It was a winding-engine for John Sparrow, and the drawing is headed ' Cockshead No. 2. Aug. 10. 1793'. The framing is entirely of wood, and the beam and connecting-rod are likewise of wood. The cylinder, 14 in. diam., is carried on a wood frame, the vertical members of which pass down one on each side of the condensing cistern. The beam-gudgeon bearings are supported on a pair of beams built into the house wall and supported under the bearings by posts; diagonal struts extend from the tops of the posts to a trestle frame carrying the crank-shaft bearing. The plug-

---

[1] B. & W. Colln.: Letter Books, 1788 May 14. ' Observations on the drawings of the engine for   winding coals for Messrs. Reynolds & Co.'

rod is of wood with an iron link connexion to the beam, and the air-pump rod is jointed to its lower end.

On the drawing of a very similar engine for Hawks & Co., dated Dec. 7, 1795, we find on the boiler steam-pipe the notes ' to be very well wrapped '—an early instance of clothing pipes.

Up to the year 1788 the smallest engine made for sale was a 4 h.p. The annual premium for this was 20 guineas, and the price of the engine material including flywheel and shaft was £224 10s. ; the woodwork was estimated to cost £50, and the ' putting together ' £40, while a suitable boiler of copper cost about £40, so that the total cost of engine and boiler, exclusive of brickwork, amounted to £354 10s.[1]

Rennie at this time was pressing the firm to undertake the production of smaller engines : he had many inquiries for 3 h.p. and 2 h.p. engines, and mentions one for a 1½ h.p. engine. Watt, however, was opposed to the manufacture of these small engines ; clearly there was little to be made out of them and the royalties would be very small ; it would seem too that he did not think his engines suitable for small powers. In 1793, in connexion with an inquiry for a small engine, he makes a note that he ' doubts the propriety of using smaller engines ' than 4 h.p.

Following the entry into the concern of younger men—the sons of Boulton and of Watt, the erection of the Soho Foundry and the return of Murdock from Cornwall—we find a change in the spirit of the management. The term of Watt's patent was approaching its end, when the payment of royalties would cease, and the firm—or rather firms, for as will be shown elsewhere the Soho Foundry and the Engine Works at Soho Manufactory were separate concerns—had embarked upon the manufacture of engines, and it was realized that a good deal might be done in the production of engines of small powers. James Watt, junior, in particular, was keen on the production of a small engine that could be sold at a lower price than the standard design of engine with beam. In this he was seconded by Murdock, who planned a self-contained engine without a beam and with a slide valve worked from the engine shaft. Southern was sceptical of the new venture, and at any rate was anxious that it should have a fair trial at home before an engine was sent out. He had grown up under James Watt and saw eye to eye with him on matters of engine construction ; all his life he had been identified with the beam engine and no doubt he held, as many continued to hold long after his day, that nothing could be produced to equal it for accessibility of the parts, sweetness of running, and durability. Replying in 1799 to a letter from James Watt, junior, who was for pressing forward the construction of an engine with slide valve and without beam, Southern writes :

' While I feel the difficulty of resisting the torrent of ingenuity which Murdock's genius pours forth, I beg to be understood as neither wishing to repress improvement, nor as approving of all he proposes, which I am fully convinced is always with the purest views . . . and let me add, lest you should have misconceived me, that I have the highest opinion of Murdock both as to his ingenuity & his heart, that I am far from thinking him obstinate in his opinions, but I do think he is too sanguine in some of his expectations.' [2]

Watt himself shared Southern's doubts as to the new design ; his son in reply to Southern's letter says : ' My father's opinion of the sliding valves and Beam*less* engines agrees pretty much with yours ; as to myself, I cannot say that I have

[1] B. & W. Colln. Southern to Boulton, 1789 Sept. 8.

[2] Ibid. Southern to James Watt, junior, 1799 Apl. 22.

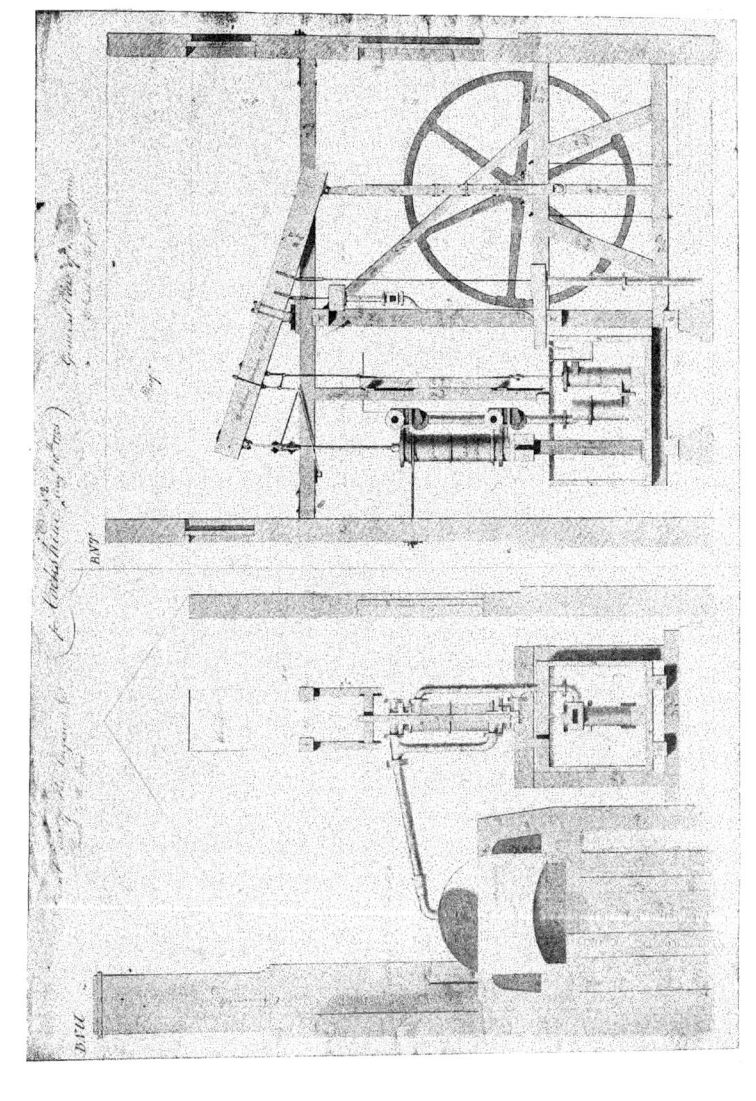

XXXVIII. ROTATIVE ENGINE WITH CRANK, 14-INCH CYLINDER, FOR JOHN SPARROW

Drawing dated Aug. 10, 1793

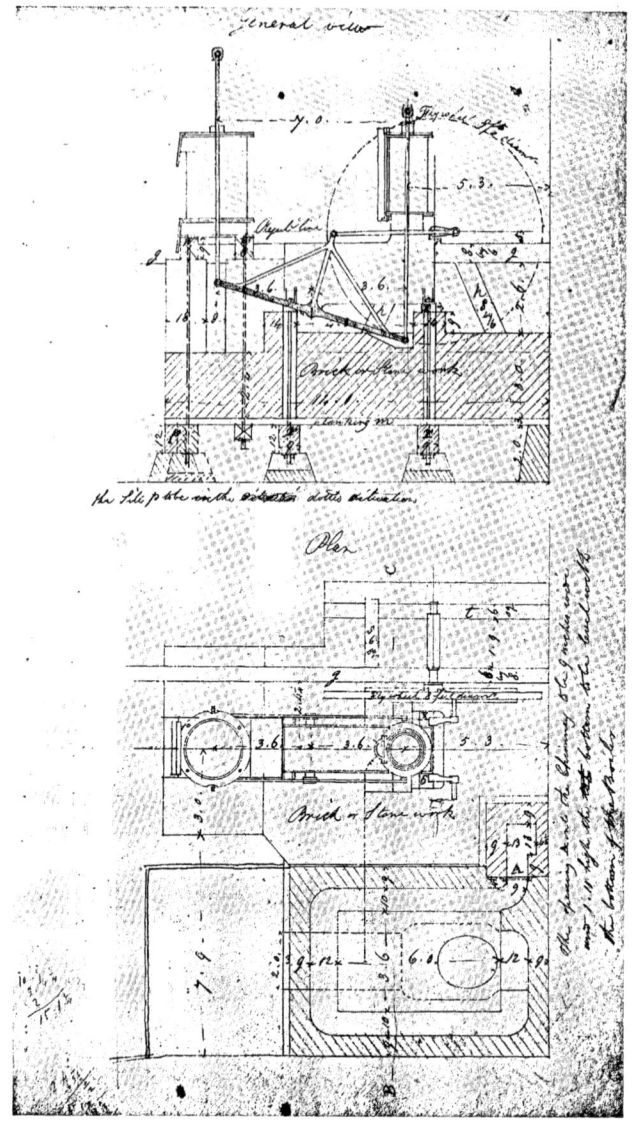

XXXIX. BLOWING-ENGINE FOR FULTON & SONS, GLASGOW, 1802

any prepossession either for or against one or the other, but am clearly for having them both tried at home first.' An order had been accepted for a 2 h.p. engine at a low price, and James Watt, junior, goes on to say that 'Symond's engine must I believe be made without a beam (on account of the price) but should certainly be made with common valves'.[1]

The 2 h.p. engine was put in hand and finished very soon after this correspondence, but, as it embodied features covered by Murdock's patent of 1799, and

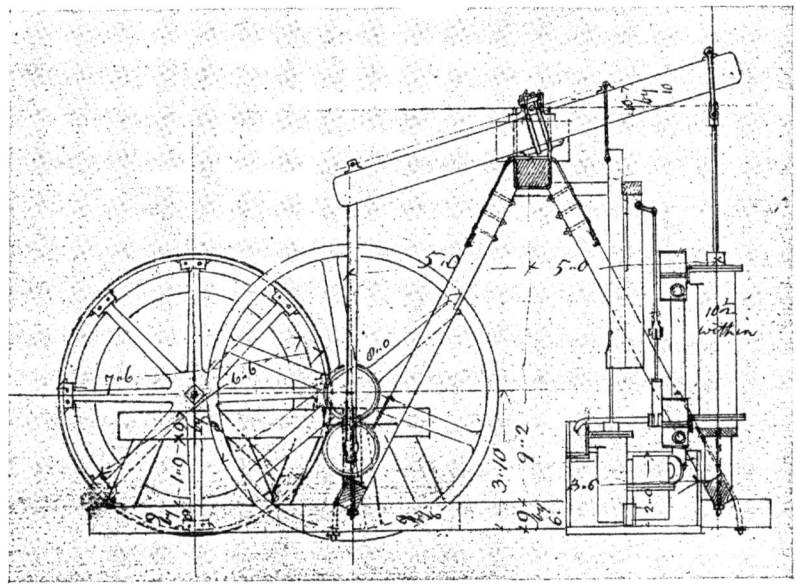

FIG. 19. Small Engine. Design of 1790

it was not expedient to send it out until that patent had passed the Great Seal, its delivery was delayed until September. This engine had the crank-shaft placed near the bottom of the cylinder, and the connexion between the piston-rod and the crank was made by a crosshead, side-rods, a bell-crank lever, and a link. In the year 1800 two other bell-crank engines of 3 h.p. were made, and in 1802 a small blowing-engine was constructed for Fulton and Sons of Glasgow on the same lines, the horizontal arm of the bell-crank being prolonged beyond the fulcrum to work two side-rods depending from a crosshead on the piston-rod of the blowing cylinder; see Plate XXXIX. Murdock was in Glasgow in August 1802 while this engine was in course of erection, and writes to Soho: 'I hope to see it work before I leave this, as there is a many people waiting to see the performance of this before the[y] give order.' However, the only other blowing-engine made on this plan was one,

[1] Ibid. James Watt, junior, to Southern, 1799 Apl. 24.

slightly larger, for the Perth Foundry Co.   This was a 6 h.p. engine, and the drawings are dated November 1802.   After this the manufacture of bell-crank engines was discontinued and the use of the beam was reverted to for small engines up to the introduction of the side-lever stationary engine, but it does not appear that engines so small as 2 and 3 h.p. were again produced.

Murdock's ideas, however, were embodied in the construction of the valve-gear of large engines.   The drawings of an engine for Soho dated 1798 (the engine was erected in 1799) show two eccentrics and the valve-stems coming out at the top of the nozzles ;   that of the Salford Twist engine, 1799, has an eccentric in the form of a cam, while the engine for Coates & Co., 1803, has a cast-iron eccentric disk, brass strap, and a wood rod.   The general drawing of this last engine is repro-duced in Plate XL.   The beam and connecting-rod, also the pillars under the beam-gudgeon bearings, are of cast iron ;   the air-pump is worked by an iron rod and the cylinder is shown supported by a framework of wood.   The boiler-feed apparatus is shown, but this will be considered later on, as also the development of the beam and of certain other parts of the engine.   It may be noted here that cast-iron beams had been used a couple of years before this.   Apart, possibly, from the method of supporting the cylinder, this engine for Coates & Co. may be taken as representing the normal practice in rotative engines at the period of Watt's retirement from Soho.

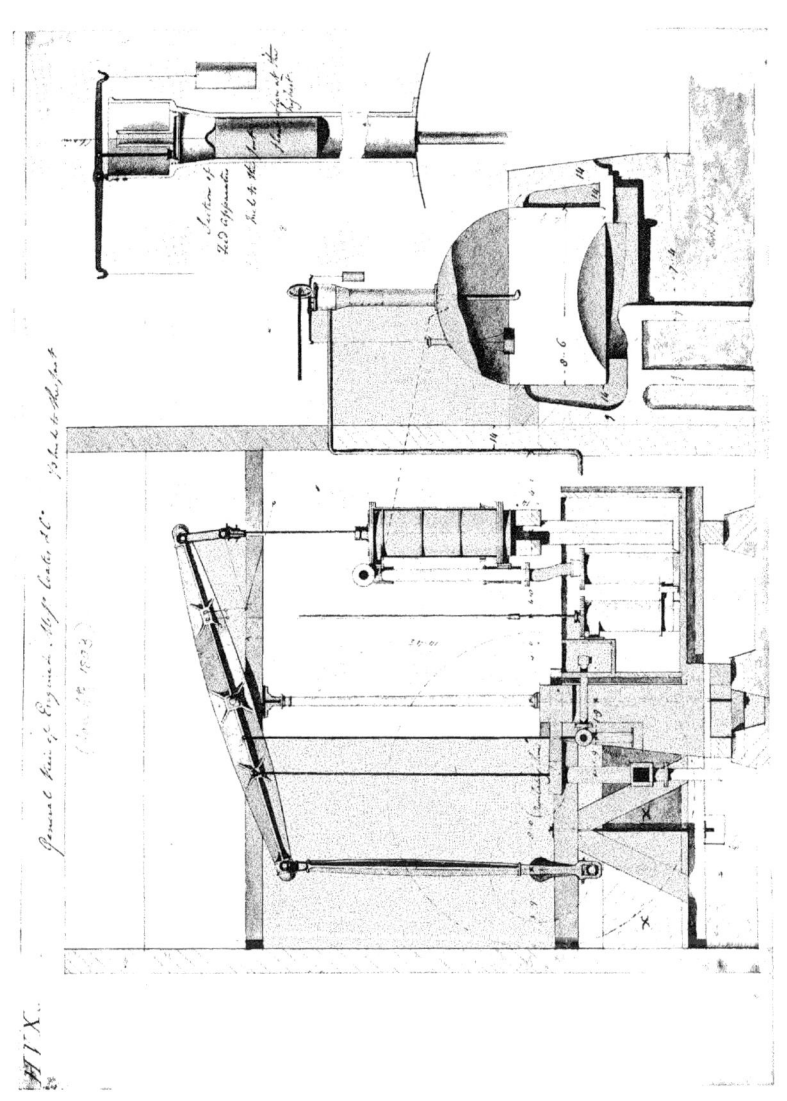

XI.

NL. ROTATIVE ENGINE FOR COATES & CO.

Drawing dated Jan. 6, 1803

XLI. THE LAP ENGINE AT THE SCIENCE MUSEUM, SOUTH KENSINGTON

XV

THE VALVES AND THEIR
WORKING-GEAR

IT has been stated, earlier in this volume, that the valve used by Watt in the first engine set up at Soho was a modification of the form found in the Newcomen engine. This construction (Plate XVII) is indicated also on the drawing of the Bow engine (Plate XIX) and in those of the condenser for the Broseley and Bloomfield engines (Plate LXXII), which are designs of the year 1775 ; but Watt was already engaged in devising a new valve which was fitted to the Soho engine before the end of the year. We do not know its form, but it is clear that it proved unsatisfactory, for very soon after we have a ' Drawing of conical valve and nozzle for Soho engine ', dated January 1776. This drawing shows a valve in the form of a short cylinder or thick disk acting as a plug in a circular passage in the valve box, the entrance to which is flared or coned. The same construction, but without the flared opening, is shown in the Bedworth drawings, May 1776, and a few months later in the drawing of Colevile's engine (see Fig. 20). Then in April 1777 another new nozzle was fitted to the Soho engine, and now we have arrived at the drop valve ; for there can be no doubt that it was the successful working of this last form that led to the provision of the new nozzle for the Broseley engine, shown in Plate XLII. It will be seen that the valve is still a short cylinder or thick disk, like Colevile's, but its lower edge now beds itself upon a conical seat or flared entrance to the passage in the valve box. This drawing is dated May 1777 ; within twelve months or so the other engines, Bloomfield, Bow, and Bedworth, had been provided with new nozzles, and from this time onwards until within a few years of the termination of the Boulton and Watt partnership all the engines built for customers were fitted with drop valves worked by a plug-tree or plug-rod hung upon the engine beam and moving up and down with it. It should be stated, however, that the term ' drop ' valve is here used in a broad sense, and that in some of the engines the valves were inverted and were closed by an upward movement.

Murdock returned from Cornwall to Soho in 1798 and soon after we find some radical changes being made, including, in the rotative engines, the introduction of the eccentric for working the valves from the engine shaft, and of the slide valve.

It is proposed to consider, first, the drop valves and the boxes or nozzles in which they were mounted ; secondly, the working-gear or mechanism by which they were actuated from the plug-tree ; thirdly, Murdock's rearrangement of the valves , then the introduction of the eccentric for working them ; and, finally, although perhaps, strictly speaking, it is not within the scope of this volume, the introduction of the slide valve will be dealt with briefly.

THE VALVES

THE drop valves employed in the first instance were plain copper disks working against conical seats formed in the cast iron of the valve box or nozzle. They made line contact only, for it seemed to Watt that, in view of the possibility of grit and dirt getting on to the seat, surface contact was inadmissible.[1] In practice, however, it was found that the copper disks caused grooving of the annular surfaces of the seats; in fact the valve dug a bed for itself in the cast iron.[2] To get over this difficulty Watt fell back upon surface contact; he coned the edge of the disk and brought its lower surface below the coned surface of the seat. This was in the year 1779; the injection valves had been made in this manner previously. At the same time he fitted the nozzles with separate valve seats of brass.[3]

Originally the valve disk was mounted on a separate wrought-iron stem, and was held up against a shoulder on the stem by a conical collar of copper soldered in place. This mode of fastening was a source of trouble, the collar got loose and the disk could not be lifted off its seat.[4] Apparently this is what Watt refers to, in one of his letters, as the ' Bloomfield disease of a bitt in our throat'. The first expedient to get over this difficulty was to rivet the collar on the stem; then the shoulder on the stem was placed under the disk instead of above it; this is shown in the drawing of the nozzle for the Trevaskus engine, Plate XLIV. Later on, the disks and stems were cast in one piece, but even in 1797 we find the built-up construction still in use, as in the case of the nozzle for the Hebburn engine, Plate XLVI.

In his first experiments Watt seems to have carried the valve stems out of the nozzles through stuffing-boxes, for we find him telling Boulton, in regard to the valves, that a ' better way of hauling them up is contrived, free of stuffing-boxes'. The plan adopted was a rocking spindle carrying, inside the nozzle, a lever coupled to the valve by links, and passing through the wall of the nozzle. Apparently a simple cone joint was used at first, the spindle being held against its seat by a set-screw that engaged its outer end and was carried by a light bridge fixed to the nozzle; but after a few years' experience Watt found that he had to fall back upon the use of a stuffing-box. The drawing, dated December 1780, of ' Nozles for a 58 Inch Cylinder', Plate XLV, in the section of the upper nozzle, shows the spindle with a conical shoulder working against a corresponding seat in the nozzle casting, and there is no provision for packing the joint. Plate XLIV, the nozzle for Trevaskus engine, the drawing of which bears the date, ' Novʳ. 1781', shows the same conical shoulder working against a seat in a removable bush in the nozzle casting, but here the spindle has a plain cylindrical part between the shoulder and the tapered part which receives the actuating-lever, and the boss on the casting is extended outwards beyond the collar, leaving an annular space between it and the spindle adapted to receive packing. It seems, however, to have been the practice with the earlier construction to wrap a little oakum round the necks of the spindles. The drawing, dated October 1783, Plate XLIII, of the upper

---

[1] When the valves leaked the defect was cured by tinning their edges (Boulton Papers. Watt to Boulton, 1778 Aug. 13).

[2] B. & W. Colln.: Letter Books. Watt to Harrison, 1779 Aug. 16.

[3] Ibid.: Letter Books. Watt to Wilson,

1779 Aug. 28: ' I am making movable brass seats for the regulators, which can be easily taken out and new ones put in when worn.'

[4] Boulton Papers. Watt to Boulton, 1778 Aug. 13.

nozzle and spindle of Rochfort engine (a small pumping-engine, 13-inch cylinder, made by Wilkinson to go to France) does not show a stuffing-box for the rocking spindle, but it does show a groove in the spindle, no doubt intended to receive packing ; the spindle works in ' Steel Collars ' and its cone is ground in ; the plan adopted for preventing endways movement of the spindle is clearly shown.

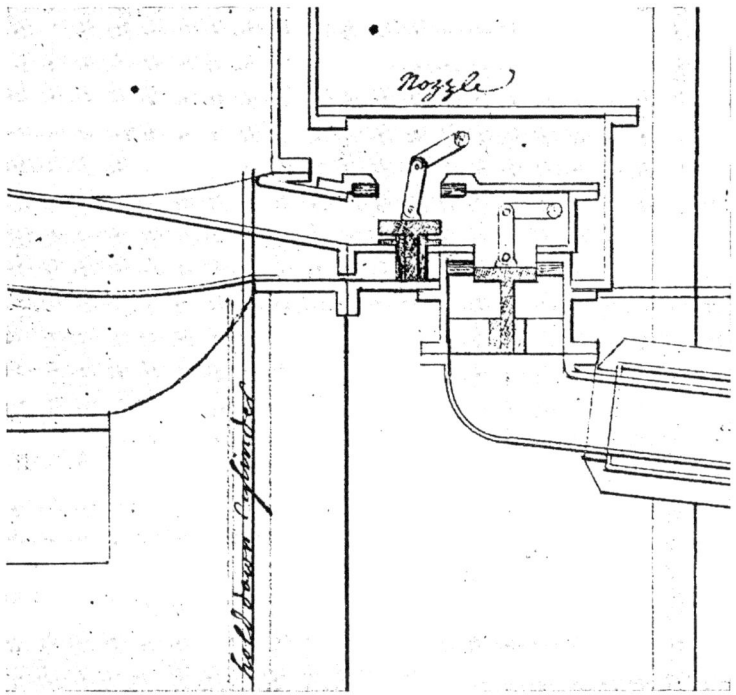

FIG. 20. Valves for Colevile's Engine

The simple lever and link connexion between the rocking spindle and the valve, Fig. 20, was speedily relinquished for a lever with an arched end as in Plate XLII, and this in turn gave place to a toothed sector and rack. Plate XLIV represents the upper and lower nozzles of the engine for Trevaskus mine in Cornwall. The engine had a 45-inch cylinder, and was built in 1782. The sectors, the racks and their guides, as well as the removable valve seats, are clearly shown. At first it was the practice to fix the seats by running in lead, and we have Boulton writing from Truro that, ' the brass rings which are inlaid in ye nossells for the valves to beat upon should not be put in with lead as it soon becomes a calx like white lead and as soon as the steam begins to pass between the iron and brass it cuts it away like a file.'

At the date of the drawing we have just been considering the drop valve had fully established its position, and was being applied by other engineers to the Newcomen engine. In 1779 it had been fitted to the old engine at Bedworth, and Boulton had expressed the view that it should be patented, if possible.[1]

As the engines increased in size, larger valves had to be used and the power required to open them increased proportionately. To meet this difficulty Watt devised a pressure-relieving arrangement for the exhaust valves. The first application of this contrivance was to one of the Poldice 63-inch engines in 1778. The drawing selected for illustration, Plate XLV, representing the nozzles for a 58-inch cylinder engine, is dated December 1780, so that it may be taken as embodying the results of some experience with the original construction ; however, in regard to the pressure-relieving device itself there is no essential difference, but this drawing shows the rack-and-sector actuation, and removable brass seats for the valves ; whereas the earlier drawing has the link connexion, and has not the removable seats. The main exhaust-valve consists of a cylinder working up and down over a disk that is fixed on a pillar projecting upwards from a bridge in the mouth of the eduction pipe ; the lower edge of the cylinder beats upon a seat secured in the nozzle, and its upper end is closed in, except for an opening in the central dished part ; this opening forms the seat for a pilot valve jointed to the end of the rack and having its stem guided in a bridge, across the opening of the main valve, and passing through a slot in a bar fixed across the top of the same. When the rack is moved upwards, it pulls up the small valve ; steam is then allowed to pass into the main valve, thus the pressure on its upper surface is nearly balanced and it readily follows the pilot valve when this catches under the cross-bar. The drawing shows the pilot valve opened and about to raise the main valve from its seat. Plate XLVI, 'Nozles for Hebburn Colliery', shows an exhaust valve on the same principle designed in 1797. The valve is now actuated from below, and the disk, in the form of a basin, is made in one casting with the valve seat ; the valve stem has a shoulder, to engage under a bridge across the opening in the top of the main valve, and another shoulder against which the pilot valve is held by a nut. Both valves are shown in the raised position. It will be observed that, here, the valve seats are clamped between the sections of the nozzle, being formed with flanges for that purpose.

We do not know the date of the invention of the Hornblower valve, from which the Cornish double-beat valve was evolved, but it seems probable that it was suggested by this contrivance of Watt's.

However, Watt himself was not altogether satisfied with it—no doubt for one thing it was expensive to make—so in 1783 we find him taking up again an idea that he had conceived seven years before, i. e. inverting the valves and causing them to open downwards ;[2] on this plan the pressure of steam, say on the exhaust valve, instead of resisting the opening, would tend to throw the valve open. This invention is covered by his patent of 1784, and the Wheal Towan and Wheal Crane engines, built in 1784 and 1785, had the exhaust valves made in this way; the Soho engine had such a valve fitted to it in 1785. The erectors did not like this construction, and it required considerable attention to keep the gear in order; nevertheless engines continued to be fitted with it for a number of years. Calvert's

[1] B. & W. Colln. Boulton to Watt, 1779 Mar. 5.
[2] Doldowlod Papers. Watt's Journal, 1783

Apl. 6 : 'made drawing of a nozle with inverted valves'.

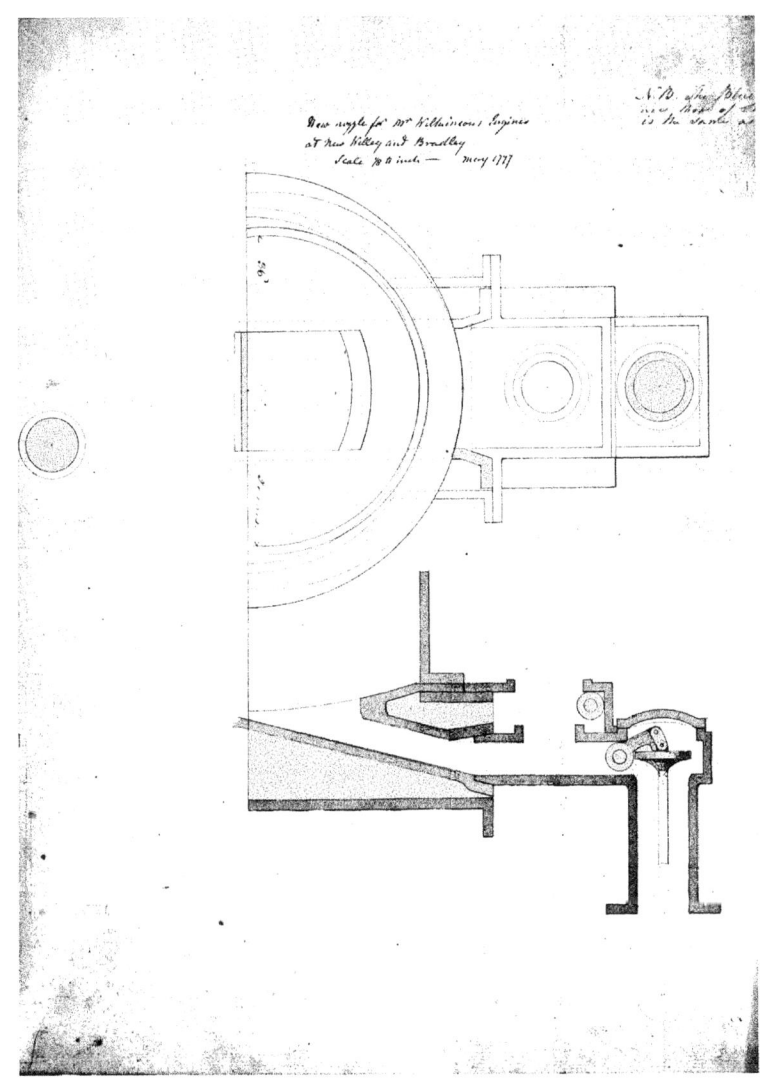

. XLII. NEW NOZZLE FOR BROSELEY AND BRADLEY ENGINES, MAY 1777

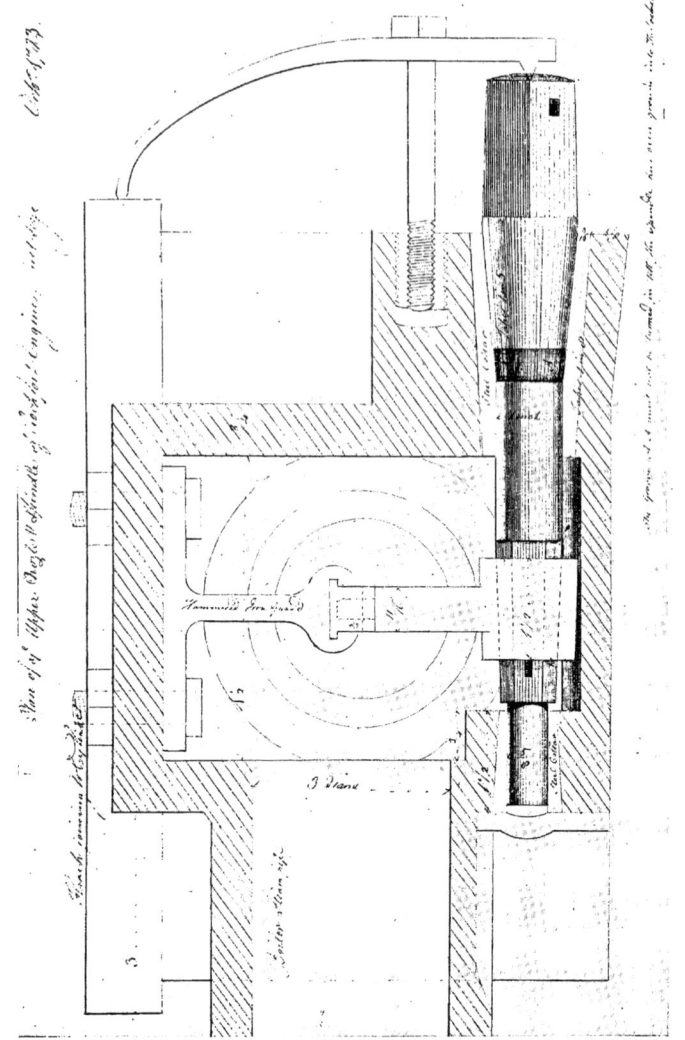

XLIII. UPPER NOZZLE FOR ROCHFORT ENGINE, OCTOBER 1783

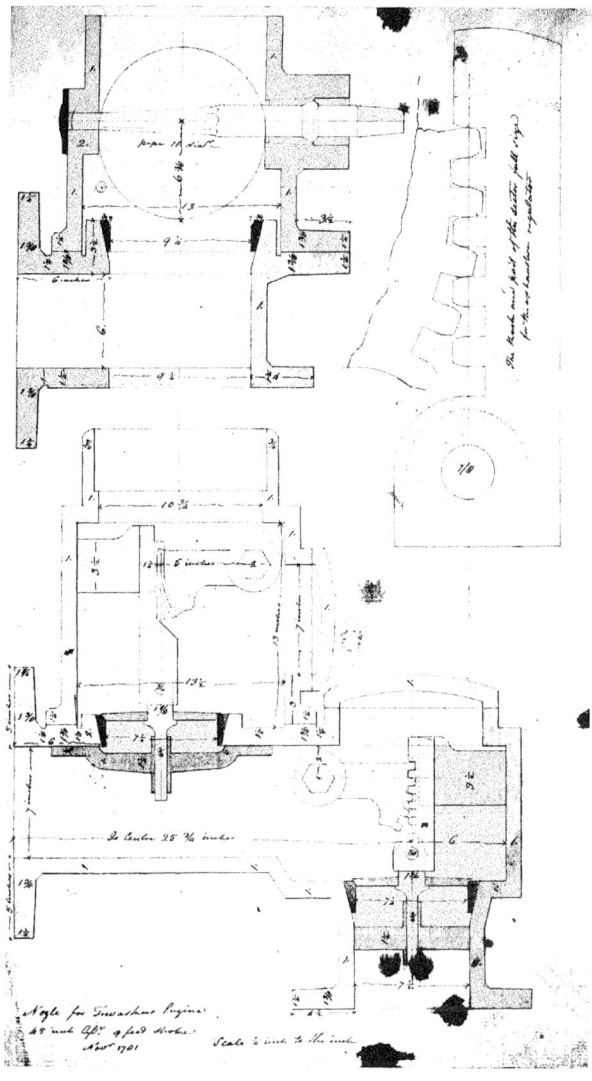

XLIV. NOZZLES FOR TREVASKUS ENGINE, NOVEMBER 1781

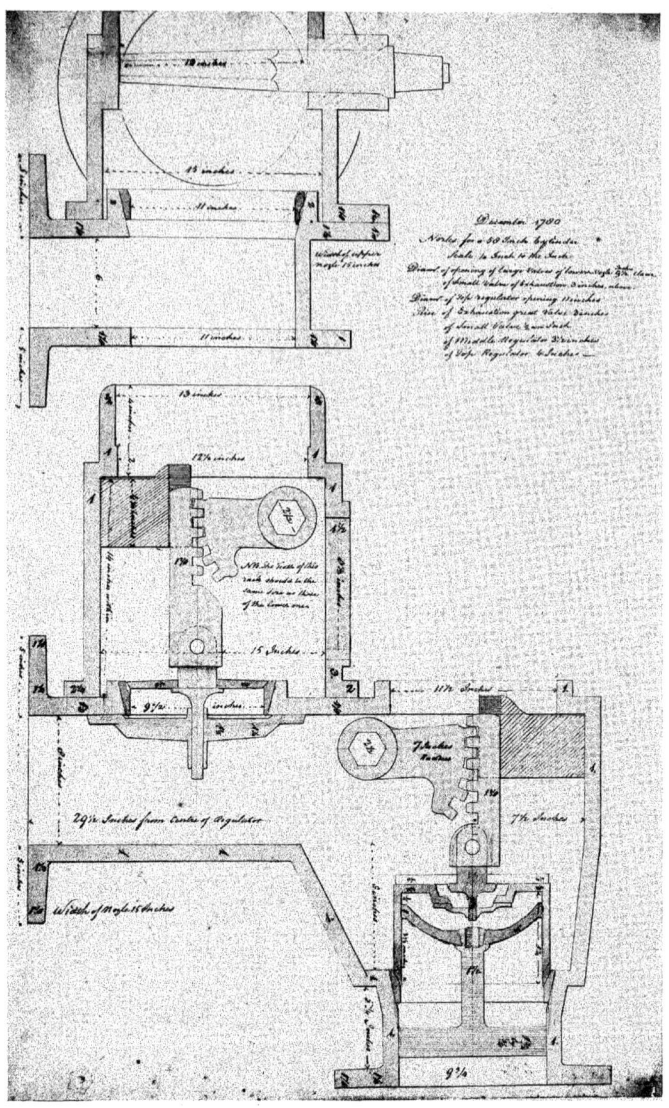

XLV. NOZZLES FOR 58-INCH ENGINE, DECEMBER 1780

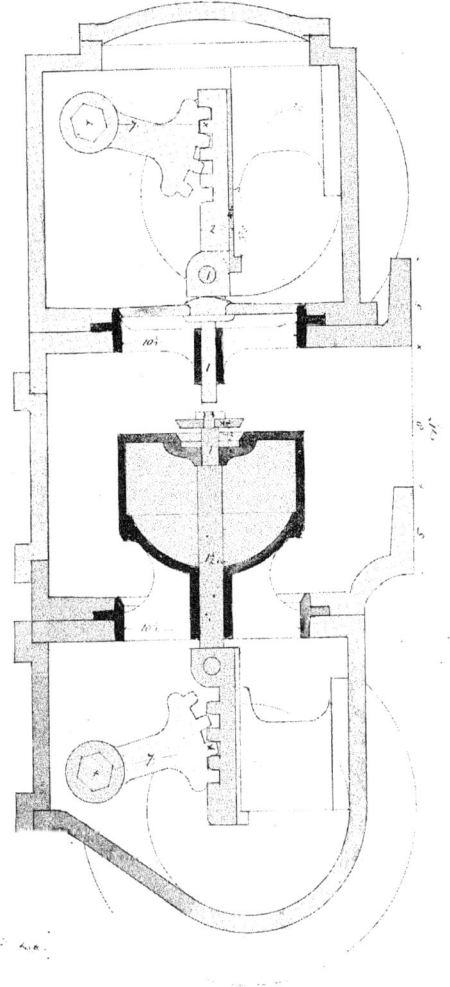

XLVI. NOZZLE FOR HEBBURN ENGINE, OCTOBER, 1797

engine at the Peacock Brewery, London, had a new exhaust valve on this plan put in in 1797, but, as we have seen, the Hebburn engine of the same date has the pilot-valve arrangement.

In the first engines, those with the open-top cylinder, there were but two valves, the steam and the exhaust valves, and they were placed at the lower end of the cylinder. Upon the introduction, in 1777, of the closed-top working-cylinder and separate jacket, a third valve, the 'top regulator', was added to control the admission of steam to the upper end of the cylinder when the engine was underloaded. In the Soho expansive engine, 1777, the top regulator and the steam or equilibrium valve were placed at the top of the cylinder and the exhaust valve at the bottom, but the usual practice was to place the equilibrium with the exhaust at the bottom of the cylinder. The two valve boxes, or nozzles, were connected by a perpendicular pipe through which the steam passed from the upper end of the cylinder to the lower. At first this pipe was made with a flange joint at each end, but in 1780 the lower joint was made as a spigot and socket.[1] In the double-acting engines both nozzles had steam and exhaust valves and were connected by a pair of perpendicular pipes, one to take the exhaust from the upper nozzle and the other to supply steam to the lower nozzle.

The valve nozzles were the most complicated castings in the entire engine and no doubt the patterns were costly, so that it is not surprising to find that the same pattern was used for different engines. For example, the Wanlockhead 36-inch, Hallamannin 40-inch, Wren's Nest 36-inch, and Broseley 38-inch (reconstruction) all had nozzles from the same pattern. For many years the nozzles were built up of separate boxes ; in the Trevaskus engine, Plate XLIV, the upper and lower nozzles are each formed of two separate castings. This building-up was not so objectionable with the single-acting engines, but with the double-acting engines the increased number of parts and consequently of joints was a source of difficulty. Accordingly one of the first things undertaken when the Soho Foundry began working was the casting of upper and lower nozzles each in one piece ; for the smaller engines they went farther and made the upper and lower nozzles and the side pipes in one casting.

THE WORKING-GEAR OR VALVE GEAR

HAVING considered the valves themselves, and the nozzles in which they were inserted, we now turn to the mechanism by which they were worked.

As in the Newcomen engine, the valves were operated by pegs on a plug-tree, a vertical bar or beam of wood suspended from the engine beam and moving up and down synchronously with the piston, acting upon arms secured to cross-shafts or arbors. In the Newcomen engine there were two valves to be opened and closed, the steam admission valve and the injection valve or cock. Plate XLVII, reproduced from a drawing in the Boulton and Watt Collection, shows part of a Newcomen engine put up by Smeaton at the New River Head in 1769. (It is the same design as the portable engine by the same engineer illustrated in Farey's *Steam Engine*.) The arbor for controlling the injection is shown immediately above, and that for working the steam valve immediately below, the injection cock ; the piston is at the bottom of the cylinder just about to begin its upward movement, the steam valve being open and the injection cock closed. The injection

[1] B. & W. Colln. : Letter Books. Watt to Turner, 1780 Mar. 20.

A a

arbor has four arms, a forked arm to take over the handle of the injection cock, a handling-arm standing in the path of a peg on the plug-tree, a detent arm to engage a pivoted latch, and an arm carrying a weight. As the plug-tree reaches the top of its stroke a peg thereon strikes the latch clear of the detent arm, the arbor is then carried round by the fall of the weighted arm (the motion of which is restricted by a leather strap) and the forked arm swings round the injection handle to open the cock. On the downward movement of the plug-tree another peg thereon presses upon the handling-arm and turns it down; the consequent rotation of the arbor causes the closing of the injection cock, the raising of the weighted arm, and the re-engagement of the detent lever with its latch. The steam arbor carries a tumbling lever, adapted to play on each side of the vertical to an extent limited by a leather check-strap, a downwardly projecting lever coupled by a link to the arm of the oscillating valve, and two curved arms standing in the paths of pegs on the plug-tree, one on each side. As the piston nears the end of its upward stroke, one of the pegs catches the upper curved arm and turns the arbor round until the tumbling lever has passed the vertical; this lever then falls over sharply, carrying with it the arbor and causing the closing of the valve. On the downward movement of the piston, near the bottom of the stroke, the lower curved arm is caught by its peg and the mechanism is restored by a similar action to the position shown. In the construction described, the lever on the steam arbor is jointed directly to the link connecting it with the valve. A more usual arrangement was to have the pin on the link engaging between the links of a forked arm on the arbor; this fork with the weighted lever made a combination in the form of the letter Y, hence this arbor was termed the Y-shaft. In no part of the Newcomen engine did greater divergence of practice occur than in the valve gear. In the Watt engine too there were considerable changes in the gear in the period now under review; and indeed we may say generally that in the development of the steam engine down to the present day no organ has undergone greater change than the valve gear.

Watt at first used the Newcomen gear or a fairly close modification of it, but very soon began scheming improvements and about the middle of the year 1776 he writes, too optimistically as it turned out, that ' the clockwork of the engine has been simplified to such a degree that I believe we shall never think it worth while to aim at any further improvements'. This letter seems to mark the invention of the gear with the curved arm or wiper on the cross-shaft or arbor acting upon a straight arm on the rocking spindle in the valve box. The engines for Bloomfield, Broseley, and Bedworth when first erected had gears of a different form, but they were all very soon altered to this type. At first Watt tried to work the two valves from one arbor, but we find him writing in reference to the Wheal Busy engine, and after experiments on the Soho engine, ' I find it will be best to have two Y-shafts and 2 loggerheads [i.e. tumbling bobs], that is one to each regulator', and in fact it seems that all the gears of this type had an arbor, or as Watt, following the Newcomen practice, very frequently terms it, a Y-shaft, for each valve.

As the wiper simply pressed upon the valve arm, it is obvious that it could operate in one direction only—the action of the wiper opened the valve; movement in the other direction to close the valve and cause the arm to follow the wiper on its inoperative stroke was due to the weight of the valve itself and of the attached parts, and, in the case of the equilibrium valve, to the preponderance of steam

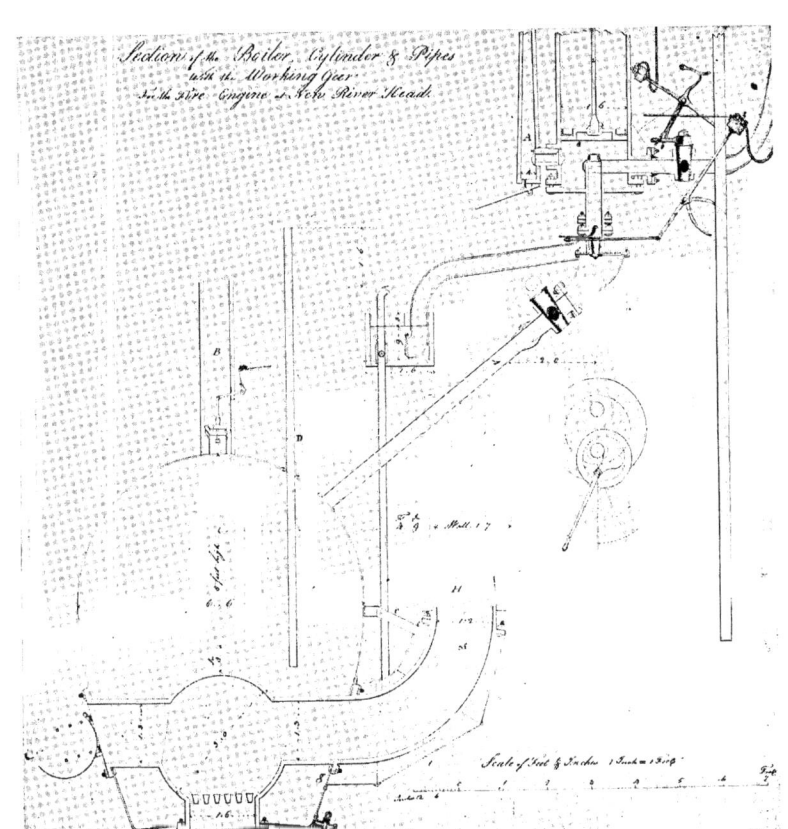

XLVII. NEWCOMEN ENGINE AT NEW RIVER HEAD

pressure on its upper surface. This type of gear, which we have termed the 'wiper gear', continued to be made for some years. In 1783 a new arrangement was devised, in which the connexion between the arbors and the valve spindles was made by links and levers. This form we have called the 'linkage gear'; it continued to be made, for the rotative as well as for the pumping-engines, after the year 1800; but just before that date we shall find actuation by an eccentric on the engine shaft, in the case of the rotative engines, beginning to take the place of actuation from the beam.

In dealing with the 'wiper gear' we shall refer to the cataract which was used on occasion as an accessory to it.

THE WIPER GEAR

It is desirable to say, as briefly as possible, what the gear was required to do. The first engines, with an open-top inner cylinder enclosed in an outer cylinder, had but two valves, one to admit steam from the jacket space to the lower end of the inner cylinder below the piston, and the other to allow the steam to escape from the cylinder to the condenser. When the outer cylinder was replaced by a jacket, put on in segmental panels, and a lid was fitted to the working cylinder, a pipe had to be provided to convey steam from the top of the cylinder to the bottom; at the top of this pipe was mounted a third valve which was worked to admit and cut off the supply of steam from the boiler. This valve was called the 'top regulator'; the valve that admitted steam below the piston was the 'steam regulator', or 'middle regulator', and the remaining valve was the 'exhaustion regulator'.

In order to avoid confusion in what follows, the valve that admits steam from the steam pipe to the engine is termed the steam valve, and, inasmuch as it serves to put both sides of the piston in equilibrium, the 'middle' or 'steam' regulator is called the equilibrium valve.

In addition to these valves, there was the injection valve to be opened and closed at each down-stroke of the engine.

When the piston was at the top of the cylinder, in order to make the down-stroke, the steam valve (if one was provided), the exhaust valve, and the injection valve required to be open; and the equilibrium valve shut. Before the piston reached the bottom of the cylinder the first three valves had to close, and then the equilibrium valve had to open for the up-stroke. Watt laid down the rule that the exhaust valve was to be shut when the piston had nearly completed its downward movement and a little before the equilibrium valve was opened, and that the injection valve was to be opened an instant before the exhaust valve and to be shut soon after the piston began to move down.

It was requisite to contrive the gear so that the valves could be worked either by hand or automatically at will; moreover, it was necessary to be able to close both valves, equilibrium and exhaust, at once, in order to stop the engine, and to have both open at once for blowing through at starting.

The earliest existing drawing of the wiper gear is that for Colevile's engine, an engine with open-top working cylinder and two valves only; it is reproduced in Plate XLVIII. The upper set of mechanism is for the equilibrium valve, and the lower, shaded, set is for the exhaust and injection. The upper mechanism consists of an arbor with a 'loggerhead', or tumbling bob J, a bent handling-arm K, and a horn or curved wiper L which bears on the upper surface of a straight arm secured on the valve spindle. The lower mechanism has a similar arbor, handling-arm, and wiper, and, in addition, a detent engaging a pivoted catch provided with a tail N standing in the path of a peg on the plug-tree ; to the detent is suspended a rod H carrying a weight ; the wiper bears on the under side of a straight arm on the exhaust valve spindle.

The parts are shown in thick lines in the position which they occupy when the piston is at the bottom of the cylinder, with the equilibrium valve open and the exhaust and injection valves shut ; the exhaust valve is on the same side of its spindle as the arm, the equilibrium valve is on the opposite side of its spindle.

As the piston, and with it the plug-tree, moves upward and approaches the top of its stroke, a peg on the plug-tree acts upon the tail N of the catch and frees it from the detent ; the weighted rod H then pulls round the arbor, the wiper raises the arm on the valve spindle and opens the exhaust valve, and by means of a cord it opens also the injection valve. The parts then assume the position indicated by thin lines. Upon the descent of the piston a peg in the plug-tree acts upon the handling-arm, first along the part A, B, and then along the part B, G, and restores it to the position shown ; its movement is accompanied by that of the wiper which allows the injection valve to close ; the exhaust arm follows the wiper and the exhaust valve closes. The downward movement of the handling-arm also raises the detent and the weighted rod H, and the detent re-engages with the catch, now free, as the peg has moved clear of its tail.

In short, the exhaust valve, with the pressure in the cylinder above it, and the vacuum in the condenser below it, is opened by the descent of a weight, which weight is held by a self-acting catch until this is tripped by the upward movement of the plug-tree ; it is closed by its own weight and that of the arm on its spindle upon the downward movement of the plug-tree, and at the same time the weight for opening is raised and re-engaged with its catch. It is interesting to note that Watt used the trip mechanism to open the valve, whereas in modern practice it is used to close the valve.

If it should happen that the peg E does not depress the handling-arm far enough to allow the valve to close, an additional peg F may be inserted in one of the holes of the inner row in the plug-tree, which, acting upon the arm nearer to its axis, ensures sufficient movement. The extent of the opening of the valve is adjusted by a wedge or other device to limit the descent of the weighted rod H.

Turning now to the equilibrium valve, and commencing with the piston on the down-stroke, the valve closed, and the tumbling bob J, the wiper L, and the handling-arm all in the raised position ; as the plug-tree reaches the lower end of its stroke a peg thereon acting upon the handling-arm turns the arbor until the tumbling bob has passed the vertical ; the bob then falls over and its weight together with that of the arm and wiper, by the action of the wiper on the arm of the valve spindle, causes the opening of the valve. As the plug approaches the top of its stroke another peg acts upon the handling-arm and raises it with the tumbling bob and the wiper ; as the wiper moves up the arm on the valve spindle follows it and the valve closes by its own weight and the difference in pressure on its surfaces.

In a note on the drawing Watt says that at Wheal Spirit (i.e. Wheal Busy) the loggerhead had been found unnecessary—' for when the plug-tree raised the horn off the lever the regulator shut, and when the engine returned and the plug allowed the handle K to descend, the arch L lay upon the lever, but could not open

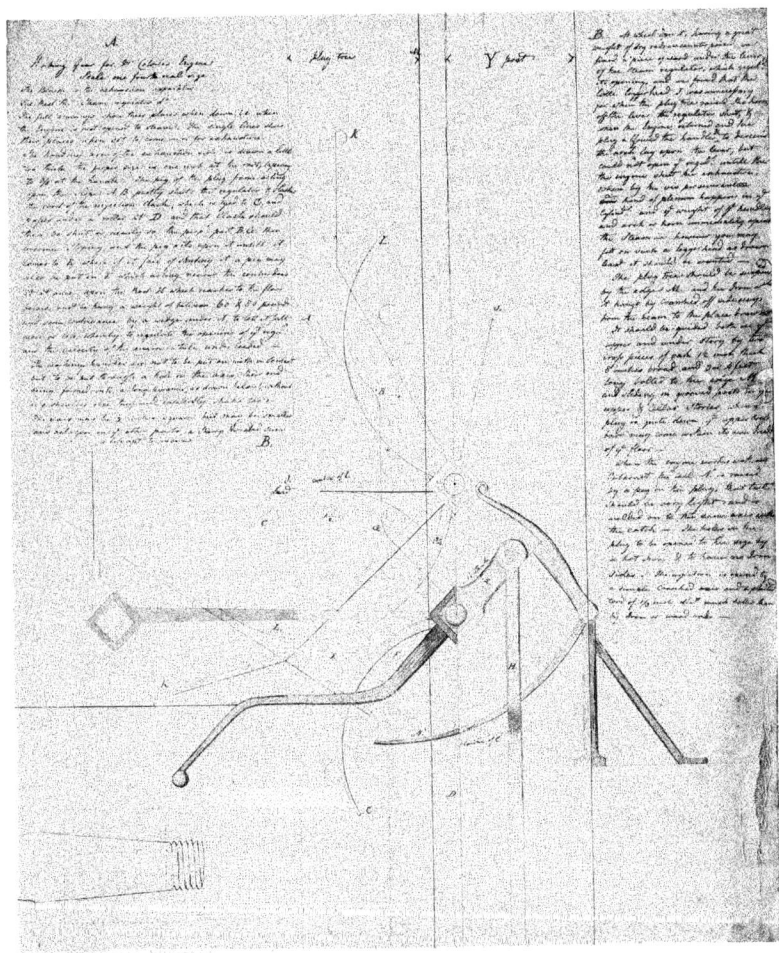

XLVIII. WORKING-GEAR FOR COLEVILE'S ENGINE

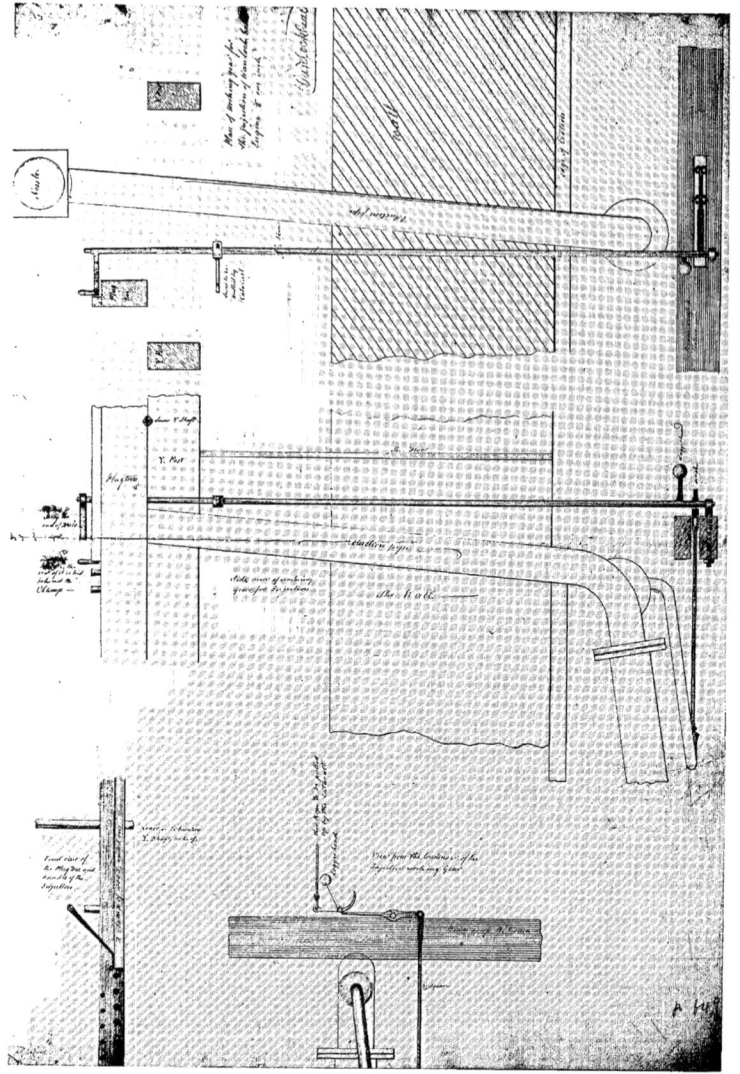

XLIX. INJECTION GEAR FOR WANLOCKHEAD ENGINE, 1779

ye regul^r. until the engine shut the exhaustion, when by the vis perseverantiae a kind of plenum happens in ye Cylind^r and ye weight of ye handle and arch or horn immediately opened the steam.' He goes on to say that a loggerhead as drawn may be fitted to the gear, lest it should be wanted, but in the drawings of some gears made soon after this, e.g. Birmingham Navigation, Ketley, and Jary's, the loggerhead is not shown.

The small figure at the bottom of the drawing shows the manner of securing the handling-arms in their arbors ; the ends are tapered and square in section.

The upper valve was worked in the same way by an arbor with a wiper taking under an arm on the valve spindle. When the piston was nearing the bottom of the cylinder, with the valve open, a peg on the plug-tree pressed down the arm on the arbor and lifted a weight suspended from a lever projecting on the other side of the shaft ; at the same time the wiper moved down and allowed the valve to close. On the ascent of the plug-tree, as it approached the top of its stroke, another peg engaged under the arm, turned it and the wiper upwards, thereby raising the arm on the valve spindle and lifting the valve off its seat against the pressure of the steam ; the movement was then continued by the weight acting upon the lever. The descent of the weight, and hence the lift of the valve, was regulated by stops, or by a cord. In some cases, the Birmingham Navigation engine for instance, the upper working-gear had a detent and catch, like the exhaust gear.

The gear for the three valves, top, equilibrium, and exhaust, having been described, it now remains to deal with the actuation of the injection valve. The Chacewater (Wheal Busy) engine at first had the injection ' worked by string tied to the plug wch pulled it open a little before the engine went quite out and shut it immediately when begun to come in ', but this was soon altered and the valve was opened by a rope connected to an arm on the exhaust arbor and closed by a peg on the plug-tree.[1] From the notes on the drawing of Colevile's gear it appears that a cord was fastened to the end C of the exhaust wiper and passed under a roller placed at D and that it was connected to a cranked axle.

However, the standard arrangement for some years, found for example in the Wanlockhead engine, 1779, Plate XLIX, is the following : a shaft extending lengthways of the engine and at right angles to the arbors has, at one end, a handling-arm standing in the path of a batten or chock of wood fixed on the plug-tree, and, at the other, a tumbling bob and a wiper bearing upon the tail of a lever to the opposite end of which is jointed a link descending to the valve. A lever on this shaft served to release the catch of the exhaust detent. When the engine was required to work at full speed, or rather to make the full number of strokes per minute, the injection valve was opened by the plug-tree ; when the work to be done demanded fewer strokes, it was opened by means of a cataract.

The foregoing account is sufficient to give a general idea of the working-gear used in Watt's early pumping-engines. Variations in design were being made constantly, and, moreover, there is evidence that the erectors did not in all cases follow the drawings very closely. The mechanism, it will be understood, was not made at Soho, but at the place where the engine was being put up. In the first engine he was sent out to erect, i.e. Wanlockhead, Murdock took it upon himself to arrange the gear so that the steam valve was worked from the exhaust shaft. Jabez Hornblower, who was erecting an engine at Donnington Wood, when he heard of this plan was very much struck with it and asked permission to construct

[1] Doldowlod Papers : Watt's Journal. 1778 Aug. 24 ; Sept. 24.

his gear in the same way ; this Watt refused to grant. As late as 1786, Southern, writing to the erector, Malcolm Logan, in reference to an engine to be put up in Holland, says : ' Mr. Watt has directed the method of the working-gear, I know he will be displeased with any variation of consequence.' ·

It would weary the reader, even if the material were available and space permitted it, to follow out the variations in design and to trace out the development of this gear up to the time when it was displaced by the linkage gear. It will be. better to select a particular gear designed after Watt had had some years' experience and to indicate the main features of difference as compared with Colevile's gear. For this purpose the mechanism of the engine known as ' Mr. Wilkinson's French engine' answers very well. The engine, a single-acting blowing-engine with a 42-inch steam cylinder, was constructed entirely by Wilkinson ; the drawings of the gear are exceptionally complete and no doubt embody the best constructions known to Watt at the date, 1782. Referring to Plate L, the equilibrium valve is actuated in the same way, but the arbor is not provided with a tumbling bob. The exhaust-valve gear has a wiper, handling-arm, detent with suspended weight, and catch, as before, but whereas in Colevile's engine the equilibrium and exhaust valves were actuated quite independently we now have an interlock between them, by which the opening of the exhaust valve is rendered dependent on the closing of the equilibrium valve. For this purpose the catch for the exhaust detent is released, not by a peg on the plug-tree, but by a trigger in the form of a bell-crank lever, one arm of which stands in the path of a cam or projection on the equilibrium arbor. The rotation of this arbor, which takes place as the piston reaches the end of the up-stroke, brings the projection into contact with the trigger and thereby lifts the catch ; the weight suspended on the detent then pulls round the exhaust arbor and, by the action of the wiper on the valve lever, lifts the exhaust valve. The piston now begins its down-stroke, and as the plug-tree reaches the lower limit of its travel a peg thereon presses down the handling-arm, and the detent, with its suspended weight, is lifted into engagement with the catch, while the valve lever and the exhaust valve drop under the influence of gravity.

Another feature of difference is the manner of working the injection valve. The opening movement is now effected from the spindle of the detent catch ; this spindle in turning to release the exhaust-valve gear, by means of a lever and link, pulls up a latch which releases a weighted bell-crank arrangement coupled to the injection valve. This valve is closed, soon after the piston begins its descent, by means of a chock on the plug-tree acting upon an arm on the bell-crank spindle ; the bell-crank, turning in the opposite direction, closes the valve, pulls up the weight, and re-engages, by its upwardly projecting arm, with the latch. The connexion between the detent-catch spindle and the latch for the injection has a screw adjustment to ensure that the injection valve shall be opened in advance of the exhaust valve, and the cam or projection on the equilibrium arbor is adjustable round the arbor to provide that the trigger may not be raised and the exhaust valve opened until the equilibrium valve is shut.   Plate LI shows the lower nozzle and working-gear in plan ; the upper arbor is cut away, just beyond the point at which the handling-arm is fastened into it, to expose the lower arbor with its handling-arm and detent.

The upper working-gear, Plate LII, has mounted on its arbor, in addition to a wiper, and an arm adapted to be acted upon by a chock on the plug-tree, a detent retained by a pivoted catch. · When the exhaust-valve arbor turns to open its valve

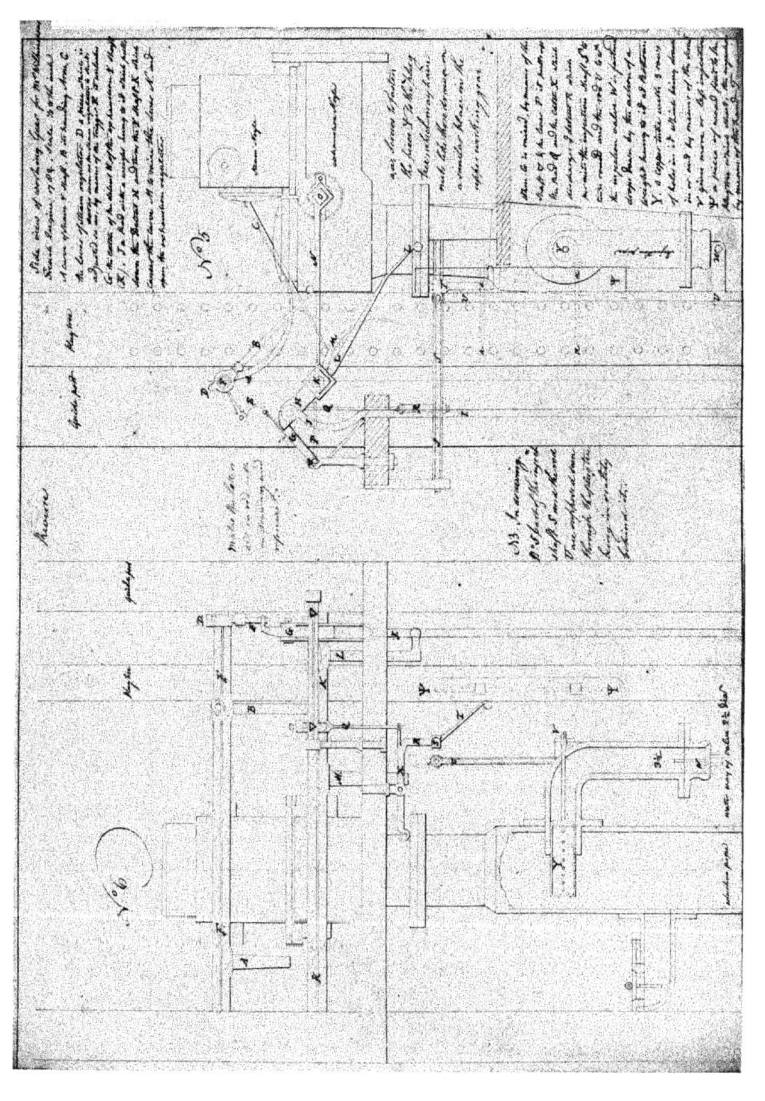

L. WORKING-GEAR FOR 'MR. WILKINSON'S FRENCH ENGINE', 1782

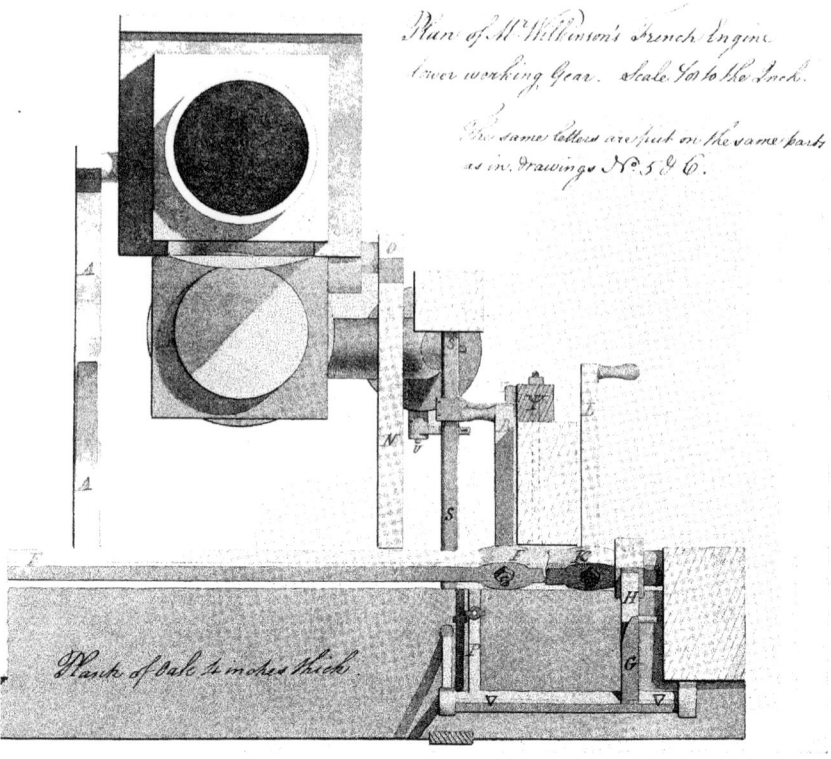

Plan of Mr. Wilkinson's French Engine lower working Gear. Scale 100 to the Inch.

The same letters are put on the same parts as in drawings No. 5 & 6.

Plank of Oak 4 inches thick.

LI. PLAN OF LOWER WORKING-GEAR FOR ' MR. WILKINSON'S FRENCH ENGINE ', 1782

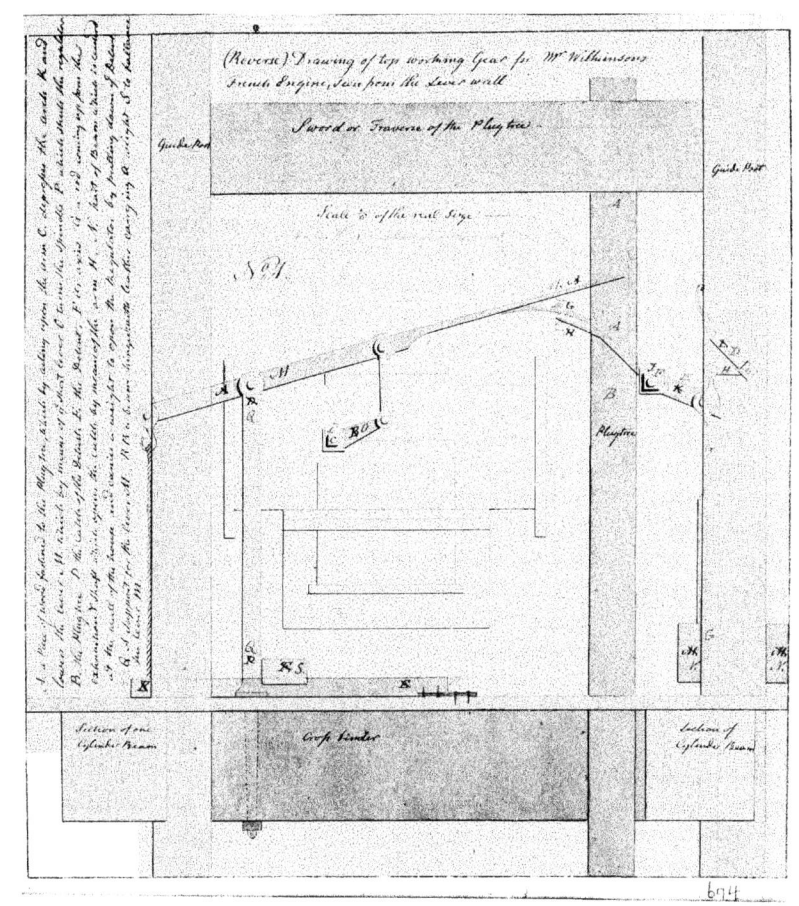

LII. UPPER WORKING-GEAR FOR 'MR. WILKINSON'S FRENCH ENGINE', 1782

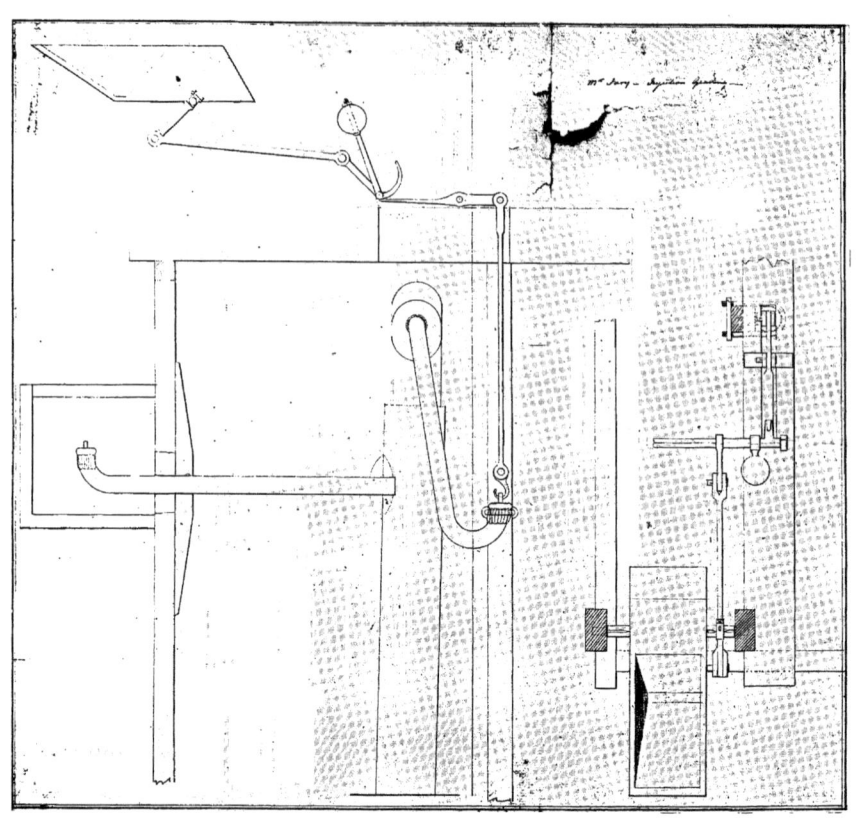

LIII. CATARACT AND INJECTION GEAR FOR JARY'S ENGINE, 1779

a projection on the arbor pushes up a vertical rod and raises the catch to release the detent which drops under the influence of a weighted lever coupled to it; the consequent rotation of the wiper lifts the valve lever and opens the top valve. Upon the descent of the plug-tree the chock presses down the arm on the top arbor, the wiper moves down with it, and the valve lever, partly counterbalanced, follows it and the valve closes under the action of gravity; at the same time the detent pulls up its weight and re-engages with its catch.

In brief, the rotation of the equilibrium arbor, to close the valve, causes the unlatching of the exhaust and injection gears and allows these valves to open; and the rotation of the exhaust arbor, to open the valve, causes the unlatching of the top gear and the opening of the steam valve.

THE CATARACT

When Watt went to Cornwall for the first time he found the cataract in use on the Newcomen engines. When and by whom it was first applied to the engine is not known. Apparently it was not known outside Cornwall, for Watt writes:

' The cataracts are very good things, and as I shall have experience of one here, shall advise you how they answer '; [1] and soon after Boulton suggests that the apparatus should be fitted to one of the Soho engines—' I think we should have a rehearsal of Cataracticus at Soho before we perform in publick, as our workmen may be au fait.' [2]

The device that was in use in Cornwall was a tumbling arrangement; a stream of water was directed into a box mounted on pivots; when the box became charged with a certain weight of water it overbalanced, emptied itself, and was then swung back to its original position by the weight of the attached parts. By adjusting the stream of water, by means of a cock, it was possible to regulate the time between successive operations of the device and consequently the period of dwell between successive strokes of the engine. The apparatus is in fact a form of water clock.

This was the form of cataract employed by the Soho firm for some years. Plate LIII reproduces the drawing of the cataract and its connexion to the injection valve for the engine for M. Jary of Nantes—the first Boulton and Watt engine put up abroad. The tumbling box is of wood and has fixed to its under side an axle turning between centres, and carrying an arm which is linked to an arm on the shaft which carries the tumbling bob and wiper; the wiper presses down the tail of a lever, the other end of which pulls up the injection valve.

The drawings for the Nantes engine are dated 1779; in the same year a new form of cataract was tried at Bedworth. Of this we have no particulars except that ' the same water always serves for it and is never changed '.[3] Boulton expressed the view that it answered very well, but it is to be inferred that some fault developed, for in the course of the following year Ale and Cakes engine was fitted with some form of air cataract, said to be very neat and convenient. The drawings of some of the Wheal Virgin engines show cataracts in the form of circular bellows; probably they were air cataracts such as were used at Ale and Cakes.

The piston and cylinder arrangement came into use a few years after this; no drawing of this form of cataract has been noted in the Boulton and Watt

[1] B. & W. Colln. Watt to Boulton, 1777 Aug. 30.
[2] Ibid. Boulton to Watt, 1777 (not fully dated).
[3] Ibid. Boulton to Watt, 1779 Jan. 6.

Collection, but in 1785, in connexion with the working-gear for Wheal Towan and Wheal Crane engines, we read of ' an air or water regulating pump similar to those of the cataracts'.

THE LINKAGE GEAR

Upon the advent of the double-acting engine a new form of working-gear made its appearance ; the wiper acting upon the valve arm is now replaced by a lever connected by a link to a lever on the rocking spindle that actuates the valve. For the Soho corn-mill engine (the first double-acting engine) we have a drawing, dated December 1782, of ' The Side View of the lower working-gear for the double 18 inch engine ' showing such a connexion, and with each arbor provided with a detent to engage a pivoted catch, and with an arm to which is suspended the weight for opening the valve. Following this, in January 1783, we have drawings dealing with ' Various ways of equalizing the different powers of the pressures on the valves at ye different degrees of opening—it being taken for granted that the pressures are, when shut 160 lb. ; when $\frac{1}{4}$ open 90 lb. and when fully open 40 lb. or nearly so.' The engine was tried at work in March 1783, in all probability with the linkage gear shown in the drawing of December 1782, but it would seem that the gear did not give satisfaction, for in the following May we find a new drawing showing a wiper gear for the same engine. Watt now seems to have come to the conclusion that it would be best to work out the application of the new form of gear to the single-acting engine before going on with the double-acting. One of the single-acting engines on hand at the time was for Whitegrit mine (in the parish of Shelve, Shropshire) ; the first drawing of a working-gear for this engine, made in April 1783, shows the wiper gear, next there is a drawing of a linkage gear, marked—' laid aside because the rod for opening the upper working gear would be in the way of the man's hand.' There are then two drawings of ' Working gear executed for Whitegritt engine 1783 ' showing a linkage gear. Other single-acting engines for which linkage gears were designed in the same year are the Wheal Maid whim engine, the Rochfort engine, and Fenton's engine.

Fenton's engine was for pumping at Rothwell Haigh Colliery near Leeds. The drawings of the working-gear, made in November 1783, are reproduced in Plates LIV and LV. Apart from the linkage connexions between the arbors and the rocking-spindles in the nozzles, which are clearly shown, the gear is very like that for Wilkinson's French engine described above. It will be observed that the linkages form toggle mechanisms. This is one of the features in the gear described in the specification of Watt's 1784 patent, wherein each valve is worked by a toggle linkage, which, when the valve is closed, is overset or carried beyond the dead centre, so that it has no tendency to turn back, while, on the other hand, when it begins to open the valve, it acts with great power. The drawing to the specification shows the arm on each valve-spindle coupled by a link to a lever on the arbor and each handling-arm worked by a pair of pegs on the plug-tree, one to close the valve and overset the linkage, the other to unlock the toggle and allow the valve to open, another feature of the specification being that the valves are inverted and ' are pushed open by the action of the steam upon them '.

The drawings of the Wheal Towan and Wheal Crane double-acting engines, 1784, show the exhaust valves arranged to open downwards by the pressure of the steam, and both steam and exhaust valves operated by toggle linkages. They show also what became a characteristic feature of the gears for double-acting

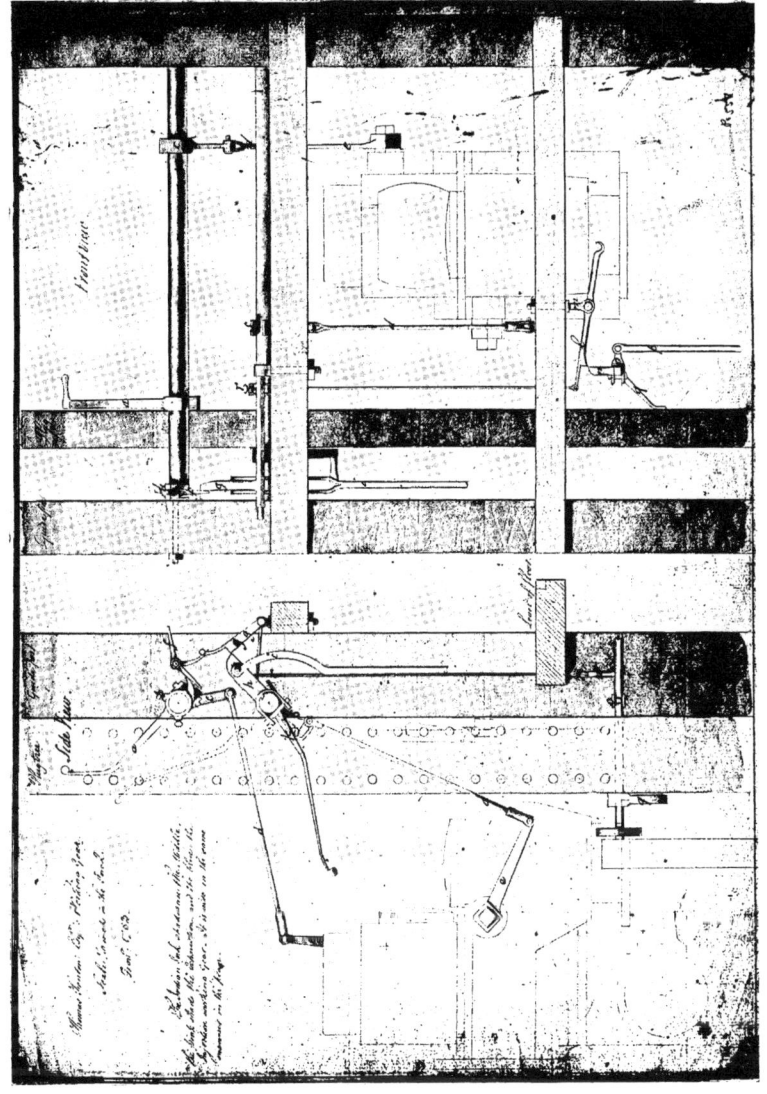

LIV. WORKING-GEAR FOR FENTON'S ENGINE, NOVEMBER 1783

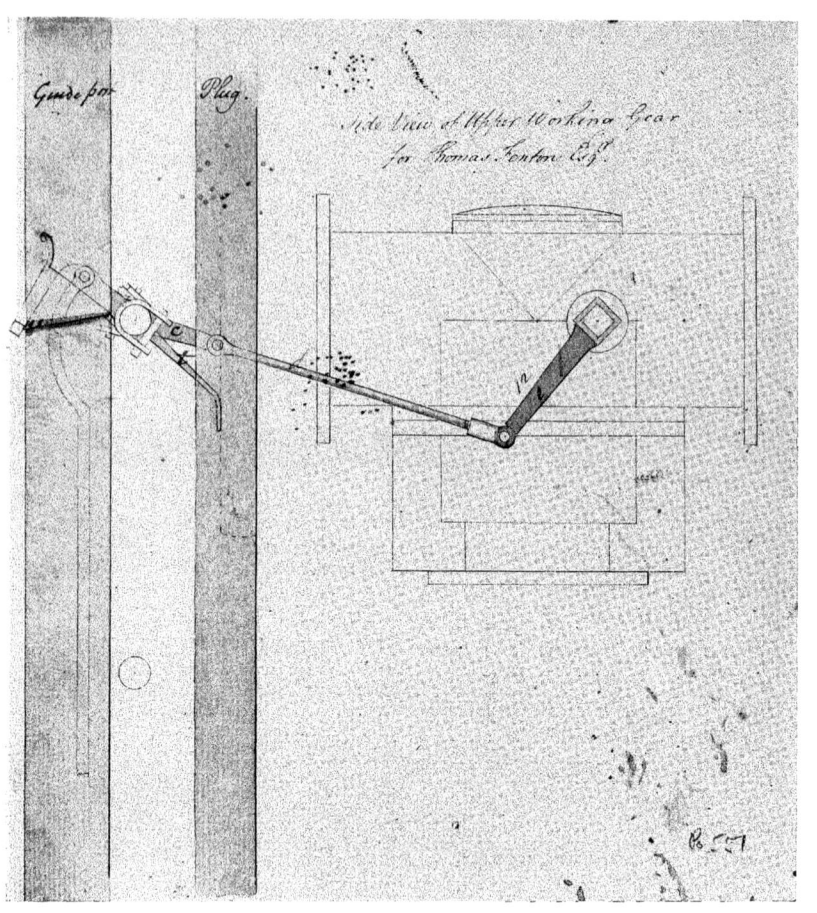

LV. UPPER WORKING-GEAR FOR FENTON'S ENGINE, NOVEMBER 1783

engines, i.e. the four valves are worked from two arbors, the upper steam and lower exhaust from one, and the upper exhaust and lower steam from the other. Each arbor has a handling-arm ; when one of these is struck by a peg or chock on the plug-tree the shaft is turned a little to carry the toggle mechanism over the dead point, the pressure of steam upon the exhaust valve then pushes it open, and, completing the movement of the shaft, causes the opening of the steam valve at the other end of the cylinder. These gears had no weighted arms, and no interlock between the two gears on each engine. The Wheal Mount engine, in the following year, 1785, with a working-gear otherwise similar had locking-catches such that the opening of one pair of valves was made dependent on the closing of the other pair. This feature of the interlock between the two arbors is one that received a good deal of attention, and several modifications of it are found in the drawings at Birmingham.

The first engine for the Albion Mills, as originally made, had no locking-catches, but the gear was reconstructed, with valves opening upwards, and a double-ended catch pivoted midway between the two arbors was arranged to engage detents on these arbors. The construction will be understood by reference to the drawing of the gear for Robinson's engine, 1791, Plate LVI, which, however, differs in one respect, i.e. the weighted arms on the arbors are replaced by springs, as will be explained later on. When the piston, moving upwards, approaches the end of its stroke an adjustable chock on the plug-tree raises the handling-arm of the lower arbor and by the engagement of its detent with the lower end of the pivoted catch causes the catch to swing and move its upper end clear of the detent on the upper arbor. This arbor is then free to turn under the action of a weighted arm and the upper steam and lower exhaust valves are opened as the piston arrives at the top of its stroke. These valves remain open during the descent of the piston, until, as it approaches the bottom of its stroke, another chock on the plug-tree depresses the handling-arm of the upper arbor ; the consequent rotation of this arbor with its arms causes the closing of the valves and the lifting of a weighted rod attached to one of the arms ; at the same time the detent engages with the hook at the upper end of the catch and the arbor becomes locked with the valves closed and the weight raised ready to reopen them. The movement of the detent over the curved end of the catch swings the catch so that its lower hook clears the detent of the lower arbor and so unlocks this arbor which, under the influence of its weighted arm, turns and causes the opening of the upper exhaust and lower steam valves.

The drawings for Liptrap's engine, 1786, show an exceptional form of working-gear. There are four arbors, one for each valve, each provided with a lever ; the lever on the top steam arbor is coupled by a leather strap to that on the lower exhaust arbor ; similarly the levers of the top exhaust and bottom steam arbors are coupled by another strap. Rennie seems to have had a good opinion of this gear, but it was reconstructed very soon. Another double-acting engine with four arbors in its working-gear was that for Thompson and Baxter, 1788.

The Hebburn 63-inch double-acting pumping-engine, 1798, was designed for expansive working and had four arbors and two plug-trees. The steam valves were closed by one plug-tree, and the exhaust valves by the other. The handling-arms for the steam valves were shaped to allow the slider or chock to pass over them ; those for the exhaust valves were not so. To open, or rather to allow the opening of, the valves, the mechanisms of the upper steam and lower exhaust valves were disengaged by one chock, and those of the lower steam and upper exhaust by another ; both these chocks could be moved out of action when the

B b

gear was to be worked by hand in starting the engine. The chock for the upper steam and lower exhaust was of a special design; it was seated in a recess in the plug-tree and pivoted at its lower end; normally it was pressed outwards by a spring in order that its upper end might be in position to engage the catch on the arbor, but it could be drawn back within the recess by mean of a link and lever. This device could be used for stopping the engine. The injection valve was worked by a lever coupled by straps to levers on the steam-valve arbors; it was opened and closed with the opening and closing of each steam valve.

As we have seen, the linkage gear was used for single-acting as well as for double-acting engines. A gear designed in 1786 for an engine put up in Holland is shown in Plate LVII. The equilibrium valve, in this case at the top of the cylinder, is worked with the exhaust valve from the lower arbor and the top valve from the upper arbor. On the descent of the plug-tree, when the piston is nearing the bottom of its stroke, a slider or chock thereon acting on the upper handling-arm turns the arbor, lifts the weight, re-engages the detent with its catch, and, by the linkage connexion causes the closing of the top valve. At the same time a peg on the plug-tree strikes a lever (not shown) which moves the catch *b* clear of the detent on the lower arbor; this arbor is then turned by its weight, simultaneously closing the exhaust valve and opening the equilibrium valve ready for the up-stroke. In its upward movement, when near the top of the stroke, the plug-tree, by means of a chock, turns the lower handling-arm, and with it the arbor and attached parts, closing the equilibrium and opening the exhaust valve, lifting the weight and restoring the detent into engagement with its catch. This arbor also by means of a wiper lifts the tail of the catch for the detent of the upper arbor which is then pulled round by its weight to open the top valve.

It will be seen that the equilibrium and exhaust valves are positively connected and that one opens when the other closes, that they are worked one way, to close the equilibrium and open the exhaust by the plug-tree, and in the other, to open the equilibrium and close the exhaust, by the weight. The top valve is opened by the weight when this is liberated by the rotation of the lower arbor, and is closed by the plug-tree. The injection valve is worked by a separate arbor in the same way as in Poli's engine.

For the engine known as Poli's engine (built for the King of Naples) the working-gear, Plate LVIII, has an arbor for each valve, and the gears of the steam, equilibrium, and exhaust valves are interlocked. The opening of the equilibrium valve is made dependent on the closing of the exhaust valve; and the opening of the steam and exhaust valves, upon the closing of the equilibrium valve. The injection valve is independently worked. In the drawing the gear is shown with all the valves closed. The exhaust valve opens by moving downwards.

The steam arbor and the equilibrium arbor each has a detent adapted to engage a pivoted catch, and each tends to turn in the direction to open its valve under the influence of a weight suspended upon an arm projecting from it. Leather straps connect the tail of the top-arbor detent catch to a lever on the equilibrium arbor, the tail of the equilibrium-arbor detent catch to a lever on the exhaust arbor, and another lever on the exhaust arbor to a lever on the top arbor.

Rotation of the equilibrium arbor to close the equilibrium valve, and shut off the communication between the top and the bottom of the cylinder, releases the catch of the top-arbor detent; the top arbor turns under the influence of the suspended weight to open the top valve and, by means of the strap κ, pulls round

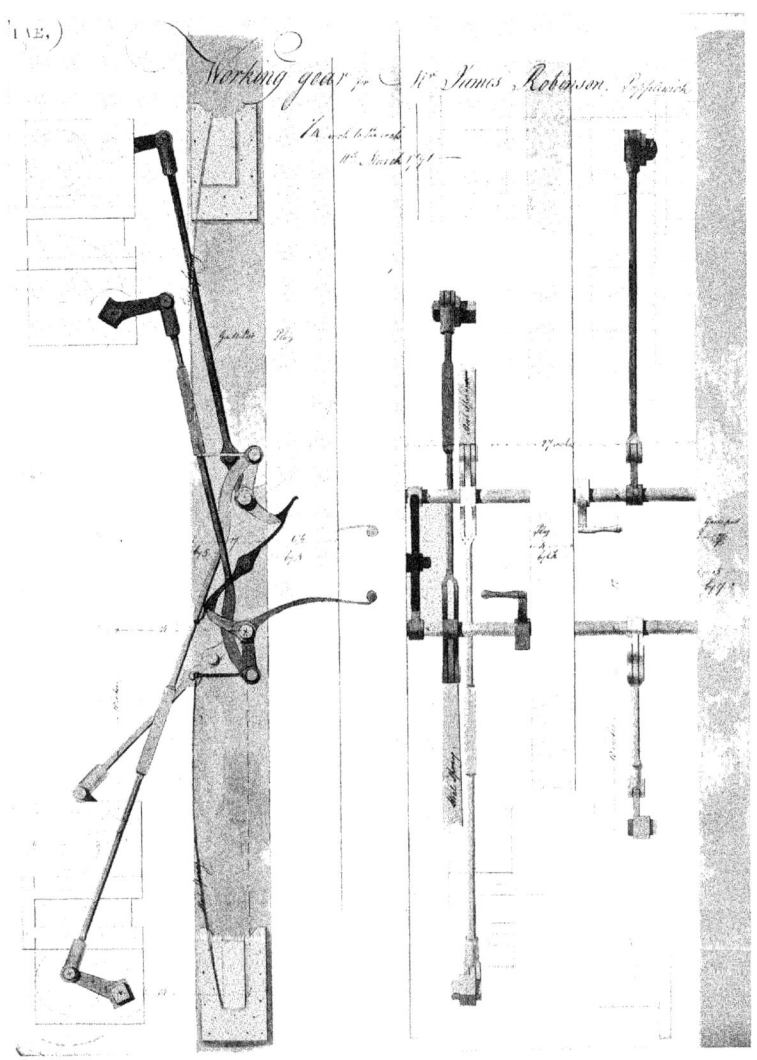

LVI. WORKING-GEAR FOR ROBINSON'S ENGINE, MARCH 1791

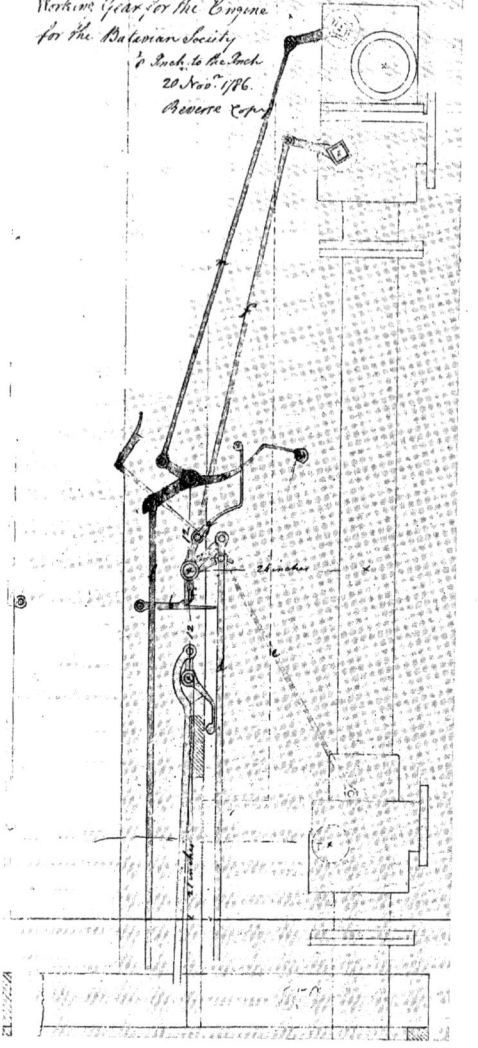

LVII. WORKING-GEAR FOR DUTCH ENGINE,
NOVEMBER 1786

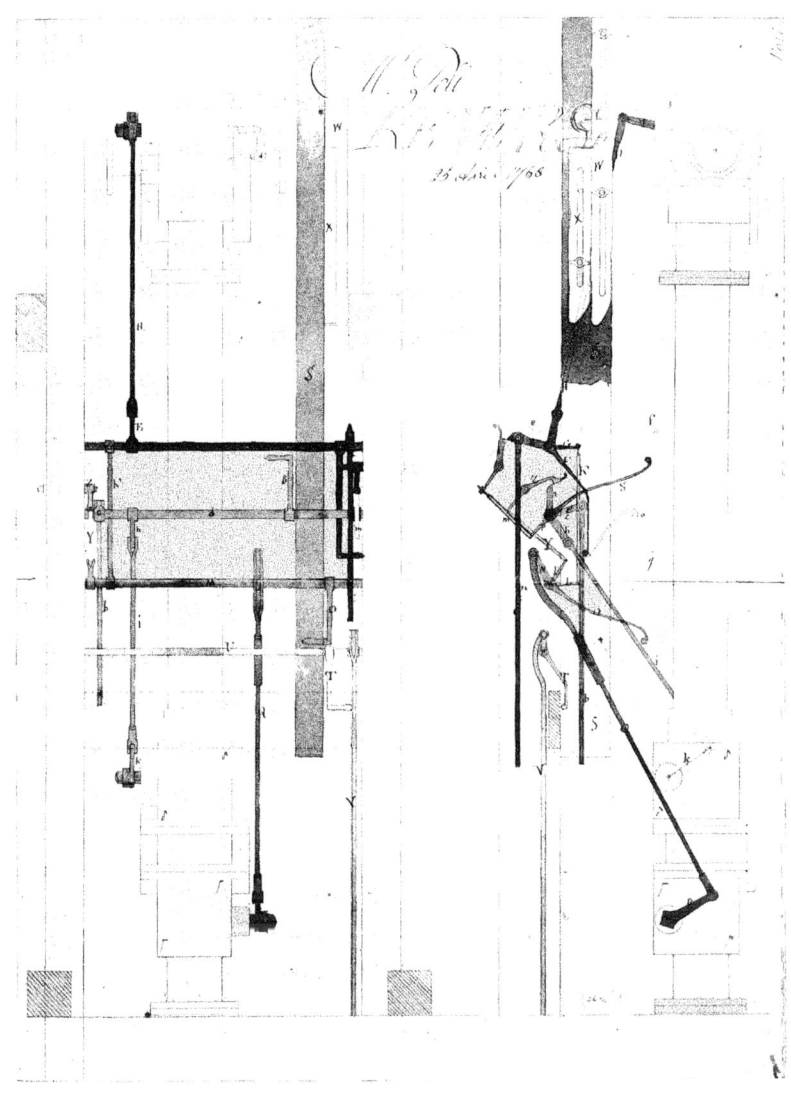

LVIII. WORKING-GEAR FOR POLI'S ENGINE, APRIL 1788

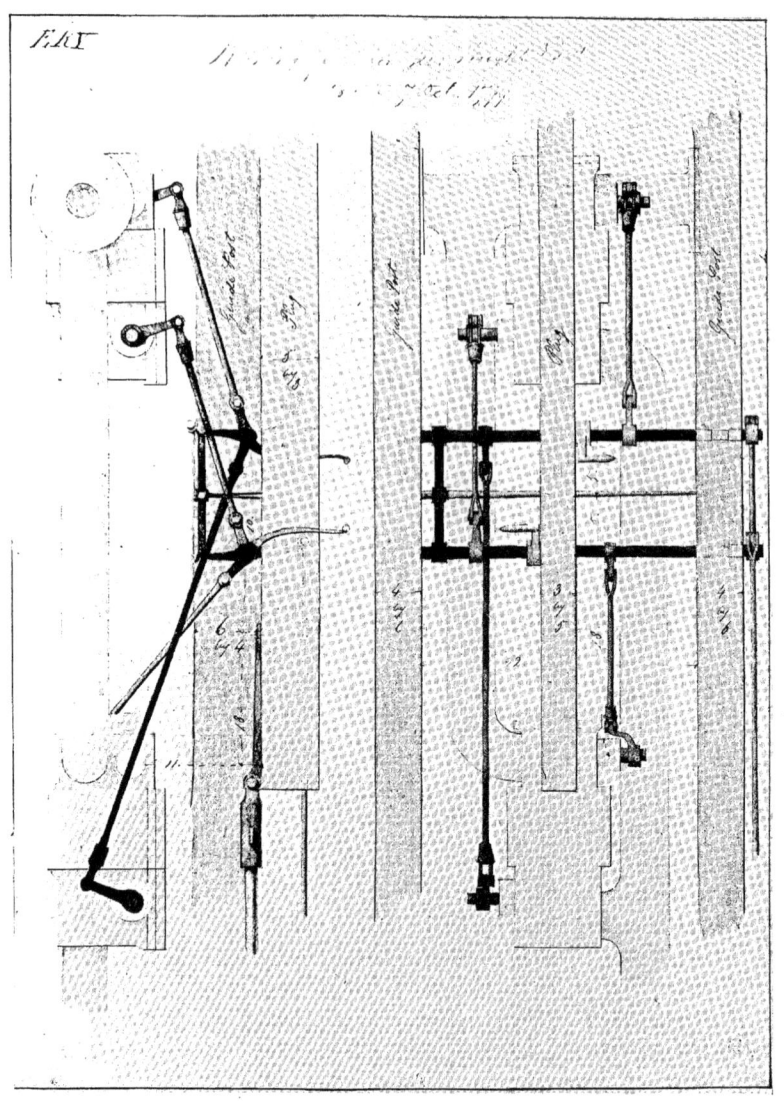

LIX. WORKING-GEAR FOR WRIGHT'S ENGINE, OCTOBER 1799

the exhaust arbor to initiate the movement of the exhaust valve; the movement of this valve is completed by the steam pressure on its upper face and the valve carries with it the exhaust arbor and the attached parts; the top of the cylinder is now in communication with the boiler and the bottom with the condenser and the down-stroke of the piston ensues. In the corresponding downward movement of the plug-tree the chock w acts upon the handling-arm F of the top valve and the chock x upon the handling-arm o of the exhaust valve, now in their raised positions, and restore them to the position shown with the top and exhaust valves shut. The rotation of the exhaust arbor, by means of the lever P and strap Y, pulls down the detent catch of the equilibrium arbor and this now turns under the influence of the suspended weight to throw open the equilibrium valve for the up-stroke. As the piston ascends a peg on the plug-tree engages the lower side of the handling-arm g and raises it to close the equilibrium valve. When the arm has been restored to the position shown, it pulls the strap m to disengage the catch of the top-arbor detent for the down-stroke.

The injection is worked by an arm T and rod v; the valve is opened by the action of a peg on the plug-tree on the underside of the arm; it is closed, as the piston begins to descend, by chock (not shown) pressing upon its upper side.

The equilibrium gear is released, to open the valve, by the rotation of the exhaust arbor; the top gear is released to open the valve by the rotation of the equilibrium arbor, and the exhaust arbor is turned to start the opening of the exhaust valve, by the rotation of the top arbor. Each gear is restored to the position shown by the action of the plug-tree.

The use of three arbors, however, was exceptional. The gear designed in 1798, for the single-acting pumping-engine at Herland Mine in Cornwall, had two arbors and resembled very closely that shown in Plate LIX; the upper arbor worked the top valve and the exhaust, and the lower arbor the equilibrium valve only.

Indeed we may say that, substantially, the gear described above for the Albion Mills engine remained the standard form until the termination of Watt's active connexion with engine designing. There is, however, one important exception to be noticed, i.e. in the course of the last ten years the weights acting upon the arbors are replaced by flat springs.

As far back as the year 1777 Watt had been experimenting with springs;[1] nothing came of the idea at that time, but about 1790 he took up the matter again. The drawing of the working-gear for an engine for James Robinson at Papplewick near Nottingham, dated Mar. 11, 1791 and reproduced in Plate LVI, shows an arrangement in which flat steel springs are connected by links to levers on the arbors and are anchored to cross-bars connecting the guide-posts of the plug-tree, the one opposite the top steam valve, and the other opposite the bottom exhaust valve. This became the standard form of gear and remained so for some years, although the erectors and enginemen did not look upon it with favour. Towards the end of 1791 the man sent out to erect an engine at Nantes wrote home:

'I have found with the working gear two long springs for opening the valves, which I think are not good things at all, for they seldom work well, and sometimes are breaking. If you have no objection, I should be glad to take them away, and apply weights, with a piston working in a cylinder cover'd with water in the cistern, and a sluice to adjust it, which is the best I have seen yet and makes the gear work very pleasant.'[2]

[1] B. & W. Colln. Watt to Boulton, 1777 May 10.   [2] Ibid. Richard Dayus to Watt, 1791 Nov. 11.

The erector's dislike to springs may have been due, in part, to an objection to change of any sort, but it is quite likely also that there was a difficulty in producing springs that would answer the purpose and last in work for any length of time. However, the spring gear cannot have been without good points for, as we have said, it remained in use for some years. Still, to the end, the workmen preferred to use weights. In 1797 some customers at Nantwich write that the engineer who is putting up their engine advises weights instead of springs for opening the valves, and ask if Boulton and Watt approve of this change.[1] Shortly after this the firm gave up the use of springs and reverted to the weights as shown in the next illustration, Plate LIX, from a drawing dated Oct. 7, 1799. Here the detent catch is in a different form and consists of a centrally pivoted vertical bar having flat ends terminating in horns ; a spring arm standing out at right angles from the centre of the bar is pressed at each end of the stroke by one or other of two small chocks adjustable on the plug-rod.

It will be noticed from this last drawing of 1799 that the wooden plug-tree and guide-posts are still retained, and that the arbors are carried by the guide-posts. Another drawing of the same year shows these arbors mounted in cast-iron brackets secured on the perpendicular pipes. Leather seems to have been used for the arbor bearings, at least in some cases ; such bearings were sent with the working-gear of the engine at Nantes, referred to above, and we find the erector asking that brass bearings may be sent instead.

In the early days Murdock, in Cornwall, was making parts of the gear of wood, for we find Boulton writing to Watt in 1784 that ' there are many parts of ye working-gear of wood which works quiet and easy and saves expense, ye detents are faced with a bit of sheet brass'.

Another note bearing upon the methods of construction is found on the back of a drawing for the ' Working gear for Horsehays ', dated 1784. This note runs :

' All the holes of the joints should be bushed with steel. The best way of doing which will be to plug them solid with a Plug of Steel welded in them large enough that the hole which must be drilled may have about $\frac{3}{16}$ of an Inch round its circumference and the steel hardened and tempered to a brown. The Pins should also be made of steel and turned to fit the Holes and then hardened and tempered.'

Whether these directions corresponded with the ordinary practice is perhaps doubtful.

Between 1780 and 1790 it became the practice, at any rate for the rotative engines, to construct the working-gear at Soho, and at one period it seems that the fitting was not well done. In 1787 Rennie, in London, was complaining that the gears were ' sent out so badly fitted that a considerable time is taken in erecting '.[2]

AUTOMATIC EXPANSION GEAR

Although nothing came of it, it is worthy of note that in 1798 Peter Ewart devised an automatic expansion gear, no doubt the first step in that direction. George Lee of Manchester was having a new engine built and was desirous of trying the expansive mode of working ; he and Ewart had discussed the matter and he forwarded to Soho, for consideration, the scheme that Ewart had arrived at.[3] The engine was the normal type of rotative built at that period—double-

[1] B. & W. Colln. Bott, Birch & Co. to Boulton & Watt, 1797 Oct. 27.
[2] Ibid. Rennie to Watt, 1787 Mar. 31.
[3] Ibid. George Lee to James Watt, junior. 1798 Nov. 7.

acting with drop valves worked by the plug-tree. Ewart's idea was to adjust vertically on the plug-tree the chocks that actuated the top and bottom steam valves ; this he proposed to effect by racks and a pinion turned by a cross-slide that was to be moved endways by the governor. He writes on his sketch that he ' is sensible that it is too much in the gimcrack way, but thinks it might answer under the judicious management of Mr. Lee who is laudably inclined to make the experiment of an expansive rotative engine '. The project did not materialize ; Southern seems to have suggested that before doing anything they should satisfy themselves that there was some real advantage to be gained by expansive working.

'SOCKET' VALVES

At the outset Watt, as we have seen, deliberately laid aside the plan of carrying the valve stems out through stuffing-boxes in the lids of the valve boxes, and the rocking-spindle arrangement of valve actuation which he then devised continued to be the only one fitted to the engines made for customers until within a year or so of 1800. When Murdock returned from Cornwall to Soho he devised an arrangement, that became known as ' Murdock's socket valves', in which the valves in each nozzle were placed concentrically, one below the other, the spindle of the one passing up through the hollow spindle of the other, and the hollow spindle sliding through a stuffing-box in the nozzle lid ; the two spindles were actuated by separate external levers. The first engine in which this plan was adopted was made for Boulton's mint in 1798, but the engine, although ready, was not set to work until May 1799. The two horse-power bell-crank engine made for Symonds was the first engine with these valves sent away from Soho ; it was dispatched in September 1799, but had been completed some time before. See p. 171. This plan seems very soon to have superseded the old one, that is to say, in engines to which drop valves continued to be applied.

THE ECCENTRIC GEAR

Up to the year 1800, not only the pumping-engines, but the rotative engines as well, had their valves worked from the beam of the engine. In that year for the first time the valves of the rotative engines were worked from the rotating shaft. This signally important change was another of Murdock's inventions. From a note by James Watt, junior, we learn that an eccentric of wood was applied to one of the small engines at Soho Manufactory in September, and then one of iron to the large engine at Soho Foundry in December 1800, and in the same month one was sent out with an engine for Sayce and Kelson of Bath ; in the following year wooden eccentrics were sent with an engine for Leith Docks in June and an iron one for an engine for Liverpool Docks in August.[1]

The drawings in the Boulton and Watt Collection, however, show that the project had been under consideration some two years earlier, although probably it had not been carried into practice. The drawings of an engine for Soho Mint dated 1798 show two eccentrics, and those of the Salford Twist sun-and-planet engine in the following year have an eccentric also. This last engine was working with the eccentric before the middle of 1802, when we read that ' Mr. Lee's eccentric

---

[1] Ibid. James Watt, junior, to Southern, 1803 June 25.

circle works very inharmoniously and must be immediately replaced by a new one'.[1]

Some, at any rate, of these early eccentrics were not embraced by straps and were merely cams, but in 1803 we find a drawing showing a cast-iron disk, brass strap, and a wooden eccentric-rod. This is for an engine for Coates & Co. of Hull. Accompanying the drawing are directions for setting the eccentric, an operation that was no doubt a difficult one at first ; in 1802 William Creighton, one of the draughtsmen, was at Bath in connexion with an engine being put up there and writes home that he could not set the eccentric properly. This was a slide-valve engine.

In applying the eccentric to the actuation of drop valves, the lever of the steam valve at one end of the cylinder was coupled directly by a link to the lever of the exhaust valve at the other end, and the rocking shafts for working them were placed at the lower end of the cylinder.

THE SLIDE VALVE

Murdock's patent of 1799 includes improvements in the valves of steam engines. His specification describes a cylindrical valve rocked on its axis, and the long D-valve, and says, in reference to the latter : ' The invention consists in connecting the upper and lower valves so as to be worked by one rod, and in making the stem which connects them hollow, so as to serve for an eduction pipe to the upper end of the cylinder, by which means two valves are made to answer the purpose of four in Mr. Watt's double engine.'

Broadly speaking, the sliding principle was not new. The valves of the Newcomen engine were flat sectors oscillating on pivots over a flat face, and this, with a cavity, was the form used by Watt in the first instance ; he even called them ' sliding valves'.[2] But it is interesting to find that there is distinct evidence that rectilinearly moving slide valves had been designed at Soho long before the date of Murdock's patent.

In the chapter on 'The Rotative Engine' reference has been made to the 12-inch engine with a trunk piston put up at Soho. The drawings of this engine show four different schemes for valves. First we have drop valves. Next we have, dated January 1783, a 'Drawing of a Sliding Regulator for a double Reciprocating Engine'. This is reproduced in Plate LX ; it shows a valve of D-shape in cross-section, and a valve box with a steam branch at the centre and an exhaust branch at each end. The drawing is finished, coloured, and shaded ; at first sight the thought presents itself that, in error, the drawing had been antedated by some years; comparison with the other drawings in the same portfolio, however, lends no support to this view. Following this is a drawing, dated June 1783 and marked ' Drawing of the Sliding valve for 12 inch Engine at Soho ', showing a piston valve with two square pistons connected by a round tube through which the exhaust is led from the upper end of the valve box to the lower. Corresponding to this drawing is the hand sketch reproduced in Fig. 21, which it will be observed bears neither title nor date. Lastly we have a drawing dated November 1783 and headed ' 12 Inch Sliding Valve', Plate LXI. Here each valve consists of a segment, of a

[1] B. & W. Colln.   James Watt, junior, to M. R. Boulton, 1802 June 23.
[2] Writing from Kinneil to Dr. Small, Mar 1,

1770, Watt states that he intends to have a new valve, not a sliding one, but with two doors (Doldowlod Papers).

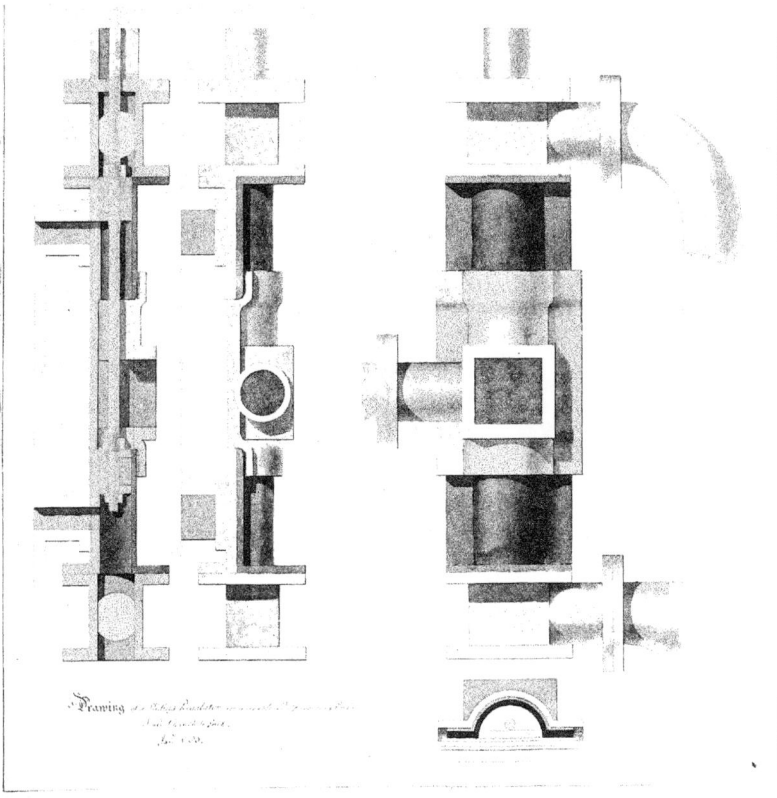

Drawing *of a Sliding Regulator on a wooden foundation board*
scale 1 inch to a foot
Jan.ʸ 1783

LX. 'SLIDING REGULATOR', JANUARY 1783

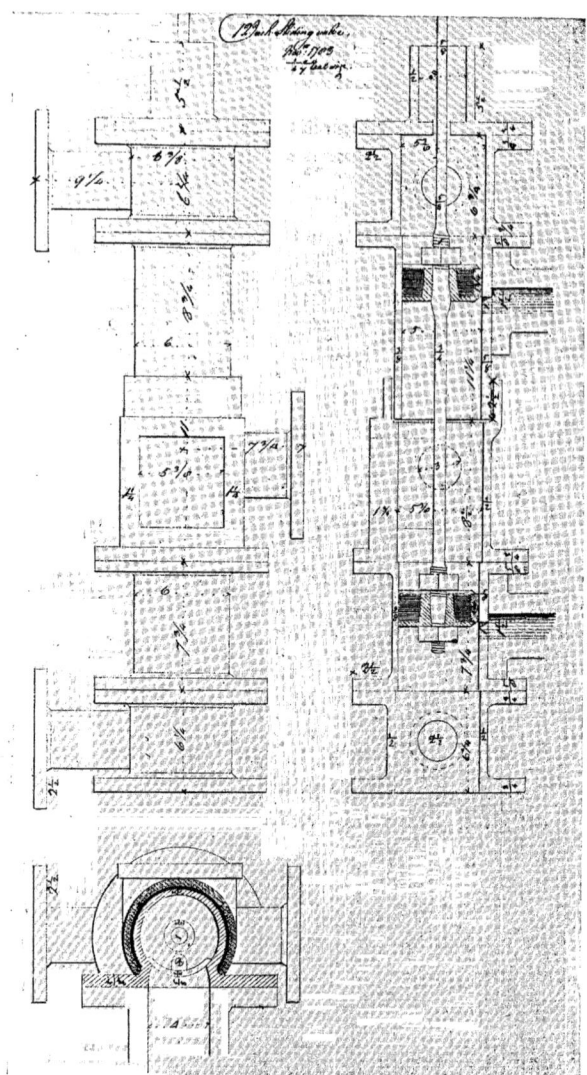

LXI. 'SLIDING VALVE', NOVEMBER 1783

size adapted to cover the port, clamped between a pair of disks secured on the valve rod and working in a cylindrical valve box having a branch at the centre and at each end. A drawing dated 1793, which seems to be for the same engine, shows the trunk piston and the cylinder unaltered, but the valves are drop valves. Nevertheless there is reason for thinking that the engine was made and tried with the valve shown in the last-mentioned drawing. Writing from Cornwall in 1786, in reference to Murdock's model steam carriage, Boulton says : ' Wm. uses no separate valves but uses ye valve piston something like the 12 inch little engine at Soho, but not quite.' This makes it clear that at the date of the letter the engine had such a valve.

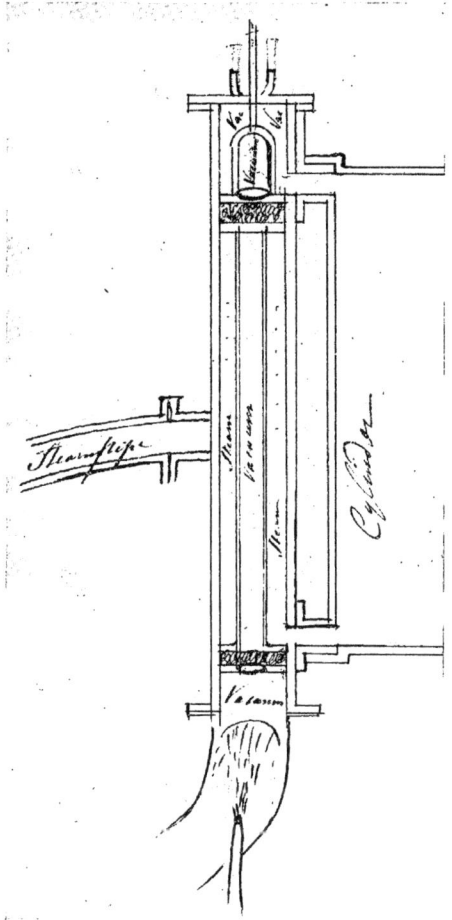

The sketch reproduced in Fig. 21 is not from the hand of Watt, possibly it may have been made by Murdock, and the idea of these sliding valves may have been due to him. The fact that, as Boulton states, he applied a similar valve himself soon after lends colour to this suggestion. The model steam carriage lately in the Birmingham Art Gallery has a piston valve with a passage along its axis. This model is commonly said to have been made in 1784 or 1785, but Murdock himself gave the date as 1792 (see p. 294).

However, the material available is not sufficient to allow a definite conclusion to be arrived at in this matter, but it is a little remarkable that when the slide valve was taken up again at Soho we find no reference to the earlier scheming. The drawings of 1783 were no doubt made by Southern, who had com-

FIG. 21. Piston valve in 1783

menced his service under Boulton and Watt in 1782, and he was still at the head of the drawing office.

It was towards the end of 1798 or early in 1799 that Murdock brought forward his idea for the D-slide. The firm had decided to embark upon the construction of small engines of a type cheaper than the beam engine, and it was proposed to fit such an engine then on order with a slide valve. Southern, however, demurred to supplying a slide-valve engine to a customer until they had had some experience at Soho with such valves. He expressed himself ' much afraid that the rubbing planes will get fluted, and should that happen, the repairs will be so *difficult* and *expensive*, that you will find the loudest complaints will be made against you'. At the same time, he thinks that for a sliding valve Murdock's design is quite unexceptionable, and that there is nothing to be hoped for in the way of improvement.[1]

Watt supported Southern in the objection to fitting slide valves on this engine, and it was made with the drop valves. Within a year or two, however, engines with slide valves were being made; for instance, an engine for Williams at Bath, 1801–2, had a slide valve worked by an eccentric, and in 1803 we find Rennie ordering ' 2 six horse sliding valve engines for the Hull Docks'. Curiously enough, seeing that the idea was to produce a cheaper engine, the slide-valve engines seem to have been more costly than the drop-valve engines; at least we learn in 1810 that for a 14 h.p. engine slide valves cost £60 or £70 more than circular valves. The difference seems very considerable even allowing for the fact that the flat surfaces had to be prepared by hand. It was not until some years later that a planing machine was at work at Soho.

[1] B. & W. Colln. Southern to James Watt, junior, 1799 Apl. 22.

XVI

GENERAL ENGINE DETAILS

CYLINDERS; PISTONS AND PISTON-RODS;
STUFFING-BOXES; PARALLEL MOTION;
ENGINE BEAMS; CONDENSERS

CYLINDERS

SETTING aside the Soho expansive engine, which in the first instance was made without a steam jacket, the cylinders of the early engines comprised a plain working cylinder flanged at its lower end, and an outer cylinder flanged at both ends, at the top to receive the cylinder lid and at the bottom to bolt on to the outer cylinder bottom, which was tied down to the foundation by holding-down bolts. The upper end of the working cylinder was open; its lower end was closed by the inner bottom from which proceeded the branch for the supply to and discharge of steam from the lower end of the cylinder. This branch was bolted to the outer bottom, which was formed with a corresponding branch for connexion to the valve box or nozzle. The inner bottoms for the first Soho engine and the Bloomfield and Broseley engines were made of sheet copper. A certain amount of trouble arose with that of the Soho engine, so Boulton had it replaced by one of cast iron, and this material was used for the few cylinders subsequently made on this plan.

The joint between the branch on the inner bottom and the inner wall of the outer bottom was inaccessible. Watt, in 1779, refers to it as ' that plagued joining of the copper bottom to the nozzle pipe which we could not get at & always went wrong '.[1]

The use of an integral outer cylinder was soon abandoned—purchasers of engines were complaining that Wilkinson, the ironfounder, was charging at the same rate for the unbored outer cylinder as he was for the bored working cylinder, and he refused to abate his prices, so the idea of dispensing with this cylinder presented itself. The first scheme was to replace the cast-iron cylinder by one of plate iron; wrought-iron plates, however, were expensive and difficult to obtain at that time, and it is not quite clear that any engine was made with a cylinder on this plan. At any rate it was very soon followed by another, that of building the outer casing of segmental panels of cast iron, and this became the standard practice of Boulton and Watt. This change necessitated a number of other changes in design. The working cylinder now became the essential structural feature; the lid was secured on its upper end; it was formed with a branch near the top, and the inner bottom, closing its lower end, rested upon and was bolted

---

[1] B. & W. Colln. : Letter Books. Watt to John Wilkinson, 1779 July 16.

C C

to the outer bottom, and its branch was led out through the wall of the outer bottom and connected directly to the nozzle. The jacket was now a mere appendage to the cylinder, and whereas formerly the working steam passed from one end of the cylinder to the other through the jacket, an external perpendicular pipe was now used, and steam was supplied to the jacket by a separate pipe. The standard construction of the upper end of the cylinder in 1779 is shown in Fig. 22 (from the printed book of 'Directions for erecting . . . steam engines'). The upper end of the casing is shown at N; it fits around a ring cast on the cylinder, and at M are shown tubes, passing through this ring and through the cylinder flange, and entering crooked holes drilled in the flange B of the lid which is formed with lugs at this part. These tubes supply steam from the cylinder jacket to the lid jacket, which is formed by a copper disk O screwed to a ring cast on the lid and under a separate ring G. The branch at the upper end of the cylinder is shown at A. The upper end of the cylinder was bored $\frac{1}{4}$ or $\frac{3}{16}$ inch larger in diameter than the working part; the term 'bell mouthed' for this formation seems to have been first used in 1779.[1]

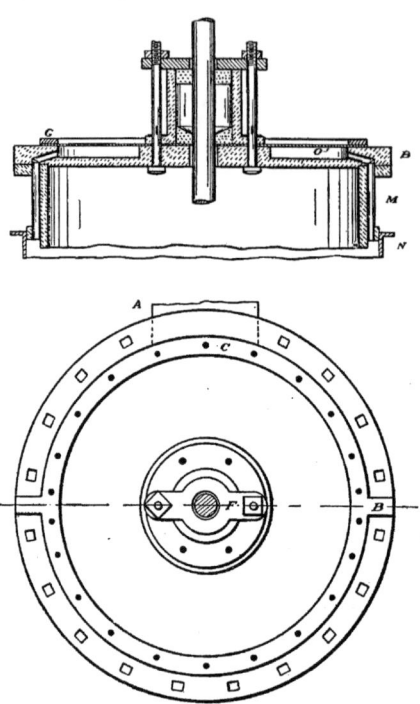

FIG. 22.   Upper end of Cylinder, 1779

The upper flange of the cylinder was faced, 'shaven flat for an inch and a half broad,'[2] and the lower face of the flange of the lid was turned, the unturned part forming the projection that enters the cylinder.

A 52-inch cylinder made in 1795 is shown in Plate LXII without its jacket. The note at the top of the drawing reads: 'The face of the flanches at the top and bottom of the cylinder to be turned flat and true, as also the side of the flanch of the cylinder lid and cylinder inner bottom that joins to the cylinder.' The barrel is 1 in. thick and the flanges $1\frac{1}{2}$ in. At the bell mouth the cylinder is $52\frac{1}{4}$ in. diam. The false cover for the lid is shown in Fig. 23; it is now made of cast iron, 'to be cast as thin as possible,' and around the neck that fits over the stuffing-box is formed an annular channel leading into a 'cup to hold grease'.

In a few instances the cylinders were cast in two lengths and the lids were made in halves, the division lying along a diameter.

[1] B. & W. Colln.: Letter Books. Watt to Turner, 1779 July 25.

[2] Ibid.: Letter Books. Watt to Turner, 1777 Nov. 6.

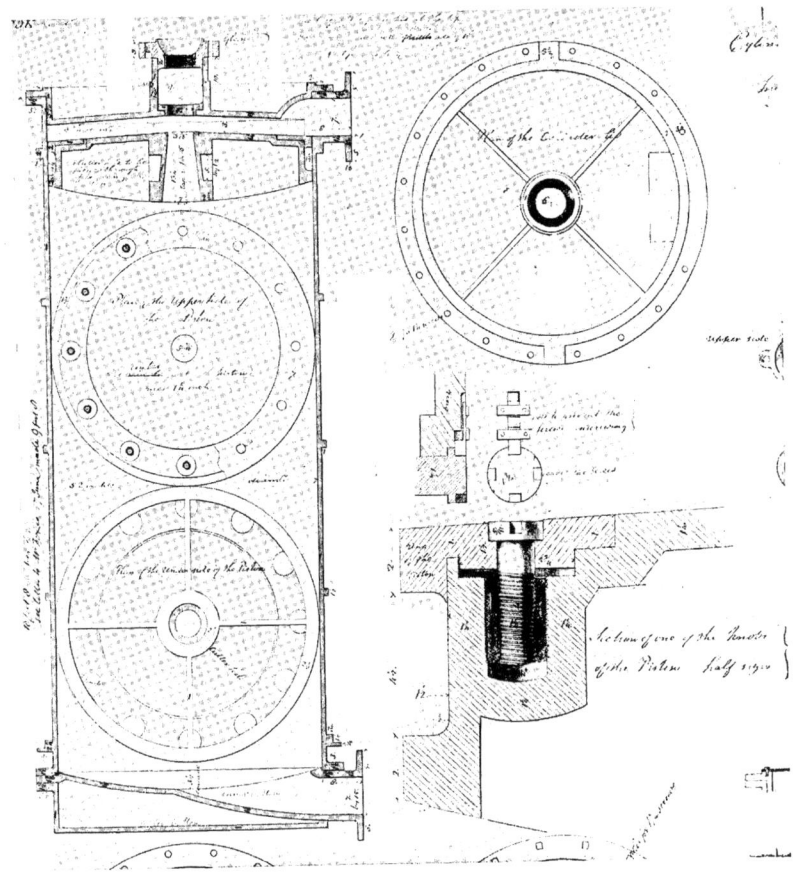

LXII. CYLINDER AND PISTON, 52-INCH, 1795

PISTONS AND PISTON-RODS

THE difficulties encountered in finding a satisfactory packing for the piston of the first Soho engine have been dealt with in an earlier chapter.

Hemp packing was in use for the pistons of Newcomen engines and answered quite satisfactorily, but there it was always saturated with water, and Watt, working with a dry piston, seems to have feared that it would be destroyed by heat in his engine. When in 1776 he set about trying it, Boulton advised him to try asbestos cloth if oakum would not do. However, the oakum seems to have been made to serve without much difficulty. The knack seems to have been to make the packing hard and to keep it thoroughly saturated with grease, and early in 1777 Watt was able to write of his piston

'It scarcely ever needs repairs after the first fortnight, during which time it must be often looked at,'[1] but it was not until after 1780 that the practice settled down to a simple hemp packing compressed by a junk-ring. In the meantime the standard construction had been that shown in Fig. 24, having a hemp packing under a ring of lead, made in sectors, and pressed down by springs. The sectors of lead are marked K, and the springs C. The springs are T-shaped and extend inwards to the eye of the piston where they are bent down to enter notches in the piston boss; this was to prevent them from moving outwards and rubbing against the cylinder, a difficulty that had been met with in practice. The springs are secured by screws to a ring D which in turn is fixed by screws H to the radial arms of the piston. To these arms are secured also

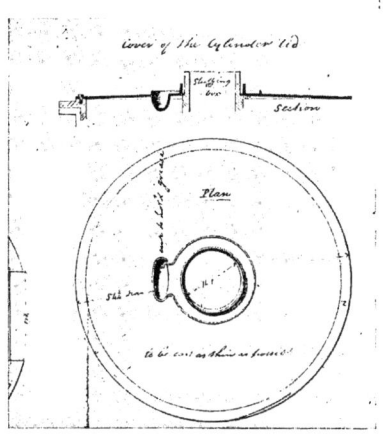

FIG. 23. ' Cover of the Cylinder lid ', 1795

the shackles F for lifting the piston out of the cylinder. The spaces between the arms were filled up with wood; this filling frequently gave trouble, pieces broke off and got into the valves. The printed book of ' Directions for erecting and working . . . steam engines' [1779] explains how the piston is to be packed: a flat rope or gasket is to be plaited from white or untarred rope yarns; this after a preliminary hammering is put round the piston and beaten down with a wooden driver and a mallet; molten tallow is then poured on, next a layer of oakum is applied and then another gasket, and so on until a depth of from three to four inches is attained, according to the size of the piston. The packing is to be beaten solid, but not so hard as to set up too great friction with the cylinder, and ' abundance of tallow should be allowed it especially at first'. When the engine is at work the cylinder cover should be lifted once every week, the spring and leads of the piston taken off and the packing beaten down with the driver and mallet, and fresh oakum or gasket added when necessary; molten tallow should be poured

[1] Ibid.: Letter Books. Watt to Hornblower, 1777 Feb. 19.

on before the leads are replaced, at the rate of two pounds for every foot of diameter of the piston.

Although in 1777 Watt seemed to be fairly satisfied, the piston was not perfect and, in the early days, at any rate, there was a good deal of trouble caused by the sticking of the leads; pewter was tried, and it was suggested that they might be made of brass or gun-metal. At one time it was the practice to put a ring of soft metal in sectors under the packing as well as above it, and the surfaces were made inclined to wedge the sectors outwards into contact with the cylinder. No definite indication has been found of the precise idea that Watt had in view in using the leads; there is evidence that in some cases they were not applied until the engine had been at work for some time and the cylinder had worn smooth. When the leads and springs were discarded the piston assumed what became the standard form of hemp-packed piston in which the packing is compressed by screwing down a junk-ring or piston-lid.

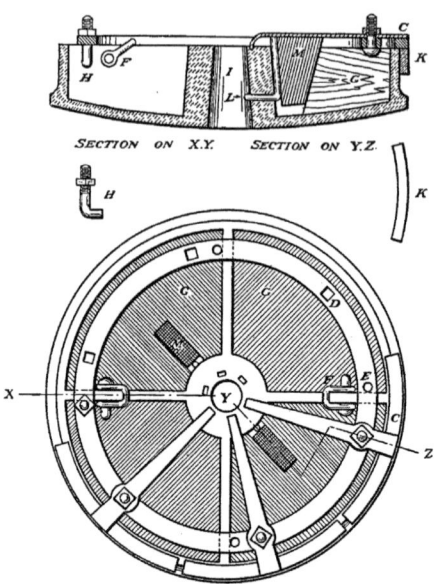

SECTION ON X.Y.   SECTION ON Y.Z.

FIG. 24. Piston, 1779

Plate LXII, taken from a drawing dated 1795 for North Downs No. 3 engine, shows in section the rim of a 52-inch piston. The junk-ring is held down by twelve 1¼-inch screws engaging in brass nuts which in turn are screwed into holes tapped in 'knobs' in the body of the piston. The head of each screw has a series of notches to take the end of a sliding latch or 'bolt', to prevent the screw unscrewing. The depth of the piston at the centre is 13¼ in., and at the periphery 8½ in. Whitbread's engine, 1784, had a 24-inch piston constructed in the same way, and the same means for locking the screws.

The piston-rod end was coned with the larger diameter below. This was a happy arrangement in two respects. It ensured a good fit (the rod and the piston, it may be explained, were made at different places and were not put together until the engine was being erected; a gauge for the taper was sent to the maker of each part) and the working pressure kept the piston tight on the rod; Watt at first thought that no other fastening was necessary, but it was found that the rod sometimes came down leaving the piston behind it. To cure this the first scheme was to screw a gland to the lower side of the piston covering the end of the rod. The plan shown in Fig. 24 was then adopted, pins were passed through radial holes in the piston boss and entered notches in the rod; access to the pins was afforded by slots cut in the wood filling of the piston; these slots were fitted with wedges M which bore upon the upturned tails of the pins. Later on, when the double-

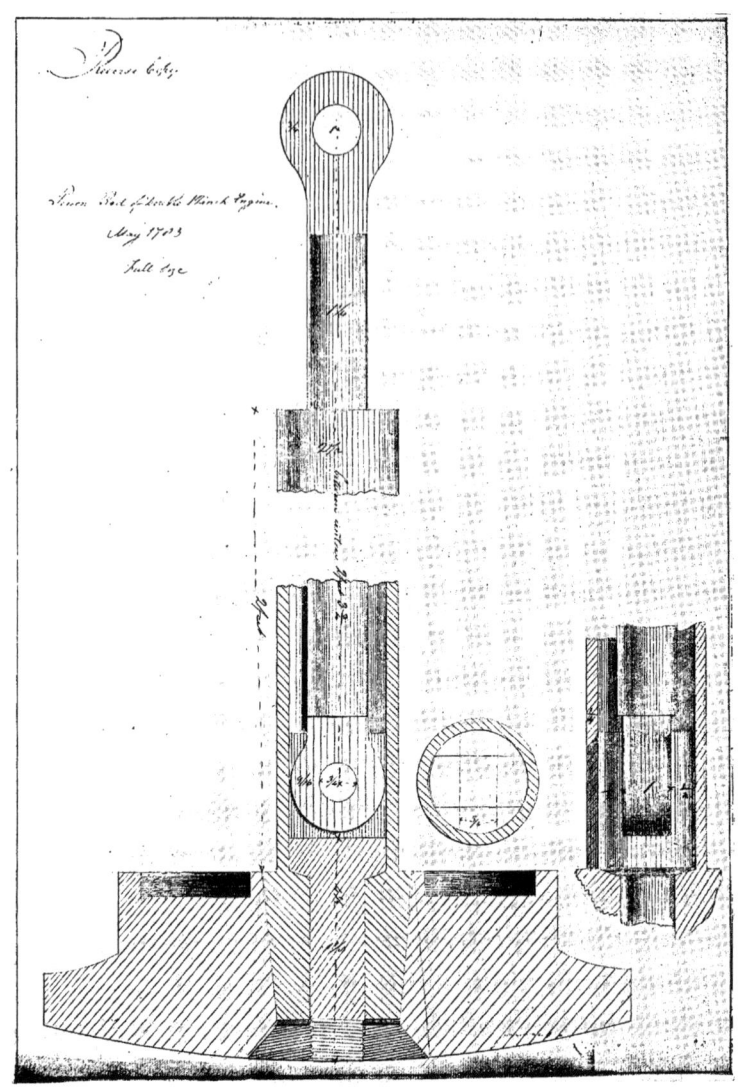

LXIII. TRUNK PISTON, 1783

acting engine came into use, these pins were replaced by cotters passing through the rod.

The 52-inch piston referred to above has a rod 5 in. diam., coned where it fits in the piston to a taper of 1 in 5 and secured by a cotter 5 in. by $1\frac{1}{2}$ in. passing through a slot in the boss of the piston. The cotter hole in the boss was formed partly in casting and was cut through to the eye after the cone was bored.

The method of securing the piston-rod to the chains is seen in Fig. 25, which shows a connexion for three chains, for two chains, and for one chain. It will be observed that in the one-chain and two-chain arrangements the cap is made in two distinct parts, and that with three chains, although the two parts are united to form the loop for the middle chain, they are not welded together to form a socket for the rod. The cotter, ' the great forelock, or cutter', is secured by a pair of smaller cotters, and these in turn by a leather thong passed through perforations in their ends.

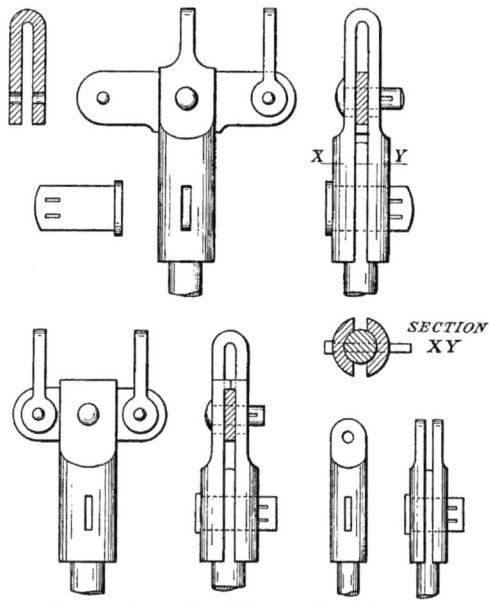

The rods for the most part were made in Cumberland or by an anchorsmith in London. Boulton in 1779 has occasion to compare the two makes. Bedworth piston-rod, he says, ' is like spunge or net work upon the whole surface and no part of it takes a pollish, this is owing to ye nature of the Cumberland iron, whereas ye Chelsea, Shadwell, and ye

FIG. 25. Connexion of piston-rod to chains, 1779

Birm^{gm} Navigation, which I saw yesterday, wears like a piece of polished cast steel without fibres or grain of any sort, from whence I conclude that piston rods composed of a number of small bars of good Swedes iron well fagotted, makes the best rods. I think it is a beautiful thing to see such a rod as ye B. Navigation engine work.' [1]

Trunk pistons. The specification of Watt's patent of 1784 shows a trunk piston, and it has been stated in Chapter XIV that such a construction is found in the drawings of a 12-inch engine, put up at Soho for experimental purposes. Plate LXIII is reproduced from the detail drawing dated 1783, and calls for no explanation. It seems that the engine was made with this piston ; we have no information as to its behaviour when at work, but another drawing of the engine, dated 1793, shows the cylinder unaltered and the piston with the trunk.

[1] B. & W. Colln. Boulton to Watt, 1779 Oct. 23.

STUFFING-BOXES

As Watt was the first to use a piston-rod working through the lid of the cylinder of a steam engine, it follows that no one could have used a stuffing-box on an engine cylinder before him, but it is perhaps a little rash to assert that he was the inventor of the hemp-packed stuffing-box. Watt never made this claim for himself, and it is likely that the contrivance was in use before his time. However this may be, he seems to have used it from the beginning, and to have met with no particular difficulty in connexion with it; at all events this was so with the earlier single-acting engines in which the pressure on the underside of the cylinder lid differed little, if at all, from that of the atmosphere.

In Fig. 26 three drawings have been brought together to illustrate the development of the stuffing-box. In the drawings for Colevile's engine, 1776, it is shown as a separate casting bolted to the cylinder cover; the upper half of the length of the box is bored out to a cylindrical form, then we have a sharply inclined cone forming the seat for the packing, and at the bottom the opening through which the piston-rod works; above the packing is an internally coned bush fitting within the cylindrical part of the box and intended to be held down by a gland. In the following year the plan of fitting a brass bush at the bottom of the box was adopted. This is the construction marked 1779, copied from the printed book of 'Directions for erecting and working the newly-invented steam engines' [1779]. Here the box is fitted with brass collars, and the gland is held down by bolts through which, above the flange of the box, are passed cotters, so that these bolts take the place of two of the holding-down screws. The box is to be packed 'with soft rope yarn, wrapt round the rod, until you have nearly filled the box, then take a collar of deal wood, two inches thick, made easy for the rod and the box; divide it in two by its diameter, lay it on the top of the stuffing, and apply the gland above it; as you go on with the packing, melt some grease and pour amongst it; and when finished screw down the gland moderately tight'. It would seem that no take-up of the packing was contemplated but that the gland was pressed down to the top of the box. Another set of instructions, printed a few years later than the above, directs that so much packing is to be used 'as to prevent the gland from touching the top of the stuffing-box, otherwise there will be no certainty that the box is tight; in which case, either air will pass through it, and injure the vacuum, or much tallow will be wasted, and will not cure the evil'.

The next step in the development of the stuffing-box was to cast the box on the cylinder lid. This we find in Whitbread's engine, 1784; here the gland is fitted with a brass bush and formed with a cylindrical extension to enter the box and, on its upper surface, with a depression or cup to contain tallow; the bush in the bottom of the box is formed with an extension to enter the eye of the cylinder cover. Substantially this drawing represents the standard practice for ordinary stuffing-boxes at Soho, at any rate up to the time of Watt's retirement.

We have now to say something about the lantern stuffing-box. With the normal form of single-acting engine in which the upper side of the piston was never exposed to less than about atmospheric pressure, as stated above, the stuffing-box had not presented much difficulty. With the double-acting engine, and with that form of single-acting engine in which the top of the cylinder was open to the condenser on the up-stroke of the piston, there was a greater difference of pressure to contend with, and any leakage which might occur would be leakage of air

inwards which, reducing the vacuum, would seriously affect the performance of
the engine. To meet this difficulty Watt adopted the lantern stuffing-box. This
he did at least as early as 1781 for the Soho expansive engine, for on the drawing
of another engine, dated Sept. 3, 1781, he gives a written direction that the stuffing-
box 'must have double brasses, and a pipe to supply it with steam as at Soho
engine'. Plate LXIV shows the stuffing-box of the Wheal Towan engine, 1784;
midway in the box, with packing above and below it, is placed a ring fitting easily
in the box and on the piston-rod ; this ring is formed with a cavity at the centre

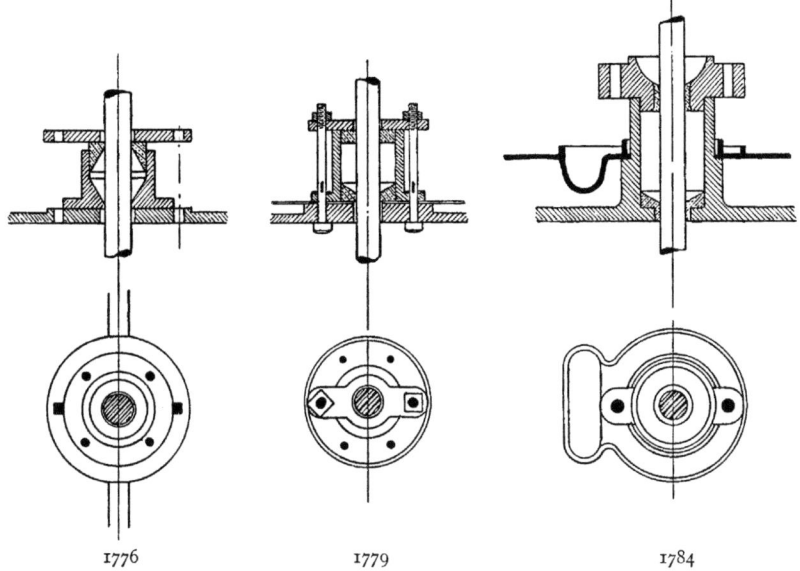

1776   1779   1784

FIG. 26. Development of the stuffing-box

and a groove at the periphery connected together by a series of radial holes ; steam
is led to the peripheral groove from the steam jacket of the cylinder cover by
inclined holes passing through the wall of the stuffing-box. It will be seen that
any leakage that may take place into the cylinder will be of steam from the ring.
We are not in position to say to what extent the lantern stuffing-box was used
by Boulton and Watt ; later on it became one of the features of Cornish engine
practice. The invention has been credited to Jonathan Hornblower, the younger,
but in view of what is said above as to its use by Watt in 1781, or earlier, this
point demands reconsideration. It was in 1781 that Hornblower took out his
patent for the compound engine, and it was in the same year that he began to
build an engine, that is to say on a working scale ; but we have no evidence of his
having used this stuffing-box until 1788 when he described it in an account of his
engine in Howard's *New Royal Cyclopaedia*. However, it is clear that the idea

of using a ring of steam to prevent the leakage of air round the piston of an atmospheric engine was in his mind in 1781, for this is one of the features of his patent specification, and from this he may have arrived very quickly at the cognate application to the piston-rod.

On the other hand, even if Hornblower was using the contrivance before Watt, it is difficult to see how he could have learned what Hornblower was doing; Watt found it sufficiently difficult to find out the main lines of the engine, and it would be even more so in regard to a detail of construction. On the whole, with our present information, the case seems to be one of independent invention.

THE PARALLEL MOTION

THE origin of the parallel motion and of its precursor, the three-bar motion, has been dealt with in the chapter on the double-acting engine (Chapter XII); it remains for us here to consider the manner in which these inventions were carried into practice. First, as to the three-bar motion—Plate LXV reproduces the drawing of 'Messrs. Stonard & Curtis's Motion, May 1785'. The old form of folded piston-rod cap is retained, but the cap is extended upwards to form eyes for a pin which carries a cross-piece; the ends of the cross-piece are round and are seated in bearings in the vertical links; these links at their upper ends work on a pin clamped in a 'saddle plate at top of motion'; the saddle-plate is formed with a rib fitted into a groove in the engine beam, and is secured by bolts passing through the beam; the lower ends of the links are jointed to the 'regulating radius's'. The joints at the middle and at the upper ends of the links are fitted with brasses adjustable by transverse keys.

It has been explained that this mechanism was soon superseded by the parallel motion; indeed, our next illustration, Plate LXVI, a reproduction of the 'Reverse Drawings of Parallel Motion for Albion Mill',[1] bears the same date, May 1785, as the drawing just described The piston-rod cap is now fitted with bearings to take over a transverse pin passing through the 'parallel bars', and the 'great perpendicular links'; these links have adjustable brasses at each end, those at the top fit over a pin clamped in the 'saddle plate top of motion'. The parallel bars are of wood with metal straps over their ends, the strap at one end fits over the transverse pin, mentioned above, that at the other end embraces the central squared part of a transverse pin on the round ends of which the inner or back links and the 'Reg. radius' rods are fitted. Like the parallel bars, the radius rods are of wood with iron straps; the strap at the front end embraces the squared part of a spindle, the ends of which turn in bearings secured to the 'spring beams'; these bearings are located well in front of the axis of the cylinder, so that the spindle does not foul the piston-rod. The inner vertical links are suspended from a cross-pin clamped in a saddle-plate—the 'Condenser saddle plate'—which serves also to secure a pin from which another pair of shorter links are suspended; these links at their lower ends are coupled to the central points of the inner links, and to the pin passing through them is jointed the upper end of the air-pump rod.

The parallel motion for Sheriff Hill, shown in Plate LXVII, is eleven years later in date than that for Albion Mills. The parallel bars and the radius rods are now of iron; the connexion of the air-pump rod to the motion has been simplified,

---

[1] For an explanation of the term 'Reverse Drawings', see p. 269.

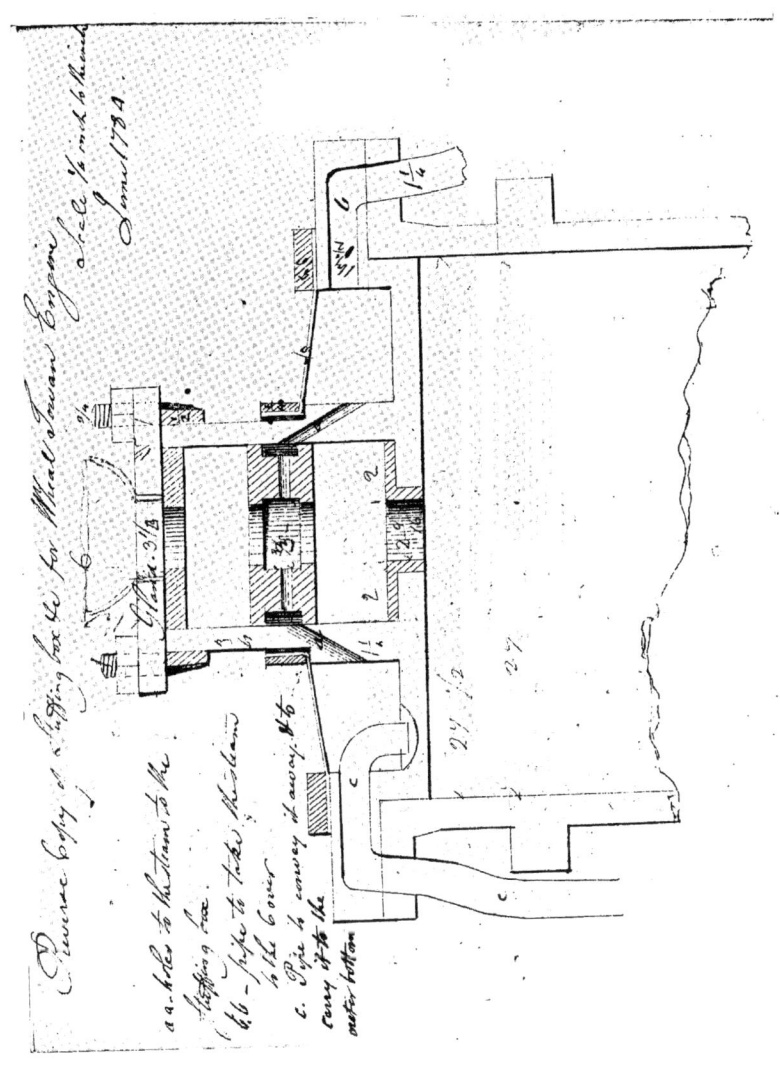

LXIV. STUFFING-BOX FOR WHEAL TOWAN ENGINE, JUNE 1784

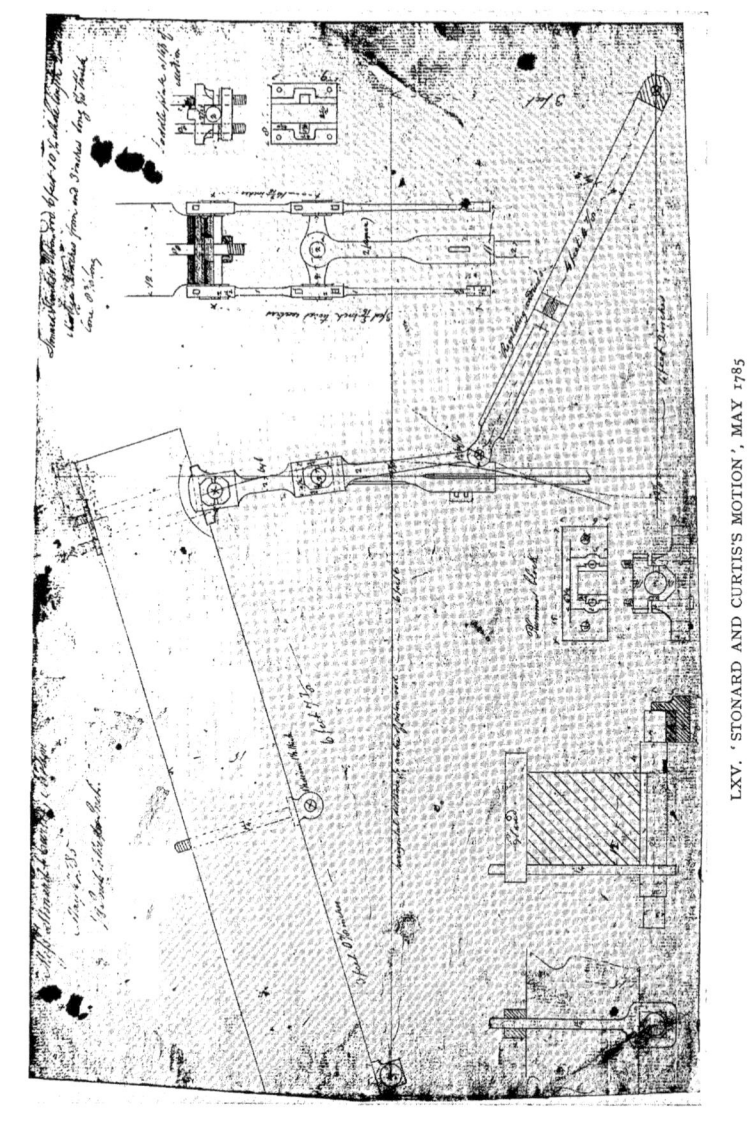

LXV. 'STONARD AND CURTIS'S MOTION', MAY 1785

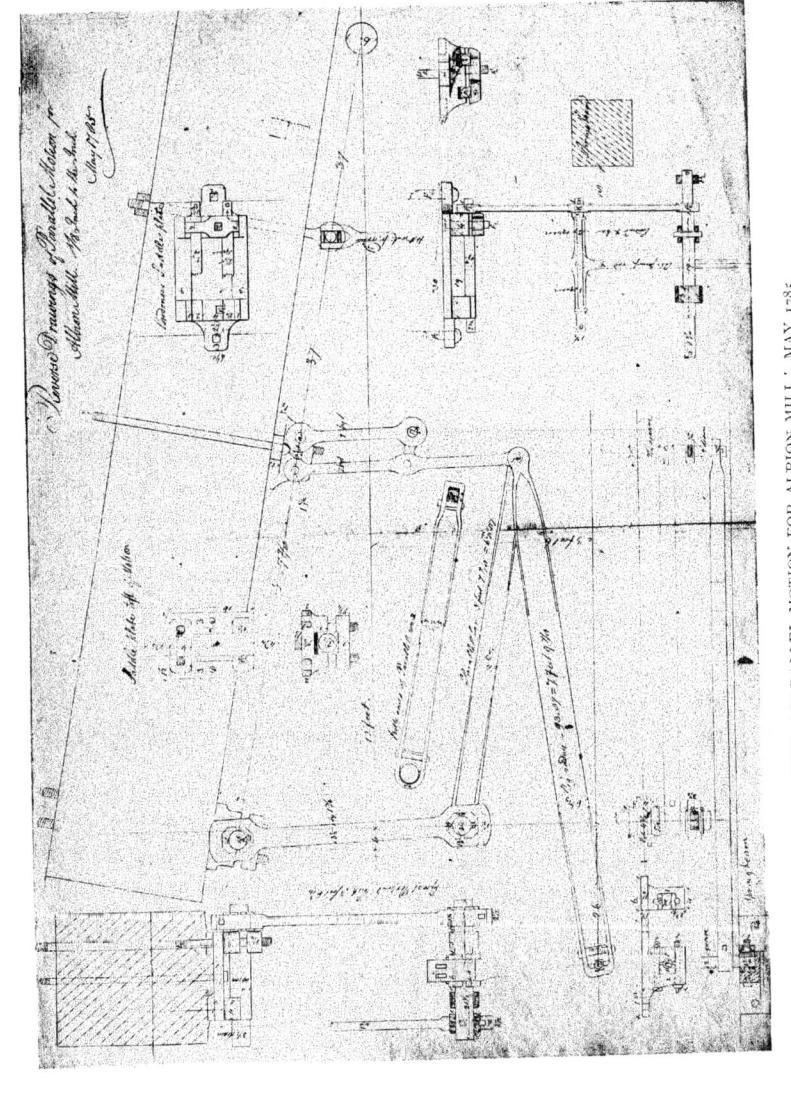

LXVI. 'PARALLEL MOTION FOR ALBION MILL'. MAY 1785

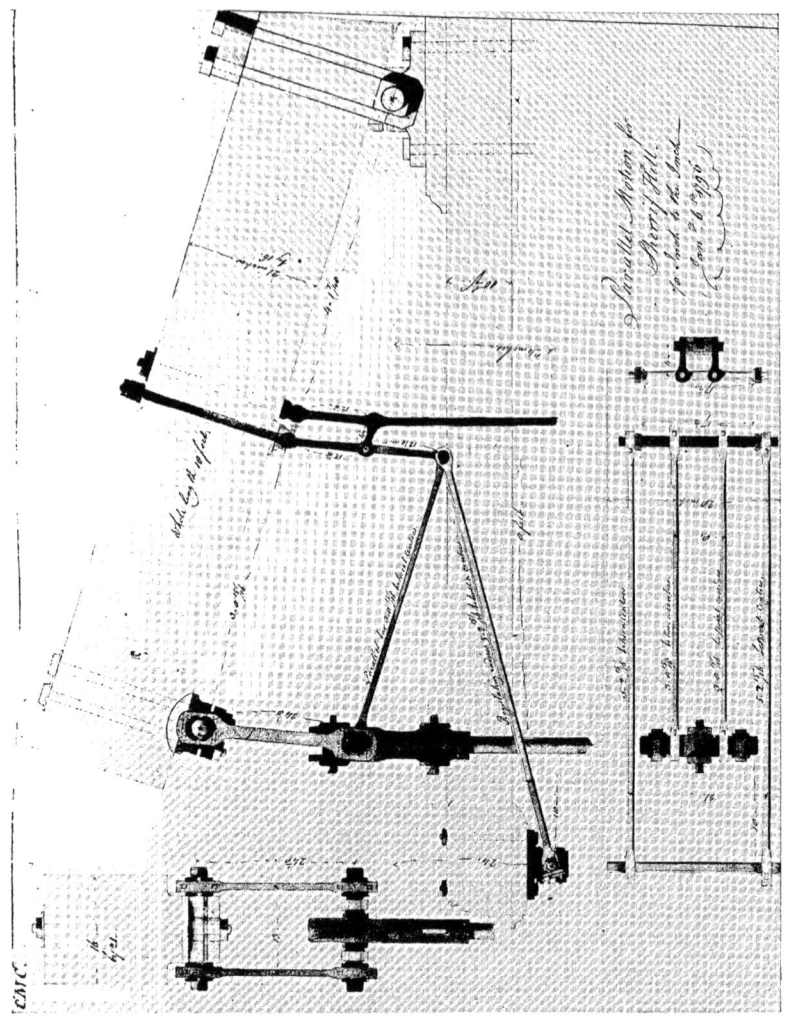

LXVII. PARALLEL MOTION FOR SHERIFF HILL, JANUARY 1796

and the common saddle-plate for the inner link and the air-pump link has been replaced by two separate fixings.

The pins in some, at any rate, of the parallel motions were made of cast iron, and the practice seems to have been a source of trouble. An engine at Cockshead Colliery, started in March 1793, had given great satisfaction at first, but in the course of the year 1794 a number of breakages occurred, and at the beginning of 1795 we have a letter with a long list of complaints. In so far as concerns the parallel motion, we learn that : ' The first thing we had broke was the lower great cast pin of the parallel motion, and the great perpendicular link, also the two little pins at top and bottom of the little perpendicular links and bended the piston rod very badly, all at same time.' Some time after they broke ' some small screw pins that hold the plumber blocks on at the inner end of the beam' ; then finally, ' the last of failiers was the great cast pin at the top of the great perpendicular links which broke and stoped us near 90 hours'. The writer goes on to say that they have replaced most of the cast-iron pins by wrought iron and propose to do the same for all ; they can make the wrought-iron pins themselves, and it is thought that they will save their cost in oil and brass in a year.[1]

It was admitted on all sides that the parallel motion was a beautiful and ingenious contrivance, but it called for increased intelligence on the part of the engine-man. Possibly this may have been the reason why some of the customers of Boulton and Watt preferred so late as 1793 to have chains and arch heads fitted to an engine.

ENGINE BEAMS

WITH very few exceptions, all the engines made under Watt's patent were beam engines, and all his beams were of wood ; at first oak exclusively was used ; then he found that in Cornwall deal logs were employed, and he sometimes varied the requirement that they should be of oak. Then again, somewhat reluctantly, he agreed to a built-up beam composed of a series of four or six logs laid side by side in pairs and held rigidly together by wooden keys and iron bolts. In the larger engines he provided a king-post tied to the lower side of the beam by diagonal braces. This arrangement prevailed until the advent of the double-acting engine ; then, as he points out to Boulton, ' our beams hitherto were made to resist a force acting downwards, but this must resist in both ways, and is above the power of a single log, and, on account of its length, above the power of any reasonable quantity of logs combined, without something in the nature of an arch or king post be applied to it'. In the same letter he says : ' The thing principally in suspense now is the working beam, which I am making some experiments to enable me to determine upon.' [2] The result was a stayed beam having two diagonal members extending upwards from the centre of the log, the ends of which members were tied together and to the ends of the log. The model beam (from the collection at the Science Museum, South Kensington) shown in Fig. 27 is of this construction, and is no doubt one of those upon which Watt's experiments were made. Such a beam was designed for the Wheal Virgin 63-inch engine in 1786, and is shown in the drawing of the Hebburn engine, Plate XXVIII. The beams were a very expensive part of the engine. In 1784 Boulton estimates the cost of Poldice beam at near

---

[1] B. & W. Colln. John Martin to John Southern, 1795 Jan. 11.

[2] Muirhead : *Mech. Inv.* Watt to Boulton, 1784 Sept. 11.

£600. Watt had been expressing doubt as to the durability of the beams, and Boulton replies : ' You are certainly mistaken about the durability of engine beams, I will venter to say that the beam now making for Poldice new engine will last 30 years with fair usage. Tingtang beam seems as good now as the first day. They ought to last long as its the most expensive part of the engine. I suppose Poldice beam will cost near 600 l.' [1] Tingtang engine had been at work about seven years at the time. The expense was an indication of the difficulty of getting suitable timbers, and as time went on the difficulty increased. The members of the firm were constantly on the watch for oak logs, and as late as the year 1800 we find James Watt, junior, saying : ' I propose going on Sunday with Southern, if he is able, to Wicknor Bridges to inspect some timber there, said to be calculated for our purpose.' [2] This letter opens with the statement that ' almost every order from Scotland is attended with one for a beam, and we shall soon be out unless we get a supply '. Another letter of the same date, quoting prices for engines in

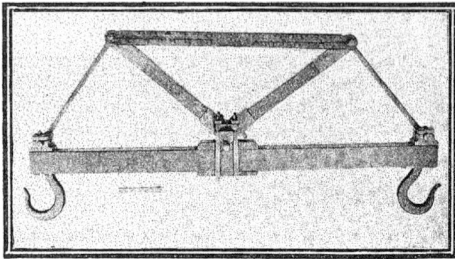

FIG. 27. Model beam, in the Science Museum, South Kensington

Cornwall, assumes the beam to be in one piece and strong enough without stays or king-posts.

The difficulty was solved by the substitution of cast iron for wood. The drawings of the Blaenavon blowing-engine in 1800 show a beam composed of two cast-iron webs ; in outline it closely resembles a single-log wooden beam, and it is likely that the design was prepared at the request of the ironmaster for whom the engine was being built. In the following year we have the drawing, dated Mar. 11, 1801, of a cast-iron beam for the Salford Twist engine ; this again has two webs, but the outline is now that which became the standard form as shown in Plate LXVIII. This figure is from the drawing, dated May 11, 1803, of the single-web beam for the engine for pumping at the Canal locks at Fazeley Street, Birmingham. It shows the cast-iron main gudgeon, 8 in. square in the body, which is wedged and keyed in the 9-inch square opening in the middle of the beam ; the oblong opening is for the purpose of receiving the upper pins of the back link of the parallel motion and of the link coupled to the air-pump rod. The end of the beam is made cylindrical to form a journal which allows of some lateral play in the connexions ; it is fitted with a cast-iron cap secured by a key. The drawing is interesting as showing how standard patterns were altered for particular purposes ; the beam is to be made to the ' 6 feet stroke beam pattern ' with the alterations indicated ; the gudgeon, to the ' 30 H. main gudgeon pattern ' also altered ; both patterns are to be restored to the original dimensions after use ; the cap at the end of the beam is the ' 45 H. Cap '.

As we might expect from such an ardent advocate of cast iron, John Rennie was early in the field in calling for cast-iron beams. In 1801 he writes that he has determined to have iron beams for three large dock engines, then in hand,

[1] B. & W. Colln. Boulton to Watt, 1784 July 8.   [2] Ibid. James Watt, junior, to M. R. Boulton, 1800 Oct. 17.

on account of the difficulty in getting good wooden ones.[1] It is clear that he did not think the iron beams would be better than those of wood ; it was merely that he could get them in iron without difficulty, whereas wooden ones were not obtainable. The Soho people seem to have hesitated in complying with his request, for a few months later he is again pressing for iron beams for the London Docks engines.[2] The substitution of a rigid material such as cast iron for the resilient material that had been used hitherto was naturally a step in which some hesitation would be felt, and in which Boulton and Watt would prefer to move slowly. As an instance of this hesitation, the drawings of the cast-iron beam for the Salford Twist engine contain evidence that the design of a skeleton beam of forged iron was under consideration at the same time. However, the cast-iron beams were found to answer perfectly well, and within thirteen years Watt was writing : 'for several years past, working-beams of timber for engines of every size have been entirely laid aside, and those made of cast iron have been employed in place of them, whereby the bending, splitting, twisting, dry-rot, etc., to which wood is subject are completely avoided '.[3]

To go back to the wooden beams, Plate LXIX shows the beam of the Tingtang engine, 1777, composed of six logs tied together by keys and bolts. Plate LXX represents a single-log beam with a king-post designed in 1785, and Plate LXXI a beam composed of two logs and with diagonal spurs designed in 1800 ; while Fig. 28

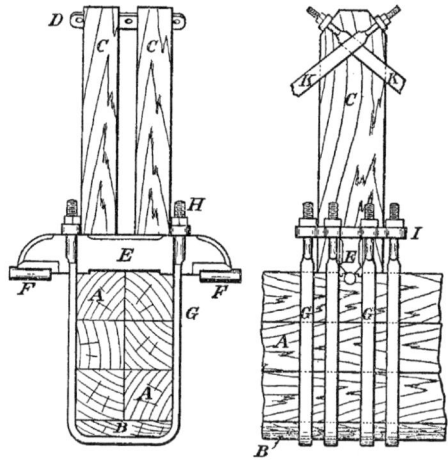

FIG. 28. Central part of beam, 1779

from the book of 'Directions for erecting . . . engines' of 1779 gives in detail the central part of the beam. All these drawings relate to beams for single-acting engines and show the gudgeon placed above the beam. For the double-acting and the rotative engines it was the practice to secure the gudgeon below the beam. Referring more particularly to Fig. 28, the gudgeon, E, was of cast iron, formed as a sort of bridge-piece extending across the beam and with trunnions, F, at its ends ;[4] it was held in place by U-straps, G, extending round the beam and passing through keeper plates or glands, I, resting on the gudgeon and clamped down by nuts on the screwed ends of the straps. The straps were let into recesses in the sides of the beam, A, and this was the main provision for keeping the gudgeon in its place in the length of the beam. To prevent chafing and injury of the beam

[1] Ibid. Rennie to James Watt, junior, 1801 June 30.
[2] Ibid. Rennie to Boulton & Watt, 1801 Dec. 28.
[3] Robison : *Steam and Steam Engines*, 143.

[4] The construction here described was in use in 1779 ; it superseded an earlier plan in which the gudgeon was formed with a wide saddle-plate to rest on the beam.

by the straps, a piece of wood, B, was spiked on to its lower face. The king-posts, C, were recessed to fit over the gudgeon ; from their upper ends diagonal stays, K, extended to the outer ends of the beam and to the lower sides thereof, where they were secured by bolts passing transversely through the beam. These bolts also served to secure the parts of the beam together in the lateral direction. The horse-heads or arches were connected to the lower part of the beam by diagonal stays intended ' to prevent the logs from sliding on one another by the difference of the direction in which the chain acts upon them when the end of the lever is up or down '. These stays were recessed into the beam to allow the king-post stays to pass over them. Extending diagonally inwards from the top of the arches to the top of the beam were the ' martingales ' ; these rested upon the tops of the arches and terminated in eyes through which passed the adjusting-screws for suspending the main chains ; their inner ends were also formed with eyes to receive bolts passing downwards through the beam. The main arches were recessed on each side to receive the arch-plates which were connected by bolts passing through the woodwork. The arches for the plug-tree and condenser-pump rods were formed with shoulders to rest on the top of the beam and were secured by screw bolts. The keys inserted between the layers of the beam, to prevent relative movement longitudinally, were recommended to be of very hard dry oak two inches thick, seven or eight inches wide at one end and four or five at the other. The ' Directions ' lay particular stress on care being exercised to avoid weakening the beam near the gudgeon, as by making bolt holes through it.

The beam chains were composed of open links, single and double alternately, united by pin joints, with their inner edges curved to correspond to the arch ; one or more chains were used according to the size of the engine ; they were made usually of wrought iron, but in 1777, in connexion with the Tingtang engine, Watt seemed inclined to advise cast-iron chains, mainly on the ground that the links were of uniform size and shape. It seems that cast-iron chains were in use in the Midlands and had been tried but given up in Cornwall.

The gudgeons were supported on plummer blocks fitted with brasses. Writing to Hornblower in 1777, in reference to the Tingtang engine, Watt says :

' The plummer blocks we make now of cast iron one foot square and two feet long, and thereby save the greatest part of the weight of the brass brasses which are lodged in a hole made to fit them in the plummer blocks. A pair of brass brasses for your engine will not weigh above 1¼ cwt. and by being equally supported will be as strong as if they were 10 times the weight.' [1]

The terms of this letter suggest that hitherto brasses had not been fitted in cast-iron plummer blocks, at any rate for engine beams. In another letter Watt writes : ' The rounded part of our cast iron gudgeons is about 12 inches long and six inches dia' ; not to increase the strength, for that may be done to any degree by the depth of the square part above the round which is 9 inches by 6. But to give a greater bearing, for if a great weight is laid upon a small gudgeon it will infallibly tear the brass it works in. As to the friction, that is lessened by contriving it so that the brass stands always full of oil.' [2]

A few years later (1780), in connexion with one of the Poldice engines, we have a drawing showing a brass in the form of half an octagon in section, bedding only

---

[1] B. & W. Colln. : Letter Books. Watt to Hornblower, 1777 Jan. 20.

[2] Ibid. : Letter Books. Watt to Hornblower, 1777 Feb. 19.

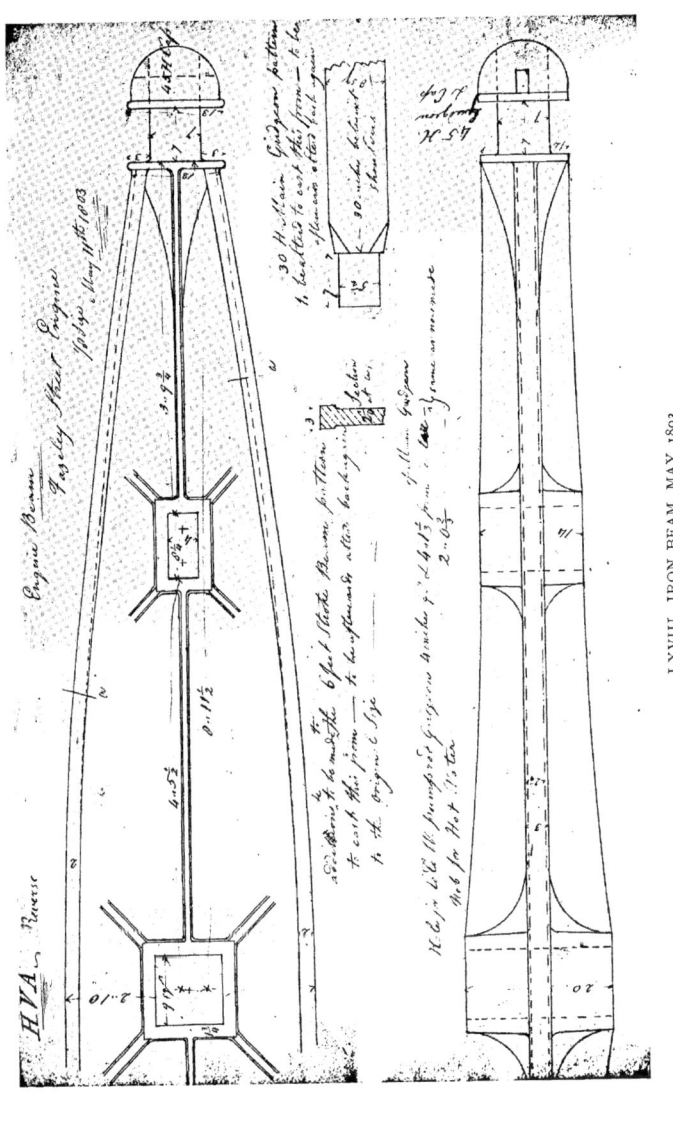

LXVIII. IRON BEAM, MAY 1803

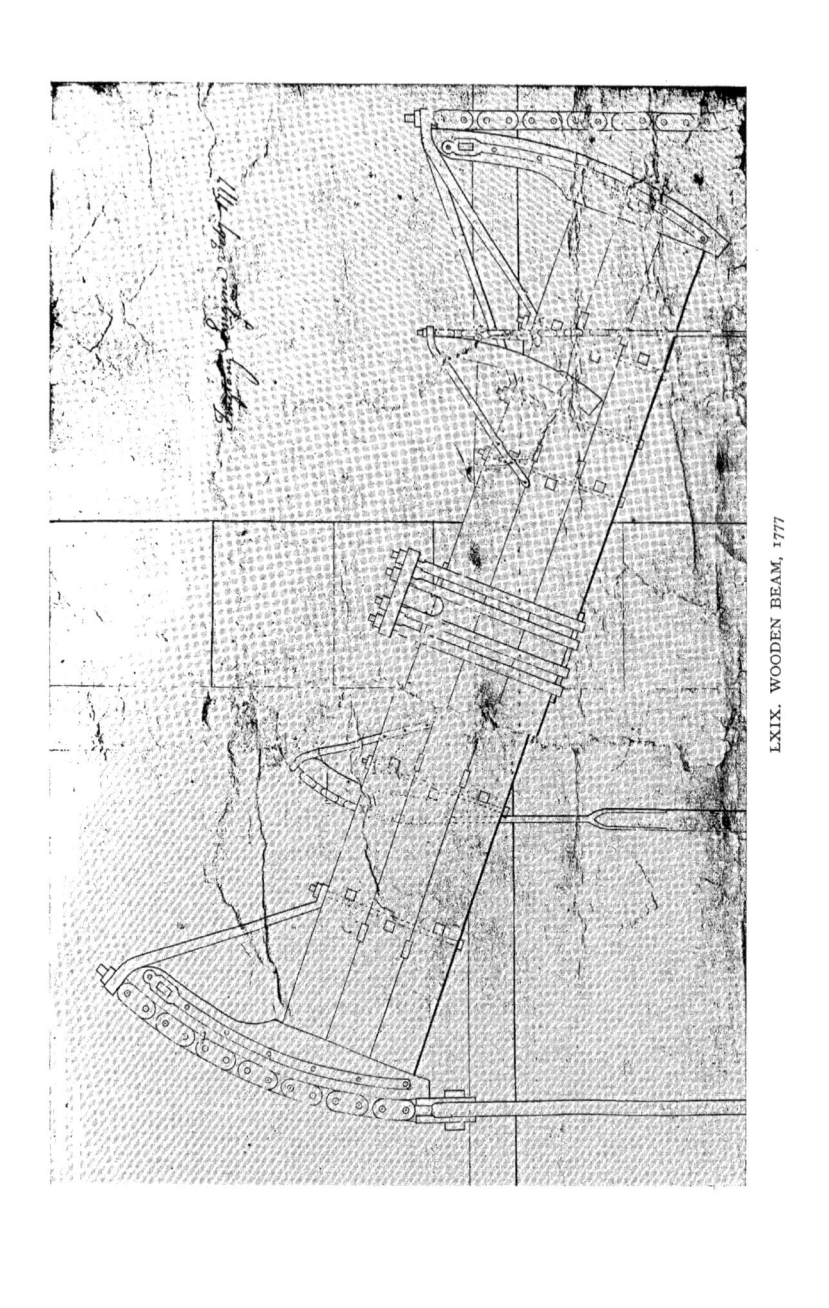

LXIX. WOODEN BEAM, 1777

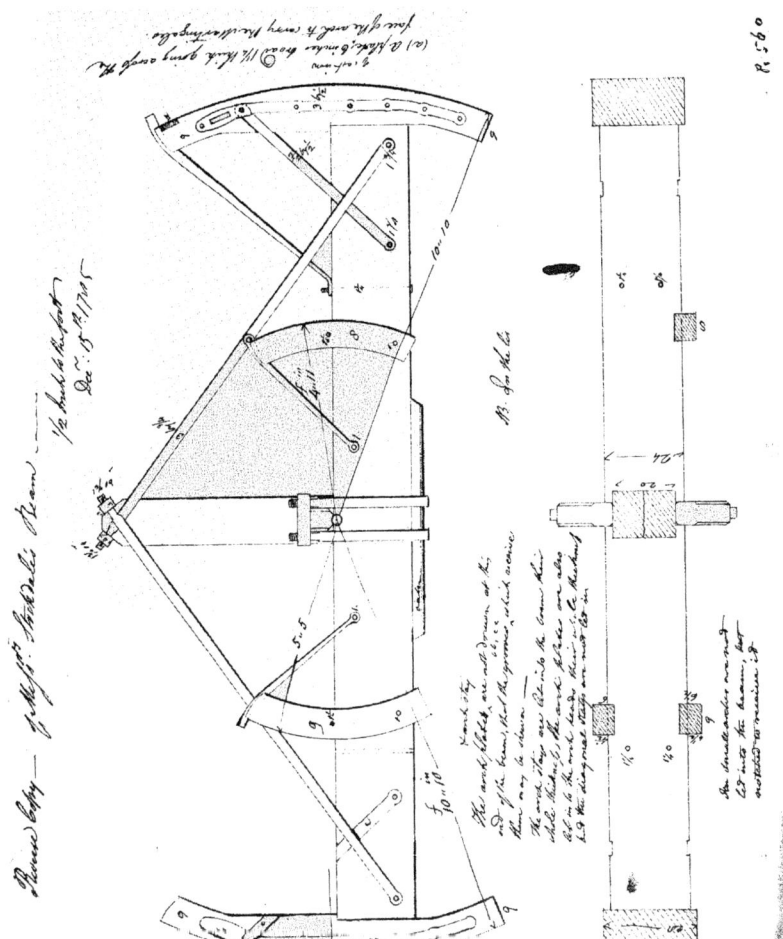

R. 560

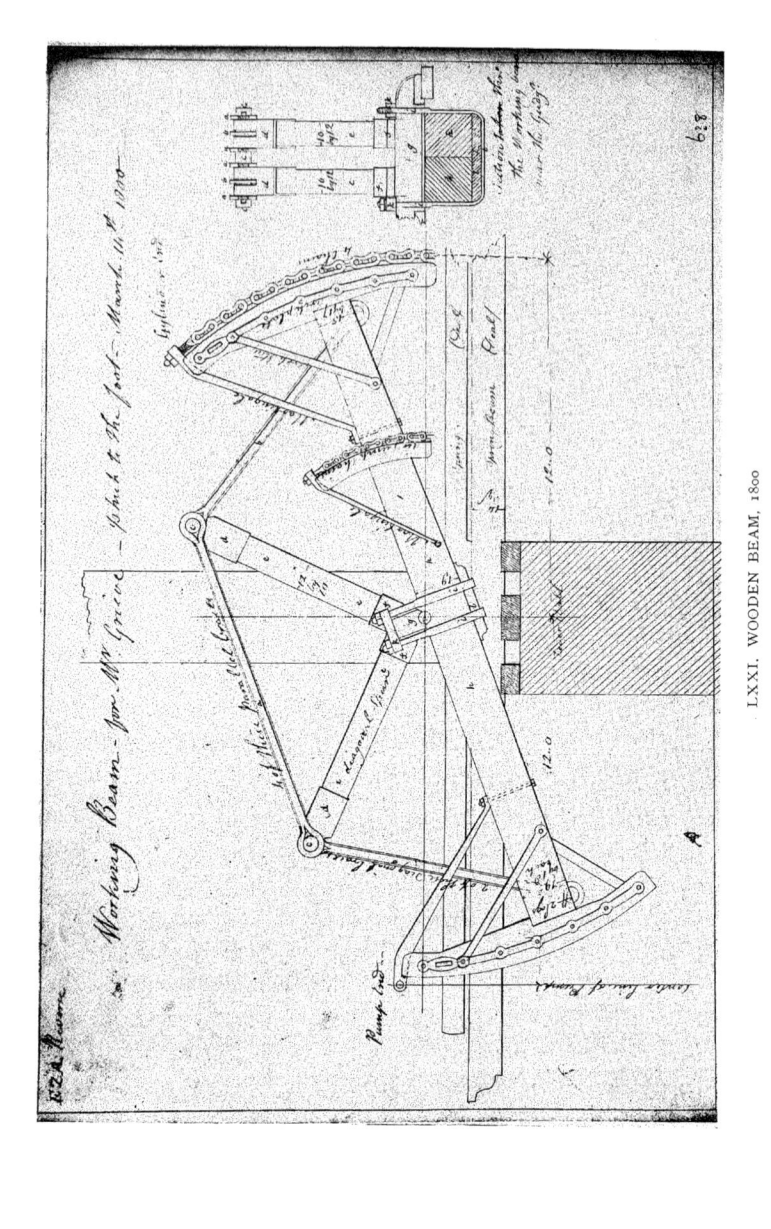

LXXI. WOODEN BEAM, 1800

on the sloping sides, with a direction that ' the brass must not touch the bottom by half an inch'.

The plummer block and gudgeon shown in Fig. 29 were designed in 1786 for a 24-inch double-acting engine for Poldice mine. It will be observed that the brass is carried up higher on one side than on the other.

The beam drawings, Plates LXIX, LXX, and LXXI, show at the upper ends of the arch-plates rectangular pins, the ' catch pins'; these project sideways from the beam and serve to limit the stroke of the engine by striking upon wooden

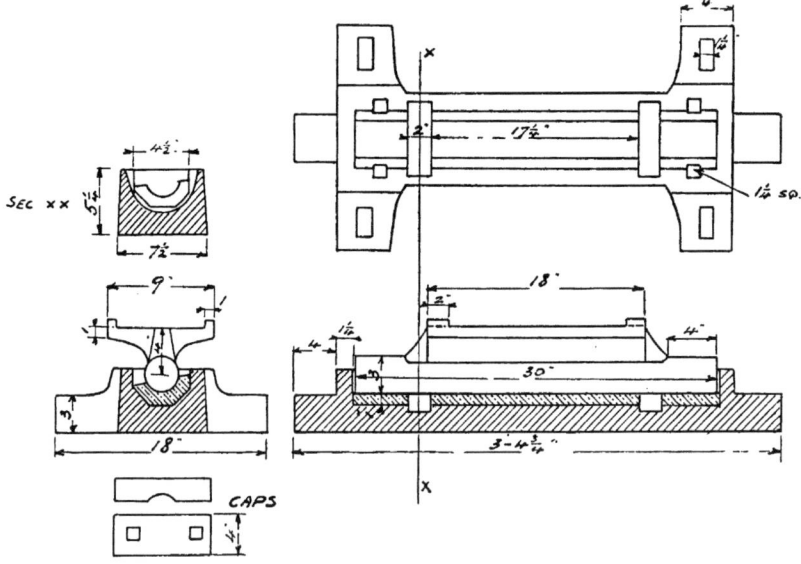

FIG. 29. Beam gudgeon and plummer block, 1786

springs, Plate LXXI, which are bolted to the 'spring beams' and adapted to have about an inch of play at their free ends. The spring beams extend, one on each side of the engine beam, from one wall of the house to the other.

THE CONDENSER

FOR a general account of the various forms of condenser that Watt tried in the early days we cannot do better than quote the draft of a letter that he wrote to Smeaton at the beginning of 1778 : [1]

' The first idea of it in 1763 or 4 was a copper spiral worm with a pump to extract the air & water, but in that form had not sufficient multiplication of surface. 2dly., a number of thin tubes going through an air tight vessel which was the first executed and but for the

[1] This draft is cancelled, and the letter actually sent contains far less information.

expence and the crust formed by bad water is still the best, being scarce any burthen upon the engine, the quantity of water and air being exceeding small. 3dly. A vessel in which the steam rose up through cold water, which was expelled when heated, by other cold water in such a manner as to require very little power to make the change ; this was also a good condenser but too complicated. In this condenser I also occasionally used an injection of cold water into the eduction pipe as do now. These two last were tryed in great in the engine at Kinneil. Some imperfections in ye engine made think that more air came in with the injection than is really the case. Therefore our first experiments at Soho were made with the thin plate and pipe condensers, but when came to execute a large order was afraid of the complexness of these methods & resolved to sacrifice part of the power to convenience, so returned to the injection. I made a middle receiver with two air pumps connected with a third which is still in use and answers very well.' [1]

The list in the first part of this letter does not include the plate condenser, and the reference to this form lower down seems to suggest that it was not tried until the engine was at Soho. This, however, is not the case. Watt, writing to Roebuck towards the end of 1765, says that he has altered the plan of the condenser from small pipes to thin interstices between plates, and that ' 16 double plates of a foot square and half an inch asunder will answer the purpose of 900 pipes of an eighth diameter'. The letter is endorsed in the handwriting of Watt—at a much later date, probably in 1808 : ' Scheme of new condenser for the iron cylinder, called the plate condenser. It came to Soho, and many experiments were tryed with it ; but it came unsoldered, and the drum condenser was substituted.' [2]

The ' drum condenser' was the pipe arrangement, the second item in the letter to Smeaton. It is shown in Plate XVI, reproducing the drawing laid before Parliament in 1775.

Although fairly satisfactory results were obtained with the surface condenser on the original Soho engine, when he had to design large engines, Watt, as he says in his letter to Smeaton, threw aside this plan and adopted jet condensation, in spite of the additional load that he saw would be thrown on the engine ; and to the jet condenser Watt remained faithful to the end. Boulton indeed in 1781 wished to reopen the question and argued the advantages of surface action, but without avail. At this date the original pipe condenser of the Soho engine still stood in the yard there.

The jet condenser was tried on the Soho engine in August 1775 ; the first arrangement did not work very well and we have no particulars of its construction. The second plan, as we know from a description by Watt himself,[3] had a separate condensing vessel with a rose under an opening in its lid, and three pumps ; the larger, communicating with the lower part of the vessel, was intended to extract water ; the other pumps communicated with the vessel about a foot above the bottom and were for the purpose of extracting the air and maintaining the vacuum. One of these air-pumps was worked directly from one side of the engine beam, and the pistons of the two were connected by a chain passing round an overhead wheel. This was the form of condenser used for the Broseley and Bloomfield engines, Plates LXXII and LXXIII. The connexion of the water-pump, c, with the lower end of the condenser, a, and of the two air-pumps, b, at points 9 in. above its bottom are clearly shown, also the rose, 'grate or cullender', placed under the lid of the condenser which is provided with an oscillating sector-valve to admit and cut

[1] B. & W. Colln. : Letter Books. Watt to Smeaton, 1778 Jan 17.
[2] Muirhead : *Mech. Inv.* Watt to Roebuck, 1765 Dec. 5.
[3] B. & W. Colln. : Letter Books. Watt to J. Wilkinson, 1775 Aug. 10.

off the condensing water. The overhead wheel for connecting the rods of the two air-pumps is indicated at G in the two drawings. It will be noticed that certain of the parts are marked with two dimensions; in the original drawings one set is in black ink for the Bloomfield engine and the other in red for the Broseley engine. By the time these engines were started, in March 1776, Watt had arrived at another plan which he informed Smeaton he was not inclined to alter.[1]

'It consists of a jack head pump, shut at bottom, with a common clack bucket, and a valve in the cover of the pump, to discharge the air and water. The eduction steam pipe which comes down from the cylinder communicates with this pump both above and below the bucket and has valves to prevent anything from going back from the pump to the eduction pipe. The bucket descends by its own weight, and is raised by the engine, when the great piston descends, being hung to the outer end of the great lever; the injection is made both into the upper part of this pump, and into the eduction pipe, and operates beyond my ideas in point of quietness and perfection.'

The condenser of the Soho 18-inch engine had been altered to this plan with an air-pump 7 in. diam. and 20 in. stroke, with the result that a vacuum of 27 to 29 in. of mercury was obtained, 'or in general only about 1½ inches below the barometer'.

But although Watt, in the letter just quoted, says that he had arrived at a plan that he was not inclined to alter, the next engine built, i.e. Bedworth, started March 1777, shows a material departure from it. Instead of a single pump with a valve in its cover to discharge the air and water, we now have three pumps placed side by side in a line parallel to the fulcrum of the engine beam; the middle pump stands above the others, and Watt explains that its function is 'to receive the hot water lifted up by the other two and by lessening the surface exposed to pressure of atmosphere, to extract it with greater ease'.[2] The pumps were all of the same diameter, so that the combined area of the two air-pumps was double that of the hot-water pump. The plan of using two air-pumps was resorted to in order to get a balanced arrangement and so to avoid 'twisting the beam'; it was used in the Tingtang engine and in Colevile's engine, but was soon relinquished. The use of the hot-water pump to remove the discharge of the air-pump, however, was continued and, as will be seen later, was a feature of the standard practice for a number of years.

The pumps of the Bedworth condenser are shown in detail in Plate LXXIV (from a drawing made in May 1776). The lower ends of the two air-pumps and of the exhaust pipe (here marked 'steam pipe') are bolted on to the 'lower box'; this box is open at the top, and the inner edges of the flanges on the air-pumps and that on the exhaust pipe are some little distance apart to leave spaces for access to the valves, the spaces are provided with lids; the valves are hung upon and beat against the ends of the box c which is fixed to the exhaust pipe and fits closely in the lower box. The upper ends of the air-pumps are formed with branches for connecting them together; they are open at the top and provided with a cover having three openings; the air-pump lids are bolted over the side openings and the lower end of the hot-water pump over the central opening; the foot valve of the hot-water pump is fitted over this central opening and access to it is obtained by a door on the side of the pump. The general arrangement of the condenser is the same as that shown in the drawing of Colevile's engine, Plates XX and XXI. The pumps are in a water cistern outside the engine-house and remote from the cylinder;

---

[1] This and other letters of Watt to Smeaton are given in full in Farey's *Steam Engine*.

[2] B. & W. Colln.: Letter Books. Watt to Smeaton, 1778 Jan. 17.

condensation is effected mainly in the long exhaust pipe which is enclosed in a water-jacket. The drawing shows two other features of the plan that Watt thought was to be final ; the one is that injection is made into the upper ends of the air-pumps as well as into the exhaust pipe ; it will be seen that the injection pipe has three limbs, a long one projecting upwards into the exhaust pipe and two shorter ones entering the upper ends of the air-pumps ; it takes in water by a mouth opening below the level of the water in the cistern ; this mouth is shown open, but there. is no doubt that it was provided with a valve operated by the working-gear. The other feature is the connexion of the exhaust pipe with the air-pump above as well as below the bucket ; a small branch is seen leading from the descending part of the exhaust pipe to the upper box of the air-pump ; the idea no doubt was to provide a passage for air direct to the upper ends of the pumps. It is likely that these arrangements were applied to the Bedworth engine, but doubtful whether they were to Colevile's or any other engine.

The correspondence for a few years after 1777 shows a constant succession of small improvements being made. The Tingtang condenser was to be on the same lines as those for Bedworth and Colevile's engines, but ' the ends of the beating box for lower valves are to be made sloping which will give the valve less motion in opening and shutting '.[1]

Coming on to the Wheal Union engine we have a distinct simplification, the use of one air-pump instead of two, and while the hot-water pump is retained a valve in the air-pump lid is used as well. Watt, in sending the drawings to Bersham, writes :

' To make it more simple and easier to all parties I have ventured upon a single air pump of 24 inches dia^r, which I request may be made light & true. The bucket is to be like a little piston, and its cone may be bored by any tool you have ready so as to be 2 inches dia^r at upper side, it does not signify tho' part of the hole be cylindrical as I would not have it more than three inches dia^r below. There are to be two valve holes in the lid with projections round them that we may get at to file flat.' [2]

Up to 1779 the round valves had been made with spindles separate from the disks. Watt did not like these valves, they were constantly getting loose on their spindles, but for small pumps he could get nothing better, so in sending the Penryn-dee drawings to Bersham he asks that the valves in the air-pump lid may have the spindles cast in one with the disks. As to those for the bucket and clack of the hot-water pump, he fears that the spindles would wear away too fast if made of brass.[3] Wilkinson in reply states that he had made the round valves for the air-pump lids of his own engine and those for Ale and Cakes and Tresavean engines with the spindle or pin and valve all in one of brass, and that it was thought that this plan would do for Penryndee hot-water pump, by taking care to have good stout metal. He adds : ' We shall screw in the brass seats with this metal *quite through* from ye outside and afterwards bore out the seat parts.'

In the standard arrangement of condenser for single-acting engines, the air and hot-water pumps were placed outside the engine-house and were worked from the outer end of the engine beam ; the condensing chamber was formed by the eduction pipe itself which proceeded from the lower nozzle of the steam cylinder through the wall of the house in a downwardly inclined direction, and then, making a sharp bend, vertically to the foot of the air-pump. The injection pipe entered

[1] B. & W. Colln. : Letter Books. Watt to John Wilkinson, 1777 Jan. 13.
[2] Ibid. : Letter Books. Watt to John Turner, 1777 Nov. 6.
[3] Ibid. : Letter Books. Watt to John Wilkinson, 1779 Sept. 20.

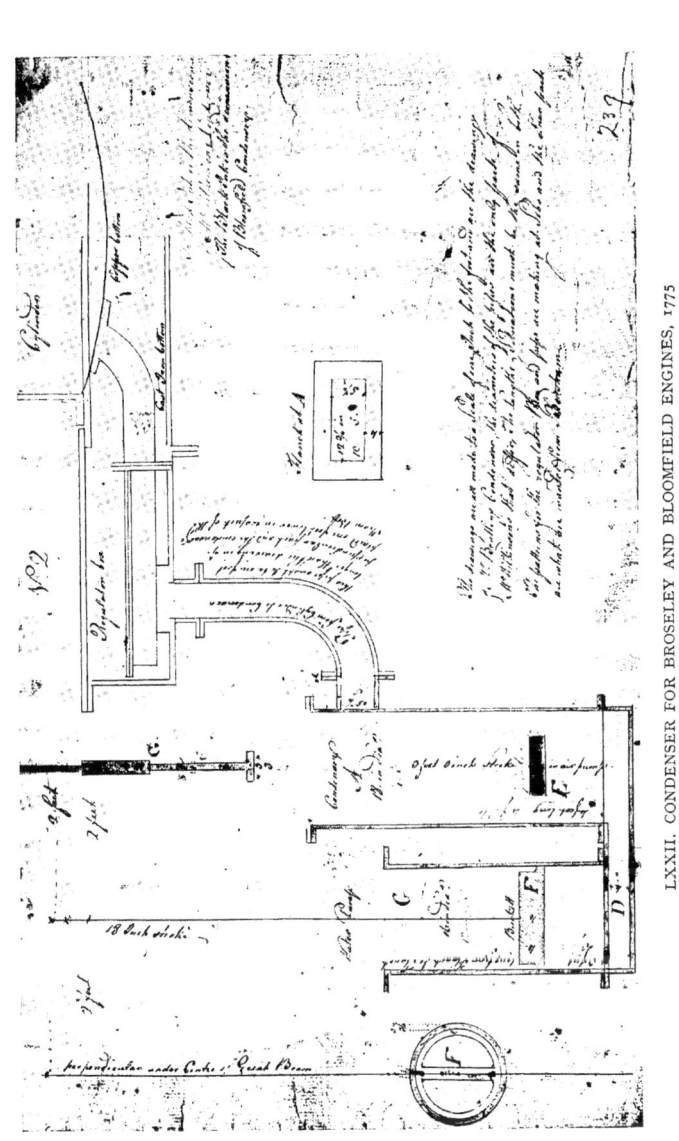

LXXII. CONDENSER FOR BROSELEY AND BLOOMFIELD ENGINES, 1775

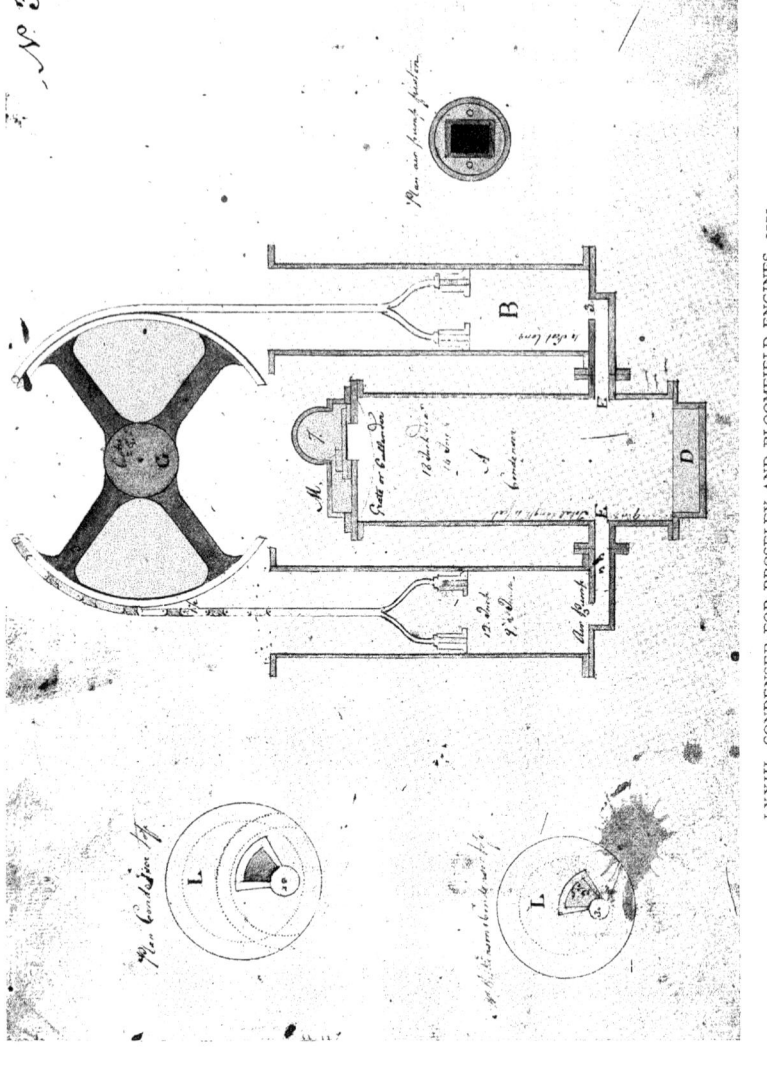

LXXIII. CONDENSER FOR BROSELEY AND BLOOMFIELD ENGINES, 1775

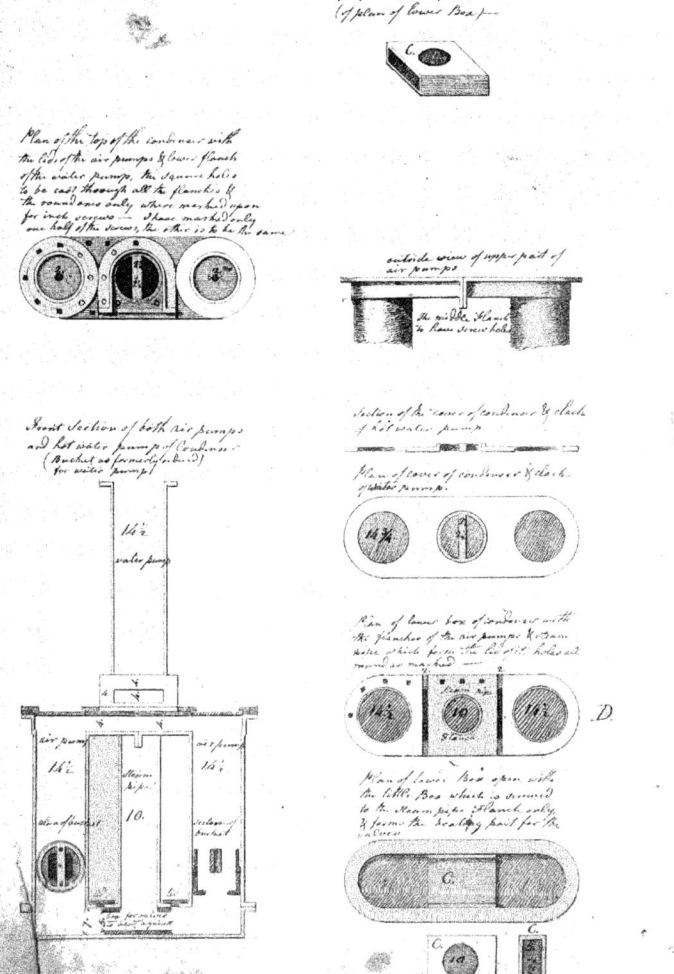

LXXIV. PUMPS FOR BEDWORTH CONDENSER, MAY 1776

the eduction pipe at the bend and the nozzle was set to direct the line of the jet upon the upper surface of the eduction pipe, towards the end nearest the cylinder, i.e. the jet met the entering steam. The vertical limb of the eduction pipe and the air-pump were immersed in the condenser cistern, which was built up of oak or deal planks laid edge to edge. The hot-water pump was mounted higher than the air-pump, and the lower end of the one was connected by an horizontal box with the upper end of the other; this is shown in Plate LXXV, a drawing of the Wheal Virgin (Elvan) 58-inch engine, dated 1780. The pistons of both pumps were fitted with flap valves, and similar valves were fitted at the foot of the eduction pipe and in the lower end of the hot-water pump. In brief, the two pumps were in series; the lid of the air-pump, however, was fitted with two circular valves which opened outwards and relieved any excess of pressure that might arise from a defect in the working of the hot-water pump, or other cause. The hot-water pump delivered at such a height that water could flow from it by gravity into the open top of the boiler feed-pipe. The free end of the injection pipe opened into the condenser cistern and was fitted with a valve, that was opened and closed by the working-gear at each stroke of the engine.

The eduction pipe, at any rate in the earlier engines, was of copper, and at first it seems to have been enclosed for a part of its length in a water-jacket; this arrangement will be found in the general drawing of Colevile's engine, Plate XX. In addition to the injection pipe, another small pipe was connected to the eduction pipe; it passed through the wall of the condenser cistern and terminated in a valve opening upwards in a small box, containing water, attached to the side of the cistern. This was the blowing-valve; its use was to allow steam to be blown through the engine at starting. It was desirable to connect this valve well down on the eduction pipe and originally it was arranged to open into the cistern, but this meant that the engine could not be started until the steam had been raised to a pressure sufficient to overcome the head of water in the cistern, which might be more than that required to work the engine; blowing into the cistern also had the effect of heating the water which was to be used for condensation. The plan of carrying the pipe through the side of the cistern and placing the valve under a small head of water was due to Jabez Hornblower, when erecting the engines at Ketley and Donnington Wood in 1779.[1]

Another appliance for clearing the cylinder, &c., consisted of a hand-lever so placed that it could be coupled to the air-pump rod; with this the pump could be worked while the engine was at rest, and the cylinder exhausted and the joints tested without the aid of steam. Plate LXXV shows the device as arranged for one of the Wheal Virgin engines in 1780.

This lever appears to have been the only means originally provided for clearing the engine of air at starting. Jabez Hornblower in his 'History of the Steam Engine' in Gregory's *Mechanics* stated that the blowing-valve 'was first applied by Mr. Hornblower at an engine on a mine called Ting Tang'. That would be in 1778, and Mr. Hornblower would be Jonathan, the father of Jabez, who was the engineer at that mine. This statement was not seriously contested, Watt's remark about it being: 'I cannot recollect whether the blowing-valve was first used at Tingtang engine or not. It was however no invention of Jabez Hornblower, being an essential part of Newcomen's engine and perfectly well known to me.'[2]

[1] B. & W. Colln. Jabez Hornblower to Watt, 1779 Feb. 22.

[2] Doldowlod Papers. Watt to James Watt, junior, 1808 Nov. 12.

The statement that the blowing-valve was an essential part of the Newcomen engine is somewhat misleading. The older engine had a ' snifting' valve on the cylinder, through which air was blown out in the course of the normal working of the engine, and this would certainly be well known to Watt; but the suggestion that that valve might be applied to the condenser, for use in starting the engine, may very well have come from Hornblower in the course of his correspondence (dating from October 1776) with Boulton and Watt in reference to the Tingtang engine. This engine, however, was not the first to which the blowing-valve was applied; the iron castings for it did not reach Cornwall until January 1778, and nearly six months earlier we find Watt writing to Dudley, the erector of the Wheal Busy engine, giving him the following directions about the blowing-pipe :

' The blowing pipe to be about 2 inches or 2½ diameter is to be fixed to the side of the eduction pipe about 2 feet under water and being 18 inches long the valve will lye 6 inches under water. As you may find it difficult to get a conical valve for it made with you, a common can lid clack faced with lead will do.' [1]

It is likely that the device had been tried at Soho just before this. Many years after, in 1808, Murdock said that he recollected one being put on the Soho engine in 1777; he was scalded by the engine-man blowing through while he, Murdock, was in the cistern.[2]

The disposition of the blowing and injection pipes and their valves is shown in Plate LIII.

In the Boulton and Watt Collection there is an undated paper, in the handwriting of Watt, headed ' Particular description of the Working gear and directions for using it ' ; this document appears to be the draught of the directions sent to France with a blowing-engine constructed by John Wilkinson in 1782 ; in it Watt discusses the arrangement of the blowing-valve, but finally recommends that no such valve be used ; in lieu thereof he advises that a hand-pump be connected to as low a point as possible of the eduction pipe.

The regular form of the air- and hot-water pumps at this period, 1779, as illustrated in the printed book of ' Directions for erecting and working the newly-invented steam engines. By Boulton and Watt', is given in Fig. 30, where we have a plan showing the two pumps and the foot of the eduction pipe and, above it, a section through the lid, A, of the air-pump with its circular valves, C, and brass stuffing-box, b, provided with a collar of wood or brass, d, to press down the packing. The plan shows the gland, B, of the stuffing-box and one of the guards, D, to prevent over-lifting of the valves ; at F is shown the box that connects the upper part of the air-pump with the foot of the hot-water pump E, the flange of which is shown at G and the clacks at H. The flange on the lower end of the copper eduction pipe, U, is clamped, by means of an iron ring, T, and screw bolts, upon the eduction-pipe foot, S, which is bolted to a flanged branch on the air-pump bottom ; R is the axis of the foot-valve, access to which is provided by an opening, Q. The eduction-pipe foot and valve are shown in elevation and section at the bottom of the drawing, also the cap of the air-pump rod. The air-pump bucket, L, is shown in section and plan ; one of the valves is removed in the plan. Three views are given of the bucket, M, of the hot-water pump, also a section of the pump showing its foot-valve, H, and a guard to prevent the valves from lifting too high ; this guard consists of a square bolt traversing a bent bar of iron. The clack valves are each forged solid

---

[1] B. & W. Colln. : Letter Books. Watt to Dudley, 1777 July 23.

[2] Doldowlod Papers. James Lawson to James Watt, junior, 1808 Nov. 14.

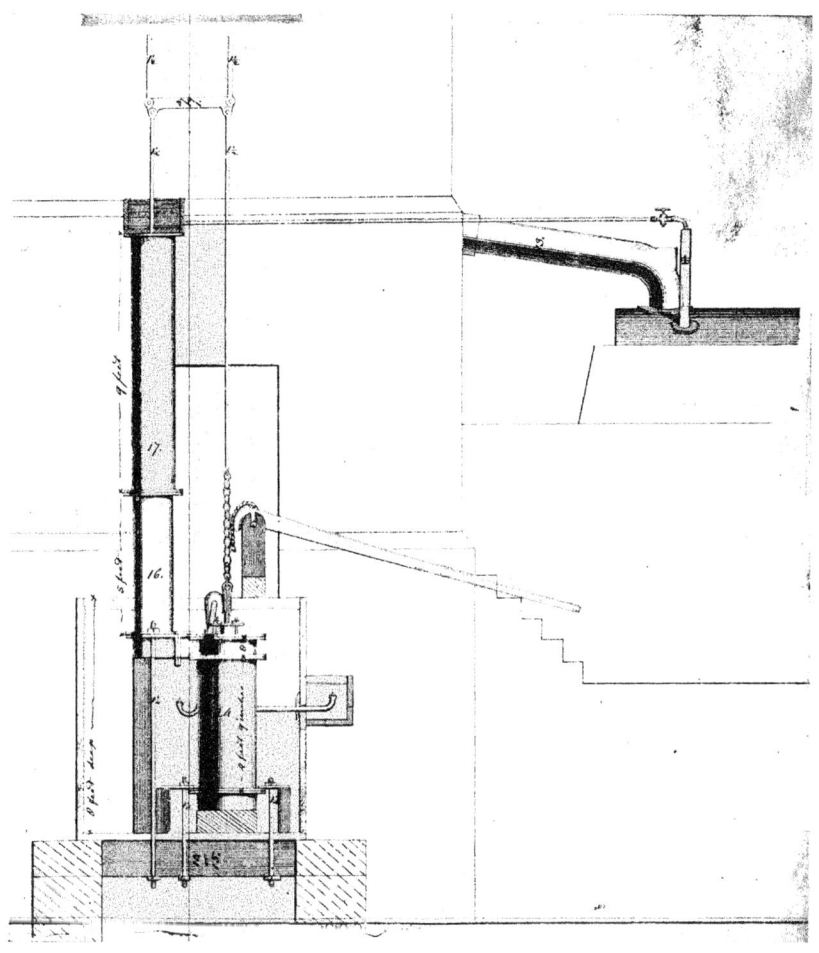

LXXV. CONDENSER, SHOWING APPLICATION OF HAND-LEVER, 1780

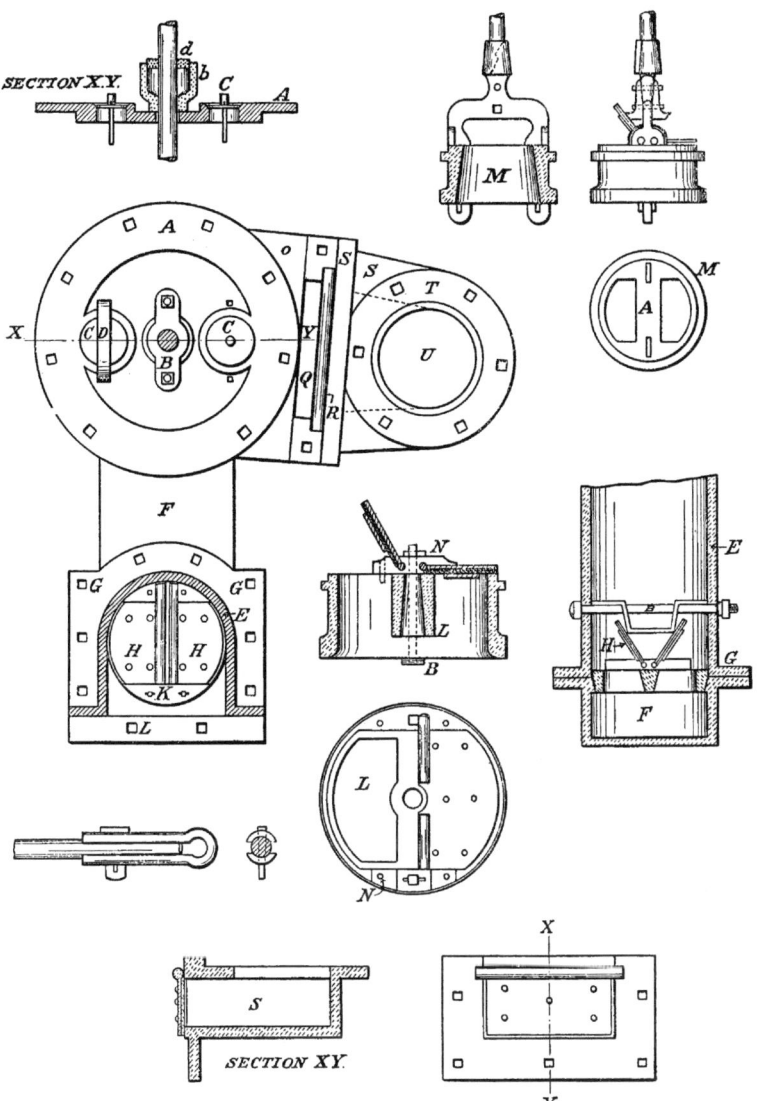

SECTION X.Y.

SECTION XY.

FIG. 30. Air- and hot-water pumps, 1779

with its pivot and faced with copper. If the requisite skill is available a copper rim may be brazed or hard-soldered in place; otherwise a plate of copper is secured by an under-plate of iron and square rivets. The valves when riveted up are to be made red hot, laid on their seats, and beaten down by a sledge-hammer and a short block of wood. The valve pivots are held down by keeper plates secured by cotter bolts. In the case of the air-pump bucket each plate, N, is held by a central bolt passing through the bucket, and a pair of steady pins. For the hot-water valves, H, the pivots are held down by keepers, K, and studs screwed into the cast iron.

When the water is corrosive all the valves of the condenser are to be made of bronze, and the valve at the eduction-pipe foot should be dished to reduce the surface requiring to be filed up and to strengthen the valve.

Later on it was found necessary in some situations to make the entire condenser of brass, and Boulton writing from Cornwall in 1785 says that all engines east of Redruth ought to have brass condensers.[1] In 1788, in connexion with the Shelve Field Gravels engine, Southern suggests that the hot-water pump might be made of wood. The air-pump lids were sometimes made of wood, also the clack valve at the foot of the eduction pipe.[2]

As stated above, the lower end of the copper eduction pipe is clamped to its cast-iron foot by means of an iron ring and screw bolts. A similar joint was used at the other end of the pipe where it joined the nozzle, and sometimes the sections in which the pipe was made were coupled in the same way. More usually, however, they were joined by spigot-and-faucet joints. Fig. 31 (after an illustration in the book of Directions) shows the manner of connecting two pipes, A, D; the faucet part, A, has an iron ring, C, shrunk upon it, to enable it to stand caulking, and the two parts are drawn together by long screws engaging rings, B, clamped one on each part. The same drawing also shows the method of fixing the blowing-pipe, E, and the injection pipe, F, in the eduction pipe, A; G in each case is a copper cup run full of solder in which is secured a faucet pipe, H. In the injection pipe is shown the block, K, of tin or brass by which the discharge end of the pipe is constricted. The drawings of Wilkinson's French engine, and of the Bradley Forge engine, show the injection pipe with a telescope rose end, but this was an exceptional usage; the usual plan was a nozzle as described above, or a mere opening in the wall of the condensing vessel.

For the rotative engines the position of the air- and hot-water pumps was changed and they were now worked from the cylinder end of the beam. The eduction pipe consequently became short, so it was necessary to provide a separate condensing-vessel; this was placed in the cistern and the injection was led into it. For some years the injection was turned on and off at each stroke as in the single-acting engines. By 1791, engines were working with a constant stream of injection water, but this does not seem to have yet become the established practice, for we have one of the erectors asking whether the engine he is about to put up is 'to work with a constant stream of injection, or with two jets at every stroke, but [he says] I think we have found the double mill engines in London work as well, or better, with a constant stream, and with a deal less noise'.[3]

We have seen that the use of a hot-water pump to remove the whole discharge of the air-pump became a feature of the standard practice. The value of this plan

[1] B. & W. Colln. Boulton to Watt, 1785 July 31.
[2] Ibid. Boulton to Watt, 1785 Sept. 23.
[3] Ibid. Richard Dayus to Boulton & Watt, 1791 Aug. 15.

was challenged in 1788 by John Southern, Watt's assistant and draughtsman, who contended that ' the hot water pump has no effect towards augmenting the degree of vacuum'. He admits that a pump is necessary to lift water for the boiler, but points out that a pump for that purpose only would be much smaller and cheaper than the hot-water pumps in use.[1] It appears from his letter that already some engines had been made without hot-water pumps, indeed the drawing of Cotes's engine, 1784, Plate XXXV, has the air-pump delivering by a clack

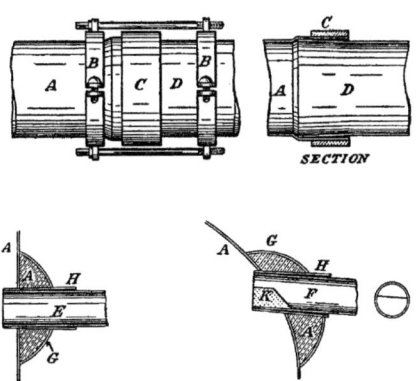

Fig. 31. Eduction pipe details, 1779

valve into a hot well formed by partitioning off a part of the cistern, but this departure had been looked upon hitherto as a makeshift. Southern's reasoning on this point seems to have convinced Watt, and from this time the use of the hot-water pump was given up and the air-pump was arranged to deliver into a hot well from which the boiler supply was drawn by a small pump. The drawings of the Cockshead engine, Plate XXXVIII, Hebburn engine (1798), Plate XXVIII, and Coates's engine (1803), Plate XL, all show this arrangement. The drawings of Coates's engines show also the cold-water pumps for supplying the condenser cistern. The air-pump of Cotes's engine (1784) is worked from a separate arch on the engine beam; the usual practice in the rotative engines was to work it by the plug-tree of the working-gear, but the separate actuation continued to be used occasionally up to the end of our period.

[1] Ibid. Southern to Watt, 1788 Aug. 15.

XVII

ROTATIVE ENGINE DETAILS AND ACCESSORIES

SUN-AND-PLANET GEAR

THE genesis of the sun-and-planet gear has been discussed in Chapter XIII. It remains to consider some of the applications in practice.

The drawing selected for reproduction, Plate LXXVI, is that of the gear of the engine put up to drive a tilt hammer at Wilkinson's Bradley Forge. This was the first engine, outside Soho, to which the sun-and-planet gear was applied, and the drawing is dated July 1782 ; it shows, in addition to the gear, the hollow shaft, the cam-ring for working the hammer, and the central part of the flywheel, but these parts will not be considered here. The sun-wheel is 48 in. diam., and has 36 teeth, so that the pitch is about 4 in. ; it consists of two rings, 3 in. wide, bolted together with the teeth stepped ; the shape of the teeth is shown to a large scale in the top figure of the drawing. The outer ring is bolted to a disk formed on the end of the journal section of the cast-iron shaft ; the periphery of the disk is formed with recesses to receive joggles on the rings.

The planet-wheel is half the size of the sun-wheel and consists of two toothed disks, separated by a disk of wrought iron engaging between the rings of the sun-wheel ; the inner disk is formed with a gudgeon to receive one end of a link which at the other end embraces the shaft ; the outer face of the other disk is formed with a ─┼─-shaped recess to receive a corresponding joggle cast on the end of the connecting-rod, to which the planet-wheel is secured by bolts passing through both disks.

For the first Albion Mills engine the two toothed cast-iron rings appear to have been separated by rings of wrought iron put on in segments. The job was badly executed by Wilkinson and gave considerable trouble. Boulton, writing to Watt in March 1786, reports that the rings were ' badly made of bad iron and are obliged to be remade as they are broke to pieces '. He has the new rings turned both on the sides and faces and cottered firmly in place. At the same time there was trouble with the sun-wheel gudgeons working loose in their shafts, and Boulton thought ' we had better have given Mr. Wilkinson 40 or 50£ rather than they should have been so badly fitted '.[1]

At about the same period we find the two rings of teeth cast in one piece and separated by an integral ring extending to the pitch line. Then it appears that in 1787 it was decided to give up the use of stepped gearing and to use ordinary toothed wheels. About the same time we have an indication of a method of securing the wheels on their shafts, i.e. ' all the wheels and also the crank, are made to wedge upon squares, first wood wedges are driven in, as many as can be & then thin iron wedges into them, and it has been found that they do not give way '.[2]

---

[1] B. & W. Colln. Boulton to Watt, 1786 Mar. 15, 22, and 27.   [2] Ibid. Southern to Wilson, 1787 Mar. 22.

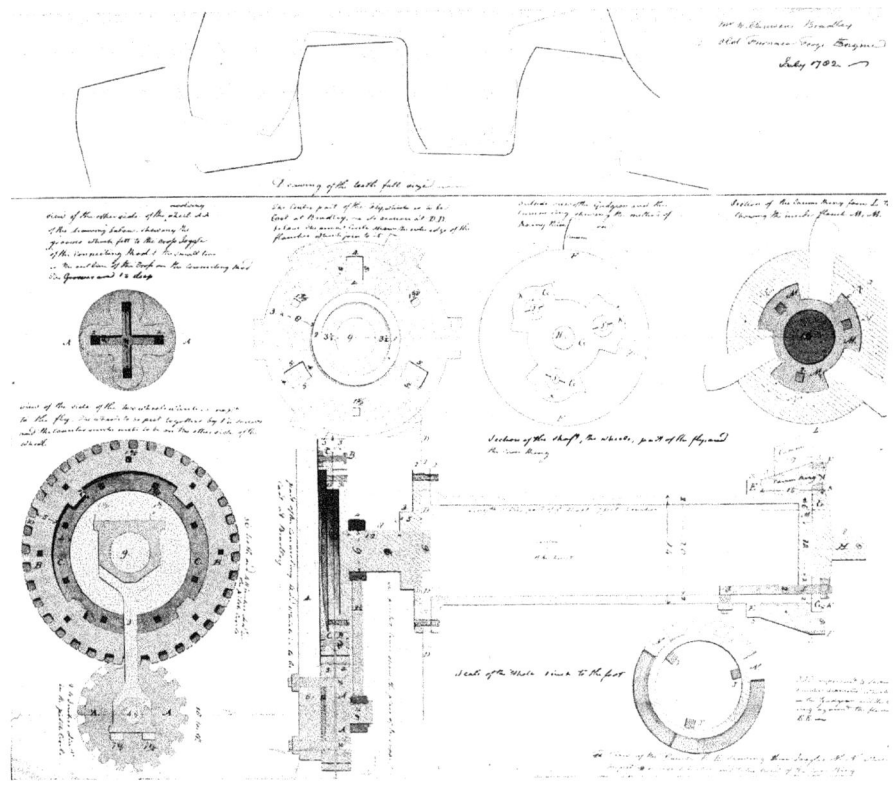

LXXVI. SUN-AND-PLANET GEAR AND CAM-SHAFT FOR BRADLEY FORGE ENGINE, JULY 1782

CONNECTING-RODS

By reference to the drawings of the Bradley Forge engine, Plates XXXIV and LXXVI, it will be seen that the upper end of the wooden connecting-rod terminates in two plain straps which carry an ordinary joint pin, while its lower end consists of a bar of cast iron enlarged at the bottom to carry a rib of cruciform shape ; this rib fits into a corresponding groove in the planet-wheel which is secured to the rod by four bolts, one at the end of each rib.

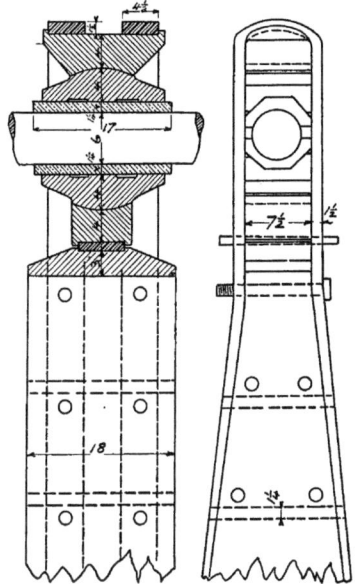

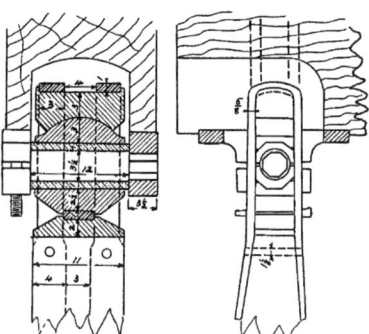

Fig. 32. Connecting-rod, upper end, 1791   Fig. 33. Connecting-rod, upper end, 1792

Reynolds's portable winding-engine, Plate XXXVI, has a rod of cast iron, following the carpentry design, with a strap of wrought iron at its upper end terminating in an eye fitted with a bearing. The design of another engine for Reynolds in 1789 shows a ball joint at the upper end of the rod.

The Soho drawings show that the top of the connecting-rod was receiving considerable attention in the years 1791 to 1794. Figs 32 to 35 have been prepared to show the development which took place in that period. It was found necessary to hang the rod in such a manner that it could be swung slightly to one side or the other to suit the position of the sun-and-planet gear at its lower end. Fig. 32, from a drawing dated 1791, shows the swivelling arrangement composed of two members, each in two parts, the inner member fitting over the bearings for the pin, and the lower part of the outer member tightened by a wedge bedding on a plate on the end of the rod, the whole being embraced by two iron straps secured to the rod ; the pin was fixed to the beam. The arrangement is faulty

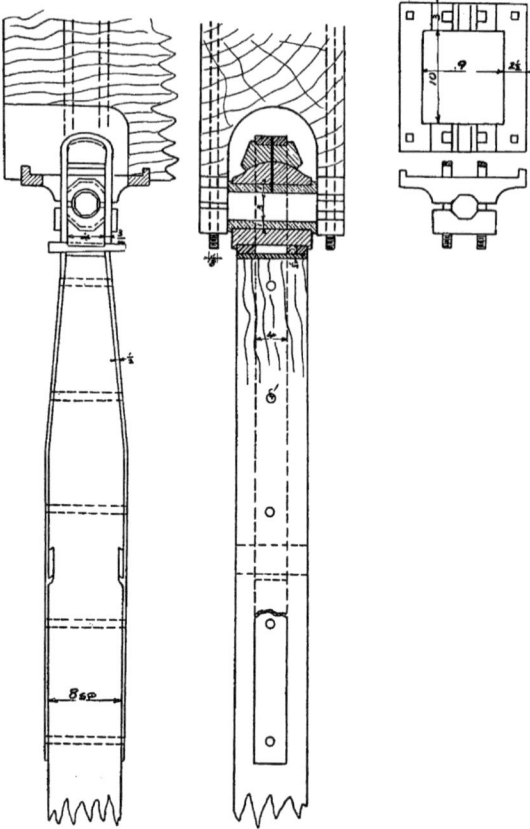

FIG. 34. Connecting-rod, upper end, 1792

in that the same wedge is used both to lock the swivel, to maintain the rod in its proper plane, and to tighten the bearings for the pin. Fig. 33 (1792) still has the one adjusting-wedge, but it now bears on the lower part of the inner member of the swivel joint and the lower part of the outer member is dispensed with. Fig. 34, another design of the year 1792, shows the introduction of two side wedges instead of one central wedge ; this facilitated the setting of the rod sideways, but it left the adjustment of the bearing bound up with the swivel arrangement. In Fig. 35 this difficulty is met ; the pin is now fixed in the rod end and is mounted in bearings on the beam, which bearings are capable of independent adjustment ; it will be seen that it has become unnecessary to make the large cavity at the end of the beam required by the earlier designs, that the pin can be firmly secured to the rod, and also that the straps at the rod end are in the form of closed loops.

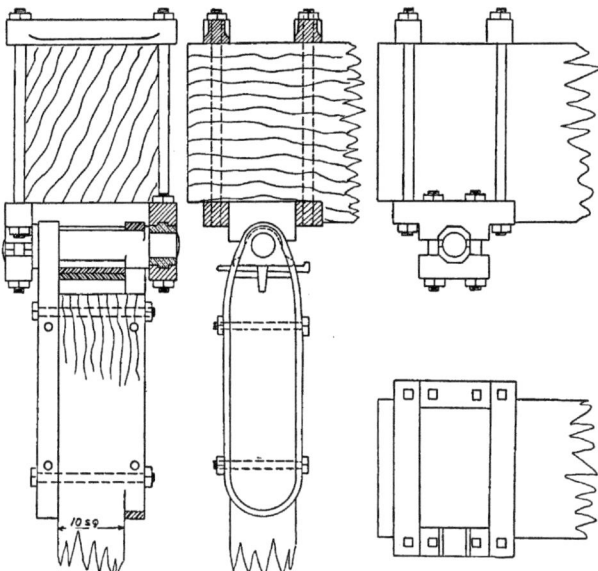

FIG. 35. Connecting-rod, upper end, 1794

This design is of the year 1794; it was the final design for wooden rods and remained in use until such rods were superseded by iron rods.

It had been suggested by Boulton in 1784 that the rods should be made of cast iron, and it has been mentioned above that a winding-engine design of 1788 shows the application of this material, but the use of cast iron did not become at all general until the year 1800. Plate LXXVII is a design of that date. The body of the rod is of cruciform section and terminates at the top in a socket in which is cottered a shank having a hemispherical head ; this head is seated in a corresponding recess in a casting formed with gudgeons to engage in bearings on the beam. The lower side of the protuberance on the casting is of spherical form to bed in a corresponding recess in the end of the rod.

CRANKS

In 1792, when Pickard's patent had run its course, Boulton and Watt began to make use of the crank in place of the sun-and-planet gear. It cannot be said that they made a sudden change in their practice, for up to the end of 1796 they had applied the crank to half a dozen engines only and the number of crank engines made up to the year 1800 was not considerable. There is little to say as to the Soho practice in respect of this part of the engine, but in the first instance the pin was cast on the crank ; then in 1794 we find a separate pin drawn up by a concentric bolt, and very soon after we find the pin secured by a cotter passing through the boss on the crank.

FLYWHEELS

The flywheel selected for illustration, Fig. 36, after a drawing of 1791, is 18 ft. diam. The rim, composed of six segments 8 in. by 3 in., is mounted on a six-armed spider made in halves bolted together at the boss which is made to go on a square shaft. On one side the arms have ribs extending from the boss outwards ; they terminate in T-heads having a central notch on the outer edge. The ends of the segments are recessed to receive the T-heads and are each formed with a projection ; the projections on the abutting ends of two adjoining segments fitting the notch in the T-head. The parts are secured together at each joint by six rivets passing through holes cast in the arms and drilled in the segments. The flywheel of Wilkinson's Bradley Forge engine, 1782, had a six-armed spider with T-heads, but in this case the rim was double, one set of segments being fixed on each side of the arms.

It will be interesting here to indicate how Watt reasoned about the flywheel for the experimental tilt hammer at Soho. Under the date Nov. 28, 1782, he says :

' The fly is 7 feet diameter and weighs 5 cwt or 570 lb. not comprehending the wood of the arms, nor the camm wheel and other parts which act at a lesser radius. The fly makes 50 turns per minute, consequently its circumference moves with a velocity of 1,050 feet in a minute or 17 feet in a second. 17 feet in a second is due to a fall of 4 feet and consequently the fly when moving with that velocity should raise its own weight 4 feet high before it was brought to rest. Now the hammer (weighing 120 lbs. besides the handle) is the 4·6 part of the weight of the fly, conseqⁱʸ should be raised 4·6 times 4 feet = 18·4 feet, but as the hammer is raised upwards with a velocity greater than gravity would give it downwards, it will require a power more than double its weight to give it that velocity, consequently would be raised less than 9 feet even though the machine had no friction, or in other words would make 6 strokes of 18 inches each, but as it strikes the rabbat very strongly it possibly is raised with a much greater velocity and conseqⁱʸ more force employed in giving it the blow.' [1]

FRAMES

In the Boulton and Watt pumping-engines, as in the earlier atmospheric engines, the cylinder, condenser, bearings for the beam gudgeons, &c., were held together in their correct relative positions by the masonry of the engine-house and foundations. Usually the beam was pivoted in one of the walls of the house with the pump end projecting outside it, but sometimes, as, for instance, in the New River pumping-engine shown in Plate XXXVII, the entire length of the beam was within the house. This was the plan adopted for the rotative engines.

[1] B. & W. Colln. ' Blotting and Calculation Book, 1782 and 1783.'

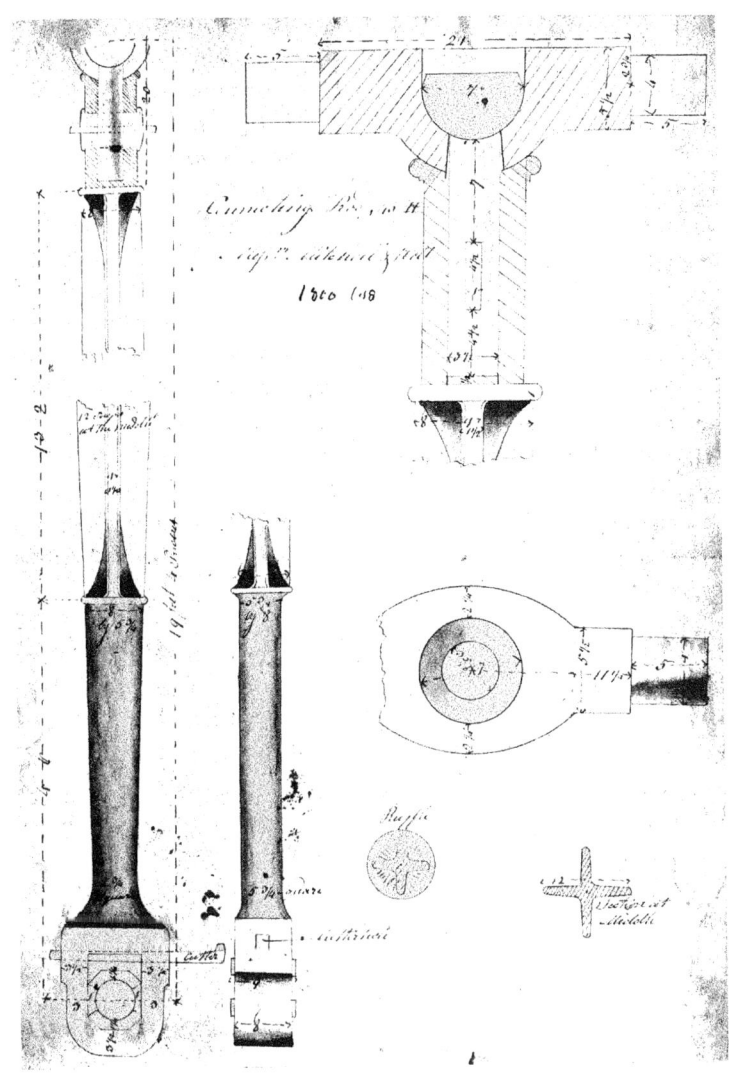

LXXVII. CAST-IRON CONNECTING-ROD, 1800

LXXVIII. ENGINE FRAME, 1788

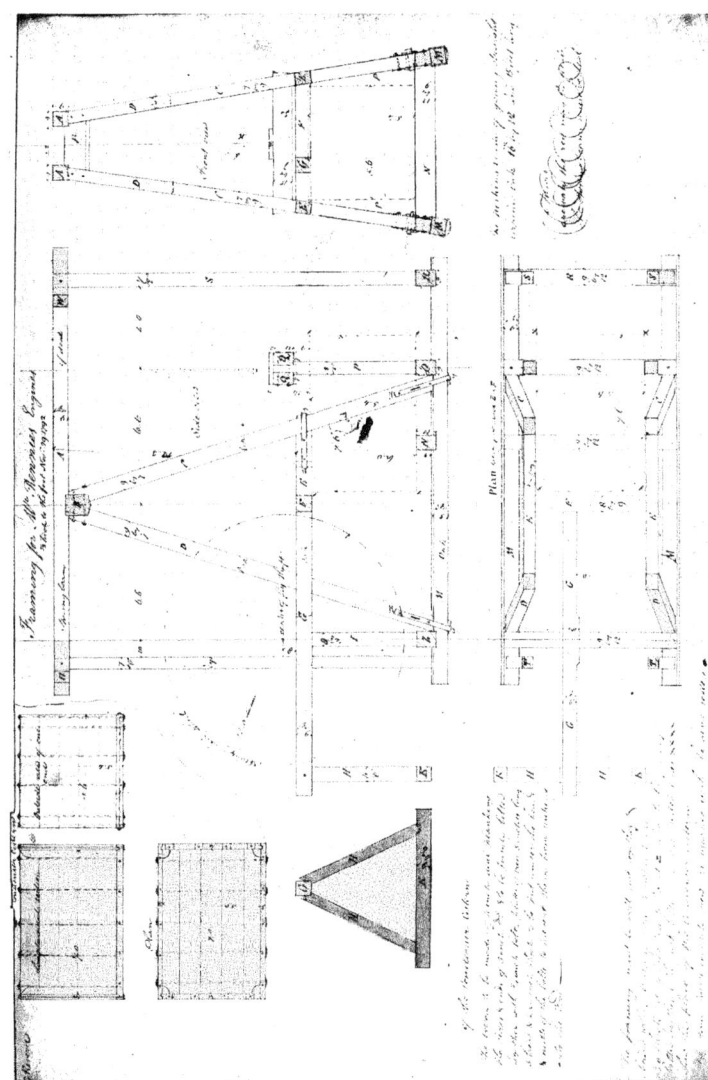

LXXIX. ENGINE FRAME, 1792

Wilkinson's Bradley Forge engine is an exception, in that it follows the lines of the pumping-engines, but Cotes's engine, 1784, Plate XXXV, shows the engine enclosed in the house, and a framework of timber supporting and connecting the various parts. The arrangement of the framing is perhaps more clearly seen in the drawing of the Cockshead engine, 1793, Plate XXXVIII, but here the usual arrangement of building the spring beams (the horizontal members that carry the gudgeon bearings of the working beam) into the walls of the house at each end is shown, whereas in Cotes's engine they terminate at the bearings.

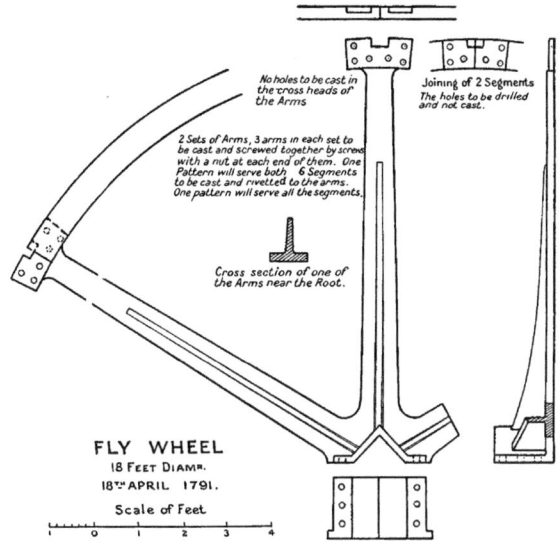

FIG. 36. Flywheel, 1791.

To show the frame construction in detail a design of the year 1788 has been selected for reproduction in Plate LXXVIII. The spring beams, built into the house at each end, are supported under the bearings of the working-beam gudgeons by the ' main framing' composed of two ' uprights' united by a ' cross-piece' and, at their lower ends, by a sill ; to the left of the main framing is the ' back framing' likewise made up of two uprights united, at the bottom, by a sill, and by the ' bearers that carry the cylinder'. The bearing of the engine shaft is carried on the ' headstock of fly framing', which is supported by three ' spurs' on a sill ; the headstock and the sill are tied together by bolts shown by thick lines. A ' diagonal spur' extends from the headstock to the ' cap piece' of the main framing. The condenser cistern, shown in dotted lines, fits between the uprights of the back framing and is carried on sills that rest on the sills of the main and back framings.

With this arrangement the house is still an essential part of the engine frame. The winding-engine for Reynolds, 1788, Plate XXXVI, and the canal engine, 1790, Fig. 19, p. 171, show self-contained engines designed to be portable. With the extending use of the rotative engine a demand arose for a self-contained fixed engine. Plate LXXIX shows the frame of such an engine from a drawing of 1792. The spring beams are supported, at each end, by a pair of uprights and, centrally, by pairs of diagonal members, the lower ends of which are united to a pair of longitudinal sills. The cylinder rests on cross-beams supported by posts on a cross-piece resting on the sills. The beam for the engine-shaft bearing is supported at its outer end on a triangular frame ; at its inner end it is secured to a cross-piece which lies between two members extending between the diagonals that support the spring beam, and centrally it is supported by a post resting on a cross-bar uniting the sills. The drawing indicates that the diagonal members and the sills are to be of oak, but deal is to be used for the spring beams. The joints of the framing are to be bound with straps of iron 4 in. wide and held with 1-inch bolts.

But although a self-contained form had thus been arrived at, the older form continued to be used, indeed to be used most frequently. In 1795 we find Southern asking ' are we to make the framing of the Spanish 8 horse engine in the common way to be erected in a house, or with four diagonal uprights as in an independent framing '.[1]

THE GOVERNOR AND THE THROTTLE VALVE

WRITING in the year 1814, Watt said : ' The application of the centrifugal *principle* was not a new invention, but had been applied by others to the regulation of water and windmills, and other things, but Mr. Watt improved the mechanism by which it acted upon the machine and adapted it to his engines.' [2]

There can be no question that Watt was the first to apply the centrifugal governor to the steam engine, but it is interesting to note that the idea had occurred to another inventor. In Stuart's *Anecdotes* we read that ' A Mr. Clarke of Manchester suggested the application of this fine mechanism to the regulation of the flow of steam from the boiler into the cylinder '.[3]   A man of that name, probably the same individual, was associated in an attempt to apply the Newcomen engine for the production of rotary motion.

We do not find any very definite evidence that the governor had been applied to regulate the speed of water- and windmills, but there is no doubt that it was in use to adjust the feed to and the distance apart of millstones before Watt took it up. In fact it is likely that the idea was suggested to him by such an application at the Albion Mills.

Boulton, writing from London in reference to this mill, informs Watt that he finds many new mechanical schemes employed, including one ' for regulating the pressure or distance of the top mill stone from the Bed stone in such a manner that the faster the engine goes the lower or closer it grinds & when the engine stops the top stone rises up & I think the principal advantage of this invention is in making it easy to set the engine to work because the top stone cannot press upon the lower untill the mill is in full motion ; this is produced by the centrifugal force of 2 lead weights which rise up horizontal when in motion & fall down when ye motion is decreased, by which means they act on a lever that is divided as 30 to 1, but to explain it requires a drawing '.[4]

[1] B. & W. Colln. Southern to James Watt, junior, 1795 Feb. 11.
[2] Robison : *Steam and Steam Engines*, 155 (Appendix by Mr. Watt).
[3] Stuart : *Anecdotes of Steam Engines*, 1829, II. 360.
[4] B. & W. Colln. Boulton to Watt, 1788 May 28.

It is known that the Albion Mills engines were fitted with governors, and it has been assumed that they were so fitted in the first instance, but it is reasonable to suppose that had this been the case Boulton would have been quite familiar with the device and that his letter would have been couched in different terms.

The earliest indications of engine governors met with in the Boulton and Watt Collection date from the end of the year 1788, just five months after the date of the letter quoted above.   We have first a drawing headed 'Centrifugal Speed Regulator' and marked in pencil 'Whirling Regulator' ; this is dated Nov. 8, 1788.   Then in the 'Drawings Day Book'[1] under the same date is the entry, 'Drawing of centrifugal engine regulator for no. of strokes', and a further entry that this drawing was taken to Soho on the same day by Southern.   It should be explained that at this time the drawing office was in Watt's house at Harper's Hill.   Then, in the early days of December, we find a number of entries showing that Southern had attended at Soho 'about the Whirling Regulator'.   On the 11th he starts a new drawing which he finishes on the 13th and takes to Soho on the 17th.   This drawing, dated '13 Dec. 1788', is marked 'M. Boulton Esq. LAP', i.e. it is for the Lap engine, now in the Science Museum, South Kensington, then under construction for Boulton's Manufactory.   This interesting drawing is reproduced in Plate LXXX.   It represents the first governor actually applied to a steam engine and the construction is so clear that no explanation is called for.

The next governor drawing we find is dated September 1790.   In the meantime the Lap engine governor had been working successfully and Boulton had been mentioning it to clients of the firm.   One of these, Peter Drinkwater of Manchester, writes towards the end of 1789 :

'When I had the pleasure of seeing your Mr. Boulton here, he politely observed that there were several little matters of later invention which you meant to annex to the engine and on that account expressed a wish that the engine when putting up, or going, might not be too much subjected to public inspection.   I accordingly engaged it should not.   Among these inventions one, I understand, is of a nature solely calculated to secure more effectually an equable motion under different degrees of heat from the fire ; a property so extremely essential in preparing cotton to work into fine yarn, that I would on no account have you deny the use of this instrument.'[2]

By this time the governor had become fairly well known and at the end of 1790 we find Rennie pressing for four or five to be sent to London, as 'several of our engine proprietors are very anxious about governors'.[3]

In 1793 there is evidence that other engine-builders were imitating Boulton and Watt.   One such case is reported from Leeds, and Southern in his reply mentions that there is no patent for the governor or the throttle valve.

'It is told to you, which you need not tell again, that we have no patent for the governor nor throttle valve, so that Messrs. Sayner and Son do not infringe ; for which good reason you may let them go on their own way.   No great things are apprehended from them, nor their throttle valve, nor governor.   Let them go on and say nothing about it.'[4]

It is likely that Watt had considered the question of patenting and had come to the conclusion that it was not expedient to do so.   He seems to have been aware

[1] This is a thick folio volume in a modern binding, and contains a record of the work done in the drawing-office over a number of years.
[2] B. & W. Colln.   Peter Drinkwater to Boulton and Watt, 1789 Nov. 21.

[3] It will be observed that the term ' governor ' had already come into use.
[4] B. & W. Colln.   Southern to Lawson, 1793 July 26.

of the patents granted in 1787 to Thomas Mead and in 1789 to Stephen Hooper for centrifugal regulators for millstones, &c.[1] It is clear that the contrivance of Hooper had come into use, and likely that that of Mead had done so also. Whether it was one of these that Boulton saw at the Albion Mills is not known.

In the Boulton and Watt Collection there are some letters from the Dearmans of the Eagle Foundry, Birmingham, apparently in reply to a request from Watt for information on the subject. In the first letter J. P. Dearman says :

' I have requested G. Warde to give us the names of the patentee of the governor if there is any such person. . . . Two or three years ago there was a person travelling thro' this country, threatning to prosecute the users of the regulator, of the name of Varlo and Son, from Ounslet or Hounsleet near Leeds. I am told the regulator for millstones has been used about us here 20 or 30 years ago.' [2]

About a month later, in January 1794, S. C. Dearman sends a printed description of Hooper's patented invention with an illustration. His letter is of great interest, as is likewise a letter from his correspondent Warde. Dearman writes :

' I have made several enquiries as opportunities offered about the governor balls and find there are several in the country about here like Averns at the windmill put up by a travelling drunken millwright who pretended to the invention and received a few guineas according to the bargains he was able to make for the job. What his name was or what is become of him I can find no traces. It was afterwards reported that he had worked for the inventor and patentee and, on leaving him, took that unjust way of putting some money in his own pocket.' [3]

The letter from Warde to Dearman, a copy of which was retained by Boulton and Watt, runs :

' I met accidentally the person alluded to respecting the governor. His name is Kingsford, and is a considerable mealman between Canterbury and Deal, somewhere I believe near Hythe, he is not the patentee, but acting as a nominee or executor for him. I have known them some years in common use applied as Avern, the baker in Birmingham uses them in the windmill. I mentioned this to Mr. Kingsford who asked me if I could date my knowledge of them, which I could do only by guess ; he thought if I did not mistake, the date of which I had an idea would antidate the patent. When I wrote first to you I also wrote to the millwright (at Peterbro) who first shewed it me to know how far back he could trace the use of them, but have got no answer.'

It would seem then that the Dearmans and Warde all had the idea that the governor had been in use in connexion with millstones before the dates of the patents to Mead and to Hooper.

The next illustration, Plate LXXXI, reproduces a drawing ten years later than that of the Lap Engine governor. It will be seen that the governor is now provided with stops, a and b, the one ' to prevent the balls opening too wide asunder ', and the other to prevent them ' from coming too close together '. It is driven by a rope,

---

[1] No. 1628 of 1787. Thomas Mead of Port Sandwich, Kent, Carpenter. A regulator on a new principle for wind and other mills for the better and more regular furling and unfurling the sails on windmills without the constant attendance of a man, and for grinding corn and other grain, and dressing of flour and meal, superior in quality to the present practice, and for regulating all kinds of machinery where the first power is unequal.
No. 1706 of 1789. Stephen Hooper of Mar-

gate, Gent. Certain new constructed machinery for regulating the power and motion of wind and other mills, as also the process of grinding and dressing therein, and for regulating all other machinery where the first motion is un-equal.
[2] B. & W. Colln. J. P. Dearman to Watt, 1793 Dec. 24.
[3] Ibid. S. C. Dearman to Watt, 1794 Jan. 28.

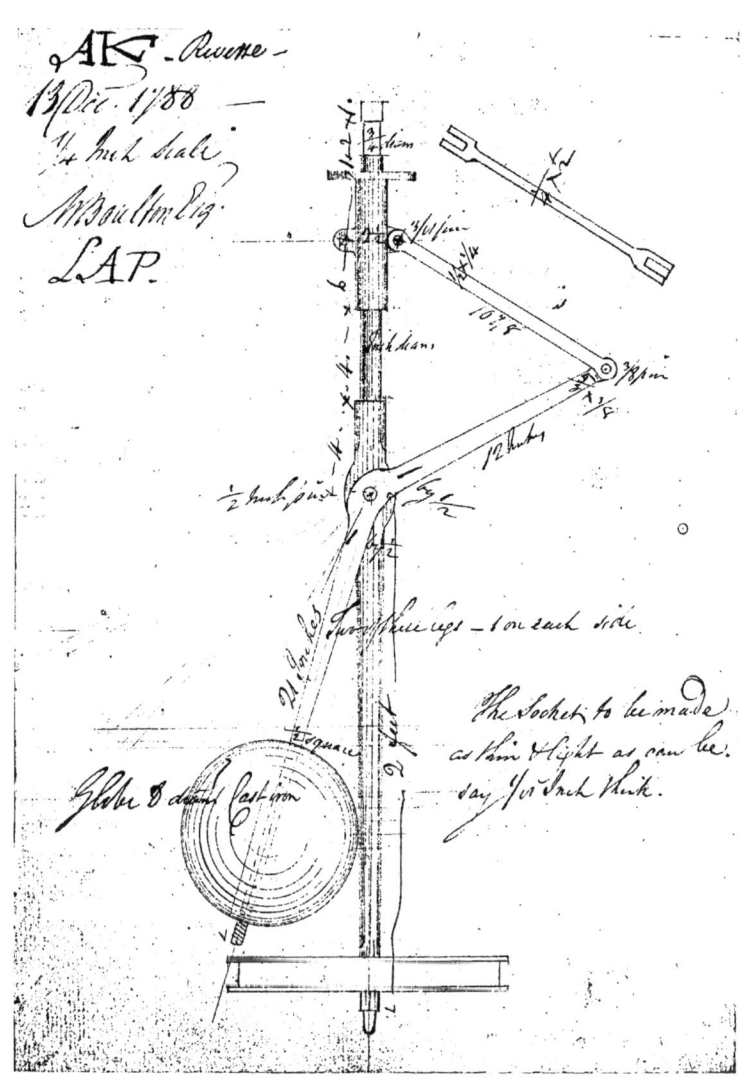

LXXX. GOVERNOR FOR THE LAP ENGINE, 1788

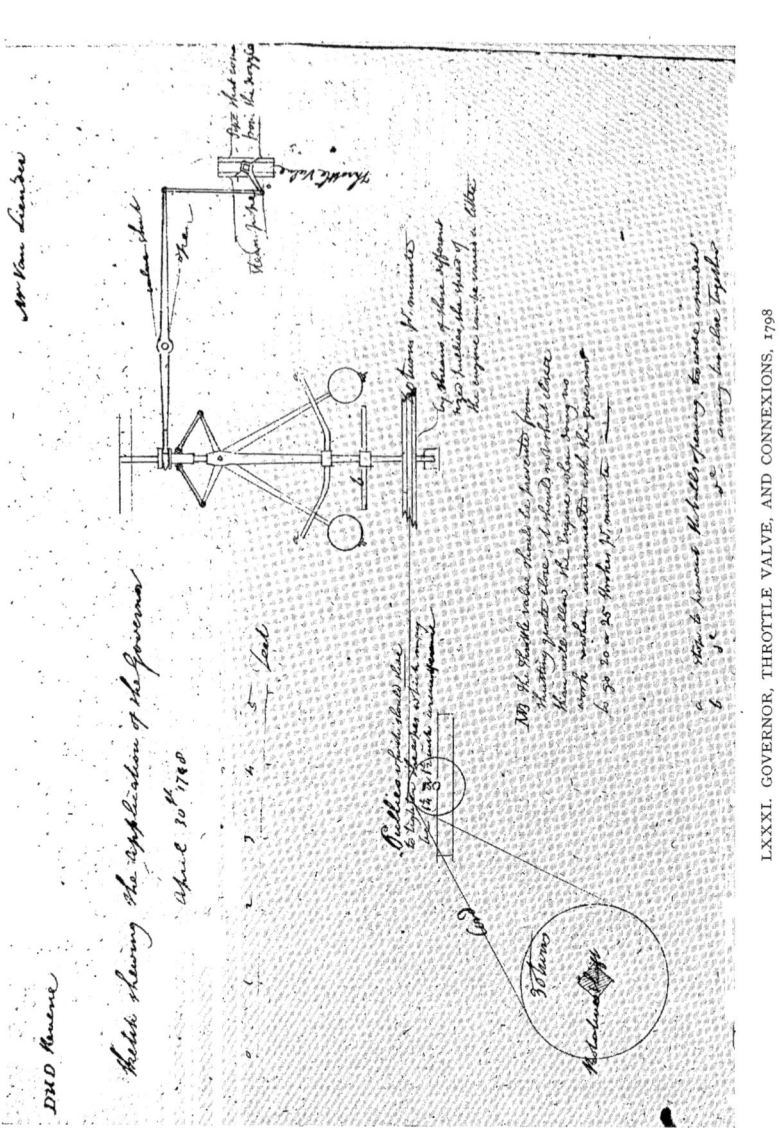

LXXXI. GOVERNOR, THROTTLE VALVE, AND CONNEXIONS, 1798

1¼ or 1½ in. circumference, passing over tension pulleys carried on a slide ; the pulley on the governor spindle is a stepped pulley, to allow the speed of the engine to be varied a little. The drawing shows the connexion from the governor sleeve to the throttle valve, interposed between the 'pipe that comes from the nozzle' and the 'steam pipe' and there is a note as to the adjustment of the valve : 'N.B. The Throttle valve should be prevented from shutting quite close, it should not shut closer than will allow the engine, when doing no work, & when unconnected with the governor to go 20 or 25 strokes per minute.'

The throttle valve was associated with the governor from the beginning, indeed, some months before the governor is first referred to, and a few weeks before Boulton informs Watt of the centrifugal device applied to the millstones at the Albion Mills, we find a full description of a throttle valve in connexion with a coal-winding engine for Reynolds at Ketley. It seems clear that the valve was designed originally for regulation by hand. In the directions sent to Ketley with the drawings of the engine we find the valve described thus : 'It consists of a ring of cast iron 3 inches thick, having a spindle like one of the valve spindles and two thin sheets of copper rivetted together and to the square part of the spindle, and wood between them to fill up.' [1]

How long this built-up construction of valve disk remained in use is not clear, but in his description of the engine written in 1814 Watt gives the single disk with an eye along its diameter to receive the spindle. He mentions also that in some cases the valve is regulated by hand.

LUBRICATION

'THE proper grease for the piston and cylinder stuffing box is melted tallow, and for the chains, gudgeons etc., common Spanish olive oil (called Sallet oil) which for some uses may be thickened by dissolving some tallow or butter in it, by means of heat. Linseed oil should never be used as grease, as it dries and creates more friction than would have been without it. Hog's lard, or train oil, if applied anywhere about the cylinder, or where it is hot, will thicken like linseed oil. When the oil or grease about the great chains, or any of the working parts, grows clotted or very thick, it should be scraped off before new grease is added.' Thus the printed book of 'Directions for erecting and working the . . . engines', issued in 1779. It would seem that it was found that the engine-men were negligent about the cleaning off of the old dried grease, for in a printed sheet of directions issued subsequently to the book great stress is laid upon the necessity of attending to this matter.

The usual weekly allowance of tallow for the piston was two pounds per foot of diameter. In 1775 we find Boulton suggesting the use of blacklead dust in the cylinder of the Soho engine, and a mixture of blacklead and tallow came into use to some extent, particularly for new cylinders or newly packed pistons ; it was found, however, that the blacklead had a tendency to wear the cylinders. In 1792 Boulton and Watt were using what they called 'pomatum', which may have been this mixture of blacklead and tallow. The engine at Gregory Mine was out of order and Southern wrote to one of the erectors to go and set it right, 'sending him likewise directions for the use of the pomatum'.[2] For one of the engines in Manchester in 1795 Lawson, who was in charge of the erection, tried tar instead of

[1] B. & W. Colln. : Letter Books, 1788 May 14. 'Observations on the drawings of the engine for winding coals for Messrs. Reynolds & Co.'

[2] B. & W. Colln. Southern to Boulton & Watt, 1792 Apl. 12.

grease. In reporting the experiment to Soho he seems very pleased with the results ; Watt, however, remarked that tar had been tried before and found to injure the cylinder, and in any case coal tar should not be used.[1]

The Soho drawings of our period show few indications of provision for the supply of lubricant, in the way of oil-cups and oil-holes. As will be seen from the section on stuffing-boxes, designs of the year 1784 show the piston-rod stuffing-box gland formed with a well ; this was to be charged with tallow once in six hours. As to the piston, Watt, at first at any rate, seems to have thought it sufficient to pour melted tallow over the hemp packing every time the piston was packed, which was once a week ; the operation of course necessitated the taking up of the cylinder cover.   As early as 1777 Boulton inquires of Watt, in reference to the Bow engine with which there was a little trouble, whether he may drill holes in the top of the cylinder in order to pour in grease or suet. Although they are not shown on the drawings, no doubt it became the practice to make such holes and to fit them with plugs. It appears that in Cornwall a grease funnel with a movable plug was introduced at an early date by Richard Michell,[2] whose name will be found to occur frequently in the chapter on ' Boulton and Watt in Cornwall'. It is interesting to note that in 1769 Watt was using a pump to circulate oil through the cylinder of the engine at Kinneil.

The beam gudgeons were arranged to work in oil from an early period. Watt, in the correspondence relating to the Tingtang engine, 1777, says : ' As to the friction, that is lessened by contriving it so that the brass stands always full of oil.' Reference to Fig. 29 will show that the brass is seated in a well in the plummer block.

JOINTS AND JOINTING-MATERIALS

In an account of the invention of the iron cement, written in 1814, Watt states that the joints of the cylinder and other parts of the Newcomen engine were generally made tight by being screwed together upon rings of lead covered with glazier's putty. This method, he says, ' would not answer Mr. Watt's purpose. He at first made his joints very true, and screwed them together upon pasteboard, softened by soaking in water, which answered tolerably well for a time, but was not sufficiently durable.'[3]   The book of ' Directions for erecting and working the newly-invented steam engines by Boulton and Watt' [1779] gives a good deal of information about the pasteboard joints. To begin with, it is explained that the pasteboard is ' not such as is composed of paper pasted together', but ' such as is used for the boards of books', i.e. millboard. The boards, about $\frac{3}{16}$ in. thick, are cut into segments; these are thinned at the ends to overlap each other, are soaked in warm water until soft, then laid to dry on a board and afterwards soaked in linseed oil. In making the joints the prepared segments are coated on each side with a putty composed of linseed oil and whiting (not white lead) and are laid between the flanges of the joint. Watt, in the account quoted above, says ' he at first made his joints very true'. This is not correct; at first the castings were used as they came from the foundry, and the directions tell us that the pasteboard segments are to be pared down to ' such thicknesses as the different parts of the joint may require (if it be more open in some places than in others)'. In some

[1] B. & W. Colln. Southern to Lawson, 1795 Mar. 25.
[2] Information supplied by his descendant,

Mr. W. A. Michell of Redruth.
[3] Robison : *Steam and Steam Engines*, 158.

cases metal packing-pieces were required, as the use of more than one thickness of the pasteboard was to be avoided. The pasteboard joint was used between the inner and outer cylinder bottoms, between the inner cylinder bottom and the lower end of the cylinder, and in other parts of the engine. The joint of the cylinder cover was preferably made with pasteboard covered with putty on its lower side only, or a soft rope coiled spirally might be used with the putty, but this was not considered so good a plan ; a plaited rope was not to be used, as it had been found that air-tight joints could not be obtained with this packing. Pasteboard and putty could be used for the joints of the condenser, or, where under water, sheets of lead and Russia duck covered with putty might be employed. Caulking is prescribed for some of the joints, and the screws for the cylinder joints were to be lapped round with rope yarn and putty both under the heads of the screws and under the nuts ' so that each screw may be airtight of itself '.

To go back now to the account written by Watt in 1814. The pasteboard joints being found not sufficiently durable, Watt ' endeavoured to find out some more lasting substance ; and observing that at the iron founderies they filled up flaws by iron borings or filings, moistened by urine, which in time became hard, he improved upon this by mixing the iron borings or filings with a small quantity of sulphur, and a little salammoniac, to which he afterwards added some fine sand from the grindstone troughs. This mixture being moistened with water and spread upon the joint, heats soon after it is screwed together, becomes hard, and remains good and tight for years, which has contributed in no small degree to the perfection of the engines.'

Watt goes on to say that Murdock, much about the same time, ' without communication with Mr. Watt, made a cement of iron borings and salammoniac, without the sulphur. But the latter gives the valuable property of making the cement set immediately.'

We learn of the iron cement for the first time in the year 1782 when Watt writes from Cornwall to inform Boulton that he is experimenting on a new cement for joints, as he had lost faith in putty which always failed in the long run in all the vacuum joints. He says that he has several compositions in view:

' one, a fine powder of iron in a metallic state mixed with substances which may dispose it to rust and some mucilaginous matter which may give it elasticity and keep it in place till it rusts, and a new substance which is neither soluble in ∇ [water] $\mathcal{V}$ [acids] nor oils but can be softened by water to the consistence of caoutchouc & which when mixed with earthy substances becomes as hard as stone, and can be had cheap, being the product of an English vegetable. What continued heat may do to it experiment must determine ; in other respects it would make a hole of an inch wide quite tight in one minute after applied and adheres to any dry substance most viciously. Caustic alkalies or acids destroy it or weaken it.' [1]

In regard to this new substance, ' the product of an English vegetable ', we have no information ; nor is there anything to show that Watt, in the first instance, had in view the use of sal ammoniac.   There is a widespread idea that the combination of iron borings and sal ammoniac was due to Murdock ; indeed, as we have seen above, Watt admitted that he had independently invented it.

Murdock's cement had come into use in Cornwall by the middle of 1784 when we find Boulton writing: ' The method of using salammoniack in the makeing of joints seems to answer so well that an engine being once well put together may

[1] Boulton Papers.   Watt to Boulton, 1782 Apl. 10.

be carryd about as one solid piece without spoiling the joints.' [1] About twelve months later he writes, again from Cornwall : ' When you send anything down here, do send a little of the new cement that one may compare the joints made with it against those made with ye salarmoniak.' [2]

It would seem that Watt had abandoned the experiments of 1782 without arriving at a satisfactory result, but, upon hearing of Murdock's success in 1784, had again taken the matter up and produced the ' new cement' mentioned in Boulton's letter, which would differ from that of Murdock in having sulphur as one of its ingredients. It has been stated above that there is nothing to show that Watt, in the first instance, had in view the use of sal ammoniac. He was well acquainted with the effect of this substance in producing rust on iron (in the book of Directions of 1779, in connexion with the fixing of studs in tapped holes in cast iron, there is a direction that the end of the screw must be ' wetted with common salt or salammoniac to rust it in ' ; the application of the salt to the joints of boilers is also mentioned), but he may have considered it unsuitable for the cement, or indeed he may have overlooked it altogether until he heard of what Murdock had done.

From this time it is clear that the iron cement came into regular use. At first it appears to have been prepared as required during the erection of an engine from borings sent with the cylinder from Bersham Ironworks, but in 1788 we have Southern writing to the manager at Bersham : ' It will be unnecessary in future to furnish any engines with gun borings, as the cement is made at Soho where it is mixed with the turnings &c all at once.' [3]   For the next ten years or so there is evidence that the cement was made in considerable quantities at Soho ; nevertheless engineers continued to prepare it for themselves in some parts of the country. In 1801 the engineer of the United Mines, in Cornwall, writes that ' we make all the joints of the cylinders & nosels with strips of Holland duck smear'd over with rust made of fine iron borings, salarmoniac, and sulphur brimstone, *i.e.* 5 or 6 thicknesses of duck in a joint of the large engines '.[4]   The writer adds that all joints surrounded by water, as in the condensing cistern, are made with soft putty and duck.

THE COUNTER

WATT had applied a counter to the Soho engine for the trials made in 1774-5. This was not the pendulum arrangement that afterwards became so well known in connexion with the Boulton and Watt engines, but a ratchet-and-pawl device worked from the beam. Joseph Harrison, in 1775, said that ' the method took in counting the strokes was by fixing up a machine that had a communication with the engine beam, that moved one division every stroke of the engine '.[5]

When it became a question of counting the strokes for the purpose of determining the premiums payable on engines supplied to customers, it seemed that this form of counter was open to objection and that it was desirable to have a contrivance that could not be tampered with, or, as Boulton writes early in 1777, ' an unalterable counter would be as good a check as can be desired '.[6]   Soon after,

[1] B. & W. Colln. Boulton to Watt, 1784 July 8.
[2] Ibid. Boulton to Watt, 1785 July 31.
[3] Ibid. : Letter Books. Southern to Gilpin, 1788 Mar. 3.
[4] Ibid. R. Michell to Boulton, Watt & Co.,

1801 Aug. 25.
[5] Doldowlod Papers. Evidence on Watt's Engine Bill, 1775.
[6] B. & W. Colln. Boulton to Watt, 1777 Feb. 1.

Boulton was in London and engaged in testing the Chelsea and Shadwell engines, and now we find that he has got the pendulum counting-apparatus, for he writes to Watt:

' We have got an excellent Counter that goes true, & can't committ a mistake, for 30 Years. It may be hermeticaly seald up in an Iron Box & it hath no external communication whatsoever. You may even exhaust ye Box of air & I think it will be proper to annex one, on our own acc^t., to every great Engine we make, & wch shall only be opend when we visit the Engine and a Book to register the strokes & ye day of ye Month. The Machine moves, it being fixed upon any part of the Beam, but gravity keeps a Bar, on wch is a dead scapement perpendicular. [Sketch.] Ye scapem^t wheel is a contrite. It scapes at 20 degrees but the Chelsea Beam moves 30 degrees when a 6 foot stroke.' [1]

The sketch and the relevant part of the letter are reproduced in Plate LXXXII.

Boulton does not say how he got the apparatus or from whom he heard of it, but there is some reason for thinking that it had been made for the purpose of a pedometer by Wyke and Green of Liverpool, and that it was brought to his notice by Henderson who had come to Soho from Liverpool.

About four months after this letter, under the date 4 Sept. 1777, Wyke and Green send in a bill to Boulton and Watt for ' 1 long frame & wheel for counting the strokes of fire engines &c. 1l. 5s. 0d.' With this bill is a letter that runs:

' Mr. L. Handerson wrote us for a set of wheels & pinions suitable for the above machine, but thought it most proper to make the frame likewise (at least for the first) that we might make the wheels & pinions more proportionable to each other, & the wheels pinions & pivitts &c. at proper distances from each other to answer the next on both sides of them. Now we have made this, can soon make you more as wanted if you find this to your mind, or if any alteration in the plan is thought necessary please to advise & it shall be observed. . . . P.S. Please to give our compliments to Mr. Henderson, advise him we have two of the pocket walking machines very near compleated.' [2]

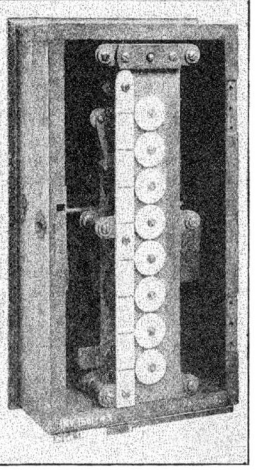

FIG. 37. Engine counter at the Science Museum, South Kensington

At the same time Boulton was having some counters made at Soho, and we find him writing to Watt: ' I have many engine counters now made. Pray shall I send you any of 'em. These have cost in work & material 2½ guineas each, but I shall get them for 2 guineas in future.' [3] However, no more seem to have been made at Soho, at that time at any rate, for another order was given to Wyke and Green which they were very slow in filling and it was suggested that an order should be given to ' Holliwell of Birmingham (who made some counters while at Soho) '.[4] This does not seem to have been done and a little later Wyke and Green write to say that they have two nearly ready and eight more in hand. Then they got other

[1] Ibid. Boulton to Watt, 1777 May 12.
[2] Ibid. Wyke & Green to Boulton & Watt, 1777 Sept. 4. The name of John Wyke is a well-known one. He had been in business in Liverpool for a number of years as a watch

and watch-tool maker. He issued a very fine and interesting catalogue of tools, a copy of which is in the Patent Office Library.
[3] Ibid. Boulton to Watt, 1777 Sept. 13.
[4] Ibid. Walker to Watt, 1778 June 10.

orders, for in 1781 they say : ' We have a few more of the pedometers you ordered nearly ready to send you . . . we have not been favoured with the new pattern of counting machines you promised Mr. Wyke when at Birmingham, which please to send us with your directions.' [1] At that date the apparatus was supplied at the rate of two guineas each, and they seem to have been producing some at Soho, for we find Boulton and Watt buying six dozen enamel dials in 1781 and ordering twelve dozen in 1782. They were also using gilt dials.

It would seem that in some cases Wyke and Green made the complete apparatus, while in others they supplied only the pendulum and the wheels, and these were put into their frames at Soho. The engine counters would differ only from the pedometers in requiring a larger train of gearing and a greater number of register-ing dials.

These counters are supposed usually to have been secured on a vertical side of the engine beam, but according to Farey [2] they were placed on the upper surface of the beam with the pendulum horizontal, its length lying across the length of the beam, and the axes of the pendulum and wheels standing vertical.

A counter now at the Science Museum, South Kensington, is shown in Fig. 37.

THE INDICATOR

NOWADAYS, and indeed for over a century past, the term ' indicator ' is, and has been, applied to an instrument with a moving card and a pencil that traces out a diagram of the varying pressure of the steam in the engine cylinder through-out the stroke. This was not the original signification of the term ; the instrument devised and called by that name by James Watt was a pressure gauge. It had a cylinder and a piston controlled by a spiral spring, and the rod of the piston carried a pointer working over a fixed scale. When this instrument was applied to the cylinder of an engine the variation of pressure therein during a stroke could be observed, and, by watching at the same time the movement of some reciprocating part of the engine, the pressures at different points in the stroke could be noted. So far as is known there are no extant drawings or examples of this form of the instrument, but of a modification, in which the piston-rod is connected to a lever that carries the pointer, examples are to be found, for instance, at the Science Museum, South Kensington ; see Fig. 38. The date of the invention is not definitely known, but it must have been soon after 1790. The Boulton and Watt Collection contains a number of letters, ranging from 1794 to 1796, that refer to experiments made with this instrument, and there is a paper endorsed ' G. Lee's experiments with Indicator, 1796 ' and headed ' Experiments made with a steam engine constructed by Boulton & Watt at Salford Cotton Mill '. This is set out in tabular form and there is a column headed ' Indicator in Cylinder Top ' and divided into a column for ' Depression ' and another for ' Elevation '. The experiments were carried out for the most part in 1795, but there is one in December 1793 and another in December 1794. Watt described the instrument in the following terms :

' The barometer being adapted only to ascertain the degree of exhaustion in the condenser where its variations were small, the vibrations of the mercury rendered it very difficult, if not impracticable, to ascertain the state of exhaustion of the cylinder at the different periods

[1] B. & W. Colln. Wyke & Green to Boulton   [2] Farey : *Steam Engine*, 521.
& Watt, 1778 Sept. 7 ; 1781 Apl. 20.

of the stroke of the engine ; it became therefore necessary to contrive an instrument for that purpose that should be less subject to vibration, and should show nearly the degree of exhaustion in the cylinder at all periods. The following instrument, called the *Indicator*, is found to answer the end sufficiently. A cylinder about an inch diameter, and six inches long, exceedingly truly bored, has a solid piston accurately fitted to it, so as to slide easy by the help of some oil ; the stem of the piston is guided in the direction of the axis of the cylinder, so that it may not be subject to jam or cause friction in any part of its motion. The bottom of this cylinder has a cock and a small pipe joined to it, which, having a conical end, may be inserted in a hole drilled in the cylinder of the engine near one of the ends, so that by opening the small cock, a communication may be effected between the inside of the cylinder and the indicator.

'The cylinder of the indicator is fastened upon a wooden or metal frame, more than twice its own length ; one end of a spiral steel spring, like that of a spring steelyard, is attached to the upper end of the frame and the other end of the spring is attached to the upper end of the piston-rod of the indicator. The spring is made of such a strength, that when the cylinder of the indicator is perfectly exhausted, the pressure of the atmosphere may force its piston down within an inch of its bottom. An index being fixed to the top of its piston-rod, the point where it stands, when quite exhausted, is marked from an observation of a barometer communicating with the same exhausted vessel, and the scale divided accordingly.' [1]

The substitution of a pencil for the pointer and the provision of a moving board or tablet to carry a card may not seem to constitute an invention of very great moment, but undoubtedly the conception of an instrument that could trace a pressure diagram was a very important and distinct step in advance. Tradition assigns the invention to Southern,

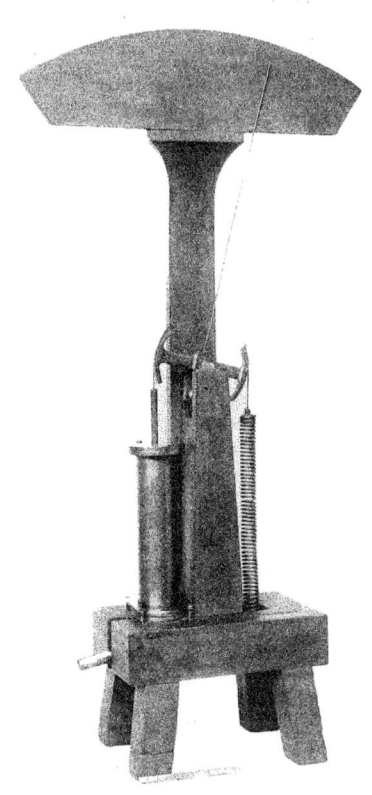

Fig. 38. Indicator or pressure gauge. At the Science Museum, South Kensington

and although the Boulton and Watt papers do not furnish us with any definite statement that he was the inventor, they contain references which make it clear that tradition is right in this instance. The invention originated in 1796 ; on March 14th in that year Southern wrote to Lawson, one of Boulton and Watt's outdoor men, ' Tell Mr. Lee I have contrived an instrument that shall tell *accurately*

[1] Robison : *Steam and Steam Engines*, 156.

what power any engine exerts.' In the same year, 1796, Aug. 13th, we have a hand-sketch of an indicator diagram. It is in a letter from Southern to Lawson, then at Manchester, referring to a diagram that Lawson had just sent to Soho, which he could not understand. Southern writes : ' I have received the Indicator's letter or rather scrawl which is not to be decyphered but by conjecture, without some further information. It is absolutely necessary to know at what part of the line the indicator points to when the piston is *exactly* either at the top or bottom.' It appears that Lawson had rigged up a card-carrier that moved in one direction only, so that the diagram was a continuous line instead of a closed curve. Southern goes on to tell him :

' Great accuracy should be observed at the moment the piston is at the top or bottom, & a person should be stationed opposite the connecting rod to call out the instant it passes the centre of the sun wheel, another person being stationed to observe the indicator pencil at the same moment. The indicator should be placed above and below to see whether any & what difference there is. Please to repeat the experiment with all the attention you & Mr. Lee can command, & send sketches & full description.'

Southern then proceeds to show how a closed diagram may be obtained :

' It would be better if instead of drawing the board uniformly forward, a pair of wheels was applied so as to make one revolution for a double stroke of the engine and crank fixt upon one end of such a length as to give the stroke you wish for the board to move. The exactness of the beginning and ending might be ascertained very nicely, and as the pencil would go over and over again the same track or nearly, the mean might be taken with more precision.'

It is clear that we have here the genesis of the indicator diagram whence the power of the engine can be determined readily.

In the following year we find a Derbyshire firm writing to Boulton and Watt to request that they will send over a man to determine the power of an engine, as ' we were informed that you had a machine for calculating the power of engines. If so, we suppose it will completely shew the power of this engine.' [1]

The earliest indicator diagrams found in the Boulton and Watt Collection were taken in the year 1803. There is first a ' Diagram made by the Indicator at the 6 horse engine at Soho, Jan. 18, 1803 '. Then there is a set taken off the Foundry engine, Mar. 13, 1803, on slips of foolscap paper, 12" × 6", with the actual diagrams 9⅜" long. Another card has a set of curves, in ink of different colours, of diagrams taken in March 1803, at Houldsworth's Mill, Manchester ; these show the gain due to expansive working, and the effects of changes in the valve setting. The engine had the valves worked by eccentric. In December of the same year a set of diagrams was taken off an engine at Wisbech.

Lawson's attempts in 1796 were made, no doubt, with a makeshift adaptation of the Watt instrument. It seems likely that by the year 1803 a complete instrument on the new plan had been made ; certainly in 1806 we read that : ' Mr. Wedgwood wishes you would make him an indicator for his own use that he may try his engine under different parts of the machinery.' [2]

The letters quoted above make it clear that the instrument was becoming known, but it seems that Boulton and Watt tried to keep the knowledge of it to

[1] B. & W. Colln. Coke and Billingsley, Pinxton China Works near Alfreton, 1797 Mar. 16. Thompson of Ashover had put down an engine for them and it would not do the work required of it.
[2] Ibid. Wm. Murdock to Boulton & Watt, 1806 Feb. 5.

themselves as far as possible. John Farey, a keen observer of all things connected with the steam engine, was ignorant of it until 1819, when he came across one in Russia. Ten years later, in his evidence before a Parliamentary Committee, he gave the following account of the instrument and the manner in which he became acquainted with it.

' Many years ago Mr. Watt invented and applied a small instrument which he called an Indicator, to his steam engine ; it indicates what extent of plenum and vacuum is alternately formed within the cylinder, in order to impel the piston when the engine is at work. It is of very important use in giving engine makers true knowledge how to make good engines ; and it was of very great use to the inventor just as a hydrometer is to a distiller. He kept it a profound secret for many years, and in 1814 when he published an account of his other inventions he gave only an imperfect description of a part of this one, without any hint of parts which are essential to the successful use. A complete instrument afterwards fell into my hands in Russia, where it had been made by some of the people sent out from this country with Mr. Watt's steam engines. At my return to England I made one and also shewed several other engineers how to make such for themselves and since that time every one of those persons has very greatly improved his practice by the light it has enabled him to throw upon an obscure part of the operation of steam in an engine. One person who had made an indicator from a sketch that I drew for him, has since printed a description of it in a public journal.' [1]

In the Memoir of John Farey in the *Proceedings of the Institution of Civil Engineers* (1851-2, Vol. XI, p. 100) it is stated that on his return to England from Russia Farey employed McNaught to construct the indicators for general use, ' and thenceforth he was constantly employed to use the instrument in disputed cases of the power of steam engines '.[2]

The publication referred to in Farey's evidence seems to be the *Quarterly Journal of Science*, which in Vol. XIII (1822), p. 91, has an ' Account of a steam engine indicator' in the form of a letter signed ' H. H. junr., Glasgow '. The writer described the indicator by the aid of an engraving, and states that the instrument ' employed in our works was made from a description of the instrument given us by Mr. Field of London '. He adds that Mr. Hutton, of the Anderston Foundry near this city, now makes the instruments upon a very convenient plan. Probably ' H. H.' was Mr. Hutton himself or his son. Although it seems that soon after this indicators were being made in various parts of the country, the use of the instrument did not spread very rapidly. In 1835 John U. Rastrick tried without success to buy or borrow one in Birmingham, and had to apply to James Watt, junior, for the loan of one.[3] Watt, in reply, explains that ' we do not make indicators for sale ', and adds, ' I much fear that all we have are at present out in the hands of our erectors '.[4] However, Rastrick obtained the loan he desired, but after all he did not use the instrument. He wanted to apply it to the air cylinder of a blowing-engine, and the engine was not working.

The indicator as made at Soho at about this date is described at some length in a document, in the Boulton and Watt Collection, endorsed, ' Description of

[1] Report from the Select Committee on the Law relative to Patents for Inventions, 1829, p. 138.
[2] There are some remarks by Farey on the construction of the Boulton & Watt indicators in the *Proceedings of the Institution of Civil Engineers*, 1842, II. 109. He said that ' in the Boulton and Watt indicators the motion allowed to the piston by the spring was very short, but recently indicators have been frequently made without the knowledge of their true principle '.
[3] B. & W. Colln. John U. Rastrick to James Watt, junior, 1835 Oct. 13.
[4] Ibid. James Watt, junior, to Rastrick, 1835 Oct. 13.

Mr. Watt's Indicator . . . as altered from the copy sent to Mr. Watt to Paris 27 May, '39' :

' The Indicator . . . has a cylinder full 1⅛ inch in diameter, so that its area is exactly 1 square inch, the length is 10⅜ inches. The spring has a tension of 1 lb. avoir. per ¼ inch, and is capable of extending downwards about 4 inches, and of being compressed 1 inch or more. This range being considered ample for the purposes of low pressure steam. It is however customary to fit the cylinder with two pistons having springs differing in strength, one for low, the other for high pressure purposes.   The scales of each being formed from the extension of the respec-

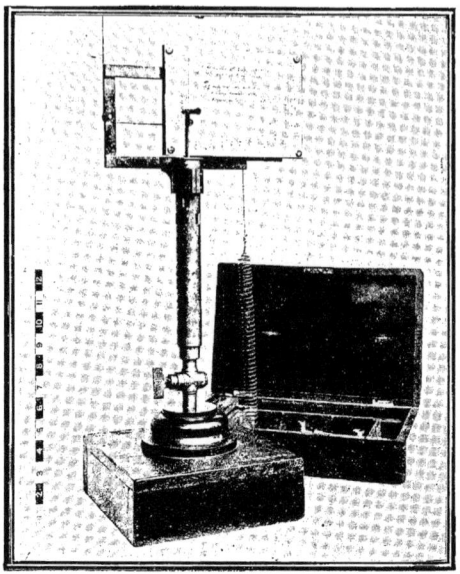

FIG. 39.   Indicator at the Science Museum, South Kensington

tive springs by a given weight. The upper end of the spring is attached to a washer under its cover, its lower end to the piston, the rod of which moves freely through the centre of the spiral, and also through the washer and cover. The piston is solid, fitting the cylinder tolerably tight without any packing, it being customary to pour a little oil in previous to immediate use, to render it airtight for the moment of experiment. The piston rod is formed of a tube, for lightness and with a view to lessen its momentum when in action ; on its upper end is fixed a brass tube containing a small & weak brass wire spring to push the pencil with which it is connected forwards at pleasure, the pencil being of a soft description to produce as little friction as possible while its point is passing over the paper. To the top of the indicator cover is fixed an iron frame by means of a thumb screw. In this frame is a sliding board or pannel, made either of wood or iron (the latter is best, not being subject to warp with the heat) which moves freely in grooves . . . and upon which the paper is fixed for receiving the figure produced by the pencil during the period of the engine stroke. This sliding pannel has attached to its middle at the back two cords, also very pliable, one of which is fixed to any moveable part of the engine, such as the parallel motion or working gear, where the motion of the part does not exceed that of the length of the figure required, which is most generally 5 to 5½ inches.

The other cord has a weight attached to its end for the purpose of overcoming the friction of the sliding board and producing the return motion during the exhaustion part of the engine's stroke. The first mentioned cord moving it during the period of the admission of the steam. The motion of the board is therefore perfectly horizontal, in one direction receiving its impulse from the part of the engine to which it is attached, and in the other from the preponderance of the weight. The bottom of the indicator cylinder is fitted with a stop cock which terminates in a tube whose diameter is usually made to fit the hole of the grease cock in the top of the engine cover.'

The account goes on to describe the manner of using the instrument, and the diagrams produced by it. The mean pressure is obtained by dividing the diagram into ten sections, and it is stated that : ' The friction of the engine alone has been found not to exceed 1½ lbs. per square inch of its piston. It is therefore customary to deduct this amount from the total average, which leaves the net power employed in performing the work.' The indicator was sometimes applied to the lower end of the cylinder as well as the top, mainly when there was a doubt as to the correctness of the valve setting. The paper when prepared was accompanied by drawings of the indicator, and representations of diagrams taken from the engines of the steamships *Salamander* and *Hydra*, but these have not been found.

An indicator now at the Science Museum, South Kensington, is shown in Fig. 39. Some writers have credited Southern with the rotating barrel for the card, but this cannot be correct. The only drawings of indicators noted in the Boulton and Watt Collection are dated 1830, and they show the reciprocating tablet; again, the description quoted above mentions only the tablet arrangement and does not in any way suggest the barrel. Now Southern died in 1815 and, had he contrived the barrel device, it is very unlikely that the firm would still be making the older form in 1830. It seems that this feature of the modern indicator was invented by John McNaught prior to the year 1830. We have seen that McNaught was employed by Farey to make indicators, and he had set up in business in Glasgow before 1833. His name first appears in the Glasgow Directory for the year 1832–3. In 1847 Maudslay and Field were making indicators with the rotating barrel; their instrument is shown in *The Indicator and Dynamometer*, by Main and Brown, published in that year.

XVIII

BOILERS. BOILER SEATINGS
BOILER FITTINGS. FIRING

BOILERS

ALTHOUGH, as will be seen later, Watt did concern himself with firing and the proper combustion of the fuel, he never gave much consideration to the form of the boiler itself.   For his first engines, boilers of the old beehive or haystack form were used ; then when the Cornish orders began to come in, and he found that most of the Cornish engineers preferred the long boiler with internal flues, he gave them drawings accordingly. As to the internal flues, Watt, in his notes to the reprint of Robison's articles on Steam and Steam Engines in the *Encyclopaedia Britannica*, says : ' The Conveying of flame through flues in the inside of the water had been practised by others before my time, and was common in the Cornish engines. The inventor is unknown, but a person of the name of Swaine was a great propagator of the practice.' He goes on to state, however, that he ' somewhat improved the form and adjusted the proportions '.

In course of time the long boiler, with and without internal flues, acquired favour outside Cornwall, particularly for the larger engines ; indeed so early as 1777–8 Wilkinson's engine at Wilson House had a boiler 12 ft. long by 6 ft. wide and 6 ft. high ; in 1787 Rennie proposed to use a boiler 12 ft. long and 5½ ft. wide with a flue 30 in. by 22 in. for a 26-inch double-acting engine. Nevertheless the haystack boiler continued to be made up to the end of Watt's connexion with Soho and later still. Under the name ' balloon boiler' it survived in Staffordshire until about 1850.

A good example of the haystack, or, as it was termed at Soho, a round boiler, is shown in Plate LXXXIII, ' Mr. Henry Goodwin's Boiler, its seating & fireplace', reproduced from a drawing made in 1788. It consists of a vertical cylinder, 3½ ft. high and 5¼ ft. diam., with a hemispherical top and the bottom dished upwards. The man-hole and the safety-valve are shown. The construction of the seating and fireplace will be considered further on.

One of the early long boilers, for Poldory Mine in Cornwall, is shown in Plate LXXXIV ; it is 15 ft. long by 7 ft. wide, and has two round flues of copper, 20 in. diam., which are secured to the end plates by means of flanged rings. The vertical sides of the boiler are tied together by stays above and below the flues. The boiler for Chacewater 63-inch engine, 1778, and the boiler for Poldice, 1780, were of the same form. One of the Wheal Virgin boilers, 1780, 19 ft. long by 8 ft. wide, was narrower at the bottom than at the top, and had two 22-inch flues, one set somewhat above the other. A boiler for Scorrier mine, 1782, had shouldered sides, that is to say, the semi-cylindrical top projected on each side beyond the vertical sides. A boiler designed in 1782 for another Cornish mine, Ale and Cakes, Plates LXXXV, LXXXVI, has a single flue of rectangular form, 22 in. wide by 44 in. high, made of iron plates. The shell is 19 ft. long by 8 ft. wide and 10 ft. 5 in. high, and the sides are tied together above and below the flue. A boiler for Poldice mine in 1792, 22 ft. long by 8 ft. 4 in. wide, had two elliptical flues, 4 ft. high and 2½ ft. wide.

Our next example, the boiler for Tate's engine, 1794, Plates LXXXVII, LXXXVIII, has a single elliptical flue, 3 ft. by 4 ft., and the uptake to this flue is now formed in the boiler itself instead of in the seating. Stays extend between the sides of the boiler, and between the sides and the uptake, and the flue. Plate LXXXIX shows two methods adopted for securing the stays to the plates. Another arrangement is shown in Plate LXXXVIII.

The long boilers referred to above all have flues, but it must not be assumed that this was the invariable practice; for instance, the boiler of Coates's engine, 1803, see Plate XL, has no flue. And again, although the examples illustrated show the vertical sides running direct into the curved top, there are cases in which a shoulder appears in each side of the boiler, i.e. the diameter of the top is greater than the width of the vertical sides. The boiler for Scorrier mentioned above is an example of this. Few examples of the inwardly curved sides, characteristic of the wagon boilers made at Soho later on, have been found in the period up to the year 1800 ; and the term ' wagon boiler' has not been noted.

To return once more to Tate's boilers, Plate LXXXVII ; above each boiler is seen a dome or steam-chamber, 3 ft. diam. and 2 ft. high ; one of the domes is shown with a branch for the steam-pipe, for the other boiler the branch was formed on the flat cover (not shown) of the dome. Each cover had an opening for a safety-valve. The dome was termed the ' water trap'.[1] From the first there were difficulties, from time to time, in preventing water from being carried over with the steam. The book of Directions (1779) says that the feed-pipe is to be fitted with a valve ' to prevent the water ever being forced up through it by steam', or, to ' prevent boiling over'. Originally this valve was placed at the lower end of the pipe ; it is so shown in the Poldory drawing, Plate LXXXIV, but very soon it was found that the more convenient position was at the upper end of the pipe.

The ' water trap', at the date of our drawing, 1794, had been applied to a number of boilers, but without, in all cases, curing the evil, the fact being that the boilers were too small for the engines, or had too little steam room. Watt puts his finger on this defect, in two of his letters written from London at this period. In the first letter he says : ' We are plagued still at Stephenson's with boiling over, in spite of a water trap, & can see no remedy but a larger boiler, or other water—I believe in general we must enlarge the dimensions of boilers, as by overloading engines & one thing or another they generally contrive to find them too small, *especially in double engines*.'[2] In the second letter he writes : ' I find upon inquiry that the boiling over happens more or less at all the engines, except one or two which are light loaded, but nowhere to the degree which it does at Stephenson's. Some means must be devised of preventing it & applied to all

[1] Rennie, writing in 1792, refers to it as a ' steam trap': ' It may be well to order a steam trap, their boiler having rather too little steam room ' (B. & W. Colln. Rennie to Watt, 1792 Jan. 7). No suggestion of what we now understand by a ' steam trap' has been found during the period of Watt's connexion with Soho, but a drawing dated 1813, showing a float apparatus,

is headed ' Design for a . . . to let off condensed water ', and in Watt's account, written in 1814, of the Albion Mills engine, in Robison's *Steam and Steam Engines*, reference is made to the use of an inverted siphon for clearing the steam-jacket during the working of the engine.

[2] Southern Papers. Watt to Southern, 1794 Feb. 4.

indiscriminately.'[1] Nothing seems to have been done at the time, and a few months later we find the chief erector in London suggesting the raising of the water trap ' considerably higher by an additional piece '.[2]

A good deal of variation is found in the size of the boilers, i.e. in the extent of heating surface allowed per cubic foot of water to be evaporated. Watt went into this matter in 1783. At first he seems to have considered merely the relation between the area of the ' surface exposed to fire ' and the contents of the cylinder of the engine, and the proportion he determined upon was 4 square feet of heating surface for every cubic foot of content of cylinder. This was for single-acting engines ; for double-acting engines the proportion was doubled. For a double-acting engine loaded only to 5 pounds per inch of piston-area he brought down the ratio to 7 to 1 ; and for a single-acting pumping-engine at Polgooth, ' as the engine is required to go very slow in summer,' he thought it sufficient to allow but 2 square feet of heating surface for each foot of cylinder content.

In calculating the size of the boiler for the Albion Mills engine Watt assumed an evaporation of 7 cubic feet of water per square foot of heating surface per hour. This figure he deduced from the performance of the boiler at the Chelsea Water-works, by calculation from the size of the cylinder and the number of strokes per minute. Certain boilers on the Cornish mines gave 9, 10, and $10\frac{1}{2}$ square feet per cubic foot evaporated per hour. A small boiler at Soho with a heating surface of 26 square feet was found to evaporate, with ease, 3 cubic feet per hour, i.e. it had about $8\frac{1}{2}$ square feet to the cubic foot per hour. The figure that became the standard, at any rate for some time, was 8 square feet of heating surface per cubic foot of water required to be evaporated per hour.[3] This applied to round as well as to long boilers.

Both Boulton and Watt favoured copper as the material for the boilers. It was considered more satisfactory in every way than iron ; it was more easily worked, and, arising mainly from the demands of the brewer and the distiller, the craft of the coppersmith had attained a high pitch of perfection, while the production of copper sheets and plates was being carried on to a considerable extent. It appears that in 1777 ' bottoms and tops in one piece of 10 or 11 feet diameter' were being made at Bristol by Freeman & Co.[4] Although the first cost was higher an old copper boiler could be sold at a good price ; a London copper-smith charged 12d. per pound for new boilers, and gave $7\frac{1}{2}d$. for old ones. This was in 1777 ; ten years later the price of new boilers had gone up to 14d. a pound. Iron boilers at the same period cost £40 per ton, so that, allowing for the fact that for the same power the copper boiler was made of smaller dimensions and thinner plates than the iron, the copper cost about double the price of the iron one ; on the other hand, it would last twice as long. The drawback of the copper boilers was that it was necessary to clean them out once a week ' as a sediment will be deposited on the bottom which will be more injurious to a copper than to an iron boiler '.[5]

The fact is that good iron plates were difficult to get ; when Watt started at Soho they were all worked out under the hammer, and if English iron was used

[1] Southern Papers. Watt to Southern, 1794 Feb. 7.
[2] B. & W. Colln. R. Dayus to Boulton & Watt, 1794 July 18.
[3] Ibid. ' Blotting and Calculation Book, 1782 & 1783.'
[4] Ibid. Boulton to Watt, 1777 May 8.
[5] B. & W. Colln. : Letter Books. Southern to his father, 1787 Apl. 13.

the result, usually, was poor. We frequently find a good deal of stress laid upon the necessity for forging them from ' Russia Slabs '.

In the early days of the Soho firm iron boiler-plates were ordered from Job Parsons of Burton-on-Trent ; the largest plates are 36 in. by 21 by $\frac{1}{2}$ in. thick, and 26 in. by 21, $\frac{1}{4}$ in. thick. At about the same time ' rolled plates ', probably for boiler tops, were got from Samuel Palmer of Lichfield. Then in 1779 Boulton and Watt go to Nicholas Ryder of Marston Forge, Northwich, in spite of his high prices—31s. 6d. per cwt. for plates and 23s. 6d. for rivet iron (120 pounds to the cwt.). The first order to him included plates 4 ft. long, 14 in. wide, and $\frac{1}{4}$ in. thick. Some of the orders, however, continued to be sent to Parsons; a set of plates supplied by him in 1785 ranged from 32 to 39 in. long. In 1786 there was trouble with boiler-plates made for a London job by a firm at Rotherhithe, ' they will not bend even hot,' says Boulton, ' they being made of his own bad iron.' Rennie, of the same lot of plates, says ' they are pretty well hammered, but wretched iron of his own making, indeed it is not better than cast iron '. John Wilkinson had commenced making boiler-plates by this time, but his wrought iron was not so good as his iron castings, and in 1796 we find a boiler maker repudiating the idea that he used Wilkinson's plates, and protesting that he used no other than Mr. Homfray's. Possibly Wilkinson's plates were made by rolling, and the defects may have been due in part to the want of experience in the new process. At any rate by the year 1795 rolled iron boiler-plates were obtainable, for we find one of the customers of Boulton and Watt declining to have his boiler made of such plates.

' Having had an opportunity of a conversation with the man, a pan-smith who attends the engine at Lawton and was the maker of the boiler now in use there, he seems perfectly satisfyed that Rolled Plates will not answer for the boiler, therefore shall be obliged to you for drawings and directions to make one here of hammered plates.' [1]

The common and indeed, in the early days, the only plan was to put the boiler together on the ground near where it was to stand, and the work seems to have been done by the local smiths. Accordingly, in the printed book of ' Directions for erecting and working the newly-invented steam engines, By Boulton and Watt ' we find directions for making the boiler.

' In making the boiler you should use rivets between $\frac{5}{8}$ths and $\frac{3}{4}$ths inch diameter. In the bottom and sides the heads of the rivets should be large and placed next the fire, or on the outside, and in the boiler top the heads should be on the inside. The Rivets should be placed at two inches distant from the centre of one rivet to the centre of the other, and their centres should be about one inch distant from the edge of the plate. The edges of the plate should be evenly cut to a line, both outside and inside. It is impossible to make a boiler top truly tight which is done otherwise. After the boiler is all put together, the edges of the plates should be thickened up, and made close by a blunt chissel about $\frac{1}{4}$ inch thick in the edge impelled by a hammer of three or more pounds weight, one man holding and moving the chissel gradually, while another strikes. All the joints above water should be wetted with a solution of sal-armoniac in water, or rather, in urine, which by rusting them, will help to make them steam tight. After the boiler is set, it may be dryed by a small fire, under it, and every joint and rivet above water painted over with thin putty, made with whiting and linseed oil, applied with a brush. A gentle fire must be continued until the putty becomes quite hard so as scarcely to be capable of being scratched off by the thumb nail, but care must be taken not to burn the putty, nor to leave off fire until it become dry.'

The precautions taken to make the boiler tight were not always attended with

[1] B. & W. Colln. Bott, Birch & Co., Nantwich, to Boulton & Watt, 1795 June 1.

success; in 1780 we read that lime had been put into a leaky boiler with good results. No special directions are given as to the punching of the holes, but in 1779 Watt, writing to M. Jary who was having an engine put up in France, said :

' There is sent with your goods a machine to pierce the holes in the boiler plate—having mark'd the places of the holes with a small brush & chalk & water, you are to put the plate under the punch and then strike on the head of the punch avec une grand marteau de 28 a 36 livres, selon l'épaisseur de plaques.' [1]

The rivets appear to have been made on the spot ; as to their size, Henderson in 1781 says, ' it is a good rule to have them twice the thickness of the plates they are to fix together.'

By the year 1790, however, boiler making had become a distinct branch of industry, and Wilkinson was carrying it on both at Bradley and Bersham, but apparently his work was not altogether satisfactory. Thomas Horton of West Bromwich seems to have turned out better boilers, and to have been preferred by Boulton and Watt. In the three years September 1796 to September 1799, he made for them fifty-five boilers of an aggregate weight of just over 100 tons. The usual price was 36s. per cwt., but four of these boilers were charged at 37s., and one at 38s.[2] Boulton and Watt sold them all at 42s. per cwt. The Soho firm seem to have been doing some boiler work themselves in 1794 ; but in 1796 Southern puts forward the suggestion that it would be a good plan ' to employ some proper person ', at Leeds and also at Manchester, ' to make boilers for B. & W., as it will be impossible to send them from hence except in pieces, which will be very troublesome & the cause of many complaints.' He thinks that 10 or 11 shillings per cwt. over the cost of the plates will be enough to give per cwt. for the net weight delivered.

At Wilkinson's Bradley Works in 1792 the practice was begun of testing the boilers, before they were sent out, by filling them up with water ' to prove their tightness '.

BOILER SEATINGS

Coming now to the boiler seatings, the printed book of ' Directions for erecting &c.' [1779] directs that :

In building the boiler setting lime mortar should be used only towards the outside ; the parts exposed to fire or flame should be laid in a mortar composed of loam or sand and clay. Long pieces of rolled iron should be laid in the brickwork to prevent it from splitting, and pieces of old cart tyre or like iron bars placed under the boiler, between it and the bricks, to preserve it from being burnt out. Four cleaning openings should be left at convenient places to afford access to the flues ; normally these are to be closed by nine-inch brickwork. One of these openings may be over the fire door and another in the chimney behind the damper. The damper is to be fitted in a groove in the brickwork, and may be provided with a counterpoise. Immediately above the brickwork of the boiler setting an opening must be left in the chimney on the side next the boiler ; this is fitted with a sliding door which may be opened more or less to admit air to moderate the chimney draught.

The boiler top above the setting is to be covered with a three-inch layer of horse or cow dung, then with a one inch layer of lime mortar, and then with two courses of bricks standing

---

[1] B. & W. Colln. : Letter Books. Watt to Jary, 1779 July 26.

[2] In 1822 Thomas Horton's business was being carried on by his sons ; William Murdock, junior, went to see them about a contract for boat-engine boilers. They offered to supply the boilers at £32 per ton, made of plates for which they paid £18.

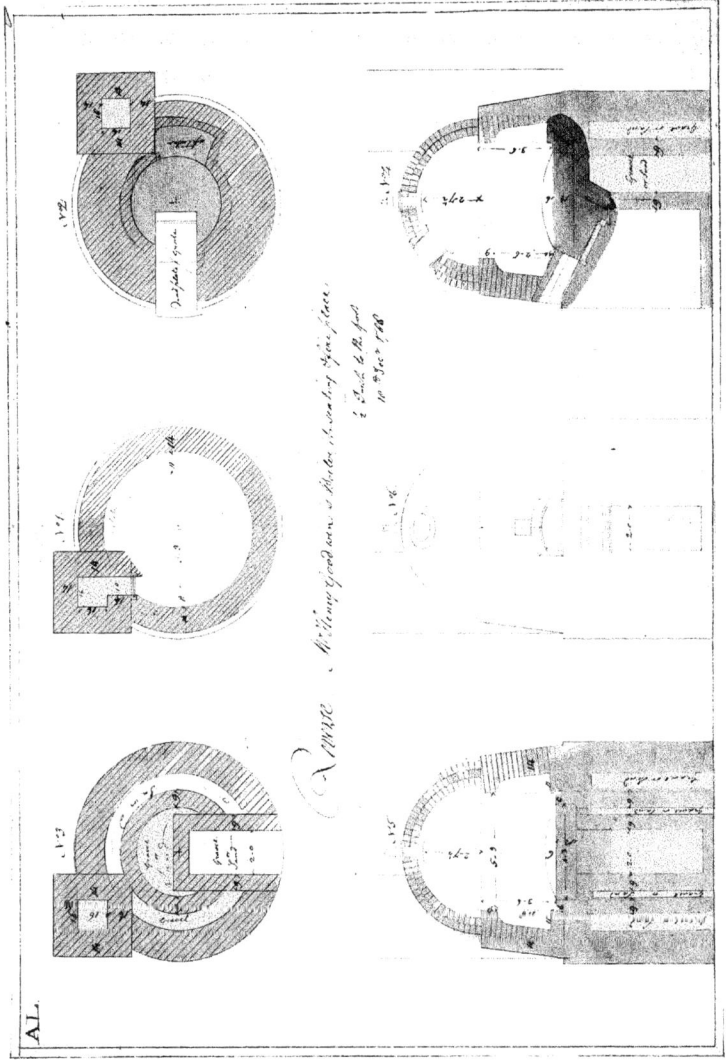

LXXXIII. GOODWIN'S BOILER, SEATING, AND FIREPLACE, 1788.

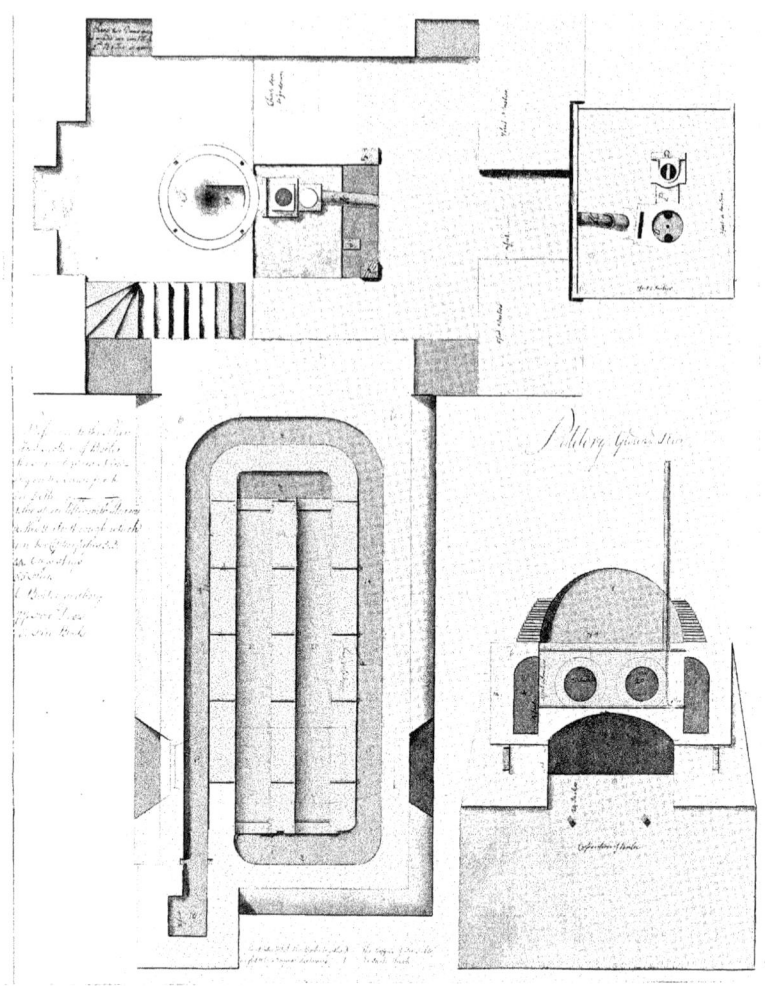

LXXXIV. POLDORY BOILER

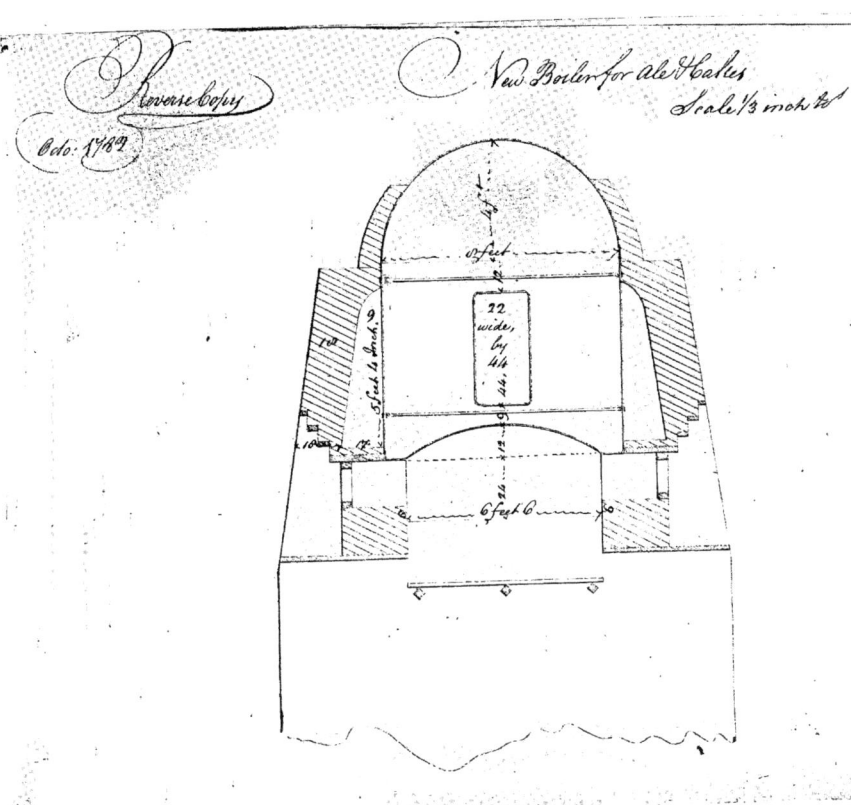

LXXXV. ALE AND CAKES BOILER, 1782

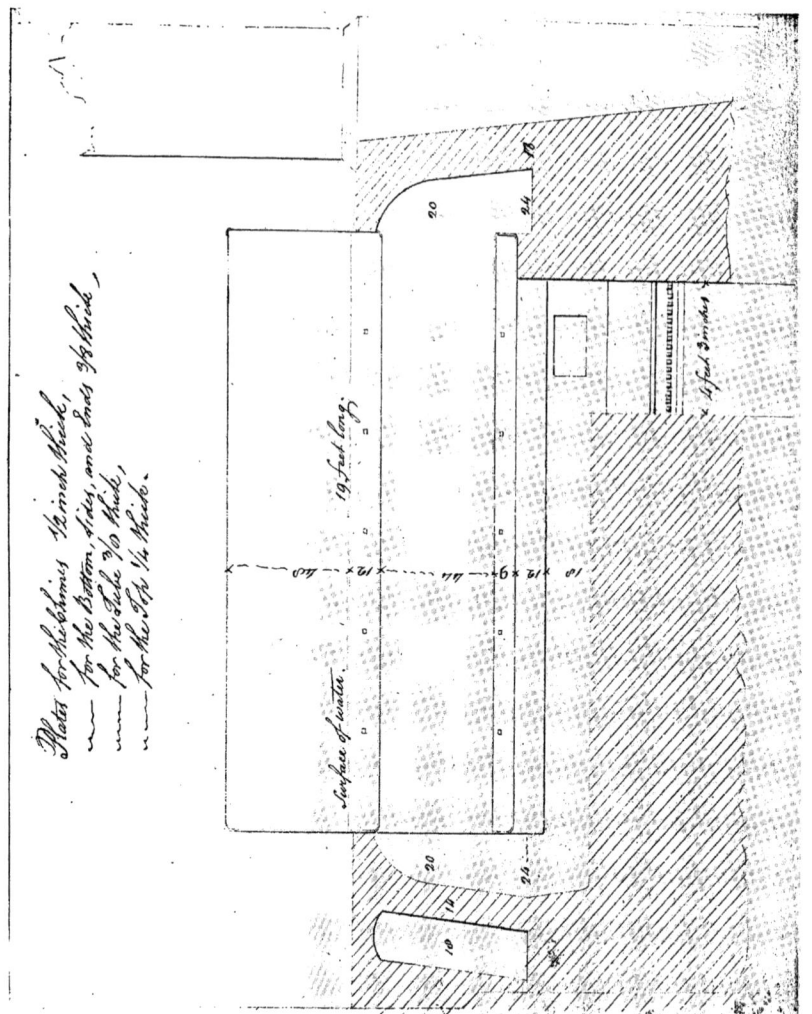

LXXXVI. ALE AND CAKES BOILER, 1782

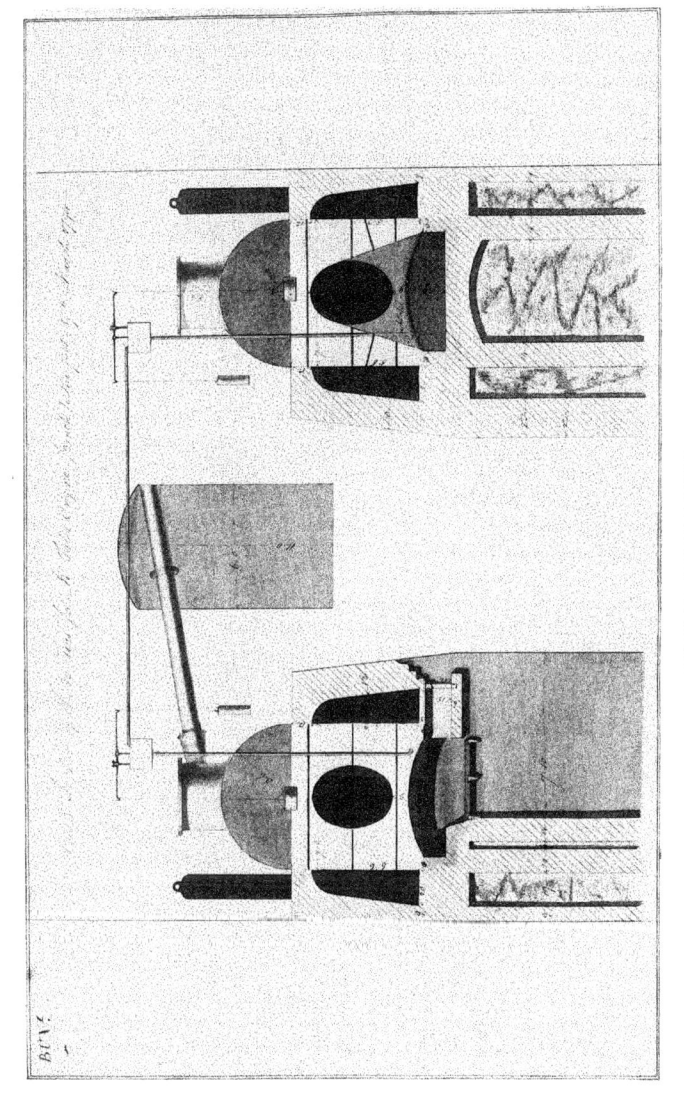

LXXXVII. TATE'S BOILER, 1794

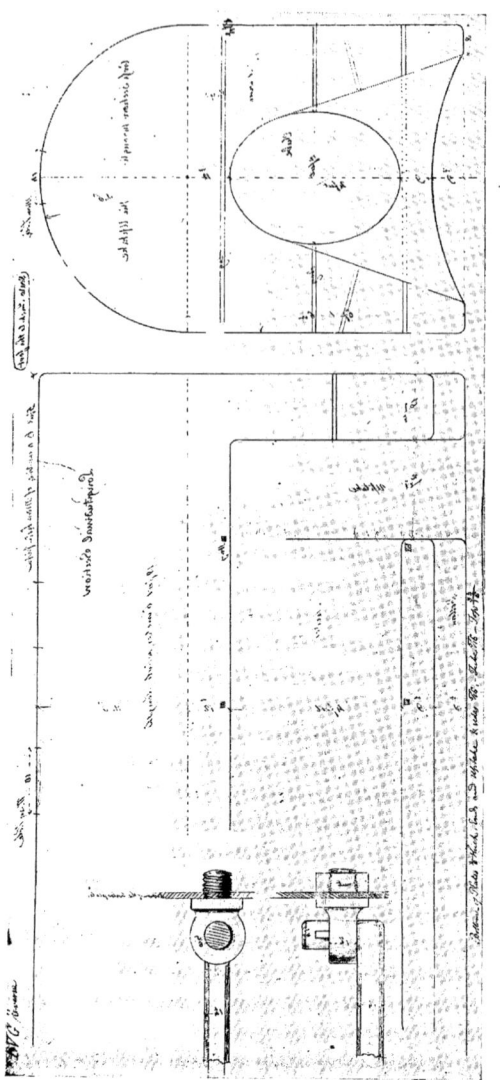

LXXXVIII. SECTION OF TATE'S BOILER, 1794

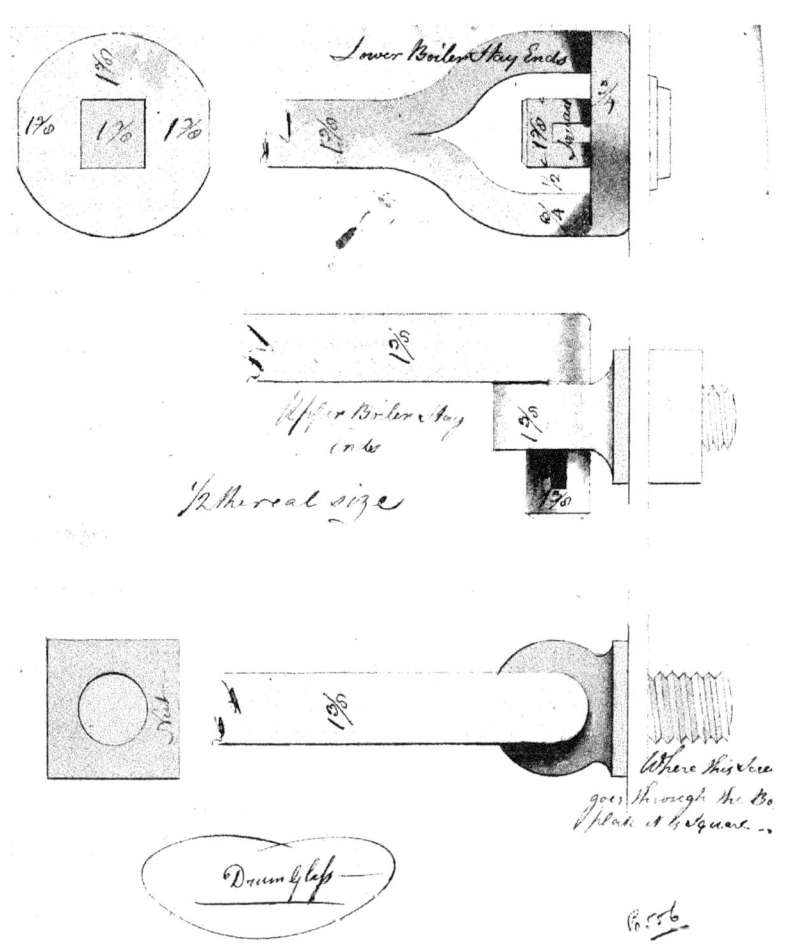

Lower Boiler Stay Ends

Upper Boiler Stay ends

½ the real size

Drum Glass

Where this screw goes through the Bo plate it is square

No 556

LXXXIX. METHODS OF SECURING BOILER STAYS

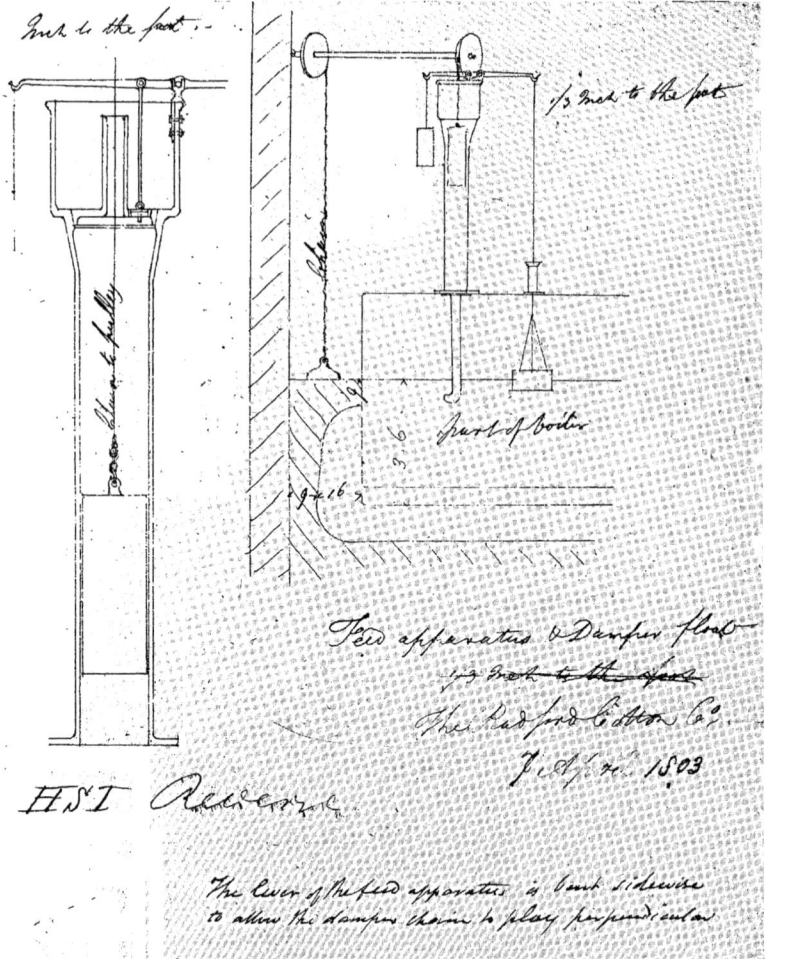

XC. FEED APPARATUS AND DAMPER FLOAT, 1803

on their ends. This covering 'makes the top steam tight, and also defends it from cold and rain, so that a boiler house is not necessary. The mortar employed must be such as stands water.'

The use of the layer of dung was a recent improvement in 1779. The fire doorway is to be two feet wide by one foot high, and the door of two leaves of boiler plate hinged at the sides and overlapping about an inch at the centre.

A seating for a round boiler is shown in Plate LXXXIII (Goodwin's boiler, 1788). It has a deep foundation composed of concentric circular walls, the spaces between which are packed with gravel or sand. The dead-plate and fire-grate are shown inclined, and opposite them is an uptake leading into the flue round the boiler; adjacent to the uptake, but separated from it by a brick partition, is the flue leading to the chimney. The top of the boiler is cased in with brickwork.

The Poldory drawing, Plate LXXXIV, shows the seating for a long boiler with two flues. The course of the flame is: along the bottom, up to the flues, and then in a circuit round the boiler to the chimney at the firing-end. There is a fire-door on each side of the boiler. The seatings are shown also in the drawings of the Ale and Cakes boiler, Plates LXXXV, LXXXVI, and of Tate's boilers, Plate LXXXVII, and again in Plate XL, the engine for Coates & Co. In all these cases the fire-doors are at the sides; in the last two it will be noticed that they are on one side only.

BOILER FITTINGS

THE feed to the boiler was effected by gravity through a vertical pipe supplied from the top of the hot-water pump, and according to the book of 'Directions &c.' of 1779 the pipe was to extend 7 ft. above the surface of the water in the boiler, and regulation was effected by a cock in the pipe from the pump. Then we are told that the boiler should be fed in a regular manner, for 'When there is too much water in the boiler, the engine will not work regular, and if there is too little, the sides of the boiler will be burnt by the flame in the flues'. In course of time an automatic regulation was devised. This is shown in Plate LXXXVII (Tate's boilers, 1794), Plate XL (Coates's engine, 1803), and, in detail, in Plate XC, a sketch of 'Feed apparatus and Damper float' for the Radford Cotton Company, 1803. A float, lying on the water in the boiler, is suspended from a counterbalanced lever by a wire passing through the boiler shell; the lever is pivoted in a bracket on a cistern formed at the top of the vertical feed-pipe, and is coupled to a valve fitted to an opening in the bottom of the cistern. It will be understood that the pipe is carried up to a height corresponding to the pressure at which the boiler is to be worked, and that when, by the lowering of the float, the valve is raised the water flows in from the cistern by gravity.

These drawings of 1803 show also an automatic regulation for the damper in the boiler flue—a feature not found in the 1794 drawing of Tate's boilers. A float, standing in the column of water in the feed-pipe, is coupled by a light chain to a pulley on one end of an overhead shaft; the damper is suspended by a chain from a pulley at the other end of the shaft. If the pressure of steam in the boiler falls the level of the water in the feed-pipe falls correspondingly, the float descends, and, turning the shaft, raises the damper; the resulting increased draught intensifies the combustion on the fire-grate, and so restores the pressure in the boiler.

Watt in his experiments in 1775 on the first engine at Soho made use of a glass tube for indicating the level of the water in the boiler, but in practice the old plan of two gauge cocks was the one adopted. In the book of ' Directions &c.' of 1779 we read that : ' The gauge pipes may be fixed into the boiler top in some convenient place ; the lower end of the longest should reach within 6 inches of the top of the flues, and the shortest should be 4 inches above it.'

The pressure gauge was of the mercury barometer type ; a tube of iron of U-shape with a float emerging from its open end opposite a fixed scale. The boilers were usually worked at a pressure sufficient to support a column of mercury from one to two inches high ; but the pressure might be increased somewhat when the engine was overloaded. It was considered inadvisable, however, to work with strong steam, if it could be avoided, ' as it increases the leakages of the boiler and joints of the steam case, and answers no good end.' On the other hand, unless the steam was kept at above the pressure of the atmosphere there was a danger of air leaking in at the joints of the boiler and spoiling the vacuum in the cylinder of the engine.

As to the construction of the safety-valves employed we have little information. The usual form seems to have been a dead-weight valve in a closed casing, through the top of which emerged a handle fixed to the valve.

Whether it was the regular practice, in the early days, to fit a stop valve in the steam-pipe between the boiler and the engine it is impossible to say ; the drawings of some engines show such valves, those of others do not. Where a valve is shown it is at the boiler end of the pipe, indeed, within the boiler, in the form of a disk adapted to close the end of the pipe and provided with a stem passing through a cross-bar and out through the wall of the pipe, at a point where it has a sharp bend ; the outer end of the stem is screwed to take a nut and the valve is opened and closed by means of a spanner.

FIRING

In the account of the trials of the 18-inch engine at Soho at the end of 1774 we find numerous references to alterations made in the fire-grate and flues. At the first the grate measured 24 in. by 20 ; it was enlarged to 24 in. by 30, and two courses of bricks were built in the throat of the chimney. It then burned 112 pounds of Wednesbury coal in 1 hour and 40 minutes. On taking out the two courses of bricks the same weight of coal lasted 2 hours and 25 minutes. Next the throat was left quite clear and the effect of thick and thin firing was tried.

When Watt was in Cornwall in 1777, about the starting of the Wheal Busy engine, one of the first things he did was to reduce the area of the fire-grate by one-third, from $4\frac{1}{2}$ ft. by 3 ft. to 3 ft. square. He found in Cornwall a very different method of firing to any that he had seen before. The practice there was to maintain a very large body of fuel on the grate and to stoke at long intervals. He thought this method excellent for caking coals, and that it was far better than the London plan of thin fires. A single fire, he said, will burn quite regular for six hours.[1]

A little later we learn of the boiler for the Tingtang engine that ' the bed of fire is 22 inches deep 30 inches broad and 7 feet long and ye upper side of ye bed is 22 inches from ye boiler lagging '.[2]

---

[1] Boulton Papers. Watt to Boulton, 1777 Sept. 13.   [2] Doldowlod Papers. Watt's Journal.

The London firing does not seem to have pleased Boulton. In 1779 he found the fire under the boiler at Chelsea Waterworks 'bad managed, it being spread thinly over the bars and ye whole fireplace, and hath many air holes in the fire'. In the following year he says of the boiler at York Buildings Waterworks:

'I never saw such a fire, not one ray of light from ye grate ; he takes out 2 bars of a side every morning, and when he has raked out ye Cynders over them he puts 'em in again and never rakes ye fire all day, so that no heat is perceptable under ye grate. It is ye Cornish fire reversed.'

But whatever the other merits of the Cornish method, it seems to have shown itself less economical than the system of thin firing, and in 1779 we find Wilson promising to 'give the London method of firing a fair tryal'.[1]

According to the book of 'Directions for erecting and working &c.' [1779] the coal should be thoroughly watered just before it is thrown on the fire in order to prevent it from being swept into the flues by the draught of the chimney, and

'the fire should be kept of an equal thickness and free from open places or holes, which are extremely prejudicial and should be filled up as soon as they appear ; if the fire grows foul and wants air by clinkers collecting on the bars, they must be got out with a poker, but the fire should be as little disturbed in that operation as possible, and the greatest care taken not to make any coals or coaks fall through, which are not thoroughly consumed ; it is very common for a fourth of the whole coals to be wasted in this manner by mere carelessness. When the fire is newly made, the damper should be raised a little, so as to let off the smoke freely, but should be let down to its proper place so soon as the smoke is gone off. The air door in the chimney should be always open more or less ; it prevents the flame from being sucked up the chimney, and very considerably increases the effect of the coals. Once a month the boiler and flues ought to be cleaned, or oftener if the water be very subject to incrust the boiler. Every morning the ashes ought to be taken out.'

In 1785 Watt obtained a patent for a furnace 'whereby greater effects are produced from the fuel, and the smoke is in a great measure prevented or consumed'.

The invention consisted in 'causing the smoke or flame of the fresh fuel, in its way to the flues or chimney, to pass together with a current of fresh air through, over, or among fuel which has already ceased to smoke, or which is converted into coke, charcoal, or cinders, and which is intensely hot, by which means the smoke and grosser parts of the flame, by coming into close contact with, or by being brought near unto the said intensely hot fuel, and by being mixed with the current of fresh or unburnt air, are consumed or converted into heat, or into pure flame, free from smoke'.

Two arrangements are described in the specification ; in the first, which is obviously regarded as the better plan, the hearth consists of an arch of brickwork which supports a column of fuel in a shaft formed in the brickwork of the boiler setting ; the fuel is fed at the top of the shaft, and the main air-supply is supposed to pass down through the interstices of the fuel column, while the auxiliary air-supply is admitted by horizontal passages just over the arch and at the foot of the fuel column. In the other arrangement the smoke of the main fire, on one side of the boiler, is burnt by means of an auxiliary fire on a small grate fired from the opposite side of the boiler.

In announcing his invention to his friend De Luc, Watt says : 'I have some

<hr />

[1] B. & W. Colln. Wilson to Boulton & Watt, 1779 July 22.

hopes of being able to get quit of the abominable smoke which attends fire-engines. Some experiments which I have made promise success. It is not on Mr. Argand's principle, but on an old one of my own, which is exceedingly different.'[1]

As he says in this letter the plan was one he had thought of some years previously. Boulton in a letter written in 1781 refers to a boiler in which the fuel was put in at the top. A visitor had informed him that a person in London was going to take out a patent for a boiler, ' in which the fire was put in at the top. I then shew'd him one, viz. : ye little copper boiler wch you ordered for ye 4 in. cylinder.'[2]

A month after his letter to De Luc, Watt writes :

' We had a first trial yesterday of a large furnace to burn without smoke under the big boiler, at Soho, that used to poison Mr. B's garden so much ; and it answered very well, as far as we could judge from a wet furnace, and without the engines being at work.'[3]

The new plan was adopted for the boilers of the Albion Mills and early in 1786 Watt informed his partner that the furnace had been tried ' which answers very well in giving no smoke '.[4]

However, the scheme did not succeed and the plan then adopted was to have a dead-plate and fire-grate sloping downwards from the fire-door at an angle of 25° to 30° to the horizontal as shown in Plate LXXXIII, Goodwin's boiler, 1788. Watt describes the method of working in the following words :

' The fire is lighted as usual, and a small quantity of air is admitted through one or two openings in the fire door, so as to blow directly on the blazing part of the fire. The fire is made at first principally near the dead-plate, and the fresh coals with which it is to be supplied, are laid upon that plate, close to the burning fuel, but not upon it. When it needs mending, the burning coals and those upon the dead-plate are pushed further down without being mixed and more coals are laid upon the dead-plate, but never should be thrown on the top of those already on the fire, as that would instantly send out a volume of smoke. In this situation they are *gradually* dried, and any smoke which issues from them is consumed by the current of air from the fire door in passing over the bright burning fuel. The opening, or openings to admit the air, are regulated, so as just to admit the quantity which consumes the smoke ; more would be prejudicial. I originally constructed these furnaces in a somewhat different manner ; but the above method has been found the most convenient, and, when properly attended to, answers the purpose perfectly with free-burning coal, but is more difficult to manage with coal that cakes.'[5]

It would seem that it was this plan that was tried on a boiler at Manchester in 1790, with results, in the direction of smoke prevention, that gave rise to a good deal of astonishment and discussion.[6] In spite of this the inclined grate did not meet with favour there, and six years later we learn that there was no desire to destroy smoke, and that the flat grate was in general use for Boulton and Watt engines in the Manchester district.

The subject of flash boilers is not in any way connected with the work of Watt, but two references thereto in the Boulton and Watt Collection have been noted, and it seems desirable to record them here.

[1] Muirhead : *Mech. Inv.* Watt to De Luc, 1785 Sept. 10.
[2] B. & W. Colln. Boulton to Watt, 1781 Aug. 9.
[3] Muirhead : *Mech. Inv.* Watt to Mrs. Watt, 1785 Oct. 9.
[4] B. & W. Colln. Watt to Boulton, 1786 Jan. 27.
[5] Robison : *Steam and Steam Engines*, 120 *n*.
[6] Muirhead : *Mech. Inv.* James Watt, junior, to Watt, 1790 July 11.

Dudley, an engineer who erected some of the first Boulton and Watt engines in Cornwall, writing to the firm from Marazion Mar. 26, 1778, says:

' Jabez Hornblower last Saturday attempted to move his new invented engine . . . the method of his working is a dry boiler (similar to that Mr. Payne tried at Wednesbury and in Derbyshire) being kept in and (*sic*) hot state and a small stream of water being conveyed in through the boiler's top, by means of a cog wheel disperses the water in a centrifugal motion round its sides.'

Some years later (Dec. 22, 1783) Boulton, in writing to Watt, mentions Mr. Richmond's project for an instantaneous steam generator : ' a red hot boiler with water dropping into it, he said would send forth steam as strong as lightening.'

XIX

THE APPLICATIONS OF THE ENGINE

BLOWING-, FORGE- AND WINDING-ENGINES

U P to the year 1782, and apart from the attempt to work forge hammers to which reference will be made later, the application of Watt's inventions was restricted to engines for pumping water and for blowing air for furnaces. The pumping-engines were applied for the most part in mines, but a number were built for public water-supply, for raising water at canal locks, &c. The great field for the mine pumping-engine was Cornwall, for although their first engines were set up at collieries in the Midlands, the Boulton and Watt engine was not taken up extensively in coal mines where, the fuel being so cheap as to render it unnecessary to embark upon any expense in order to save it, the old form of engine continued to be used. In this connexion it is curious that although a Boulton and Watt engine was erected at Byker Colliery, near Newcastle-upon-Tyne in 1778, nearly twenty years had to elapse before another engine of that construction was set up in the Northumberland and Durham area. The history of the Byker engine is obscure ; it is possible that it was badly constructed, and this may have given Boulton and Watt a set-back in that district.

With the introduction of means for converting the reciprocating motion of the beam into rotary motion of a shaft, of the rotative engine in short, the field of application was extended very widely. In the first few years after 1782 we find the Boulton and Watt engine, with the sun-and-planet gear, applied to working forge hammers, to winding coal and ores in mines, to supplying power in breweries, starch manufactories, and bleach works, to driving oil mills, corn mills, metal-rolling mills, glass-grinding machinery, fulling mills, woollen mills, and cotton-spinning mills. The first engines for cotton-spinning were set up in Nottingham. Manchester was rather slow to take up the new engine ; in 1791, although it appears that two engines had been erected there recently, it was reported that the Boulton and Watt engine was not known, and that engines of the old construction were being erected ; of these Bateman and Sherratt were the principal makers.

In respect of most of these applications it may be said that the particular use to which the engine was put did not affect the design, and that the concern of Boulton and Watt ended at the engine shaft. In others, however, we find the application more closely bound up with the construction of the engine, or at least we find Watt entering upon the design of the machinery to be driven by the engine. In particular is this the case with engines for working forge hammers, and winding-engines for mines. On these two subjects and on that of blowing-engines it is proposed to offer some remarks, but it must be understood that these remarks are based entirely upon material that has been collected for the main subject of this book, the steam-engine inventions of James Watt, and the development of the engine at Soho up to the time of his retirement in the year 1800. Other

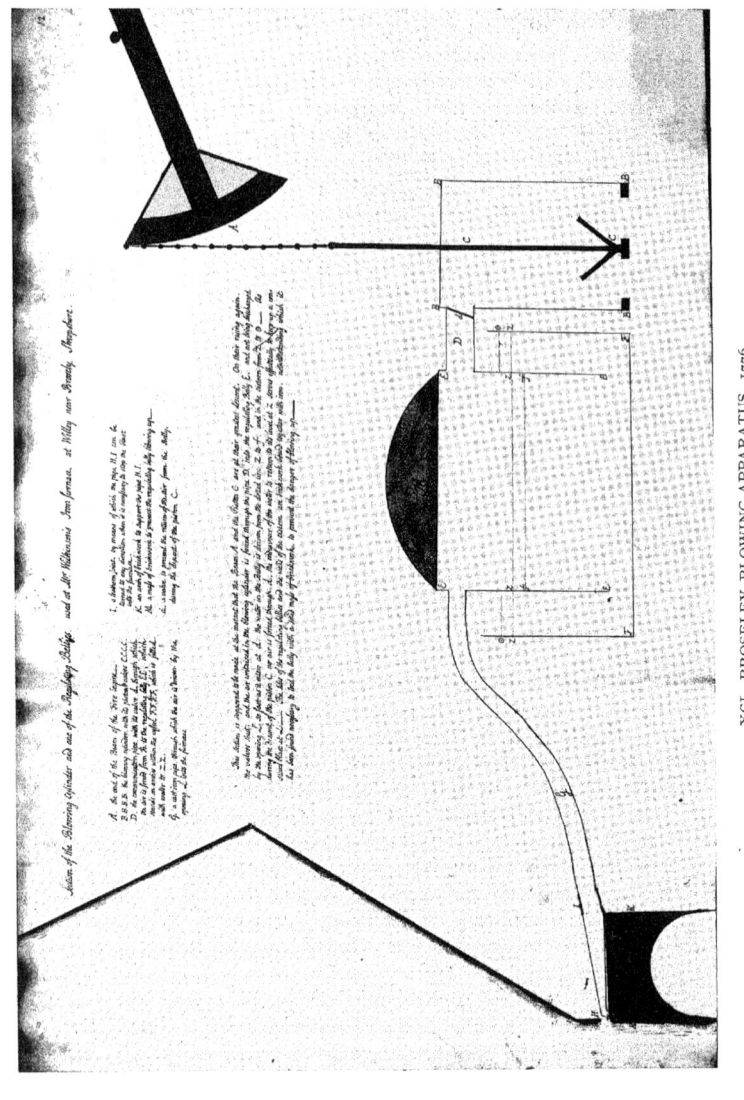

XCI. BROSELEY BLOWING-APPARATUS, 1776

Egerton MSS., British Museum

engineers were at work building steam engines for these and other purposes, so necessarily it follows that the treatment here is not exhaustive ; still it is thought that the material, culled mainly from the Boulton and Watt Collection, will form a useful contribution to the history of these particular applications of the steam engine.

BLOWING-ENGINES

ONE of the first Boulton and Watt engines erected outside Soho was, as we have already stated, a blowing-engine for the great ironmaster John Wilkinson at New Willey Furnace, near Broseley, Shropshire, which was started early in 1776. The information we have about Wilkinson's engine is very meagre—the little that is found in the Boulton and Watt Collection concerns the steam end only—but we know that the engine was a single-acting beam engine with a steam cylinder 38 in. diam., and Watt, in a letter to Smeaton in April 1776, states that the blowing-cylinder was 72 in. diam., the water column $5\frac{1}{2}$ ft. high, and that 14 strokes were made per minute. We are able to supplement these particulars to some extent from the anonymous contemporary account referred to in describing this engine in Chapter IX. It appears that Wilkinson had previously had a blowing-cylinder worked by an atmospheric engine ; this had a 49-inch steam cylinder supplied from two boilers, and was barely able to furnish sufficient blast ; the new 38-inch engine supplied more blast than was required and did it with one boiler only. From a sketch accompanying this account, see Plate XCI, it is seen that the air was forced on the upstroke of the blowing-piston which was formed with flap valves opening upwards, and that the air branch at the top of the cylinder had a flap valve opening outwards. The pressure-regulating apparatus consisted of a cistern containing water in which was placed a smaller cistern, ' the regulating belly' supported on arches and loaded to resist the air pressure ; the delivery pipe from the blowing-cylinder and the pipe to the furnace entered the regulating-belly at the top. During the forcing stroke of the blowing-piston the level of the water was lowered inside and raised outside the regulating-belly ; on the suction stroke the water sought to regain its normal level and the blast was maintained more or less uniform. The sides of both cisterns were constructed of brickwork bound together by iron. The account states that : ' There are two Regulating Bellies ; one 22 feet 9 inches by 10 feet ; the other 33 feet 9 inches by 9 feet ; each of them 3 feet 3 inches deep. The column of water in the cistern surrounding the Regulating-Bellies 5 feet 6 inches deep. The two communication pipes which pass from the blowing cylinder to each of the Bellies 15 inches diam. The discharging pipe at the nose, through which the air is driven into the furnace 3 inches diam. There are leather joints in two parts of the pipes ; and so strong is the blast, that some air will pass through the substance of the leather. The blast is equal to 4 pounds on a square inch.' [1]

It is probable that Wilkinson used the blowing-cylinder and regulating-apparatus that had been laid down for the first engine, and that Watt had nothing to do with the design of these parts. Smeaton had previously used cylinder and piston blowing-machines at Carron, but his apparatus was worked by water wheels, and he used four cylinders to maintain a uniform blast. No earlier instance is known of the direct application of the steam engine to a blowing-cylinder than this of

[1] British Museum, Egerton MS. 1941, ff. 5-20.   *Ironmaster*, 1914, 23, 24.
See also Dickinson, H. W. : *John Wilkinson,*

Wilkinson and no doubt the regulating-bellies, particularly necessary in the case of a single-acting blowing-cylinder, were now used for the first time. Farey describes this water-regulator but gives a much later date for its introduction.[1]

As we have seen in Chapter IX the steam side of the blowing-engine was not at first an entire success, but the plant as a whole was no doubt a considerable advance on former practice, and within the next five or six years Wilkinson put up four more Boulton and Watt engines for blowing, and others had been erected at Coalbrookdale, Ketley, Donnington, Wood and Rotherham,while one had been sent to France. In 1793 a blowing-engine was put up in South Wales, at the Neath Abbey Ironworks ; it was double-acting both for steam and air ; the steam cylinder was 40 in. diam., the blowing-cylinder 70 in., stroke at both ends 8 ft., 10 to 12 strokes per minute and pressure 2¾ pounds per square inch. Of the same year we have the drawings of a blowing-engine made to go out to Cadiz ; the blowing-cylinder has now a number of small round inlet valves in the cover and bottom ; the delivery valves are still hanging flaps. Then, in the years 1798 to 1801 engines were erected at works in the South Wales iron district, Penydarren, Dowlais, Sirhowy, Rhymney, Blaenavon, and Tredegar.

A drawing dated 1790 and marked ' Shelve Field Gravels ' shows an exceptional form of blowing-engine for the Salop Mine Company. The outer end of the beam is coupled by a chain or rod to a bellows ; the beam is slung from above by a pair of links, and the upper end of the piston-rod, coupled directly to the beam, has a roller running upon a vertical guide ; at the lower end of the cylinder are two cocks, side by side, one ' to let the steam in ', and the other ' to let the steam out '. The cylinder is 14 in. diam.

Although outside the period with which we are dealing in this book we cannot refrain from referring to two other blowing-engines of exceptional form intended for supplying air to foundry cupolas. Just before the year 1800, with the object of producing a cheaper and more compact apparatus than the beam engine, the bell-crank engine was designed. These blowing-engines were modifications of the bell-crank engine ; the first, made for Fulton & Sons, Glasgow, is shown in Plate XXXIX. It will be seen that the blowing-cylinder is vertical and parallel to the steam cylinder, and that its piston is worked by a crosshead and side-rods from an extension of the bell-crank lever, the upwardly extending arm of which is coupled to the crank. This engine was started in August 1802 and soon after an order for another, slightly larger, was received from the Perth Foundry Co.

FORGE-ENGINES

Very soon after his blowing-engine had been set to work at New Willey, John Wilkinson decided to set up a forge, and in May 1777 we find him approaching Boulton and Watt for an engine to work the hammer, or ' to raise a stamp of 15 cwt. 30 or 40 times a minute '. The original 18-inch single-acting pumping-engine was now lying idle at Soho and Watt at once set a man at work to apply it to working a hammer, a ' stamp hammer of 60 pounds weight ', he termed it. Within a week the apparatus was at work going at the rate of about 60 strokes a minute ; it excited great interest in Birmingham, so much so that large orders were anticipated. However, the scheme fell through ; the accounts make it clear that the action was

---

[1] Farey : *Steam Engine*, 286.

rather violent and possibly in the end the apparatus knocked itself to pieces. It was spoken of as ' the devil', and Watt in reporting its working to Boulton remarks that ' It has demolished all the fixtures many times already, and I suppose must be made wholly of cast iron'. Still at the time the project must have seemed promising, for Watt informed Wilkinson that a larger one for him ' shall be set about next week'. However, the earliest drawing that has been found is dated nearly a year later, Apl. 26, 1778 ; this is headed ' Double Battering engine for Mr. Wilkinson' and shows two separate beam-engines with 15-inch cylinders ; the connexion between the outer ends of the beams and the hammer is not indicated ; what seems to be intended for a sun-and-planet gear is pencilled in, but this is clearly of later date, for the first mention we have of this gear occurs nearly four years later. The specification of Watt's patent of 1784 shows a direct connexion by means of a link between the engine beam and a helve, and it is possible that this was the plan adopted. As to whether Wilkinson had the apparatus made and set to work we have no evidence, but the fact that we find Watt writing to him in August 1779 : ' I expect we shall want one of the devils,' suggests at least that one was then being made. However, this is the last we learn about these particular ' devils', so that if made the apparatus cannot have been a success.

Wilkinson, however, returned to the charge a couple of years later. Watt was then at work on the problem of producing rotary motion in the reciprocating engine (his patent for contrivances for this purpose is dated Oct. 25, 1781), and in July 1781 Boulton writes to inform him that Wilkinson proposes to erect at Willey a two-cylinder engine with cylinders 27 in. diam. for 7 ft. stroke to work one hammer of 16 cwt. and another of 2 cwt. at the rate of 120 strokes a minute, that he intends to make the hammer framing all of cast iron and the anvil block 10 tons in weight, and that he prefers the tilting to the lifting hammer. It may be as well to explain that the forge hammers with which we are here concerned had the hammer head at the end of an horizontal arm or helve mounted on pivots. In the lifting hammer the helve was pivoted at the end remote from the hammer head, and the cams acted on its underside to lift or throw it up against a spring beam. In the tilting hammer the cams acted to depress the end of the helve which was pivoted between the free end and the hammer head. The two-cylinder engine referred to was one of Watt's rotative engine schemes ; it comprised two separate engines working cranks set at an angle on the same shaft. Boulton was anxious to meet the views of Wilkinson and to use this form of engine ; he offered Watt, who was then busy in Cornwall, to carry the scheme through without troubling him. Watt, however, objected to the cranks being applied to the engine, he feared litigation with Pickard the holder of the crank patent, and no doubt preferred to work out a scheme himself ; but when in February 1782 he had completed a set of drawings for Wilkinson's engine and hammer he was not at all satisfied with what he had put on paper. The engine had the eccentric motion on the middle of the shaft for the production of rotary motion, and this had never been tried in practice ; and as to the hammer he confessed that he had never examined a forge attentively ; so in sending the drawings to Wilkinson's agent he writes :

' As the whole of this is a new thing and I have not been able to satisfy myself in what regards the hammer & its framing, I would advise that a modell 3 times the size of the drawing be made immediately and every thing made of the same material it is to be in great, the hammer ought then to be about 16 pounds weight.'

Whether this project was carried into practice we cannot say, but there is evidence that Wilkinson was experimenting with forge hammers at Willey. Towards the end of November one of the Soho men had been over at Broseley and returned with the story that, ' Mr. Wilkinson's new tilting forge does not answer yet, and they cannot work a hammer above 6 cwt. in the lifting one.'

In the meantime another set of drawings had been finished and this no doubt was more or less closely followed in the apparatus set to work in April 1783 at Wilkinson's works at Bradley, Staffordshire. In this case the engine had the sun-and-planet gear. However, before this engine was started an experimental hammer had been tried at Soho. In September 1781 Watt had made drawings for a new type of engine with two cylinders to which it was intended that a hammer should be applied ; work was begun on the engine very soon but nothing was done in regard to the hammer until the beginning of November 1782, some months after the second set of drawings had been sent to Wilkinson, and probably just after the scheme had been approved and was about to be put in hand.

Watt, following the usual policy of trying the new schemes at Soho before allowing customers to proceed with them, then had the hammer castings rushed through and the hammer was at work before the end of the month, November 1782.

No drawings for this hammer have been found but we need not hesitate to say that it was made on the same lines as shown on the drawings for Wilkinson to be referred to below ; however, we know that the hammer weighed 120 pounds and made six blows for each stroke of the engine which worked at the rate of 25 strokes a minute. After one of the first trials Watt informed Boulton that ' it strikes a good blow, and forges iron very well ', but he intended to replace the wood cams then in use by steel ones, to increase the weight of the hammer to about $1\frac{1}{2}$ cwt. and to run it at the rate of 250 or 300 blows per minute with a lift of 9 or 10 inches. The engine had the sun-and-planet gear and but one of the two cylinders was used. Wilkinson came to inspect the apparatus, the working of which answered the expectations formed.

We may turn to a consideration of the drawings of ' Mr. Wilkinson's Bradley Old Furnace Forge Engine', Plates XXXIV and LXXVI. The engine was single-acting with a cylinder 42 in. diam., 6 ft. stroke, and had the sun-and-planet gear. It will be seen that the gear and the fly wheel are mounted on the end of the shaft beyond the inner shaft bearing ; the body of the shaft is a cast-iron cylinder 24 in. diam. and 2 in. thick, and the outer bearing is formed by the gudgeon H. The three cams are driven into the cam-ring which fits over the end of the shaft and abuts on a flange E formed with joggles N that enter recesses in the end of the ring. The cam-ring is held on the shaft by the disk G on the gudgeon, fitting into the shaft and formed with joggles to engage joggles in the shaft and recesses in the cam-ring. The gudgeon itself is held on by bolts J having their inner ends bent at right angles to pass through the wall of the shaft. From the general drawing it will be seen that the stem of the helve lies parallel to the shaft and that the cams engage its under surface between the hammer head and the fulcrum. The framing shown consists of two cast-iron standards supporting the pivots of the helve, the spring beam, and, at the top, a large cross-beam.

As we have said the hammer was set to work in April 1783, and it seems to have answered very well. Towards the end of the month Watt in a letter to Smeaton informed him that the hammer weighed $7\frac{1}{2}$ cwt., had a lift of 2 feet, made six blows for each stroke of the engine, and had struck 300 blows per minute ;

however, as only 90 per minute were required, they were going to alter the ratio from 6 to 4½ and to increase the weight of the hammer to 10 cwt.

That Wilkinson's forge was a success is proved by the fact that other iron-masters speedily followed his example and applied the steam engine to work their hammers. In August 1784 Watt writes to Boulton that 'Hallen is getting a forge-engine erected by Engᵣ Burnard from Bradley, & Parkur at Tipton another by somebody else, on what plans I know not. John Onions's goes pretty well, but is perpetually breaking in its most essential parts.' These engines would be atmospheric and the last named had a crank. At the same time a Boulton and Watt engine was in course of erection by Richard Reynolds & Co. at Horsehay; it had a 26-inch cylinder, a parallel motion and the sun-and-planet gear, and was started at the end of November. In the following year forge-engines were being put up in the London area, one at the Union Ironworks, and another at the King and Queen Ironworks, both in Rotherhithe.

WINDING-ENGINES

THE Newcomen engine in conjunction with a water-wheel had been applied by Smeaton for winding in coal mines, notably in the Tyne district, and it had been tried working directly by Oxley at Seaton Delaval, Northumberland, before 1770 ; but prior to the introduction of the crank in 1780 it is safe to say that there is no instance of the successful application of the steam engine directly for winding. How soon after that event engines with cranks were so applied cannot be stated with any degree of precision, but we have seen, in Chapter XIV, that the raising of coal and ore in mines was among the first of the applications of the Boulton and Watt rotative engine.

As to the 15-inch cylinder engine put up by John Wilkinson, on his colliery at Bradley, Staffordshire, in 1782 or 1783, it is not clear whether, as built, it had the swashplate arrangement, or the sun-and-planet gear, and in regard to the winding apparatus, apart from the fact that the gin-wheel or drum was mounted on a vertical shaft and was 15 ft. diam., we have no information ; no doubt it was designed, as it was constructed, by his own people on the lines of the horse gins in common use. It appears that Wilkinson, early in 1782, had refused his agents permission to buy any more horses for the mine, and they worked out the applica-tion of a steam engine as best they could. About the same time Sir James Lowther visited Soho and expressed an inclination to instal steam winding-engines in his mines in Cumberland. On this occasion we find a letter from Boulton to his partner in which he gives his ideas on the subject.[1]

In July 1784 a whim-engine was set to work in Cornwall. It was an 18-inch single-acting engine with sun-and-planet gear, and appears to have been designed for Dolcoath mine, although it was actually erected on Wheal Maid, one of the Consolidated Mines group. This engine worked at Wheal Maid until September 1793, it was then moved to Baglan Colliery in Glamorgan,[2] and it has been stated that it was brought back to Cornwall and erected first at Herland mine, and later at Dolcoath,[3] the mine for which it was originally intended.

Boulton was in Cornwall at the time this engine was being erected and sent

[1] B. & W. Colln. Boulton to Watt, 1782 Jan. 19.
[2] Ibid. 'Mr. Wilson's Account of Cornish Engines. Aug. 1794.'
[3] Trevithick, F. : *Life of Richard Trevithick*, I. 91.

Watt accounts of the progress made. At first things seem to have gone badly and Murdock, who had charge of the erection, ' was quite out of heart of it & began to think the money thrown away.' However, Boulton went into the matter and under his direction successful working was soon attained. The construction of the winding apparatus as first made is not explained, but the drum seems to have been reversed by a gear train, for Boulton says ' You may depend upon it that no contrivance of putting out & putting in gear will ever answer so well as changing the motion of the engine ' ; he altered the working accordingly and then says ' There is no difficulty in stopping (by the assistance of the brake) exactly at the point you w^d wish to stop and then make one revolution back to lower the kibble and then leave it to itself, it sets to work again '.[1] A fortnight later he is able to say : ' In short nothing can be better in every respect than the whole of that machine. They drew 200 of kibbles 103 fathoms high in 11 hours & burnt 8 bushels of coal.' The brake was foot-actuated and straight on the face ; Boulton suggests that it be curved to sit on the wheel.[2]

The plan of reversing the engine was employed in all succeeding winding-engines by Boulton and Watt. No reversing-gear was used at this period, as Watt remarked in 1795, they ' changed the motion by stopping the engine & setting the fly agoing the contrary way '.

Another feature with which Boulton finds fault is the use of chains to connect the piston-rod to the beam (it will be remembered that the engine was single-acting) ; he thinks jointed rods should be used, ' for now if you attempt to stop ye engine suddenly the chain doubles up & then receives a great shock.'

A few months later the apparatus was out of action in consequence of the breaking of a balance rope. Watt had designed a double spiral barrel for the balance rope, but, possibly on account of cost, the mine people had adopted the plan of hanging to each winding-rope 100 fathoms of old rope which was led into a trunk in the shaft. Boulton now arranged ' to make a double spiral wim shaft & take a ballance wt. into a vacant shaft exactly as you directed in the beginning, all which must be proportioned to a rope 6 lb per fathom, to a kibble containing 3 cwt. of ore & to a depth of 110 fathoms ; the cage [drum] is 15 ft. diam^r. The drawings you made were adapted for Dolcoath 160 fath^m.' [3]

Within a month the new plan was in successful operation. ' You may now work with a ¼ inch of steam & may draw 100 kibbles without using the brake. They drew 400 kibbles yesterday, the same the day before.' [4]

Boulton fitted to this engine a winding indicator consisting of a vertical board in the engine-house with ' a pulley at top of such a diameter & moving with such a velocity as to give an inch of cord for every fathom the kibble ascends in the shaft . . . & thus by hanging on two little weights to represent kibbles the engine man sees exactly whereabouts the kibbles in the pit are & least he should not be attentive a bell shall ring within & another without when ye kibble comes within — fathoms of the top.' [5]

He had seen something of the sort at Newcastle, but his apparatus when finished was, he says ' a far better, cheaper, & more conven^t thing than any of those at Newcastle '.[6]

[1] B. & W. Colln. Boulton to Watt, 1784 July 8.
[2] Ibid. Boulton to Watt, 1784 July 22.
[3] Ibid. Boulton to Watt, 1784 Sept. 25.
[4] Ibid. Boulton to Watt, 1784 Oct. 21.
[5] Ibid. Boulton to Watt, 1784 July 8.
[6] Ibid. Boulton to Watt, 1784 July 22.

The double spiral balance drum which, as Boulton says, Watt directed at the beginning, became the standard practice for some years with Boulton and Watt. It is shown in Plate XCII, a drawing of the Wheal Maid engine and winding-gear, dated Nov. 16, 1785, and possibly made for some reconstruction of the plant. The engine shaft drives by bevel-gear a vertical shaft which in turn is geared to a spur-ring on the winding-drum, marked as of 15 ft. diam. The drum is on a vertical shaft and above it is the balance drum. This was the plan adopted by Messrs. Wilks at Measham in 1787. Southern, in submitting the drawing of it to them, writes :

' When the skips in the pit meet, the rope of the spiral (& its weight) is fixed to the shaft & as one rope descends & consequently overbalances the other, the rope of the equalizer begins to wind upon one of the spirals according as it turns one way or the other. The weight fastned to the end of this rope (which is called the equalizing weight) getting gradually farther and farther from the centre of the axle as the ropes in the pit get more unequal in length, it becomes proportionally more powerfull, and when the skip is quite wound up the equaliz^r weight is also wound to the end of the spiral, or upon the greatest circumference of it. . . . It is to be observed that the rope & weight of the spiral must either have a pit for themselves of 15 yards deep or else be conducted into a corner of the pit that belongs to the skips.' [1]

In the following year Southern, in explaining the arrangement to a customer in South Wales, advises him to have a man from Cornwall who had had experience with it, or else to send the man who was to make it over to Cornwall to see one at work.[2]

Between the years 1788 and 1792 three winding-engines were put up at the coal pits of John Christian (or John Christian Curwen) at Workington. The first engine had a 14-inch cylinder, 4 ft. stroke, the others had 21-inch and 22-inch cylinders and strokes of 5 ft. ; they were all double-acting engines with parallel motion and sun-and-planet gear.

The drawings of the winding-gear for the first of these engines are reproduced in Plates XCIII, XCIV. The engine shaft has at its outer end a cast-iron bevel-pinion in gear with a ring of teeth on the drum, which is mounted on a vertical shaft that is shown supported on a ' Platform of Stonework to carry the barrel &c. steadily '. The winding-drum is a plain cylinder 15 ft. diam. with the double spiral balancing-drum below it. In the detail drawing the winding-drum is shown below the spiral, but this is corrected by notes opposite each end of the shaft, at the bottom : ' This to be the upper end,' and at the top : ' This to be the lower end,' and although the ring of teeth is shown on the winding-drum there is a note that ' The teeth are not to be in this rim, but in that marked A, B above (see the draw^g of the engine) '. The teeth are mortice teeth 6 in. wide and 1½ in. thick and 1½ in. high at the outer end. The shaft is of wood with at each end a tapered part on which are driven two rings of wrought iron ; it terminates in iron gudgeons. The winding-drum consists of an iron rim supported on a pair of wood rings, and each ring is connected to the shaft by two pairs of arms, one pair crossing the other at right angles and each pair clamped to the shaft by U-bolts. In the drawing one pair of arms for each ring is seen in section with the U-bolts embracing them. The two sides of the balance drum are built up of wood and full dimensions are given for the formation of the spiral grooves. The general drawing shows the pithead framing, but the end view is incorrectly drawn; and there is a note that the two pulleys

---

[1] B. & W. Colln. : Letter Books. Southern to Wilks, 1787 Mar. 19.

[2] Ibid. : Letter Books. Southern to Morris, 1788 Jan. 9.

instead of being shown at the same height, should be one above and the other below the position indicated.

The double spiral balancing-drum was used also with an engine for the Duke of Devonshire's Ecton Mine in Staffordshire (in 1788), but in this case Boulton and Watt advised the application of taper ropes. The Duke's agent, Cornelius Flint, experienced some difficulty in procuring the rope, but finally its manufacture was undertaken by John Hall of Leek. After about six years' work Flint was able to report :

'With respect to our taper roapes, they answer (past my first expectations) indeed, as far as I am able to judge, the common shape of a roap at Ecton, considering the depth we are at &c, would not have answered. The length of the roapes in the pit is 202 fathoms, which is the depth of the pit. That length is tapered from 5 inch in circumference at the small end to 7 inches round at top, but no increase of size for the extra length required to be put upon the wheel out of the top of the pit. We have the ropes made by Mr. Samuel Goodwin of Leek, his present price that he charges is 5½ per pound.'[1]

It is hardly necessary to say that the ropes used were of hemp, but it is a matter worthy of note in this connexion that John Wilkinson had made ropes of wire, whether or not his idea was to use them for raising coal from his pits we cannot say ; he asked Boulton and Watt to try them and Watt reported : ' We have tried your wire ropes and consider them as being too stiff to answer.'[2] For what purpose they had been tried does not appear, but this was some years before the firm had any concern with winding-engines.

The correspondence with Flint discloses one of the methods of balancing previously in use :

' We have a method in this neighbourhood of ballancing the weight of Roaps &c. by a simple wheel with weights at different distances placed to the roap, the same to fall on a stage in the pit so that the weight on the ballance wheel is more or less according to the greater or less extension of the roap in the pit.'[3]

The winding indicator used at Ecton mine was worked by screw gear. There is a sketch and description of it in the Boulton and Watt Collection.

The portable coal-winding engine for Reynolds & Co. (1788) has been described and illustrated in Chapter XIV (see p. 168). It had a plain cylindrical drum, 5 ft. diam., on an horizontal axis and was not provided with a balance arrangement.

From a letter written by Southern in 1793 to Lawson, who was then in charge of the erection of engines in the Leeds district, it appears that Boulton and Watt had by that time changed their practice in the matter of balancing. He says :

We find it best & most approved to have a small barrel above or below the large one to wind up a cast iron chain as the basket descends below the middle. When the baskets meet the chain is at the bottom of its pit, and a small rope from it to the little barrel, as the basket descends the chain is wound up, & its weight increases. The chain accumulates at the bottom of the pit, and is so long as not to rise from it much when the basket is at the top. . . . The method is found to be cheap & to answer perfectly & the balance pit is not ¼th the depth of the basket pit.'[4]

By this time the Coalbrookdale firm were making winding-engines, using the Newcomen engine with a crank, and a drum on an horizontal shaft geared to the

[1] B. & W. Colln. Cornelius Flint to Boulton & Watt, 1794 Nov. 1.
[2] Ibid. : Letter Books. Watt to J. Wilkinson, 1780 Apl. 25.
[3] Ibid. Cornelius Flint to Boulton & Watt, 1788 Mar. 26.
[4] B. & W. Colln. J. Southern to J. Lawson, 1793 Dec. 11.

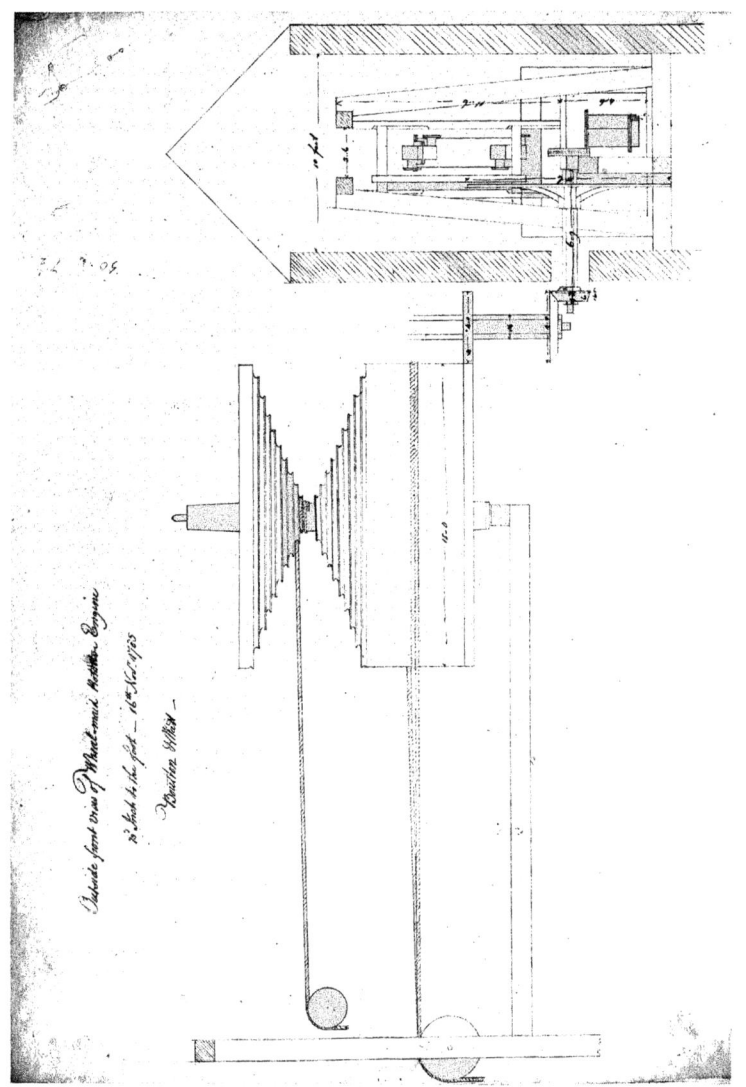

Lithographic front view of Wheal Maid Matthew Engine

to Suit to the feet — 16th Nov 1785

Boulton Watt —

NCII. WHEAL MAID ENGINE AND WINDING-GEAR, 1785

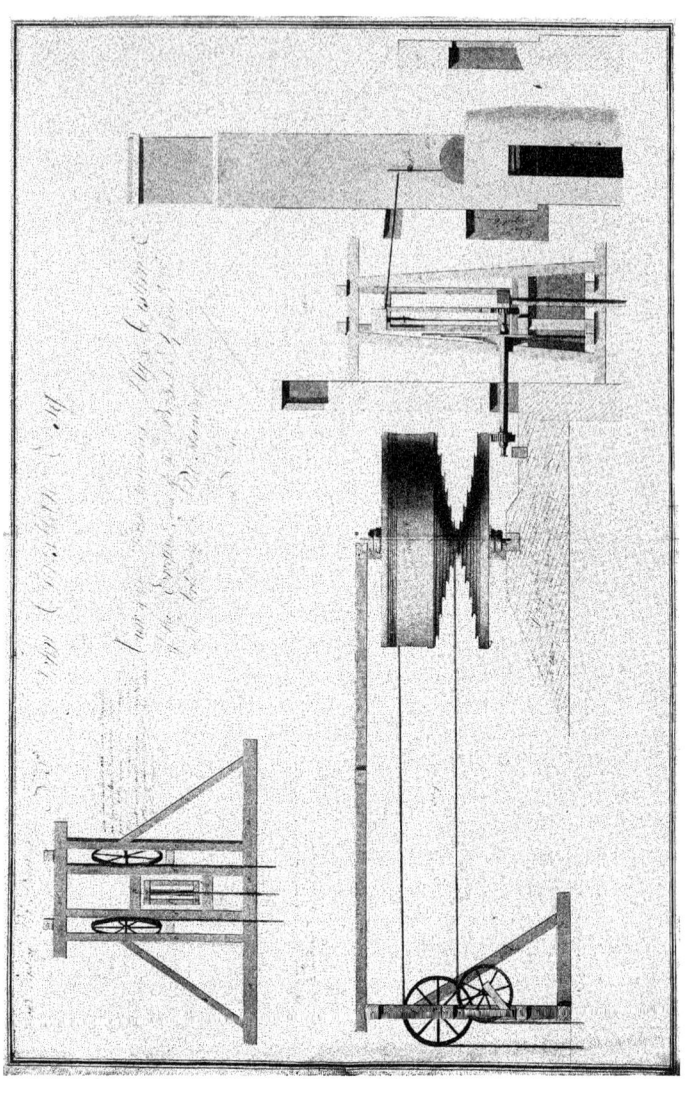

XCIII. WINDING-ENGINE FOR JOHN CHRISTIAN, 1788

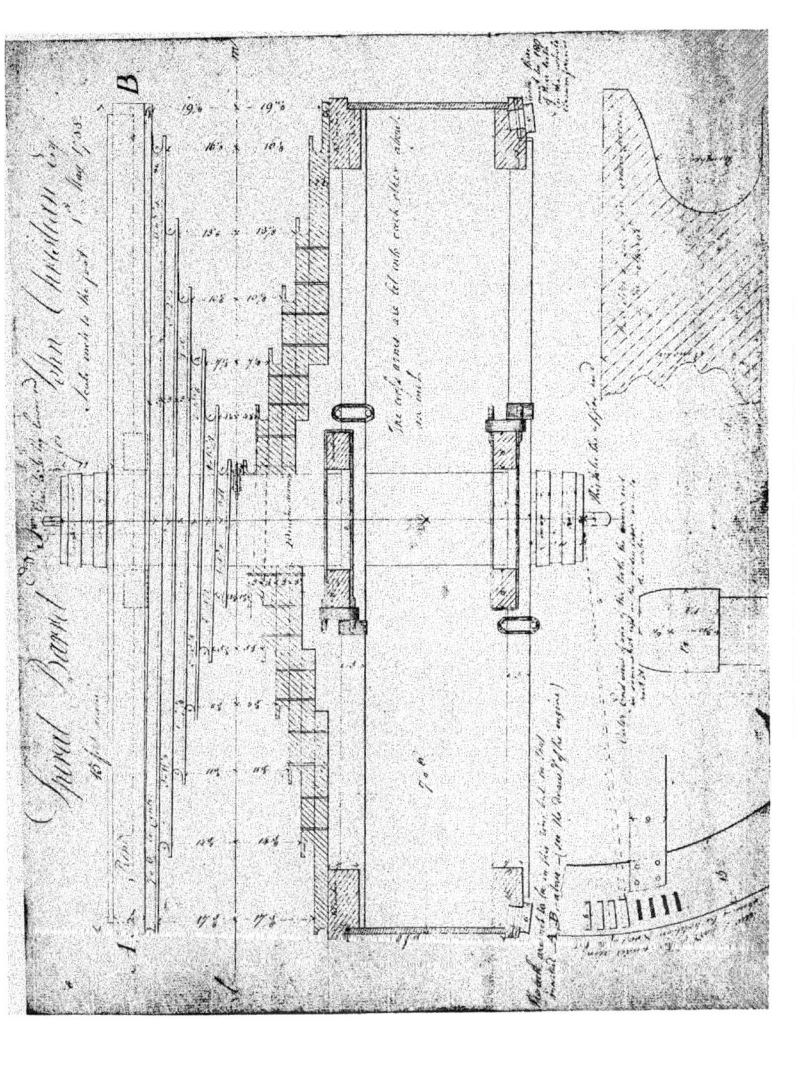

XCIV. WINDING-DRUM FOR JOHN CHRISTIAN, 1788

engine shaft. The drum was in the form of two cones base to base, a form that had been ,used previously by Smeaton. In reference to this apparatus, Southern, in the letter quoted above, remarks : ' We do not much approve of the method of draw^c coals at Coalbrookdale by the engine, which is a double cone barrel. The cone is supposed to balance, but in great depths it is very inadequate.'

Another firm, Blair, Jeffreys, and Cameron of London, were now supplying winding-engines in Northumberland and Durham. Cameron had been in the service of Boulton and Watt for some years, and after leaving them had embarked in business as a steam-engine maker with, first, Blair and later Jeffreys also as his partners. He had obtained a patent for a particular form of engine, and in 1789 he patented another invention for raising coal, ores, and water, the essential feature of which was the use of a steam engine with a crank to work a winding-drum. In conversation with Boulton in 1789 he said that he had erected five coal-winding engines, and that one of his Newcastle engines raised 300 tons of coal per day from a depth of 95 fathoms. In the course of a few years Jeffreys became the sole member of the firm, and then in 1795 we find him suing for infringement of Cameron's patent a man who was using a Newcomen engine with a crank for raising coal near Newcastle.

Jeffreys was not able to make good his case and seems to have given up engine-building ; we need not pursue this matter further, except to mention that Cameron devised a balancing-arrangement consisting of a weight at the end of an arm on a rockshaft which carried a toothed segment in gear with a pinion on the end of the rope drum. This information is taken from a letter by Thomas Barnes,[1] the manager of the Walker Colliery near Newcastle-upon-Tyne, who was corresponding with Boulton and Watt about winding-engines.

In another letter Barnes refers to the double cone drum as ' the method introduced here by Mr. Smeaton to balance the rope' and sketches one 13 ft. 6 in. long, 6 ft. 6 in. diam. at the middle, and 2 ft. 8 in. at the ends. He says it is ' by no means uncommon here to move the corf at the rate of 10 feet per second . . . the rope employed will be 5¼ in. circumference (shroud laid) and will weigh 7¾ lbs. to the fathom', and adds that they had never tried taper ropes.[2]

From his first letter in April 1794 it appears that Barnes was in want of two engines, one for Walker Colliery, 107 fathoms (Ord, Pearth & Co.), and the other for a new colliery, Benwell, 98 fathoms (Surtees & Co.). The corf at the first colliery weighed 6 cwt. ; at the second, 5 ; in each case the engine was to raise 50 corves per hour.[3] He had heard that there were Boulton and Watt winding-engines at work in Cumberland and he went over to see them, but the winding there was at

---

[1] Ibid. Thomas Barnes to Boulton & Watt, 1794 Dec. 16.

Barnes was a man of some note. His father died when he was sixteen years old and he then took on the entire management of Walker Colliery which he had retained ever since ; in addition he had the management of Benwell and seems to have acted in a consultative capacity for other collieries. He was at this time thirty years of age. Southern writes of him as ' a plain man, a man of very clear strong judgement, he is intrepid, cautious and prompt. He understands fire engines very well, and recommends B. & W.'s upon principle, his con-

viction of their superiority being grounded both in theory and experience. He has, I daresay, full influence upon his employers, and his opinion I understand is highly respected by others. His time is extremely engaged, though I do not understand that he has the main management of any Collieries except Walker and Benwell.' (B. & W. Colln. Southern to J. Watt, junior, 1795 Sept. 8.)

[2] Ibid. Thomas Barnes to Boulton & Watt, 1794 Oct. 13.

[3] Ibid. Thomas Barnes to Boulton & Watt, 1794 Apl. 24.

a slower rate than the Newcastle people wanted, at least he says that the corf moves only 18 inches for each foot of piston stroke, ' while with us the corf must go 18 inches at least for every 6 inches of the piston stroke '.[1]

Towards the end of 1795 Barnes refers to the plan of raising two corves at the same time. He is ordering an engine for Cowpen Colliery to raise 11 cwt. of coal at once, 100 fathoms, 35 times per hour, and says :

' The method they propose is something similar to that at Hebburn where 2 20-peck corves are brought up by each rope of the engine. The one corf hung a little below the other, and both being raised to their highest level before the engine stops, the lower corf is taken off the hook by one banksman, and the upper one (being lowered to the proper level for striking) by another.'[2]

In connexion with the first engines ordered for Walker and Benwell, Boulton and Watt seem to have sent Southern up to Newcastle to settle the final disposition of the winding-gear on the spot in conjunction with Barnes, and we have an interesting letter from him in reference to the Benwell engine. He found the business had been very tedious indeed owing to the slackness of the workmen, both in understanding and execution, and to the fact that they were employing machinery to produce new results. He goes on to describe very briefly the mode of operation of the machinery set up :

' The corf at bottom & the engine sets off, the engine checks itself before it comes to meetings, (which is necessary on account of the great velocity of the corf) it afterwards resumes its speed till it comes near the top when it again slackens its velocity till it arrives at top, when a brake is put upon it & the valves shut *by the engine*. The corf & the engine there remain till the banksman (or man who lands the corves) returns from the place where he has been with the last corf. He brings the corf down (if it be too high) to his pleasure by laying his hand against a rail that assists him in striking the corf, and having exchanged the empty one for the full one, he pulls a rope by which he takes off the brake, and away goes the engine the reverse way, without a single touch from the engine man, so that we hereby save one man, for otherwise it would be absolutely necessary to have a man constantly at the working gear. In the engines here for drawing coal they have a fireman, a brakesman and a banksman.'[3]

It should be mentioned here that Smeaton in his water-wheel winding-apparatus had provided for an automatic reduction of speed when the up and down corves were passing each other (it will be understood that this was necessary because the corves were hanging free in the pit), and for the cutting-off of the water-supply as the corf reached the bank.

Early in the following year James Watt, junior, inspected the Benwell engine. He was charmed with it and wrote to Southern :

' I will not longer defer expressing the pleasure I received from seeing your contrivances at Benwell. The first time we tried the engine, it performed most admirably, executing all its evolutions with the greatest regularity & precision, first at the rate of a corf in 50 & then in 40 seconds, which does not now appear to me at all too quick. I think you are entitled to great praise for the simplicity of the different modes of checking the velocity & stopping the engine, which I will not disgrace by a comparison with the other modes here in use. The first day I thought them *infallible* and appointed a number of gentlemen to see the performance last Monday. The day unfortunately turned out to be very rainy, the ropes had been taken off the barrel & were with difficulty adjusted & it took us two or three hours to get it fairly started owing to the bungling stupidity of the pitmen. The rain too prevented the break from acting properly, by smoothing the surfaces, which we at length corrected by a little ashes. In

[1] B. & W. Colln. Thomas Barnes to Boulton & Watt, 1794 Dec. 16.
[2] Ibid. Barnes to Southern, 1795 Dec. 14.
[3] Ibid. Southern to M. R. Boulton, 1795 Oct. 10.

short it was a very unfair trial. Yet the company waited with the greatest patience to be drenched through with the rain, and all testified their approbation of the engine & gear, neither of which had any blame laid upon them, indeed it was impossible the engine could perform better than it did. I had the whole set to dine with me afterwards and made them all most gloriously drunk. Whether it was in consequence of this, or of the performance of the engine, I know not, but so it is that I have received already orders for 6 twenty horse engines.' [1]

The plan of shutting the valves automatically when the corf reached the top, as mentioned in Southern's letter, does not seem to have been retained for any length of time. In 1798 at the request of James Watt, junior, Barnes drew up a description, with a sketch, of the method of working and braking then in use at the King pit, Walker. The throttle valve was closed automatically when the corf approached the surface, so as to reduce the speed to 4 or 5 ft. per second. Then

' on the appearance of the corf chain at the pithead, the engineman shuts all the working valves and prepares the engine for the retrograde motion by means of a hand lever placed just above his left shoulder, and, if this does not bring the corf to a full stop when at proper height for striking, which is generally the case, he touches the brake lever lightly with his foot, & gently removing it again the corf descends upon the sledge. During the time the banksman is unhooking, the valves are opened again by the engineman, he removes his foot from the break and the engine goes off with the throttle valve close until the corf is on the point of lifting.' [2]

The brakes of the engines in the Newcastle district were applied to the flywheels, the peripheries of which were faced with wood. In the engine Barnes was describing the brake band embraced the lower half of the flywheel ; it was anchored to a fixed beam at one end and coupled to a foot-lever at the other and consisted of a single piece of iron, 5 in. wide and $\frac{3}{8}$ in. thick, lined with elm plank.

[1] Ibid. J. Watt, junior, to Southern, 1796 Feb. 12.
[2] Ibid. J. Watt, junior, to Southern, 1798 July 11.

XX

HOW THE ENGINES WERE MADE

THE ENGINE-YARD AT SOHO MANUFACTORY; SOHO FOUNDRY

IN the first place, let us consider briefly the conditions under which the Newcomen engine was built in the year 1775, when Boulton and Watt commenced their partnership. Perhaps it is hardly necessary to say that there were no engine-building works in existence, and that engine-fitting as a separate craft had not yet arisen. The engines were put together on the mine, or, in case they were to be employed for pumping water for the supply of water-wheels, at the mill. The erection was supervised by an engineer, a man such as Budge, or one of the Hornblowers in Cornwall. The engineer, in some instances, entered into a contract to supply an engine, but more usually he seems to have been paid for his services, and the engine parts and materials were purchased directly by the mine or mill-owner, who also paid the workmen employed on the job. The principal parts bought in a condition ready for use were the cylinder, cylinder bottom, and the working barrels of the pumps. The bored cast-iron work—cylinders and pump barrels—all came from one of four ironworks—Coalbrookdale, New Willey, Bersham, or Carron.[1] The smaller castings were sometimes obtained from the same place as the cylinders, at other times locally. The boiler, and all the wrought-iron work, was made on the spot, and the wooden beams also. Accordingly the engineer had under him a staff of smiths, carpenters, plumbers, and masons for building the engine-house.

This, then, was the position of affairs when Boulton and Watt commenced erecting their engines, and it was substantially on these lines that the firm carried on work for a number of years; for although in the petition to Parliament for the extension of the period of Watt's patent it had been urged that great sums of money would require to be expended in the erection of a manufactory for the production of the new engines, as a matter of fact Boulton and Watt did not embark in any considerable engine-making establishment until a number of years later.

Taking the period say about 1780 the valves and nozzles were the only parts of the engine made regularly at Soho. The cylinder, its cover and bottom, the

---

[1] In 1762 a concordat was arranged between Abraham Darby for Coalbrookdale Company, Isaac Wilkinson for Bersham Company, and John Wilkinson for New Willey Company for similar prices to be charged to customers for all fire-engine materials, cylinders, and bored articles at one price, pipes or articles not bored at another, on equal deliveries, London excepted. *Victoria County History: Shropshire*, Vol. I, p. 464.

The Carron Ironworks commenced work in 1760, and in 1770 Smeaton planned a cylinder-boring machine for this works. In 1773, and again in 1776, the Carron Company supplied engine cylinders to Sir Roger Newdigate for his colliery at Griff near Nuneaton. Castings were made at other ironworks, and there were in existence some iron foundries, i.e. distinct from blast furnaces for smelting, but probably none of these had apparatus for boring engine cylinders.

piston, the air-pump, and the condenser were made at Bersham, then in the hands of John Wilkinson, who made nearly every one of Boulton and Watt's cylinders up to the year 1795. The nozzles themselves were cast at another of Wilkinson's works—Bradley, a few miles from Birmingham—and brought to Soho to be fitted. Beam gudgeons, plummer blocks, and sundry other castings were supplied from the same works, while the cylinder jacket, or outer cylinder, together with some smaller work were cast at the Eagle Foundry, Birmingham. Fire-bars and other furnace fittings and small parts were sometimes cast in the locality where the engine was to be erected. The copper eduction pipes were made in London, and the wrought-iron piston-rods either at Seaton in Cumberland or in London.

These parts were ordered by Boulton and Watt from the various makers on account of the mine- or mill-owner; Boulton and Watt sent the necessary drawings and instructions, and kept a staff of men to supervise and do part of the work of erecting. These men were paid by the person for whom the engine was being erected. It was one of the great difficulties of the firm to get together, and to retain, a suitable staff of erectors.

The course pursued in erecting an engine in the year 1778, when we may consider that sufficient experience had been gained to render the procedure regular and systematic, and that at least a few men had been trained for the work, is described in great detail by Watt himself in connexion with an engine for pumping brine at Thirlewood or Lawton Saltworks, Cheshire (Salmon's engine).[1] His account is too long to reproduce in full, and we must content ourselves with an outline of the main features.

When Watt arrived at Lawton on Feb. 4, 1778, the engine materials were on the spot, and he found the engine-house ready for putting up the cylinder. Next day the beam was got up and the martingales at the cylinder end put on; also the cylinder bottom was set out for drilling. On the 6th the inner bottom was drilled for the holding-down screws, and the brick platform or foundation for the cylinder was begun; this was completed the following day, when also the windlass was fixed on the upper floor. On the 9th the upper flange of the cylinder was drilled and the small holes in the inner bottom were drilled and tapped; the inner bottom was then screwed down and the inner cylinder put in place. Next day an attempt was made to raise the outer cylinder; but, in consequence of a failure of the rope, this was not accomplished until the day after, when also the inner cylinder was levelled, the two cylinders set nearly parallel, and the lower part of the nozzle fitted; on this day too the boiler seat and the ash-hole were marked out. On the 12th the setting parallel of the cylinders was completed, the bottom joints of the outer cylinder made with dung and blood, the upper part of the nozzle fitted, the condenser cistern moved into its place, and the copper eduction-pipes tried together. On this day Watt remarks: 'We had a good deal of chiseling at the flanches of inner cylinder opposite the holding down screws before would suffer front joint of outer cylinder & bottom to come fair and we had to cut off the outer side of ye washers under the pillars to a quarter inch broad, otherwise kept the bottom joint open.' On the 13th both nozzles were screwed on 'for good', the joints being made with pasteboard and putty, the boiler setting was begun, and half the upper floor of engine-house laid. Next day the two cylinders were set perpendicular, and the piston filled up with wood; the piston-rod, it was found, did not fit the piston, being 3 in. in diam. at the base of the cone instead of $2\frac{5}{8}$ in.; the boiler seat was

---

[1] Doldowlod Papers : Watt's Journal.

finished and the chimney commenced. On the 16th, the next working day, the boiler was put on its seat and the flues built up to the arch, the piston was put in, the screws of the stuffing-box lengthened, the guide-posts fixed up, holes bored in the plug-tree, and staples made for the Y-shafts. The following day the working-gear was fitted up, and on the day after more work was done on the gear, and the air-pump was screwed together. The 19th saw the joining up of the eduction pipe, ' by pouring lead with a small mixture of tin into copper bosses surrounding the joints', the drilling of the steam-pipe flanges, more work on the working-gear, the completion of the boiler setting, and progress on the chimney. The next day we have two men making screws, another soldering up the hot-water pump, another putting the arches for the air-pump chains on the beam. On the 21st certain defective joints in the eduction pipe were unsoldered and made afresh, and the beam was adjusted. The next week was taken up in work connected with the pump end of the beam, and in fitting and fixing the steam-pipe, fixing down the condenser in its cistern, hanging the chains for working the air-pump, fitting the brake or lever for working that pump by hand, and fixing the manhole screws in the boiler. Later, there is an entry that : ' Manhole screws·with T-heads would not answer, ordered bolt-head screws and fixed nuts.'

On March 3rd fire was put under the boiler, and the piston packed, and on the day after steam was turned on to the engine which was found ' in general very tight'. The succeeding days were taken up in minor jobs, and then on the after-noon of the 11th we read ' sett the engine agoing, raised water to top of pitt trees, drew much air, and the condenser top let in much water, about a gallon pr. stroke'. On the 12th the engine was christened in the presence of ' a great lot of people'. The engine went off very well, but not so the pumps ; ' the upper lift soon begun to draw air, sett on ye jack-head but it proved very leaky .... with difficulty got a little brine to top of bank'. Although the engine worked, it was not doing what it should do, so the following day, in addition to overhauling the pumps, the air-pump and condenser were attended to, and Watt gave instructions for an alteration of the gear for working the exhaust valve. He left Lawton on March 14th.

It will be seen that the work of erection began on February 4th, and that the engine was set going on March 11th. There were some minor difficulties with the pumps, but the engine itself went off very well, and it continued to do well, for in August 1779 we find that Salmon had written ' a letter full of praises of his engine'.[1]

It has been stated above that by the year 1778 the procedure in erecting the engines had become regular and systematic. This is borne out by the fact that in the following year Watt set about the production of a hand-book bearing the title : ' Directions for erecting and working the newly-invented steam engines. By Boulton and Watt.' This, the first book in the English language devoted to the steam engine, was not published in the ordinary way, but was produced for private circulation among the clients of the firm, who, it must be understood, were under no compulsion to have their engines put up by the Soho erectors. It seems that one hundred copies only were printed, and the book is accordingly very rare ; considerable use has been made of it in the present volume and it is reprinted in full with the accompanying illustrations in Appendix II. The title-page bears no date, but certain entries in Watt's Journal[2] make it clear that it was printed in 1779.

[1] Boulton Papers. Watt to Boulton, 1779 Aug. 13.   [2] Doldowlod Papers.

' 1779. May 25. Writing directions for putting engines together.
June 6. Writing directions for putting engines together.
June 7. Gave directions to Mr. Rollason to be printed.
June 10. Corrected proofs of directions and sent them 20 more pages.
June 11. All day at directions, wrote 19 pages.
June 12. In forenoon at the printers, ordered 50 copies on copy paper and 50 on thin post paper.
June 29. Wrote to Mr. Hornblower with copy of engine directions.
Aug. 10. Writing directions for working and managing engines.
Aug. 11. Finished engine directions which gave to the printer.
Sept. 4. rec^d the 1st. plate, the piston, from the engraver, he charges 10/6 for it & says it took 3 days.'

Some years later, after the introduction of the rotative engine, another set of ' Directions relating to the engine' was printed as a single sheet, no doubt with the intention that it should be pasted on a board and hung up in the engine-house. This also is reproduced, in Appendix III.

But apart from these 'Directions', &c., we find Boulton and Watt making free use of printing in the course of their business. For reply to an inquiry for a mine pumping-engine they had a seven-page quarto pamphlet : ' Proposals to the Adventurers in . . . By Boulton and Watt', setting forth in full the conditions for the grant of a licence to use the engine, and the manner in which the royalty was determined, with blank spaces for the insertion of such particulars as the name of the mine and the size of the engine. Printed lists of the materials for each engine came into use in 1778. These lists were filled in in writing with the names of the places from which the various parts were to be supplied and any necessary directions as to erecting, &c. For the engines in Cornwall printed forms were employed from 1780 for the periodical returns of the engine performances, and printed books were supplied to the mines for the same purpose. Curiously enough, as it may seem to us now, the firm did not make use of printed letter-headings.

When we consider the undeveloped state of the means of transport at that period, it becomes an interesting subject for study how the different parts from these widely separated localities were brought together at about the same time, for the erection of an engine, say, in Cornwall. It is clear that a good deal of fore-thought and an extensive organization were necessary, and that the commercial side of the Boulton and Watt concern had not a few difficulties to surmount. The transport of goods made in London was the simplest problem of all, as there was a frequent service of ships direct between the Thames and one or other of the Cornish ports. The goods from Soho and Birmingham were carted to the Canal, along which they were taken to Stourport on the Severn There they were transhipped and carried down the river to Gloucester, Bristol, or Chepstow, and again transhipped into a coasting vessel which completed the transit. The goods sent direct from Bradley could be loaded into the canal barge at the works. The Bersham goods were sent by road to Chester and there shipped to a Cornish port. Usually the cargo was made up with fire-bricks, in which there was a considerable trade to Cornwall. The piston-rods made at Seaton were usually shipped at Whitehaven or Workington to Liverpool, taken thence to Chester, where they were loaded on the same vessel as the cylinder, &c., from Bersham. At least this was the course that it was desired to follow, but frequently the rod did not arrive to time.

Now all this was sufficiently complicated, but matters were rendered still more difficult by the fact that Great Britain was at war with France, and the English Channel and St. George's Channel were infested with privateers. It was not enough to insure the cargoes against seizure, the captains of the ships had to be furnished with ransom money.[1] The idea was that in the event of capture the privateer, upon discovering the nature of the cargo, particularly upon learning that it comprised parts only of an engine, and not a complete machine, would decide that it was not worth his while to take his prize into a French port; then, upon the presentation of the ransom money, he would allow the ship to proceed, instead of sinking it, as he would do in the ordinary way. There is no record of such an event having taken place, and the only engine material actually captured by privateers was the outer casing for an engine for Ireland. This was taken on the voyage from Bristol to Waterford in 1782.

An interesting point that comes out in connexion with the privateer difficulty is the absence at the ports of facilities for dealing with heavy articles. There is in relation to one engine a discussion as to the particular port in Cornwall to which the engine parts are to be sent. Hayle is convenient for the site of the intended engine, but at this port there is no crane, so the cylinders have to be lifted from the ship's hold by such expedients as present themselves, and rolled ashore. On the other hand, there was a crane on the Truro river, but to get there meant doubling Land's End, and this was particularly dangerous on account of the privateers.

We find Murdock figuring in one of the privateer stories, but as a landsman only; the Frenchman tried to cut out a vessel that had just got into Portreath, and Murdock was one of the party armed with guns that drove off the boat.'[2]

Then we read that on one occasion an erector, returning from Cornwall to Birmingham, proceeded by sea from a Cornish port to Bristol. For weeks there was no news of him; inquiries were made in all directions without result, and the conclusion seemed inevitable that the ship had been taken by the enemy. However, the man turned up at last; the ship had proved unseaworthy, and had been driven by storms into a little out-of-the-way port.

The state of war caused the transport of the erectors to and from Cornwall to be a subject of anxiety on other grounds. The press-gangs were out and the men were liable to be seized on the way; precautions had to be taken to avoid this, or, if the men were taken, to get them liberated. When Murdock and Law were sent to Cornwall in 1779, they were furnished with letters addressed to important people in Bristol and Exeter begging interest in their favour.

In spite of the state of war, Boulton and Watt managed to send engines to France. One was put up near Nantes in 1791–2, and Richard Dayus was sent out to take charge of the erection.

But quite apart from the difficulties due to the war, the transport of the goods gave rise to many problems. It was difficult to get ships with hatchways big

---

[1] The sum usually given to the captain seems to have been £100.

[2] 'The Pool goods, which are arrived from Bristol, yesterday was very near being carried to France from Portreath; a privateer of 16 4-pounders came to anchor within the Gull Rock & sent out a boat to cut out Burrell who had them on board, but the country being raised & getting some small arms stood at the light house and beat off the boat, after standing some time in contest, supported by broadsides from the privateer. Wm. Murdock was one of the people that fired.' B. & W. Colln. Thos. Wilson to Watt, 1781 Jan. 18.

enough to pass the cylinders. The cylinder of the first engine, Tingtang, ordered for Cornwall was delayed for a considerable period on this account. Watt was very vexed, and was prepared to spend five or ten pounds to cut the hatchway larger; he felt that it was not fair that a smaller engine, Wheal Busy, ordered later, should be the first completed. The handling of the cylinders was generally an awkward affair; the difficulty at the Cornish ports has been alluded to above. At first, any goods sent from Bersham to Birmingham seem to have gone by road, then later they were taken by wagon to Preston Brook near Runcorn and carried from there along the canal. In 1791 John Wilkinson's clerk or agent at Bersham writes:

'The last cylinder which went to Ocker Hill was taken there by our team, and thrown over in a field near where it was erected. The same might be done with this. . . . There is no danger in throwing a cylinder from a waggon on soft earth, but it is not so at Preston Brook. I would sooner trust to throwing it off the waggon there *than to their cranes.*' [1]

It appears that the general practice was to throw over the wagon with the cylinder on it.

Then again there was a lack of care in stowing the goods in the ships. The piston-rod for the Wanlockhead engine was sent from Seaton to Dumfries by sea. It failed to arrive at its destination; there were protracted inquiries and at last the rod was found on the quay but sadly damaged. It appears that the ship had carried a quantity of lime, this had become wetted on the voyage, and the heat developed had injured the rod, which had been placed on or quite near to the lime. Possibly it was this accident that led to what became the usual practice, viz. the boxing-up of the rods for transit.

With the introduction of the rotative engine, more and more of the engine-work was done at Soho. At the mines there were to be found the necessary tools and workmen capable of making and putting together the parts, and the mine-owners were accustomed to buying the various parts and materials themselves and paying the cost of erection; but when it came to putting up engines in breweries, mills, &c., in London, and other towns, the conditions were altogether different. The proprietors wanted an engine; they gave the order, and desired to have no further trouble or responsibility. John Rennie did a great deal of work in connexion with the early rotative engines in the London district and elsewhere, but in most of these cases he was concerned also for the millwork and other machinery. The tendency was of course to throw more work upon Soho, and between 1790 and 1795 we find a state of transition; of some engines only the nozzle and working-gear were made at the works, in other cases the entire engine was put together there. A good instance of this is found in a letter written by Southern, Feb. 15, 1794, in reference to an engine ordered by Rennie. He says:

'I should wish to hear by letter whether you would have anything more than the cylinder & air pump cast at Bersham, & whether we shall furnish from Soho the other castings or send Mr. Rennie drawings, & likewise whether the Bersham goods shall come here to be fitted, or be sent from Chester to London.'

Another letter by Southern of about the same date is interesting for the light it throws upon this matter, and upon the method of charging for the work. It seems that James Lawson was erecting engines in Leeds and had written to Soho,

[1] B. & W. Colln. Gilbert Gilpin to Boulton & Watt, 1791 Apl. 29.

to complain of bad fitting and of excessive prices charged, and Southern tells him :[1]

'I wish you would send a list of particulars instead of saying generally that things do not fit. In the invoice of Messrs Markland's engine there is only 2½ guineas charged for drilling boiler steam pipes and some other matters. You will please to recollect that the cylinder & air pump, their piston & bucket did not come to Soho, so that neither the fitting the condenser to the air pump, the piston to its rod, or the air pump bucket to its rod could be done here. Neither could the nozzles be fitted to the cylinder, nor the pipes to the nozzles, but in W. F. & G. [Wormald, Fountain, and Gott] all this was done, which will account for some difference. But if they were not fitted at Soho, so neither does Soho charge them for fitting these parts. In short, if you will inspect the invoice and see what is charged for fitting, you will find it small upon the whole for what was fitted . . . the weight of the piston rod is charged neat weight and not the forged weight. Brasses we always charge the rough weight, as the filings which come off are not worth more than 1d. a pound, and cast iron goods the same which you must have known has been the custom for years both with Mr. Wilkinson & B. & W., & I believe everybody else [2]—for what are the borings or turnings worth ? But all turned iron or steel we charge net weight. . . . In relation to Soho prices they are precisely the same as [. . . [3]] great engine had theirs charged, they are less than Mr. Wilkinson charges, and they are not greater than will allow only a *moderate* profit. B. & W. lost one year by their engine manufactory 2000*l*. It was high time to raise their prices.'

In 1795 for a 63-inch cylinder pumping-engine for the Newcastle district the nozzles and working-gear only came from Soho. In this case the cylinder and air-pump were made at Coalbrookdale.[4]

Possibly in consequence of internal troubles at Bersham, and the resulting delay and falling off in the character of the work done, the Coalbrookdale Company was about this time doing a good deal of Boulton and Watt's engine-work. Some of the cylinders they made were of large size, 60 and 63 in. In 1795 the orders at the Dale were much behindhand, and Southern, accompanied by Ewart, went there to investigate the matter. As a result he suggested that some of the orders be sent to Walker of Rotherham. He notes also that the standard of accuracy is much below that of Wilkinson. This comes out in an account of a visit paid by Southern and Ewart to the works of Banks and Onions near Broseley. There they ' saw their boring mill and some large cylinders for a blowing engine, but they were not well done, nor do I think that they had aimed at accuracy. . . . I think their castings are as well done as the Dale Co's, but their notions of accuracy from what I saw appear inferior to the Dale Co's, and theirs are very short of Mr. Wilkinson's performances.' [5]

The letters of the Coalbrookdale Company to Boulton and Watt show that they had their troubles from time to time with bad cylinder castings. On Oct. 15, 1795, their agent writes that ' the cylinder for North Downs upon boreing proves a faulty one which will cause a delay of near two weeks in getting another ready ' ; and again, on May 25, 1798, ' in going through the last cut with boring your cylinder it proves faulty, and we have condemned it as unfit for your purpose'. This was a 40-inch cylinder.

One of the surprising things in the letters from the Coalbrookdale Company to

[1] B. & W. Colln. Southern to Lawson, 1793 Oct. 22.

[2] The Coalbrookdale Company, however, charged cylinders on the weight after boring. Ibid. Boulton to J. Watt, junior, 1795 Sept. 20.

[3] The name is torn off the letter.

[4] B. & W. Colln. Southern to J. Watt, junior, 1795 Aug. 30.

[5] Ibid. Southern to J. Watt, junior, 1795 May 29.

PLAN of BIRMINGHAM

SCALE

Showing in black the situation of Soho and
other places in the days of Watt.
The subsequent expansion of the City
over the area is superposed in Red

Boulton and Watt, particularly in view of Southern's opinion as to the accuracy of their work, is contained in a letter of Jan. 23, 1796, relating to the dimensions of a certain piston which they had made ; they say that the diameter of the opening at the bottom is ' nearly $6\frac{1}{2}$ inches or $6\frac{4125}{10000}$ '. The idea that an engineer could express a dimension in terms of ten-thousandths of an inch at this period is quite staggering. Possibly they wished to show Southern what they could do when they set out to be accurate.

The original plan of throwing the charge and responsibility of the erection of the engine upon the customer was maintained up to the year 1795 ; at least it is for the year 1795–6 that we first find particulars of wages paid by the firm to their erectors. During the years 1786 to 1795 the number of men employed as outside erectors varied from 9 to 18 ; the average works out to somewhat less than 12. The number of engines put up in the same period varied from 10 to 26 per year, the average being 16.[1] The old plan, however, continued after 1795, for in 1799 we find the younger Boulton putting forward the suggestion that Boulton and Watt should themselves erect their engines, and pointing out that their competitors Bateman and Sherratt, Smith, and Murray always did so.[2]

THE ENGINE-YARD AT SOHO MANUFACTORY

THE Soho Manufactory was established in the year 1765. Matthew Boulton at an early age had entered the business of his father, also ' Matthew' Boulton, who was a ' toymaker ' in Snow Hill, Birmingham. The son soon showed a remarkable business capacity and ingenuity, and he gradually assumed control of the business. In process of time he found the Birmingham premises too restricted, and sought a place where there was room for extension, and where a natural source of power was available. This he found at Soho, at the edge of Soho Park, then and for many years after a rural and most pleasant situation. In the year 1756 a lease for a term of ninety-nine years had been granted by John Wyrley, lord of the manor of Handsworth, to Edward Ruston of certain common and enclosed lands at Handsworth, with liberty to make a cut about half a mile in length for the purpose of turning Hockley Brook into a pool to drive a small mill for rolling metals.[3] This lease Boulton purchased in 1762, and upon the site he proceeded to erect the magnificent building the appearance of which is well known from the published engravings. The cost is stated to have amounted to £9,000.

The main building was in plan in the shape of the letter E ; it was used for the manufacture of silver and plated goods, except the wings, which were used, in part at all events, as dwellings for the managers and foremen. Behind it was a yard, a story lower than the ground at the front, and behind this again another yard still lower. The buildings of the remaining branches of the great Soho concern were placed in and around these yards.[4]

The different branches were run as separate companies, with Boulton himself as one of the partners ; thus we have Boulton and Fothergill for the silver and plated goods, Boulton and Scales, and, later, Boulton and Smith for buttons,

[1] Muirhead Papers.
[2] B. & W. Colln. M. R. Boulton to James Watt, junior, 1799 Jan. 30.
[3] *Birmingham Weekly Mercury*, Oct. 23, 1886,

' A Chapter of Local History '.
[4] *Birmingham Weekly Post*, Apl. 20, 1895, Local Notes and Queries, ' Reminiscences of Soho '.

buckles, steel toys, &c., Boulton and Watt for steam engines, and James Watt & Co., for the copying presses.   The Mint, however, seems to have been carried on by Boulton alone.[1]   These separate companies were each charged an annual sum for rent, and their proportion of the general office expenses.

In a very few years Boulton began to find inconvenience from the deficiency of his water-power, and was led to turn his attention to the steam engine.   How from this arose his acquaintance with James Watt, and the formation of the celebrated firm of Boulton and Watt, has been described earlier in this work. The engineering establishment at Soho commenced about September 1775 with a smith's shop furnished with two hearths.   It cannot have been a very large building, although Boulton refers to it at this date as ' formidable ' and later on calls it ' our great smith's shop ' ; it served also as the fitting shop, and was provided with at least one lathe.   Whether the patterns were made in the same building is not clear, but the firm seem to have made their own patterns from the beginning.   At an early date Watt complains of the pattern-maker's drunkenness and dishonesty (he stole the nails) and proposes to dismiss him and employ another man, or get the work done outside.   A few weeks later, however, we learn of reform, the pattern-maker has been working very diligently and is well on with the nozzle patterns.   Towards the end of April 1777 an additional lathe for turning piston-rods was set up, but some little time had to elapse before there was a piston-rod upon which to employ it.   The story of this first piston-rod is quite a chapter of accidents.   First we read that ' Dickson has forged a very good piston rod ' ; next, ' Dickson's piston rod was badly welded & we have had to take it asunder and re-do it ' ; then, ' Cleobury iron has turned out damned bad, the piston-rod for Huel Bussy was very well forged by Dixon and upon heating it to float it fell in two at the shooting—Dixon is now faggotting one out of Swedish small bars ', and finally, ' Dickson has finished an exceeding good rod without *shooting*, & I don't think it will lose 2 pounds in dressing up.'   After this, for many years, the greater number if not all of the piston-rods were forged and turned by outside firms.   By this time the engine business was promising to be an extensive one, ' so ', writes Watt, ' all our heads must go to work as hard as possible and some more fitters must be taken in, for I believe we shall get almost all the forging done out of the house & readily.'

Writing to Jonathan Hornblower, the elder, at about this time, Watt gives an account of the iron they were using at Soho :

' We work upon no other iron but the best tough scrap iron, made at Wednesbury, which costs us 20/- per cwt in the barr, or else the best gun barrel iron, which costs us the same price.'

He advises Hornblower to have uses or slabs of iron drawn at a forge near to the size for the martingales and other parts, and he recommends :

' John Wood's Wednesbury scrap iron as being the most certainly and truly tough and fitter for most engine purposes than any other.   It is made of the small scraps cut off in the Bir[m] manufactures welded up into barrs.   It will cost from 20/- to 22/- when drawn into uses.'

However, this high opinion of John Wood's iron was not of long continuance, for within a few months we find Watt stating that its use must be given up, as ' it is too weak and uncertain a fabric '.   The wrought-iron work supplied from Soho

---

[1] Another list of the branches enumerates the Mint, the Rolling Mill, Plating Co., Button Co., 　　　Latchet Co., J. Watt & Co.

XCV. SOHO MANUFACTORY

From a water-colour drawing in the British Museum

included screws and nuts, among them the screwed ends for welding on to the longer bolts, such as the holding-down bolts, the stirrup for the beam, and the martingales. In the letter to Hornblower quoted above, Watt describes how the screws were made :

' Our screws are made with solid heads (i.e., not welded on) and they are screwed in stocks and dies with square topt threads like a vicepin and afterwards brought all to one size by a screw plate. The nutts or burrs (as they are called here) are first entered by a taper tap and finished by a tapp which goes quite through them and makes them equally wide upon both sides. Such screws we find vastly preferable to the common sort, especially in cases where they are often to be unscrewed, or where strength is required ; but after all that pains in making them we cannot sell them under 6d. per lb., and hitherto have been losers by selling them at that price.'

When, later in the same year, 1777, Watt went to Cornwall to superintend the erection of his first engine in this county, he found that Soho prices for engine-work were considered too high, and he writes to Boulton ' our prices so much exceed what they are done for here that we shall never get our bills pleasantly paid, and to be paid with a grudge, and to lose by the business at the same time will never suit ' and accordingly ' we must give up the manufacturing of any parts of the large engines excepting the nozzles '.

This dictum of James Watt really defines what became the policy of the firm for many years. And although in 1778 Boulton wrote to Smeaton that :

' we are systematizing the business of engine making, as we have done before in the button manufactory ; we are training up workmen and making tools and machines to form the different parts of Mr. Watt's engines, with more accuracy, and at a cheaper rate than can possibly be done by the ordinary methods of working. Our workshop and apparatus will be of sufficient extent to execute all the engines which are likely to be soon wanted in this country, and it will not be worth the expense for any other engineer to erect similar works, for that would be like building a mill to grind a bushel of corn.' [1]

of the single-acting pumping-engines the nozzles were the only parts that were invariably sent out from Soho. Possibly even the parts they did handle may have seemed to Boulton very big jobs, as indeed they were compared to the buttons, buckles, and other products of the factory. Or it may have been that he wished to warn off all competitors.

Coming on to 1781 we find Boulton engaged about the plan of a new engine-shop, a two-story building that was completed by the end of the year and cost a little under £110. The firm also seem to have been considering the erection of a foundry in Cornwall, but speedily relinquished the idea although later on Boulton made the suggestion of setting up a workshop there. ' I think if Murdock or us had a shop with a few tools such as a machine to re bore old working barrells & a sockett plane to turn piston rods, with one or two good lathes to turn valves it wd. be well.' [2]

A perusal of the Boulton and Watt papers shows that Watt did not take any active part in the direction of the engine works, or in the improvement of the methods of production. He had an office at his house at Harper's Hill, and it was there that he carried out his work, calculations, drawings, and correspondence ; the mere bulk of the product forms abundant evidence that he worked very hard, and it is clear that frequently many days elapsed between his visits to Soho to

[1] Farey : *Steam Engine*, 330.   [2] B. & W. Colln. Boulton to Watt, 1782 Dec. 5.

inspect the progress of any new schemes on hand, and to discuss business matters with Boulton.   Possibly in Boulton's absence his visits may have been more frequent, but on one occasion we find Boulton writing from Cornwall requesting him to try to go down to Soho, or else to send his assistant Playfair, to see that matters are going forward satisfactorily.

It was Boulton who managed the works, kept an eye on costs, and sought to improve the methods of production.   Thus we find him writing, from Cornwall,. that cast iron was not a suitable material for the racks and sectors for operating the valves, and suggesting that ' it would be safer to forge them of steel & cut them down with a cutter in the mill ',[1] and then a little later on suggesting a machine for dividing and cutting the teeth.

Then again in 1782 when the nozzle for Trevaskus engine was in hand (it was to a new design), ' Joseph Harrison and Mr. Henderson are both positive that if nozzles are to be fitted up as you have ordered Trevascus to be they cannot be afforded by the workmen for less than 30ˡ. a piece, even if no bigger than Trevascus.' Boulton went into the question of what they had been paying for the fitting of nozzles in the old way, and expresses the result in terms of the diameter of the cylinder of the engine, i.e. 3s. 4d. per inch.   ' Thus a 56 inch and 58 inch diam is 9 guineas each, a 42 7 guineas, a 63 is 10 guineas, and a 66 is 12 guineas.'   He considers that with improved methods of work the new nozzles may be done for an advance in the proportion of 3 to 2.   He again mentions the machine to cut the racks and sectors and thinks that ' the flats of the nozzles may be ground by a great leaden lap, or a flat grindstone, both of which should be turned by a mill or an engine, and as there is additional drilling, I think a drill might also be turned by mill power '.[2]

A fortnight later he again writes to Watt on the subject of nozzle-fitting, in which they are much behindhand, suggests that the collars for the spindles be made of brass instead of steel as it would save time and expense, and comes back to the tooth-cutting machine and the grinding of the flats ' instead of chiselling '.[3]

Boulton next turns his attention to the castings themselves, and proposes to go over to Bradley Ironworks to see a nozzle moulded, and to discuss with the moulder the best means to avoid the considerable amount of chipping now entailed in fitting up the nozzles.[4]   Four years later Boulton regrets, this time in connexion

[1] B. & W. Colln.   Boulton to Watt, 1780 Nov. 6.

[2] Ibid.   Boulton to Watt, 1782 Apl. 6.

[3] Ibid.   Boulton to Watt, 1782 Apl. 19.

[4] ' We are desirous if possible to lessen the expense of fitting up our nozzles, which at present is very considerable ; and for that purpose we mean to try if, among other savings, something cannot be saved by the casting, either from the improvement of patterns in making some parts and feneering other parts wᵗʰ brass, or an extraordinary care being taken in moulding.   We are at present obliged to chip and file a great deal at the flanches and valve holes &c., which puts us to a considerable expense, and we think that by particular attention being paid to the patterns & moulding that this part of it would be lessened.' B. & W. Colln. : Letter Books.   Boulton & Watt to John Threlkeld

(manager at Bradley), 1782 June 13.

' We beg that particular attention may be paid. to casting the flanches straight ; we have had much reason to complain of this part hitherto, as also of the knobs & some of the flanches turning out hollow.   We think that some method could certainly be fallen on to have the knobs cast hollow ; this would save us a good deal of work, & would render the solid part more free [from] blown holes in the castings.   We have sent by the bearer three patterns of the exact cones for the valve holes, which you will please to have your cores turned by.   Hitherto we have been obliged to chisel out the iron to the proper cone, this expense may be saved by some attention paid in the casting. . . .' Ibid. : Letter Books.   Boulton & Watt to John Threlkeld, 1782 Aug 5.

with the working barrels of pumps, that neither Soho nor Bradley is furnished with a boring machine.[1]

The work seems to have been interrupted a good deal by the drunkenness of some of the workmen. In 1782 we find Boulton writing : ' Our forging shop wants a total reformation ; it is worse than ever, Peplo has been drunk ever since his wife's death almost. Jim Taylour has been drunk for 9 days past ; '[2] and again : ' Jim Taylour has been drunk for some time past on acc^t of a girl, & my patience being quite exhausted, I have taken Cartwright from the Navigation engines and set him about Trevascus nozzle with other assistants, since which Jim Taylour has returned, but I desired him to go about his business.'[3] Boulton, however, took him on again later.

The first serious dispute with the workmen that we read of is in the year 1791. On March 23rd in that year Boulton informs Watt of a strike of the engine-smiths. He seems to have settled it very quickly, for on the following day he writes : ' I have the satisfaction to tell you that the Smiths etc are all got to work this morning very peacably, and I am persuaded will be the better for yesterday's struggle. Turner seems to have got a fright, and is in terror lest he should be turned off.' The trouble seems to have arisen over piece-work rates. Roberts, the engine-works clerk, had left the firm, and his account book was not to be found, and the men were claiming prices that seemed too high. Possibly Roberts had been lax, and his successor, Foreman, paid more attention to the interests of his employers, at any rate he managed to make himself very much disliked by the workmen, and a few years later, in 1794, he was seriously assaulted by some of them. The ostensible reason was that he was thought to have instigated the enlistment in the army of one of the apprentices, but a memorandum by Southern goes to show that the cause of the prevailing ill feeling on the part of the workmen was more deep-seated than this, and he mentions at least three contributory factors. While Roberts was at Soho the prices paid to the smiths were unduly high, some of them frequently earned 10s. 6d. a day, and sometimes more ; after Foreman came the prices were reduced ; further the Sick Fund had been more closely watched to see that none evaded payment. But the most curious point is one in connexion with the provision of watchmen for the works. At the time of the Birmingham riots in 1791 it was thought proper to augment the number of watchmen for a certain time. The increased expense, £30 or upwards, fell upon the engine works, and it appears that the men had to contribute towards it ; after a time the additional watchmen were dispensed with, and the extra money was paid to the usual watch, but, says Southern, ' It is now some months ago since you saved by not paying to the usual watch all that you had paid to the extra-ordinary watch, and by this time I believe you have cleared near 15 or 20l. This circumstance has had its effect upon many of those who have been obliged to pay it, and is in my opinion one cause of the prejudice existing against the Engine Co's clerks.'

It is possible too that the wages paid at Soho were kept down to the lowest possible figure. We find Peter Ewart writing from Manchester early in 1792 : ' Mr. Foreman desired me to send 8 or 10 hands, but he does not consider that there is not a hand that is good for anything can be had here for less than 57 or 58 shillings per week.' Generally one gets the impression that the Engine Company

---

[1] Ibid. Boulton to Watt, 1786 Mar. 5.   [3] Ibid. Boulton to Watt, 1782 Apl. 6.
[2] Ibid. Boulton to Watt, 1782 Mar. 26.

paid very badly. They paid less even than Boulton himself at his Mint in the same works ; in 1800 he was breaking up his Mint establishment, and his fitters were free, men were badly wanted by the Engine Company, but there was the difficulty about the difference in the rates of wages.[1]

Apparently it was in the year 1794 that the firm took the first serious steps towards doing more of the engine work themselves.[2] The original idea was to have a boring-mill, and Peter Ewart, who was then in business as a millwright, and superintending the erection of Boulton and Watt's engines in Manchester, was brought to Soho to design and superintend the erection of such a machine. However, as time went on this scheme developed and resulted in a considerable extension of the engine manufactory at Soho itself, and in the erection of a separate establishment at Smethwick, which became known as the Soho Foundry, and was about a mile distant from the parent establishment. Soho had the disadvantage that everything had to be brought there and taken away by road ; when the question arose of dealing with large cylinders and other heavy articles this became an important matter, possibly it may have been the factor that determined the setting up of a new works; at any rate in selecting a site for it the point was kept in view, and the plot of land acquired had a frontage to the Birmingham Canal, from which a branch was cut and a dock formed within the works. The proposed extensions were not embarked upon without a good deal of consideration. For instance, there is a ' Statement of goods sent from Soho in the years 1791, 2, 3, & 4, with the proportion of work done by the different founders employed by B. & W.' The total value of the goods sent out amounted to a little under £40,000 and the total of the founders' accounts was a little over £20,000, accordingly the value of the work done at Soho was less than half the total. The founders mentioned are Bersham, Bradley, Dearman, and Seager, and the accounts cover cast iron, brass, and wrought iron, and fitting.

The Boulton and Watt Collection furnishes very little information on the subject of patterns and pattern-making. From the first the firm made their own patterns for such parts of the engines, i.e. nozzles, &c., as were sent from Soho, but in the early days pattern-making was not a distinct craft, the same man might in turn be engaged in pattern-making, fitting, and erecting. Some of Murdock's first work for the firm was at patterns. That the art was in its infancy is seen by a letter sent to the founder in 1783 with some patterns of wheel segments and racks. Boulton and Watt's clerk begs that they may be cast at once partly ' that the mouldings may be made before there is any danger of the patterns altering '.[3]

It has been stated earlier in this chapter that Watt for many years carried on his work at his private residence, and when he got the assistance of Playfair, and afterwards of Southern, it was at his house that they attended. This remained the state of affairs until 1790 when the drawing office was moved to Soho Manufactory. There was no separate drawing office at the Foundry, although this

[1] From the beginning the rate of wages seems to have been low, even for those days. In 1777 Robert Mackell of Glasgow was offered a job as millwright ; he was to engage for three years certain at wages not over twelve shillings per week. Murdock when he started received fifteen shillings per week.

[2] It was about this time that James Watt, junior, became a partner in the business, and his father dropped into the background. In 1798 Watt writing to Dr. Black says, ' In regard to the engine business I now take little part in it, but it goes on successfully.'

[3] B. & W. Colln.: Foundry Letter Book. Buchanan to J. Threlkeld, 1783 Jan 11.

was a distinct engine-making business; at least this seems to have been the case as late as 1826.

We find no code of rules for the government of the office until 1828. With Southern, and possibly another man, working under Watt's eye there could have been little or no need of a written set of rules, but as the office increased in size, and a number of young men were employed, the necessity for some standard regulations made itself felt and accordingly in the year mentioned a scheme of 'Proposed rules for the drawing office' was formulated. What strikes one at once is the long hours of attendance. Rule 1 is: 'The young men to be there at 9 o'clock in the morning, and to remain until 1 o'clock, when they go to dinner and return to the office at 2 o'clock; there to' remain until 5, when half an hour is allowed for tea; after which they are to be at the office until Belringing in the Evening.' On the back of the paper are some notes on the hours of attendance of clerks in other places. At 'Liverpool 10 till 5 is considered regular, 7 hours'. At Manchester, on the other hand, some of the factories called for 10 hours. There is a further note as to the attendance of the 'operators' at the 'Soho Drawing Room' in 1791 and 1792; they worked 10 hours in the summer, starting at 7 o'clock in the morning, with one hour and two half-hours out; in the winter they worked until eight in the evening, that is to say, 11 hours, or, as the note puts it, 'deducting irregular returns' 10½ hours. The note goes on with a very curious remark: 'Some years after, hour in the morning left out, as about nothing could be begun till near 8, as door was without a key for firemaker & postman.'

From the first Watt adopted the plan of figuring the leading dimensions on his scale drawings, indeed this was absolutely necessary when he made the same drawing serve for a 36-inch and a 40-inch cylinder. Frequently alternative constructions are suggested in writing and the decision as to the plan to be adopted is left to the founder.

The work of the earlier years shows no uniformity in the size of sheet of the drawings; the paper was cut to suit the drawing. One of the features that strikes one on going through the Birmingham Collection is the free use that has been made of the copying-press for duplicating drawings; the duplicates are of course reversed copies of the originals, and are marked 'Reverse drawing'. Usually the lettering was not inserted until the copy had been taken, but there are cases in which this was not so (see Plate LXXXVIII). The earliest use of tracing paper that has been noted is in 1808, but it is quite possible that it may have been used earlier.

Another feature of the Soho drawings is that colour is used, not so much to indicate the materials of construction, as to distinguish one set of mechanism from another. For example in the drawing of a working-gear for a double-acting engine, the parts peculiar to the upper steam valve may be coloured red, those to the lower steam valve blue, to the upper exhaust valve yellow, and to the lower exhaust valve green.

At the end of the eighteenth century getting in office supplies was a matter requiring forethought; for example, in 1792 we find Southern suggesting to Watt, then in London, 'perhaps you may have an opportunity of buying some good "Indian Ink"'.

The development of the Soho Engine Manufactory during the fifteen years 1786 to 1801 is illustrated by a paper in the Muirhead Collection which summarizes the amount of the inventory of the property, the number of the workmen, the wages paid, and the number of engines put up in these years. On Oct. 1, 1786, the inventory showed the value of the property to be £2,319 ; on Oct. 1, 1800, it amounted to £9,010. The wages paid to the workmen for the first year of this period amounted to £1,688, and for the last year £3,407. In 1786–7 from fifteen to twenty men were employed (not including the men engaged outside as erectors) ; in 1795–6 the number was from fifty to fifty-five.

For the period 1798–1803 some interesting particulars of the costs and methods of production at the Engine Manufactory are summarized later on with like information in respect of Soho Foundry. The two establishments were conducted as separate concerns, and the Engine Manufactory continued in operation until about 1850, when the Soho Manufactory was dismantled and the engineering establishment was concentrated at Soho Foundry. It is likely that after the death of Matthew Robinson Boulton in 1842, his concerns, other than the Engine Company, had languished, and that the main buildings were idle, or had been sublet. Moreover, the original lease for the land granted to Edward Ruston in 1756 was drawing to its end.

SOHO FOUNDRY

NEGOTIATIONS for the purchase of land for the new works were begun about the middle of 1795, and within a very short time building operations had commenced. James Watt, junior, writing on Sept. 14, 1795, stated that the new buildings ' are going on at full gallop & hope to have the brickwork completed this week ; should that be the case we shall make a dash at the foundery this season'.[1] The following day he wrote again to Southern who was at Newcastle-upon-Tyne : ' I told you yesterday that we should make a dash at the foundry this season, 60 feet square is the size we propose. Please to give us all your ideas upon the subject respecting the position of the furnaces, stoves, cyl'. pit, & the abstract of your observations upon the foundries in the north. You must let us have them soon, as we commence our operations immediately, not to lose this astonishingly fine season.' [2]

A month later we learn that two buildings had been roofed in, two others were being roofed, and the foundry building was so far advanced that it was expected the roof would be on in a month's time.[3] However, it was not until December that the roofing of the foundry was taken in hand, and it seems that the building measured about 100 ft. by 70 ft.

There appears to be no plan extant of the foundry buildings as originally erected, but the Wages Book from the commencement of the works, Aug. 5, 1795, to Mar. 31, 1797, has references to Forging shop, Smith's shop, Boring mill, Turning shop, Fitting shop, Carpenter's shop, Drying kiln, Foundry and air furnace.

Early in the year 1796 there was a celebration at the new works, at which Boulton made a great speech. The affair seems to have been highly successful, and Southern's remarks are : ' You will have been acquainted with the great feast of the christening of Soho Foundry I presume, before this gets to you. Upwards of 200 partook of the bounty of the founders and the day was conducted without

[1] B. & W. Colln. J. Watt, junior, to Southern, 1795 Sept. 14.
[2] Ibid. J. Watt, junior, to Southern, 1795 Sept. 15.
[3] Ibid. Peter Ewart to J. Lawson, 1795 Oct. 22.

any misfortune, except a little, or rather not a little, thieving, and it is not a little remarkable that nothing came amiss to them, knives, forks, peper boxes, mustard pots, drinking cups, juggs, table cloths, knife cloths, spoons, cocks, &c. &c.'[1]

However, it took some time to get the place into going order, and there was a difficulty in getting men. Abraham Storey, one of Wilkinson's men from Bersham, had been engaged as foreman or manager of the foundry, and the firm advertised in the newspapers for moulders and furnacemen. Towards the end of May they had secured a moulder, and begun casting, as appears from a letter from Southern to J. Watt, junior, May 27, 1796 : 'We have got a moulder & are begun to cast grate barrs, small plumr. blocks, and such like. I pressed Storer to begin, seeing how difficult it is to get anything from the other foundries ; wooden boxes are made to serve for the present for such cases as are not too heavy.' The man was probably 'Gilpin a moulder from Scotland' ; there must have been a furnaceman also. William Matthews, furnaceman at Wringeworth Ironworks near Chesterfield, had applied for this job, but he wanted his two sons to be taken on also, one was a greensand moulder and the other ' a small furnace blower'. Matthews stated, as to wages, that generally he had been paid by the ton, but he would be satisfied with a guinea a week. Having once started, things began to move along and about a month later Boulton wrote : 'A new nossell in one piece has been cast at the foundry & is a very good one, but we want more assistance, however, we shall soon have more, as I understand that Ken[k] will soon be here, as he has nearly finished our cylinders.'

The man referred to by Boulton seems to have been another of John Wilkinson's staff—John Kendrick. He had applied to Boulton and Watt for a situation in April, and he was willing to come to them for 21 shillings a week, house and fire, and two guineas for his expenses in moving from Bersham to Birmingham, and he adds : ' Hope if you think I merit it at the end of the year you will not object to give me 10 guineas.' Kendrick said that he had been a moulder in dried sand all his life, and for some years all the cylinders, bottoms, &c., cast at Bersham for Boulton and Watt had been made by him, or under his care. Needless to say this was exactly the class of man that Boulton and Watt were in want of. There are indications that other workmen from Bersham had been taken on at Soho, and at length it struck Watt that they were going too far, and he wrote from Bath, July 19, 1796, to Matthew Robinson Boulton : ' In respect of J. Wilkinson's men, I think you ought by no means to take any more of them, your doing so would be cruel and blameable, the men should therefore be positively refused unequivocally.' This may have put a stop to the recruiting of Wilkinson's men for the time, but a twelvemonth later Southern writing from Leeds, ' I have had no success in attempting to get any sort of hands that are wanted at Soho, as there is a want of such everywhere ', suggests that they should take on some more of the Bersham men. However, by this time Wilkinson was himself discharging his men ; at the end of June it was stated he was turning off most of his founders at Bersham and giving them written discharges,[2] and a little later M. R. Boulton writing to

[1] Ibid. Southern to J. Watt, junior, 1796 Feb. 2.

[2] John Wilkinson owned works at Bradley, at Broseley, and at Bersham near Wrexham, and he was at this time putting up another works at Brymbo, also near Wrexham. In the Bersham works his brother William had a small share.

In the year 1795 some dispute arose between the brothers, the exact nature of which has not been ascertained, and the matter was referred to arbitration. From September 1795 to April 1796 the Bersham concern was carried on by a trustee, William Fawcett of Liverpool ; a number of cylinders, air-pumps, and condensers

Southern (July 6, 1798) says: 'We have got an additional hand at the foundry, a moulder from Bersham, where, he reports, there is a great scarcity of orders and many of the workmen are to be discharged. We beg you, however, not to relax in your endeavours to procure any good workmen either as fitters or moulders.'

Southern, however, met with no success in his efforts to engage men at Leeds. Iron-founding was a rapidly increasing trade all over the country and moulders and founders were in great demand. A correspondent at Newcastle-upon-Tyne writes to the firm : ' Moulders are very scarce here, and indeed are only picked up from other countries.' John Kendrick of Bersham, in his application to Boulton and Watt, stated that he had been offered higher wages in Lancashire, Yorkshire, and Scotland than he was willing to come to them for. William Jessop of the Butterley Ironworks, knowing that Boulton and Watt had advertised for men, and thinking they might have had more applications than they could entertain, requested them to let him know of any moulders they did not want, and in a year or two John Wilkinson was in want of men; we read that he ' is like to snap up all the founders, being engaging all he can at exorbitant wages '.[1]

Soho Foundry had, from the commencement, more work than it could turn out. In the year beginning Oct. 1, 1797, of twenty-five engines supplied by Boulton and Watt ten were made at the Foundry ; in the following year fifteen were made, and in the year after eighteen. In addition to having plenty of work the venture was satisfactory in another respect—it paid very well. The funds for erecting the buildings and machinery were advanced by Boulton and Watt to the Foundry Company, which, as explained elsewhere, was a separate firm, and in October 1797 they amounted to £21,624 16s. 6½d. This debt, with interest at 5 per cent., amounted to £27,431 9s. 4d. at the end of September 1804, from which date it was gradually repaid from the profits of the concern until in August 1812 it was cleared entirely.[2] Not only was the original debt cleared, but considerable extensions were made as well, for at the end of September 1816 the capital expenditure amounted to over £48,000 ; of this £30,000 had been laid out on buildings and the balance on machinery. As one would expect, the bulk of the expenditure for machinery was in the fitting department, whereas the greatest item in the building account is for the foundry ; at this date a boiler department had been added to the works.[3]

To go back to earlier days, by about the middle of 1797 a cupola had been set up to assist the air furnaces, and the younger James Watt was greatly pleased with its performance. ' The cupola', he says, ' is doing its duty and melts down nearly a ton per day of 12 hours.' [4] In 1800 he writes again to the younger Boulton : ' An extension of our foundry, or an increase of our hands in that dep^t becomes

were on order for Boulton and Watt, and Fawcett seems to have done all in his power to get the orders executed. There was a sale by auction of the stock, &c., in September 1795, at which James Watt, junior, attended, and bought for his firm a number of cylinders that Wilkinson had cast for stock. In December of the same year the works was offered for sale, but found no bidders. John Wilkinson then seems to have bought out his brother and to have carried on the works alone for some time, but in a half-hearted manner. He died in 1808 at the age of 80.

[1] B. & W. Colln. Watt to James Watt, junior, 1800 Feb. 2.
[2] Doldowlod Papers : ' Short memorandums of different firms under which Mr. Boulton and Mr. Watt and their sons have carried on business.' [By James Watt, junior.]
[3] Muirhead Papers : Ledger ' Soho Foundry, Buildings and machinery '.
[4] B. & W. Colln. J. Watt, junior, to M. R. Boulton, 1797 June 10.

daily more urgent from the rapid accumulation of orders.   Perhaps a new building altogether for the green sand would be the most eligible plan.' [1]   A week later he informs the same correspondent that Abraham Storey ' has got on well with cylinders and air pumps and generally with engine goods, but Murdock complains he cannot get castings for his machinery '.

By this time they had succeeded in casting a cylinder as large as 66 in. diam. ; previously they had done them up to about 30 in. only.

There is an interesting paper endorsed ' Prices of castings as charged by Soho Foundry up to Dec[r]. 1st, 1801 ', giving a list of the castings made and the prices per cwt. under two headings : ' Foundry ' and ' Fitted '.   Cylinders, air-pumps, pistons, cylinder covers, &c., are 20s. from the foundry and 32s. fitted.   Upper and under nozzles 20s. and 21s. respectively.   Perpendicular steam-pipes, condensers, &c., 20s. and 20s.   Rotative wheels 18s. and 20s.   Steam cases, gudgeons, &c., 16s. and 20s.   Connecting-rods, cranks, shafts, plummer blocks, flywheels, &c., 16s. and 18s.   Governor balls, 15s. and 18s.   Grate bars, 14s.

A good deal of the smaller work was being done outside.   Of the same year as the list just referred to are other lists headed ' Articles proposed to be cast at Izons ' and ' Articles proposed to be cast at Dearman's '.   Izons and Dearman's were local iron-founders ; Dearman's was the well-known ' Eagle Foundry '.   These lists cover a great variety of engine parts such as condensers and air-pumps for small engines, water cisterns, covers for sliding-valve cases, fire-doors and frames, dampers and frames.

Events proved the wisdom of the development of the manufacturing side of the business ; orders poured into Soho and for some years we read of the difficulties of keeping pace with the work, the delays of outside founders, and the necessity for further extensions of the works.   The standard of workmanship appears, however, to have been comparatively low, in particular in regard to smith's work and greensand castings.   The Lowmoor Ironworks was turning out far better greensand work, and Fenton, Murray, and Wood were excelling Soho both in forgings and greensand castings.   Murdock and Abraham Storey, the foundry manager, were accordingly sent on a voyage of discovery in January 1799 to Leeds and Lowmoor.   They

' returned from their excursion highly delighted and full of panegyricks upon Murray's excellent work. Abraham is now entirely convinced of his inferiority, &, what is more, of the possibility of amendment, & he is now *actually* making trials of different substances to mix with the sand with the view of getting a better skin to the castings. . . . It seems his [Murray's] forge work is still better than his castings, and we are trying therefore to infuse a spirit of amendment into this department of the work. They were admitted into every part of Murray's manufactory, & spent two even[gs]. with him, & by virtue of a plentiful doze of ale succeeded in extracting from him the arcana & mysteries of his superior performances.' [2]

In the course of a few weeks, although there was no improvement in the foundry work, the smiths were turning out far better work.   Murray had presented Murdock with a specimen of his forge work ' no doubt with a view of exciting his astonishment and perhaps a despair of ever attaining the same perfection, for I must candidly acknowledge it was the most beautiful & perfect piece of work I ever beheld.   Fortunately however it has been the means of producing a very different spirit.   The emulation of our men has now been awakened, & they now vie with each other

[1] Ibid.   J. Watt, junior, to M. R. Boulton,   [2] Ibid.   M. R. Boulton to J. Watt, junior, 1800 Oct. 2.   1799 Jan. 17.

in adopting the tools & improvements suggested by Murdock. Our first imitative essays have succeeded completely, and we entertain great hopes of turning out the working gear in such a manner as will render filing a superfluous operation.' [1] It seems that Murdock had picked up hints for improvement in other directions, for we learn from the same letter that ' some other tools partly in imitation of Murray, & partly of the suggestion of Murdock have been introduced with the greatest success into the fitting department '.

Improvement in the greensand castings did not take effect, and in June 1802 they got an account of what was being done at Leeds from a former workman of theirs who had been induced to take service under Murray. This man was employed as a drysand moulder and he did not think very highly of the work done at Leeds in that line, but he was fully sensible of the excellence of the greensand work. Moreover, all the nozzles were now made in dry sand, as the only moulder capable of making them in greensand was dead. He said that the sand used was exactly the same as that in use at Soho, but the Soho people spoiled it by mixing with it too much coal dust. The bottom of Murray's greensand foundry, a distinct building from the drysand foundry, consisted of about four feet deep of this sand which was worked over and over again; it had not been renewed since he had been there, a period of thirteen weeks. When the mould was made, a little coal dust was strewed over it, which was all that was used, and he supposed this to be consumed by the iron, so that little or none remained in the sand which, although changed in colour by frequent use, was not substantially changed in its nature. Connecting-rods, shafts, and wheels of every description Murray made solely in greensand, but no pipes or hollow goods. Cylinder bottoms, lids, and pistons are also made in greensand, except the very large ones. A cupola about 8 ft. high and 20 in. diam. was used for a great deal of the work; it melted about 6 cwt. per hour, and the iron ran very much finer from it than from that at Soho, although they did not use one-third of the quantity of coke; the coke was supposed to come from Lowmoor.[2] This man, whose name was Halligan, was engaged to come back to Soho and young Watt obtained from him a great deal of information about Murray's works. It appears that his drysand foundry was about 20 yds. long by 12 wide, with two air furnaces and three stoves: one, 20 ft. by 13 ft., for loam work; another, 17 ft. by 13 ft., for boxes; and a third, 17 ft. by 9 ft., for cores. The greensand foundry was separated from the drysand foundry by a yard; it was of nearly the same dimensions and provided with two air furnaces in addition to the cupola. The cylinders above 20 horse-power were cast in loam; in this department there were four men and a boy, the same number of hands were employed on drysand work, and in the greensand foundry there were four moulders and six lads. There were two air-furnace men, one cupola tender, and five chippers for the two foundries, but the loam moulders chipped their own castings. Halligan considered that in both foundries they cast upon an average three to four tons a day, which appears to have been much the same as the output at Soho. The air-furnace men were paid a guinea a week and had houses rent free. The chippers had eighteen shillings a week. Murray employed eight smiths, but there is no information as to the number of fitters; a curious point that comes out, though, is that the pattern-makers were sent out to erect engines. In all he seems to have employed about 160 men and apprentices. Upon a review of all the information imparted to him,

[1] B. & W. Colln. M. R. Boulton to J. Watt, junior, 1799 Feb. 1.

[2] Ibid. J. Watt, junior, to M. R. Boulton, 1802 June 12.

James Watt, junior, remarked : ' It is high time we should reform.' He is struck by the plan adopted for paying the men, and thinks it worthy of imitation—' the men are paid upon the Saturday about 11 o'clock by the clerks who go into each shop, by which means their wives are enabled to go to market early. This regulation is much approved of by the men.' It would seem that the wives must have come round to the works and relieved their husbands of the money as soon as received.[1]

At about this time he saw the process of moulding in greensand at the engine works of Emmet at Birkinshaw near Bradford, and remarks ' one of our great defects is ramming the boxes too hard which prevents the escape of air & steam. They scarcely ram theirs at all and make few or no prick holes in them.' In the same letter he says he has seen the place the greensand comes from and sends instructions for ordering a barge load—forty or fifty tons—to be sent to Soho.[2]

About this period—say for the years 1798–1804—the Boulton and Watt Collection affords us a number of papers that illustrate the care and thought that was being bestowed upon the management, and are of distinct technical interest. Thus we have a series of papers dealing with the cost of production and fitting of the various parts of the engines, as ' Calculation of the expense of fitting up pistons, 1798.' This gives the actual costs of turning, drilling, and fitting the pistons of seven particular engines ranging in size from 12 in. to $30\frac{1}{4}$ in. diam. A similar paper is endorsed : ' Average cost of centring & turning Rot. shafts, main and out end gudgeons, pistons & covers, cylinder lids, air pump buckets & lids, rotative wheels & crank pins, July 1801, at Soho.' Then we have ' Particulars of work done by the piece at Soho Foundry 1800 '. The rates paid are based on widely different factors, and some of them seem of the rough-and-ready kind, but as any particular man was kept at the same class of work no doubt the results were satisfactory when taken over a good range of jobs. The piston-rods and air-pump bucket rods were turned and filed at $1\frac{1}{2}d.$ per pound net weight, this was simple ; the parallel motions were fitted (exclusive of rimering) at 3s. per inch of cylinder diameter and an allowance of one day for each set for repairing tools ; the price for fitting pistons with covers and bottom plates was 4s. 6d. for each pin screwing down the cover ; the cylinder glands were bored and turned, bushed, and the bushes bored and turned at 5s. per set for all sizes ; in the case of the air-pumps ' fitting the pump and bucket with valves, rod to bucket, pins to gland, and steady pins to top flanch ' was done at a charge of 3s. 3d. per inch as far as a 20 h.p. engine and 3s. for all above that size.

Then there is a long account covering both piece-work and day-work at Soho. It is endorsed : ' Account of the work done by the piece & day in ye Engine yard Soho June 1800. Calculations relative to ye fitting in general.' All the turning and drilling, the chipping and some of the fitting was done by day-work, but the valves were fitted to the nozzles, and to the air-pump and bucket, and the working-gear, parallel motion, governor, and some other parts were fitted by piece-work. The price for fitting governors was 18d. each. The cylinders and air-pumps were bored and faced at the foundry. As in the Foundry list, the names of the men are given for each job. For the following year there is another list entitled ' Arrangement of Workmen, and Distribution of work, September 14th, 1801.' This relates

[1] Ibid. J. Watt, junior, to M. R. Boulton, 1802 June 15.   [2] Ibid. J. Watt, junior, to M. R. Boulton, 1802 June 17.

to the Foundry, and covers all the steps in the production of an engine; apart from the foundry men and smiths, it includes six men for 'weighing, packing, loading and unloading goods and labouring work' and the total amounts to forty-three men and seven boys, but of these they were wanting eight men and four boys.

The subject is dealt with from a somewhat different point of view in a voluminous document headed 'Distribution of work at Soho Engine Manufactory, Dec. 3rd. 1800' and endorsed on the back: 'Specification of ye fitting of engine materials, and the shops where it is to be done, Dec. 3. 1800.' This is a complete list of all the parts of the engines, the operations performed upon them, the machines used, and the particular shops in which the work is to be done; thus we find 'Piston rod, centring, turning, filing, fitting rod to piston, viz., drilling and cutting and filing hole—Piston rod lathe (No. 4) heavy fitting shop'.

Another scheme for the 'Arrangement of Soho Engine Manufactory' is dated December 1801. This enumerates the following machine tools:

'No. 1. Large drill; an upright one with convenience for putting in and out of gear & sliding socket. Present speed about 8 p. m. Proposed speed 8. No. 2 small drill; an upright one with same convenience as the preceding; present speed 50 p. m. Proposed speed 75. No. 3 Large turning lathe with both centres & chock. Present speed 3½ p. m. proposed speed with 2 p. min. to 18 p. min. No. 4. piston rod lathe, with spindle & dead centres. Present speeds 18, 38, 60. Proposed speeds, 18, 30, 50, 80. No. 5 Nozzle lathe, spindle & dead centre. Present speed 123 p. m. Proposed speeds 35, 80, 130. Nos. 6 & 7 Parallel motion lathe, spindle and dead centre. Present speed about 65. Proposed do. 18, 50, 70. Nos. 8 & 9 Lathe in upper fitting shop; spindle lathe & dead centre in same cheeks. Present speed 88. Proposed do. 18, 50, 80, 120. No. 10 small lathe. Proposed speed 200, 300. No. 11 Lapping machine. No. 12. Pattern-makers lathe. Speed from 220 to 300 p. m. No. 13 steam case drill.'

The first part of this paper defines the work to be done by each machine, for instance, the lapping-machine is 'for lapping all brasses of plummer blocks &c.' Then we have a table setting out the separate shops with the machines and conveniences contained in each; the operations performed in the respective shops, and remarks on the relative positions desirable for these shops and the means of communication between them. The shops enumerated are: drilling shop, heavy-turning shop, heavy-fitting shop, nozzle shop, fitting shed, parallel-motion and working-gear shop, light-fitting shop, pattern shop, casters' shop, and smiths' shop.

It would seem that most of the machines mentioned were in the works, but that additions and alterations were required. There is a paper endorsed 'Memorandum of sundry articles wanted for the completion of the new & alteration of the old machinery at Soho, 1801' in which the first three entries are the following:

'1st Drilling frame, viz., the wooden frame, horizontal shaft, screw & wheel for large drill; bevil wheels for motion of small drill.
'2. Nozzle lathe. Cheeks, poppets, spindle & chock & dead centres. Iron pullies both for lathe & variation of speed.
'3. Grt. turning lathe, new poppet head with spindle & chock, same as large lathe at foundry. Endless screw for do & wheel. Wheels for quickening speed from engine shaft.'

The next entries relate to the engines for driving the machines; there was to be a six horse-power engine, bell-crank type with slide valves; a four horse-power engine to drive the drills and nozzle lathe, and a three horse-power engine to drive the great lathe. The question of the best system of supplying power had been discussed a year before in a paper headed 'Estimates of the different proposed methods of communicating motion to the machinery of Soho Engine Manufactory, Dec. 15, 1800.[1]

Progress in the extensions is recorded by James Watt, junior, in 1803.[2]

' We are also going on with new buildings & machinery both here [at Soho] and at the foundry. . . . Our own engine house here is undergoing a complete metamorphosis, and we are adding a new engine & shop to the foundry establishment ; the latter 134 feet long is now roofing in. It is intended for the fitting of the little engines, and I hope will enable us to get on with at least one per week, which will scarcely enable us to meet the present demand. Some other alterations of minor importance are going on there which I hope will be finished in the course of the summer ; and if it should afterwards be found expedient to add a shed for making boilers and a counting house, that establishment will have reached, what I should vote to be, its ne plus ultra.'

BORING-MILL

It has been stated on a preceding page that Peter Ewart was brought to Soho for the purpose of designing and laying out new machines and buildings. He and Southern inspected boring-mills at Coalbrookdale and elsewhere, and then about the middle of 1795 designs were prepared for both vertical and horizontal machines. These designs were considered by Watt, who showed a preference for the horizontal form. However, it was not until the close of 1796 that the first boring-mill at the foundry was actually set to work. On December 1st of that year Boulton wrote to James Watt, junior, ' I believe they have this morning began to bore a cylinder at ye foundry', and next day he informs him that ' Boreing goes on well' but the machine did not work satisfactorily ; the bar was not stiff enough, or was not properly supported ; and in 1797 William Wilkinson, who was making a call at Soho, made a number of suggestions for its improvement. The machine continued to work during 1797 and 1798. In the meantime Murdock had designed a new machine, and on Dec. 30, 1798, G. Lee of Manchester writes to Lawson that he has been to Soho and that ' the boring mill was not completed though Mr. M[urdock] was indefatigable'. However, some time was to elapse before the new machine got to work. The exact cause of the delay we do not know, possibly, for one thing, the first boring-bar was unsatisfactory; at any rate although the foundry was in full working order, they had to get a bar made at the Lowmoor Ironworks. This arrived at Soho in May 1799 ; it weighed over 3½ tons, and was charged at £21 per ton. About this time, or soon after, Murdock was laid up with illness, and very little progress was made, but on September 17th Gregory Watt writes to his brother : ' Murdock is much recovered and his revived vigour has infused new spirit into the men at the boring mill which we expect will be completed this week', and early in October boring with the large rod was commenced. The machine had ' a large rod' and ' a small rod' ; the latter had been in operation since July, before that a certain amount of work had been ' bored by temporary screw'. The first job dealt with on the large boring-bar was a 64-inch cylinder ; it took 27½ days. The particulars of the time are worth recording : Getting on ¾ day, centring and

---

[1] Three methods were considered. (a) By one engine of ten horse-power. (b) By three engines as set out above. (c) A greater number of small engines. It is clear, however, that the scheme contemplated new buildings, and a Memorandum of Mar. 10, 1802, mentions the ' intended site of the fitting sheds ', ' new drilling shop ', ' heavy fitting shops ', and ' new casting shop '.

[2] B. & W. Colln. James Watt, junior, to Gregory Watt, 1803 June 8.

fixing 1½, facing ½, setting cutter ¼, boring 11½, preparing to go through a second time 1, boring 8½, facing 1¼, bellmouthing 1½, getting off ¼.[1]

The essential feature of Murdock's design was the use of worm gear to turn the boring-bar, and this is covered by his patent of 1799. In his specification he refers to the worm as an endless screw, and says that the screw ' I have commonly used has been three threads, is 16 inches diameter, and the advance by one rotation is 6 inches ', he adds that the same gear may be used to turn pistons, cylinder covers, &c., and in fact it was applied first of all to a lathe. At the beginning of 1799 we learn that ' The new method of turning the larger lathes by means of the endless screw as shown on the drawings of the boring-mill has been tried upon our air pump lathe & found to answer every expectation '.[2] The gear had been tried a couple of months before this. ' This morning Murdock's screw commences its active operations in this life. M. R. Boulton, Mr. Southern, and the other man midwifes are attending the delivery ', writes Gregory Watt to his brother on Dec. 11, 1798. Possibly some defect had shown itself on this occasion which took some time to rectify. The letter of 1799 referred to above states also : ' The difficulty of making the endless screws is nearly overcome, & a method is now under trial which if it succeeds will enable us to make them very expeditiously.' There is no indication of the nature of this new method of making the worms.

A letter of later date, although it does not directly concern the Soho boring-mill, is worthy of notice. In 1802 James Watt, junior, writes in reference to the boring-mill used by Murray at Leeds : ' His cutter block is pushed forward upon the boring bar by an endless screw, which, or some similar contrivance, we mean to adopt, both to guard against the negligence of the borer & to save part of his wages.' [3]

The history of Soho Foundry, its staff, and its productions demands a book devoted to that subject solely ; let us hope that in the course of time such a work may be forthcoming. In the meantime, although it takes us beyond our period, it is not inappropriate here to deal very briefly with Soho affairs after James Watt has ceased to take an interest in them. Boulton died in 1809, Southern in 1815, Watt in 1819, Matthew Robinson Boulton in 1842, and James Watt, junior, in 1848. As we have seen, the firm were carrying on two separate establishments at Soho and the Foundry ; this arrangement seems to have continued until after the death of James Watt, junior, up to about 1850, when the Soho Manufactory was dismantled.

Shortly before the death of James Watt, junior, Gilbert Hamilton, junior, and Henry Wollaston Blake became partners in the concern and from 1850 Blake is said to have been the sole proprietor ; he was at any rate the principal proprietor. Blake was a well-known man in business and other circles in London.[4] He carried on the Soho Foundry until 1895, but it seems to have been a losing business for some years before this, at least when the stoppage occurred it was not for lack of orders, although the number of hands employed had dropped from 1,000, which it is said to have been a few years previously, to between 300 and 400. By the beginning of 1895 it seemed clear that the works could not longer be carried on without great loss, and in March the break-up began with a sale of surplus stock and stores.

[1] B. & W. Colln. ' Abstract account of cylinders bored from Dec. 31, 1796, to Jan. 11, 1800.'
[2] Ibid. M. R. Boulton to James Watt, junior, 1799 Feb. 1.
[3] Ibid. James Watt, junior, to M. R. Boulton, 1802 June 15.
[4] There is a Memoir of H. W. Blake in *Proc. Inst. Mech. Eng.*, 1898.

In May the freehold works with the plant and machinery were offered for sale by auction. From the particulars in the sale catalogue we learn that the area of the works was nearly twenty acres, and that the buildings of the engine works, mint, and gas works occupied five acres. Among the buildings are mentioned pattern shops and pattern stores, brass and iron foundries, smiths' shops, steam-hammer shop, coppersmiths' shop, boiler shop, and machine and fitting shops. In the matter of machines the catalogue enumerates sixty lathes of various descriptions, nine planing-machines, five radial drilling-machines, eighteen vertical drilling-machines, four boring-mills, three slotting-machines, four boiler-shell drilling-machines, three steam-hammers, and a hydraulic riveter.

This section must not be closed without a reference to an interesting paper in the Boulton and Watt Collection. It is a list of Christmas presents given in the year 1799. At Soho seven apprentices get sums varying from five shillings to ten and sixpence, and two fitters get half a guinea each to be laid out by the manager 'in whatever articles of cloathing, necessaries, or books they may wish to have'. Four of the men get bacon to the value of half a guinea, and six others get advances in their wages ; the lowest figures mentioned are from ten shillings to eleven, and the highest fourteen shillings to fifteen. The postman is to have two pairs of shoes. At the Foundry three apprentice moulders get five shillings each ; two apprentice fitters five shillings, and one seven and sixpence. Two journeymen moulders get a guinea each, but most of the fitters fare better ; two of them are down for five guineas, two for two guineas, one for one guinea, and one for half a guinea. Abraham Storey, the manager, receives ten guineas, as also does William Harrison. The latter was the son of Joseph Harrison, the workman who figures so prominently in the early records of Soho : he had been one of the erectors and was now in charge of one of the shops at the Foundry.

XXI

THE STAFF OF BOULTON AND WATT

T HE Boulton and Watt correspondence, particularly in the early period, contains many references to the workmen employed; some of these references are in connexion with the erection of particular engines, but many relate to difficulties in the Soho workshop arising from the drunkenness and irregularities of the men. As time went on, there is less evidence of this; no doubt the organization was improved, a steadier lot of men were got together, and it became possible to weed out the worst of them; indeed, later on, Soho men came to have a high reputation and the establishment may truly be said to have been a nursery for millwrights and mechanical engineers.

The 'father' of the Soho workmen was undoubtedly Joseph Harrison. Nothing is known of his antecedents, but it is reasonable to suppose that Boulton picked out a man who had some experience in the erection of the Newcomen engine. At any rate it was Harrison who re-erected the Kinneil engine at Soho, and in 1775, when the Bill for the extension of Watt's patent was before Parliament, he was one of the witnesses examined by the Committee of the House of Commons; he then described himself as 'a smith at the Soho Manufactory at Birmingham'. With the increase in the engine business and the consequent enlargement of the staff, Harrison became the leading man in the Soho engine-yard, but he was frequently sent out erecting, and many of the earlier engines—Bow, Chelsea, Richmond, &c.—were put up or completed by him. One gathers that he was a genial, modest sort of man, not given to making difficulties about his work. In 1779 Hallamannin, one of the first Boulton and Watt engines put up in Cornwall, was doing badly, the result mainly of faulty erection, and the firm decided upon sending Harrison to put it right. 'This', writes Watt, 'we have resolved upon as the least hurtful to Dudley, as Joseph is not given to stir up strife or affront people, and is able I expect to redress all grievances.'[1] Dudley, it should be said, was the engineer who had put up the engine. He was not one of Boulton and Watt's men.

Harrison's failing was the common one. From time to time he gave way to drinking bouts, moreover he was not a man of any education, so in 1778 when the firm decided to appoint a foreman at Soho, a man named Hall was brought in. Hall's tenure of the position was but a short one, and soon after he left Harrison seems to have been put in charge, and we find him referred to as foreman; possibly his duties were of a more restricted nature than the firm had had in view when engaging Hall. Harrison continued to be sent out erecting for a few years after this; he was erecting Shadwell and repairing Bloomfield in 1781; then, the work at Soho increasing and a number of other men having been trained for erecting, he seems to have remained at head-quarters for the rest of his life. His son, William Harrison, was in due course taken on at the works, was an erector for some years, and then 'was made foreman of the small engine manufactory in the 6-horse

---

[1] B. & W. Colln.: Letter Books. Watt to Wilson, 1779 Apl. 8.

shop at Soho Foundry, 1st Jan. 1800, and continued in that station until his death, 24th Aug. 1815'.

Other names we meet with in 1776-7 are Perrins, Peplo, Cartwright, Law, Isaac Perrins, Careless, Webb, Johnson, Dickson, and Horton. The last three were smiths. Perrins, however, was not a Boulton and Watt man, he was an independent erector; his son Isaac (of whom more hereafter) must at this time have been quite a young man acting as assistant to his father. Most of the references to Peplo are in regard to his drinking habits. Richard Cartwright was another broken reed; he it was who disclosed Watt's crank invention to Pickard's engine-man. He was with the firm as early as 1777 when he was assisting in the erection of the Bedworth engine. Another engine that he helped to put up was that at Chelsea Waterworks, and he seems to have wished to stay at Chelsea as engine-man, for we find Watt writing to Boulton: ' Remember that Cartwright is not to be trusted and wants to stay at Chelsea; if they will keep him you should let him go to them, he does not deserve his wages.' [1] Cartwright himself thought he was underpaid, for soon after this he was applying for a rise, and to Watt himself.

' Dick Cartwright insists upon having his wages increased to 18/- per week, but is such an idle scoundrel that I think we ought to turn him off, & with disgrace. I cannot bear any longer with him; he still demands pay[t]. for the day spent in detecting his theft. In fact the keeping him is an encouragement to others to treat us in the like manner.' [2]

Boulton, however, was very much against dismissing any of the men if it could in any way be avoided, so in spite of his shortcomings Cartwright remained with the firm until after 1790. His end was a sad one. In March 1797 he was applying to a firm at Bristol for a job as engine-man; this he failed to get, and we next hear of him in the following July in Salisbury gaol awaiting his trial at the next assizes, and asking Dayus to request Boulton and Watt to intercede for him. At the assizes Cartwright and another man, James Murdogh, were found guilty of stealing in a dwelling-house goods and moneys above the value of forty shillings, and they were sentenced to be hanged. The sentence, however, was commuted for one of transportation for seven years.

Careless was another man given to drink, but apart from this he seems to have been a good and diligent workman. James Law, it is clear, was the steadiest man of this early group; he accompanied Murdock when he went to Cornwall in 1779. Law remained in Cornwall about three years; he it was who on his return journey was lost sight of for some time, much to the alarm of the firm, as has been related in a previous chapter. Law proved himself a good man, and was employed in London and various parts of the country. In 1792 he was erecting an engine at Nottingham, and received a very flattering testimonial from his employer there. By this time he had his son as an assistant, and the son likewise is spoken of in laudatory terms.

Following this first group we have William Murdock (whose name has become associated for all time with those of Boulton and Watt, and to whom a special section is devoted further on), Vickers, Perkins, Williams, and Jabez Hornblower. Hornblower, the son of Jonathan Hornblower the elder, came from Cornwall to Soho in the same year that Murdock went from Soho to Cornwall; the firm had offered him work at a guinea a week, the same pay as Murdock was receiving, and he was

---

[1] Ibid. Watt to Boulton, 1779 June 26. [2] Doldowlod Papers. Watt to Boulton, 1781 Mar. 29.

for a few years one of their erectors. He worked on the Ketley, Donnington Wood, and Penryndee engines, but he seems to have had a curious twist in his character, and did not get on at all well with other workmen or with his employers. Later on he became a rival engine-maker and a pirate, and more will be found about him in Chapter XXII.

In 1781 a number of men were sent to Cornwall, James and George Taylor, Edward Bull, Robert Cameron, Malcolm Logan, and David Watson. Both Bull and Cameron set up in business later on as engine-makers, and we shall speak of them in another place. Malcolm Logan deserves a few words. In 1782 Boulton refers to him as ' a handy, active and industrious fellow ' ; he seems soon after this to have given way to drink, but to have pulled himself up in time, for in 1788 when it was a question of selecting a man to go to Naples, ' I think', writes Boulton, ' he is the properest man we have to send to Naples, as he is a handy fellow either in wood or iron, or engine, or mill, or pumps.' [1] He must in addition have been a man of quick intelligence. Later on, in 1795–6, he was at work in Spain.

The Scottish element in the staff was quite pronounced. Pearson and Buchanan, the clerks, were Scots, as also was William Playfair, and at the end of 1782 when Boulton gave a dinner to the erectors in Cornwall he pointed out that six out of the seven were Scotsmen—Murdock, Lawson, Pierson, Perkins, Logan, and Muir. To Lawson, who rose to a high position in the firm, we devote a separate section. Robert Muir remained with the firm until 1787 when he set up in business for himself as a blacksmith, coppersmith, and plumber.

The Scotsmen, at any rate those who came south, were steadier and less given to drunkenness than the average English workman of that day ; moreover, they were content with lower wages.

An erector who is entitled to more than a passing reference is Isaac Perrins. It will be remembered that one of the first engines put up by Boulton and Watt was that at Bloomfield Colliery. The proprietors had arranged to put up a Newcomen engine and an engineer named Perrins had undertaken the erection. When they were induced to put up a Boulton and Watt engine instead, Perrins retained his position and erected it ; afterwards he was concerned in the erection of Bedworth, Bog Mine, and Minsterley engines. Perrins seems to have been advanced in years ; he died in 1780 and is then referred to as ' old Perrins ', although the adjective may have been used merely to distinguish him from his sons. One of the sons, Isaac, was offered a job under the firm in Cornwall in 1780, and this he declined, but early in 1782 he applied for work and was taken on at twelve shillings a week. He was connected with Boulton and Watt for some years, and of all the men under the firm Isaac Perrins was the best known to fame, nay, his name must have been a more familiar one to the public than that of James Watt himself, for he was a prize-fighter. One morning in 1788 Watt, in Boulton's absence, dealing with the correspondence of the firm, opened a letter from a gentleman in London enclosing a newspaper cutting, and saying ' In the enclosed paper of mine is an offer from a person living with you, to fight any man in London ; would you please to inform me by a letter of what age Perrings is, what weight he has been known to carry, and something of his mode of life, that I may see whether a match cannot be made for him.'

One can picture the consternation with which the staid Watt must have perused this letter, Watt who a few years before had made it a condition in engaging

[1] B. & W. Colln.   Boulton to Watt, 1788 Jan. 25.

Southern that he was to enter into a bond to give up music! It is clear that Watt was not aware of Perrins's proclivities, for he minuted the letter ' Mr. Roberts, please inform me of the truth of this matter, that I may write to Mr. Boulton.' Boulton, however, was not in such ignorance ; five or six years before Perrins had fought ' the famous Jimmy Sargent ' and had beaten him thoroughly, and Boulton was duly informed of that event by Scales.[1] The fight that followed Perrins's challenge on the present occasion terminated quite differently, for he was badly beaten by his opponent, Thomas Johnson, a smaller and lighter man. Perrins was a big fellow, over six feet high and judging from the picture, which is extant, of the fight, he must have been a fine-looking man. His defeat does not seem to have checked his ardour for fighting ; eighteen months after we find James Watt, junior, writing to Southern,

' I advise you to get I. Perrins sent off into the country as soon as you can, for he has received a letter from Mendoza pressing him to fight Big Ben, and offering to back him with £100. If the Birmingham gamblers hear of this, as he seems well inclined to inform them, I am certain they will find it no difficult matter to work up his passions, so as to induce him to challenge his rival, in which case you may bid adieu to him for ever.' [2]

Soon after this Perrins was sent to erect engines in Manchester and towards the end of 1793 he moved his household there and took a public-house, still keeping on his work as engine erector, but about a year later, having quarrelled with and threatened Lawson, Boulton and Watt discharged him from their service. Among the points made against him on this occasion was the slovenly condition in which the Boulton and Watt engines in Manchester were kept ; but, as Perrins points out in a very outspoken manner, the firm had never paid him anything to look after the engines : ' if you had allowed me a competency to have kept them clean, I should not be afraid of durtying my hands with doing it as some of your servants are that you send here with ruffles at hands and powdered heads, more like some Lord than an engineer. It cannot be thought that I can lose my time and neglect my own business without some consideration for it.' [3] The references to the ruffles and so on is a hit at Southern, or possibly at Lawson.

In spite of his shortcomings, no doubt Perrins was a capable man as an engine erector ; the terms of James Watt junior's letter, referred to above, make it clear that the firm did not wish to lose him, and for years after he had ceased to be an accredited erector for Boulton and Watt he continued to be employed to put up their engines in the Manchester district. Some of the mill-owners preferred to employ him rather than rely upon the men sent from Soho, and he was able to keep a staff of men employed.

Another man of whom mention should be made is Richard Dayus. He was with the firm in 1786 and remained with them until 1804. In 1791 he was sent to Nantes to erect an engine for a corn mill and seems to have witnessed some of the exciting scenes of the French revolution, but to have escaped without molestation of any sort. His opinion of the French workmen that he had under him was a poor one. In 1800 he was Boulton and Watt's principal erector in London, and a little later Rennie, who thought very highly of him, says that he was getting over half a guinea a day.

It may be well to remind the reader that in the early days of mechanical

---

[1] The letter is given in Smiles, *Boulton and Watt*, 480 n.

[2] B. & W. Colln. James Watt, junior, to Southern, 1791 Apl. 14.

[3] Ibid. Perrins to Boulton & Watt, 1794 Dec. 13.

engineering fitting was not a separate craft. Most of the men we have been considering had been trained as millwrights and were able to turn their hands to erecting, to pattern-making, or to the fitting of nozzles, which, as explained elsewhere, was the only fitting work done at Soho in those days ; others were smiths and carpenters ; the smiths were usually capable of finishing up their own work, and as for the carpenters, the beam engine as then made embodied a good deal of woodwork.

Numerous examples could be furnished of the transfer of a man from one class of work to another. Murdock made the nozzle patterns for some of the first engines sent to Cornwall. Cartwright was by turns erector, pattern-maker, and nozzle fitter, but he was not so expert in the last-named branch as one of the Taylors who was recalled from an erecting job in 1779 to fit nozzles at Soho.

To turn now to the officials. The clerks, Pearson and Buchanan, as stated above, were Scots. Pearson was engaged early in 1775 and he remained at Soho as book-keeper and cashier until 1817 when he retired on a pension. Buchanan came a few years later than Pearson ; he left in 1785, much to the regret of Boulton, who had a very high opinion of him.

The first technical assistant engaged by Boulton and Watt was Henderson. Lieutenant Logan Henderson, according to Smiles, had been an officer of marines, and afterwards a sugar planter in the West Indies ; he had lost all he possessed in Jamaica, but had gained a knowledge of levelling, draining, and machinery. In the Boulton and Watt correspondence, Henderson is met with for the first time early in the year 1776. He had submitted to Boulton, from Liverpool, a scheme for a rotary engine, and Boulton in reply had sent him an account of the performance of the rotary engines that had been tried at Soho and invited Henderson to call. Henderson paid a visit to Boulton, and being satisfied that his own scheme was inferior to that of Watt, laid it aside. The visit, however, resulted in his being engaged as an assistant, and early in 1777 we find him in London with Boulton, but living at Deptford and travelling to and fro every day. While in Birmingham he lived in Boulton's house. He had charge of the erection of the Torryburn and Byker engines in 1778, and in 1781 he was in Cornwall with Watt. Boulton had thought that Henderson would have lived with Watt and been an intelligent companion with whom he could have discussed engineering matters. Henderson, however, set up an establishment of his own where he lived with a lady who is referred to as ' Miss Peggy '. Perhaps it was as well that Watt and Henderson did not live together ; as it was Watt was far from pleased with him, and his letters of this period to Boulton contain frequent references to Henderson's bad temper and sullenness, and show that there was a good deal of friction between him and Wilson, the firm's agent in Cornwall. With Murdock, Henderson got on very well ; indeed it is said that he had great influence over him, and that he, Henderson, had asserted that Murdock would leave Boulton and Watt if he did so. In 1782 Henderson returned to Birmingham, and about the beginning of the following year he terminated his connexion with the firm very abruptly. In 1784 it seems that he was concerned in a foundry in London, and then in 1790 we find him applying, without success, for the post of Engineer to the Dublin Waterworks. This is the last we know of Henderson.

Early in 1778 the increase in the work, and the trouble with the men led to the appointment of a foreman at Soho. As already stated a man named Hall was selected for the post, but he did not prove suitable for it. Watt characterized him as 'a very great blunderer' and it was found that he had been making things for himself at Soho with the firm's materials. Watt was for turning him off at once, but instead of that he was sent out as an erector and worked on the Ketley, Wren's Nest, and Snedshill engines ; after which he drops out of the firm's list of erectors. Boulton writing to Watt in 1782 says of him : 'I am sure he will be allways plunging us into mischief, & therefore we had better silently drop him. I do think it is an impudent thing of him to go about ye country erecting boreing machines & instructing founders in our work without ever consulting us how far it may or may not be agreeable to us.' The last paragraph of this letter seems to bear upon the fact that Hall had undertaken to put up a Boulton and Watt engine for Walker of Rotherham who refused to have his castings from Wilkinson.

To make up in some respect for the loss of Hall at Soho, Watt suggests to Boulton (June 27, 1778) 'I would recall Playfair who can do part of the business, and I think now that you are at home you can contrive to give him proper assistance. I must warn you that Playfair is a blunderer.' Playfair was brother to John Playfair, a professor at Edinburgh University, and son of the Reverend James Playfair of Bervie near Dundee ; like Rennie he served an apprenticeship to Andrew Meikle. He had been employed as draughtsman and clerk to Watt for a brief period before this. Evidently Watt had not a high opinion of him ; however, he was brought back, but again he was with the firm for but a limited time, and his connexion ceased in 1781, when we find Boulton writing (Oct. 23rd) that he is sorry Playfair is going, only on account of Watt not having any proper assistant in drawing. Playfair went into business on his own account in London ; he took out several patents, but finally dropped into literary work. He wrote a memoir of Watt for the *New Monthly Magazine* which James Watt, junior, found very displeasing ; there is a bundle of correspondence on this matter among the papers at Doldowlod.

JOHN SOUTHERN

WITH increasing business, Watt's work had become very heavy, and we find Boulton constantly urging him to get an assistant. After the departure of Playfair in 1781, Boulton suggested a number of young men, and among them John Southern. 'I think young Southerns would be a very likely person, he seems good-humoured and very obliging. He is now with his brother as a surgeon, but says that though he is keen in the study of surgery, yet he had rather have been employed under you as a draughtsman and assistant, and if you wish to have him I know he will gladly come. He draws tolerably neat.' [1]

John Southern was one of the sons of Thomas Southern, residing at that time at Wensley, near Wirksworth, Derbyshire, and interested in mining affairs in various parts of the country.

Boulton followed up the letter by another again recommending the employment of Southern. To this Watt, then in Cornwall, replies : 'If you have a notion that young Southern would be sufficiently sedate, would come to us for a reasonable sum annually, and would engage for a sufficient time, I should be very glad to engage him for a drawer, provided he gives bond to give up music, otherwise I am sure he will do no good, it being the source of idleness.' [2] Other letters follow,

[1] B. & W. Colln. Boulton to Watt, 1781 Aug. 31.   [2] Muirhead : *Mech. Inv.* Watt to Boulton, 1781 Oct. 1.

and in June 1782 Southern, then at the age of 24, starts with Boulton and Watt on a three years' agreement, and with them he remained until the end of his life. He seems to have dropped into his place without hesitation or friction of any kind, even Watt makes no complaints of him, and he soon proved a valuable assistant and relieved his chief of a good deal of work.

In 1793 Southern married the daughter of Thomas Dobbs of Kings Norton. Dobbs had a Boulton and Watt engine put up in his rolling-mill in 1787, and it may be that Southern's visits on this business led to his acquaintance with the lady.

At one time Southern seems to have interested himself with political affairs, and in 1794 we find Boulton writing that he had called on the Marquis of Stafford, and had learnt that from papers lately seized it appeared ' that the country was upon the brink of a ruinous attempt to overthrow its Constitution. . . . I hope to God our Southerns is out of the mess, & that none of his letters will be found amongst these papers, & I wish you would speak to him about it, for I have reasons for my fears & wish to guard him.' [1]

With this exception Southern seems to have passed a quiet, uneventful life, devoting himself to his business and to scientific research. Whether he gave up music in deference to Watt's desire is not known. There is good reason for thinking that he was the inventor of the steam-engine indicator, that is, as we now understand the term, an instrument with a pencil for tracing the diagram of the varying pressure in the cylinder of an engine (see Chapter XVII).

When in 1791 the Birmingham rioters attacked Edgbaston Hall, then in the occupation of Dr. Withering, we learn that ' Mr. Sutherne, a clerk to Boulton and Watt ' was one of the party who engaged in the defence of the house.[2]

In 1800 Southern was given in addition to his salary a percentage upon all goods produced at the Soho Engine works, or in lieu thereof £600 per annum, and later in 1810 he was admitted a partner in the firm of Boulton, Watt & Co., to receive one-sixth of the profits. After his death the firm settled £2,000 on Mrs. Southern and the children.

One imagines that Southern was not a man of robust health. In 1812 he was ill in London and in the hands of a doctor ; two years later he spent some time at Aberystwyth ; he seems to have had a fall and to have gone there to recuperate his health. He returned to his duties but died at the age of 57 years, in July of the following year, 1815, at his house at Handsworth. Southern was buried in the Kings Norton church in the family vault of his wife's father Thomas Dobbs.

JAMES LAWSON, F.R.S.

AMONG the important members of the staff of Boulton and Watt whose names have been lost sight of, not the least important is James Lawson, who started at Soho in 1779, and was for many years connected successively with Boulton and Watt, Boulton's mint, and Boulton, Watt & Co. Lawson was the son of the Reverend Archibald Lawson of Kirkmahoe, near Dumfries. When he first came to Birmingham he seems to have been in the office, for in 1781 we find him, among others, signing letters on behalf of the firm. In the following year, however, he was one of the staff of engine erectors in Cornwall, and Boulton says of him that

[1] B. & W. Colln. Boulton to Watt, 1794 May 14.
[2] Letter from Dr. Withering to Sir H. J. Gough, quoted in Chatwin, P. B. : *A History of Edgbaston*, 1914.

he can ' manage the working of an engine better than any of their engine men '.[1]
One gathers that in his early days Lawson did not show himself a very diligent
servant, at any rate the partners were not altogether satisfied with his conduct, for
in the letter just referred to Boulton says : 'I think Lawson seems improved, at least
he seems desirous of doing the best he can.' A few years later Boulton, again in
Cornwall, writes of him : ' He says that he can erect an engine himself, and that
he did most of the work at Trevaskus. He doth very little that benefits us.' [2]

It must be remembered that the men when out erecting engines were paid,
not by Boulton and Watt, but by the person or firm for whom the engine was
being erected, and possibly there were intervals when, owing to the non-arrival
of materials or other causes, no work was going forward, or there was not enough
work for all the Soho men. At any rate in 1783 Lawson had taken up mine surveying, and we learn that he had been ' a long time at Poldice taking a plan of
that mine '. During the following year Boulton's letters contain a number of
references to his work in that direction. ' Lawson is making plans and sections
of the Cornish mines, certain of the mines propose employing him to record the
progress of the workings. He can survey land or dial mines, & is expert under-
ground, & I think it would answer to the mines to pay him to keep their drawings
in course with their mine. The plans & sections which is Lawson's only employ
at present will prove of great importance as they serve to detect great mistakes
& to enable the adventurers to see clearly with their own eyes.'

It is not clear how long Lawson was engaged on the mine survey work. Boulton
in his letter of Aug. 19, 1784, takes a pessimistic view of the prospect of the work
affording him a permanent position : ' As to the mine drawings, it may pay him
for once, but I have no hopes of the mines establishing a new office (draughtsman).'

Lawson's career for the next few years is obscure, but in 1791 he was engaged
under Boulton himself at the Soho mint, and we find Boulton expressing himself
as vexed by the want of harmony between him and another man, Bouch or Busch.
Lawson was in charge of the multiplication of the dies and of one press ; he had
also ' the office of inspector of the 2 fire engines ', but, says Boulton, 'the dies
are bad and insufficient, and Lawson's press stands more hours than it works,
and breaks more arms than all the rest.' Lawson was now talking of leaving, but
Boulton wished to retain him in some capacity or other, for although he found
ground for complaint he fully recognized Lawson's skill and intelligence, and later
on, in November 1793, when he had gone back to engine erecting, Boulton having
entered into a contract to supply fifty tons of copper coin, asks him if he can leave
his business at Leeds and ' come to Soho for about a week, in order to see that
the 6 presses are in the best order. I should be much obliged to you for I fear
Busch is not so much master of that subject, besides, he has his hands too full,
& perhaps Bill Harrison would like to spend his Xmas with his mother & at the
same time could render us a little assistance which with all expenses I will pay
with pleasure.' [3]

A few months before this he had been in London assisting in preparing models
and in getting up a case for Counsel for one of the lawsuits relating to Watt's
patent.

By this time Lawson was taking a leading position as an outdoor man under

[1] B. & W. Colln. Boulton (Cosgarne) to Watt,
1782 Sept. 30.
[2] Ibid. Boulton to Watt, 1784 Aug. 19.

[3] B. & W. Colln. Boulton to Lawson, 1793
Nov. 23.

Boulton and Watt, and he was representing the firm in Leeds and Manchester. He had formed a friendship with James Watt, junior, and in the autumn of 1800 the two made a tour in North Wales, terminating in ' a fortnight amongst our Lancashire friends & engines '. At the conclusion of this tour Lawson entered into an agreement to act for the Soho firm in Scotland for five years, at a salary rising from £250 to £300 per annum with a commission of 1 per cent. on all goods sent to Scotland and ½ per cent. upon all orders procured by him elsewhere, the firm guaranteeing that in all he should receive not less than £350 per annum during the first half of the term and £400 during the latter half. James Watt, junior, was not altogether satisfied that the terms were fair to Lawson, whose services he says he would be sorry to underrate.[1] Some years later, about 1811, Lawson was appointed superintendent of machinery to the Royal Mint at a salary of £800 per annum with a residence, but he continued to transact business, in London, for the Soho firm, and received an allowance from them until his death.[2] Lawson was elected a Fellow of the Royal Society in 1812 ; he died in London Apl. 9, 1818.

PETER EWART

THE name of Peter Ewart occurs in the Boulton and Watt correspondence at a comparatively early date. Thus in 1781 we find Boulton writing to Watt : ' I have had a letter from Lady Hopton recommending a young man of ye name of Ewart, 15 years of age, but I presume he wishes to be a practical engineer & millwright. Inclosed you have a letter upon the same subject. I think he is a very likely subject from her Ladyship's description of him, and may be of use to us in time, and therefore if it is agreeable to you it will be so to me to take him. We shall find it conv^t to breed up a few young men as our business will, I am persuaded, increase.'[3] Just before, Boulton had been alluding to the necessity of getting Watt some assistance in the drawing office.

The letter enclosed by Boulton was probably the letter from Professor Robison to Watt, given in Muirhead, in which Robison solicits—

' employment for a young lad, a near relation of mine, Peter Ewart by name, who wishes to be educated as a millwright or in any good branch of the business of a civil engineer. I could not find him so proper a master as yourself, and I flatter myself that you would find him a very deserving pupil. His father is a clergyman near Dumfries, and has given the boy a very good education, but with other views. But the boy's inclinations are so much turned to mechanics, and his mind so much caught by anything of this kind, that we all agree that this is the line of business in which he is most likely to succeed. His constitution is healthy and strong, so that he is perfectly fitted for the hard labour by which he is to get his living. If therefore you can find employment for him I shall look on him as setting out in the most favourable manner..'[4]

Ewart, however, seems first to have been apprenticed to John Rennie at Musselburgh. When the latter came south to put up the Albion Mills in London Ewart accompanied him ; and it was not until about 1790 that he became connected with Boulton and Watt.

From an account of Ewart's life by W. C. Henry,[5] it appears that he was born

[1] Ibid. James Watt, junior, to M. R. Boulton, 1800 Oct. 2.
[2] Doldowlod Papers. ' Register of the loss and death of agents, &c.'
[3] B. & W. Colln. Boulton to Watt, 1781 Oct. 23.

[4] Muirhead : Mech. Inv. Robison to Watt, 1781 Oct. 22.
[5] Memoirs of the Manchester Literary and Philosophical Society, New Series, Vol. VII (1844).

at Troquaire Manse in 1767, and was educated at the Free School, Dumfries ; his father was a clergyman of the Church of Scotland, one brother was British minister at the Court of Berlin, another a physician at Bath, and a third a merchant in Liverpool. In 1788 he was sent to Soho by Rennie to erect a water-wheel and other machinery for Boulton's rolling-mill, and he was afterwards employed by Boulton in the construction of the millwork and machinery for his mint. From 1790 to 1792 Ewart was engaged in erecting Boulton and Watt engines in the Manchester district. In 1791 he informs Watt that he had taken a shop in Manchester, and was engaging some hands to begin upon the millwrighting and other work he had undertaken.[1] This business was not of long continuance, for in September 1792 he entered into partnership in a bleaching and calico-printing firm ; but this lasted only for a year. In 1795 and 1796 we find him again at Soho laying out shops and plant and devising machinery for Soho Foundry, and for the Engine Shop at Soho Manufactory. Among his work were the designs for the boring-mill, which as described in Chapter XX did not answer well. In reference to this he writes early in 1797 from Manchester to James Watt, junior, ' I am exceedingly mortified by your account of the boring mill. I cannot think that the jarring is owing to the weakness of the wheel, for every part of it is much stronger than the rod in proportion to the distance from the centre.'

To go back to Mr. Henry's account : from 1798 to 1835 Ewart was concerned in cotton-spinning in Manchester ; he was then appointed chief engineer and inspector of machinery in the Royal Dockyards. He died at his official residence at Woolwich in September 1842, in the 76th year of his age, in consequence of injuries caused by the breaking of a chain while he was superintending the removal of a large boiler. Ewart took out three patents : the first in 1813 for working looms ; the second in 1822 for coffer dams (this invention was used at the Liverpool docks) ; and the last in 1833 for spinning machines.

THOMAS WILSON

In connexion with the affairs of Boulton and Watt in Cornwall, the name of Thomas Wilson is frequently met with. He was at Chacewater in 1777 when the Wheal Busy engine (the first Watt engine set up in Cornwall) was ordered, acting as agent for the proprietors, Fentons and the Yorkshire Copper Company. Watt stayed in his house on his first visit to Cornwall, and possibly it was on this occasion that arrangements were made with him to act as the commercial or financial agent for the firm in Cornwall, a position that he continued to hold until the expiry of Watt's patent in 1800. He was an active and intelligent man, but it would seem that he engaged in too many different undertakings to make a great success in any of them. A lively account of his activities in the year 1785 is given by Boulton, who, after discussing the conditions at Chacewater and the difficulty of making the mine pay, goes on :

' I know Mr. Wilson says he has a very good opinion of her, but I will form mine only upon facts, & not opinions. Mr. W. hath now several partnerships in shares of ships, he hath always had the tickiting to attend, the assay office, the accts of the Welsh works, the several mine accounts, in wch, he, we & Wilkinson are concerned ; he hath a partnership with Ned Rogers in 2 or 3 farms ; he hath another with Lantie Atkinson in other farms; his candle trade is a very large concern ; he hath several partnerships with difft. persons in pairs of

---

[1] B. & W. Colln. Ewart to Watt, 1791 Sept. 23.

mules. He hath B. & W. business to manage, with gunpowder, candle accounts, stamps, &c., and he hath a large family. I heartily wish him success in these & all other undertakings, but it must be evident to you that he cannot bestow any attention to the mine.' [1]

For his services to Boulton and Watt, Wilson was paid a commission of 2½ per cent. on the premiums derived from Cornwall. For the twenty years 1781–1800 he received the sum of £3,485 or a little over £174 per annum on the average.[2] He died in 1820, at the age of 72 years, and was buried at Falmouth.

WILLIAM MURDOCK

It is not intended here to give a full account of the life of William Murdock— his life story has been written at some length by Smiles and other writers—but rather to supplement, and in some instances to correct, the published accounts, by the aid of such notes as have been made in the course of an investigation in the Boulton and Watt correspondence. Had space permitted it would have been eminently proper to have given a complete biography of a man whose entire life was spent in the service of the firm and whose story is so closely connected with that of James Watt.

Murdock entered the service of Boulton and Watt in the year 1777, when he was 23 years of age ; his name first appears in the books of the firm as ' Wm. Mordach ' ; he was paid at the rate of 15s. a week, and some of his first work seems to have been at pattern-making. He very soon made his mark as a man of ability and intelligence, and as early as January 1779 we find Boulton writing : ' I think Wm. Murdock a valuable man & deserves every civility & encourage-ment.' [3] Murdock was at this time engaged in repairs and alterations to the Bedworth engine. A few months later he was sent to take charge of the erection of the Wanlockhead engine, and Boulton writes of him to one of the proprietors : ' He hath a good deal of experience in our engines and is capable of putting your people to rights in any matter they may not understand, & we doubt not but he will acquit himself to your & to our satisfaction, as he is a man we have a good opinion of. Pray don't keep him longer than necessary as we want him in Corn-wall.' [4] About the same time Watt informed the same customer that he is ' a very sober, ingenious young man, who has a good deal of experience under us in putting engines together and knows all the little niceties, the omission of which might cause a bad performance in your engine '.[5]

Upon his return from Scotland, Murdock was sent to replace Jabez Hornblower in the erection of the Ketley and Donnington Wood engines ; in September he and another erector, James Law, were sent to Cornwall. As a precaution, in case of their being taken by the press-gang on the way, they were furnished with letters addressed to prominent men in Bristol and Exeter requesting protection. For the next twenty years Cornwall was Murdock's home ; there he married, and there his sons were born. From the first he assimilated with the Cornish men far better than Watt ever did, and with occasional interludes of disfavour he was well liked by the great majority of the people concerned in the mines during the whole of his stay in Cornwall. An anecdote of his early days in the county concerned with

---

[1] B. & W. Colln. Boulton to Watt, 1785 Aug. 22.
[2] Ibid. 'Statement of the Commissions paid Thos. Wilson at sundry times by Boulton & Watt.'
[3] Ibid. Boulton to Watt, 1779 Jan. 6.
[4] Ibid. Boulton to Meason, 1779 Mar. 23.
[5] Ibid. Watt to Meason, 1779 Mar. 27.

XCVI. WILLIAM MURDOCK

From the oil-painting, by Graham Gilbert, at the Birmingham Art Gallery

the beating-off of a French privateer has already been told. Temperance in the matter of strong drink was a characteristic that distinguished him from most of his fellow erectors; but, more than this, he was particularly dexterous and resourceful in erecting new engines and repairing the old ones, and he was very hard-working—'indefatigable' is the term applied to him in some of the letters. Thus Boulton, writing in 1782 from Cosgarne, says:

'Murdock hath been indefatigable, ever since they began he has scarcely been in bed or taken necessary food, for everyone seems helpless in comparison of him. . . . After Murdock had been slaved day & night on Thursday & Fryday last, he rec*d*. a letter from Wl. Virgin in the West insisting upon his coming over directly as they could not set their engine to work, & if he did not come instantly they would let out ye fire. He accordingly went on Saturday morn*g*, set ye engine to work, wch. went on very well dureing 5 or 6 hours, & then left it & returned back to ye Cons*d*. Mines about 11 at night and was employed about the engines till 4 this morn*g* & then went to bed. I found him at 10 this morn*g* in Poldice cistern, searching for pins & cotters that had jumped out, and I insisted upon his going home to bed for he had a bad cold.' [1]

Two years later Boulton, again writing from Cornwall, expresses his satisfaction in even warmer terms:

' We want more Murdocks, for of all others he is the most active man & best engine erector I ever saw, of w*ch* I had a strong proof this day. They stoped Poldice lower engine last Wednesday & took her all to pieces, took out the condenser, took up out of the shaft the greater part of the pumps, took the Nossells to pieces, cut out the iron seatings and put in brass ones w*th* new valves, mended ye eduction pipe & did a great number of repairs about the beam & engine, put the pumps down into the new engine shaft, did much work at new engine, & this day about noon both the engines, new and old were set to work again compleat. When I look at the work done it astonishes me & is entirely owing to the spirit & activity of Murdock who hath not gone to bed 3 of the nights, & I expect the mine will be in fork again by Wednesday night. I have got him into good humour again without any coaxing, but have spoke plainly to him in presence of Wilson, have prevail'd upon him not to give W*l* Virgin engines up, which he was resolved to do from the ungenerous treatment he rec*d*. from ye Capt*n*.' [2]

Although he was such a valuable man, his pay, for the first few years at any rate, was on a very low scale. When he started in Cornwall in 1779 he received 21s. a week; Law, his travelling companion, had 20s. Twelve months later he wanted an increase of pay, and thought he was entitled to two guineas a week. Boulton seems to have sympathized with his wishes, but Watt objected and thought it would be very wrong to give this sum, 'an example of that kind would ruin us by stimulating every other man we had to similar demands'. It is not very clear why Boulton and Watt should have kept the pay so low; it affected them only in so far as it increased the total cost of the engine; the erectors were not paid by them, but by the adventurers for whom the engine was being erected. Boulton managed to stave off the difficulty by persuading one of the mine companies to make Murdock a present of ten guineas, to which he added another ten on behalf of Boulton and Watt. At about this time John Budge, one of the most highly respected of the Cornish engineers, was pressing Murdock to enter into partnership with him. It was the practice in Cornwall for the mine-owners to entrust the management and supervision of their engines to an engineer, such as Jonathan Hornblower or John Budge, who made periodical visits, and might have a number of engines in his care. In this instance Budge had offered Murdock the half of all his engines. However, towards the end of 1782 Murdock

[1] Ibid. Boulton to Watt, 1782 Sept. 30.   [2] Ibid. Boulton to Watt, 1784 Nov. 8.

obtained an appointment of this nature on his own account—the seven new Boulton and Watt engines at Wheal Virgin and Poldice were placed in his charge at a fee of £6 per month. Boulton seems to have been in high glee at this event. ' Murdock will be well off, but it fixes him firm to us and our interest.' [1] This fee was· in addition to his pay for erecting engines, and no doubt, as time went on, Murdock had more work of this and other kinds, and possibly he was getting into such a position that he could make his own terms with the mine companies for erecting new engines. We find Boulton writing from Cornwall in 1784 that ' people in general are much prejudiced in fav^r of W. Murdock & he is thereby intoxicated to a great degree '.[2] And in the same letter we have an indication that he was undertaking work on his own account. ' He hath erected a new ballance bob at W. Maid wch. hangs in a joynt & rolls upon a vertical plane, but it is not so good as your old one and I dont believe he now prides himself upon it.'

Moreover, he was now taking upon himself, in erecting new engines, to depart in important features from the drawings supplied from Soho. Thus :

' The small rotative engine at Wl. Maid will be set to work in a day or two, but he hath in that varied from your drawings, for he hath hung the spear upon the *top* of the beam, i.e. ye centre of motion of the top end of the spear or connecting rod is upon the top of the beam whilst the centre of motion of the beam itself is under the beam. This he hath done in order that the revolving wheel shall always stand right to begin its stroke.' [2]

Murdock's connexion with the invention of the sun-and-planet gear and his iron cement have been referred to elsewhere in these pages. A more important invention with which his name is identified is the introduction of coal-gas as a lighting agent. His first experiments in this direction were made in Cornwall in 1792, but it was not until his return to Birmingham that anything was done on a commercial scale. The Soho Manufactory was fully equipped for lighting by gas in 1803, and the manufacture of gas-making plant became a distinct branch of the business at the Foundry, which became the training-ground of the early gas engineers.

Another project that Murdock engaged upon during his stay in Cornwall, much to the distaste of Watt and his partner, was that of steam carriages. The main facts have been published by Muirhead and by Smiles, but mention should be made of a series of letters written in 1784 by Boulton, from Cornwall, to Watt, in which reference is made to an erector, who is named ' Brown '.[3]

The partners in their correspondence frequently used private names or symbols to designate the persons that they had in view, and in this case there can be no doubt that ' Brown' was Murdock. In the first of these letters Boulton mentions that Brown is ' now taken up with his girl and the wim engine '.[4] In the second he says, ' I shall do all I can to make a pleasant Brown . . . he is certainly very usefull here, & ye adventurers in all the mines are much prejudiced in his favour ', and we have the important statement that Brown and his father were about a steam carriage (with a common engine) some years ago. Next we have Boulton inquiring about hot-air engines, and explaining, in a subsequent letter, that he had done so in ' consequence of Brown mentioning it as the thing he had thought of before he came to England '.

[1] B. & W. Colln. Boulton to Watt, 1782 Dec. 12.
[2] Ibid. Boulton to Watt, 1784 June 23.
[3] Ibid. Boulton to Watt, 1784, July 8, 22, 31, Aug. 6, 19.
[4] Smiles mentions that Murdock married in 1785 the daughter of Captain Painter, and made his home in Cross Street, Redruth.

The next letter is of particular interest, as it shows that Murdock had in view an air-cooled surface condenser, and a variable-speed gear ; apparently he intended to use the sun-and-planet gear with a set of planet-wheels of different diameters. Writing from Cosgarne on Aug. 6, 1784, Boulton says :

'I have had a little conversation with Brown about wheel car^{gs}. He proposes to catch most of the condensed steam by making it strike against broad thin copper plates and the condensed part trickling down may be caught & returned into its boiler or other reservoir. This may do some good in rain or frosty weather, & he proposes to have different sized revolvers to apply at every hill & every vale according to their angle wth ye horizon. What he wishes to do may be right, but ye means he proposes will not do. I find his father hath a w^l carriage to go without horses which I believe Logan Malcomb lately saw in Scotland, & as this was one of the first mechanical amusements wch Brown ever turned his attention to in his youth it serves to account for the furor that he is seized with. I verily believe he would sooner give up all his Cornish business & interest than be deprived of carrying the thing into execution. When a man is mad in any way it is vain to reason with him about his disorder. He doth not directly say that he hath a right to work pistons by steam, but he says he thinks or fears that the Trumpeters or somebody else will & then take out a patent for w^l car^{gs} wch is the thing he says that makes him so sollicitous to have it secured. Now as you are going to specify sundry new applications of Steam engines, Q^r. if it may not be prudent to specify the application of it to w^l cargs without making any drawing & only describe ye application of the elastic force of steam to act upon a piston or pistons in cylinders & that force applyd to the turning of the wheels by means either of cranks or by one w^l revolving round another or within another, or by any other of the methods now used to convert the reciprocating motion of a piston moveing in a cylinder (or of a reciprocating beam moved thereby) into a rotative motion, or by any other of the rotative motions invented by you & described in your specification dated . . ., 1782. I propose this by way of taking possession and saveing the expense of a patent. . . . I think a very light and cheap modell might be made takeing one of our drawn tubes of an inch diam^r to make the cylinder of.'

Watt at this time was engaged in draughting the specification for his patent of 1784, covering the parallel motion and other inventions, and he at once proceeded to incorporate in it such a description of a steam carriage as he could do in the time and space at his disposal. This procedure effectually forestalled other steam-carriage projectors for the time being, and so would allay Murdock's fears on this point; whether it was quite the method he had in view himself is another matter.

The last '·Brown' letter we have to refer to is dated about a fortnight later ; in it Boulton says : ' I think Brown is getting better into tune, but I must say something more to him before I leave upon ye Car^g. Subject, or he will relaps. We must make the best of mankind as we find it.'

Two months later Boulton, still in Cornwall, writing of Murdock in his proper name, says : ' I have also prevaild upon him to put off his determined journey to Scotland till N. D. [North Downs] engines are got to work & have quieted his mind about w^l carriages till then.' [1]

There is no documentary evidence that Murdock made a model at this time ; probably he was kept too busy with mine engines. Two years later, however, there is distinct evidence that a model had been made, for in August 1786 Wilson writes from Chacewater to Watt : ' W^m. Murdock desires me to inform you that he has made a small engine 3/4 dia. & 1½ inch stroke that he has applied to a small carriage which answers amazingly.' [2] From the terms of the letter there can be little doubt that this was, as far as Wilson knew, Murdock's first model, and that

[1] B. & W. Colln. Boulton to Watt, 1784 Nov. 8.   [2] Ibid. Thos. Wilson to Watt, 1786 Aug. 9.

it had been made quite recently. A few weeks later Murdock set off for London to exhibit his model and to apply for a patent. However, Boulton was on his way to Cornwall at the same time ; their coaches met near Exeter, when Boulton induced him to retrace his journey. Upon their arrival at Truro, Murdock unpacked his model and 'made it travel a mile or two in Rivers's great room in a circle, making it carry the fire-shovel, poker and tongs'.[1] In a later letter Boulton says : 'he hath made a very pretty working model wch keeps him in good humour, & that is a matter of great consequence to us. He says he has contrived, or rather contriving to save the power arising from the descent of the carr^e in going down hill.'[2] Boulton at the same time asks Watt to send all the engines as soon as possible, so that Murdock may be better employed than about wheel carriages.

Among the Watt papers at Doldowlod is a letter dated May 31, 1815, from John Murdock to James Watt, junior. William Murdock had broken his leg at Leamington, and was laid up there for some time, with his son John in attendance upon him. In this letter, which is in reply to an inquiry by James Watt, junior, we find the following :

'The model of the wheel carriage engine was made in the summer of 1792, & was then shown to many of the inhabitants of Redruth—about two years after Trevithick & A. Vivian called at my father's house in Redruth to consult him about removing an engine then on Hallamanin mine to Wheal Treasury Mine, where they wished the engine to work double. My father mentions this circumstance merely to bring to their recollection that on that day they asked him to shew them his model of the wheel carriage engine which worked with strong steam & no vacuum. This was immediately shewn to them in a working state. My father knows of no patent that Trevithick had for steam engines except that in conjunction with A. Vivian. However he is not quite clear about this ; he has some slight recollection of his having a patent for steam boats.'

Then in a postscript we are told : 'The model of the steam engine with one inch cylr. that works double with slide valves was made in August 1791, it is now at the foundry.'

This letter is endorsed by James Watt, junior, 'Mr. John Murdock, Leamington, 31 May, 1815. Date of his father's model of the wheel carriage engine & time when shown to Trevithick. Gives 1791 as the date of the model ; but this must be a mistake, as Mr. R. Boulton says he saw it at work in 1784.'

The occasion of the inquiry which produced this letter is not known. Trevithick had long ago relinquished his work on the locomotive, and had returned to Cornwall. It seems likely that the model referred to is identical with that still preserved, to which the date 1784 or 1785 is usually ascribed, presumably on the assumption that it is the same as that which M. R. Boulton recollected to have seen in Cornwall. Young Boulton does seem to have spent one of his school holidays with his father in Cornwall, and in the year 1784. Although, as we have seen, the earliest documentary evidence of the existence of a model is of the year 1786, it is possible that one was made and had been shown to him in 1784. That this model was identical with that under discussion in 1815 is another matter. One can hardly set M. R. Boulton's recollections of what he saw in his boyhood against those of the man who made the model, and there is no suggestion that Murdock's memory had failed in any way ; he was 61 years of age in 1815. The model formerly in the Birmingham Art Gallery has a cylinder of $\frac{3}{4}$ in. diam. by $2\frac{1}{8}$ in. stroke, a piston

[1] B. & W. Colln. Boulton to Watt, 1786   [2] Ibid. Boulton to Watt, 1786 Sept. 17.
Sept. 2.

valve, and 9¼ in. driving-wheels. The model mentioned by Wilson in 1786 had a cylinder of the same diameter for 1½ in. stroke. John Murdock's letter does not state the size of the cylinder of the carriage model made in 1792 ; the 1-inch cylinder engine is clearly a different thing, made, as he says, in 1791.

The next invention of Murdock's that we learn of is for raising water ' without pumps or great beam '.[1] This is news that Boulton hears in London and transmits to Watt. Wilson a little later sends a more circumstantial account which suggests something in the nature of air-lift pumping ; he says that Murdock

' let me see his new method of raising water ; it is by compressing air at the surface which is forced down small pipes to the bottom of the shaft & there communicates with the main pump & forces up the water, what machinery he has at the bottom I cannot tell, only there is no bucket, but there is a clack to prevent the water falling back ; his present machinery is a copper tube of 1 inch diar. & 9 inches long, in this he works a wooden plug, from the side is a very small tube, which goes from the top of his house to his well 40 feet & another larger tube by the side up which the water ascends & runs off continually in an equal stream. . . . He proposes writing to you on the business. He says pipes of 3 inches diameter will do for forcing down the air to work 17 inch pumps ; if so there must be a great saving in erecting for deep mines. He has a very small string which goes down to the bottom, which he pulls when he sets to work, I suppose to open a valve.' [2]

We learn nothing more about this scheme, so presumably it did not answer Murdock's expectations. Ten years later, just before he left Cornwall for Soho, he was constructing a water-pressure engine at Wheal Fanny,[3] and about the same time he was sending mica to Soho for use in the packing of pistons.[4]

Murdock's sojourn in Cornwall was now drawing to an end. Watt's original patent and the Boulton-Watt partnership would terminate in 1800. Watt himself had ceased to take an active share in the direction of affairs ; his son James, and Boulton's son, Matthew Robinson, had now taken over the reins, and the Soho Foundry had been commenced. Business in Cornwall was likely to fall off, and the young men felt, no doubt, that it would be wise to have a man of Murdock's experience, capacity, and energy at their disposal. So Murdock was given a post at head-quarters. He did not finally leave Cornwall until the end of 1798, but he had, before this, visited Birmingham to assist in the deliberations upon the new establishment. At the beginning of 1796 we find a letter from James Watt, junior : ' P. Ewart leaves us this evening. The last three days have been spent in scheming between him, Murdock, Mr. Southern and A. Story, they have concluded upon the different lathes, boring and drilling apparatus and drawings of all are made.' [5] Peter Ewart, as we have explained, had been brought to Soho for the purpose of designing the machinery for the new works which were planned ; Storey was the manager of the foundry.

From the early part of 1797 James Watt, junior, was calling for the presence of Murdock at Soho, but there were still many jobs to finish. It had by this time become known in the county that he was leaving Cornwall, and the Cornishmen had begun to realize what his departure would mean to them. An engine was to be put up at Wheal Jewel, and at first the adventurers declined to sign the agreement unless a clause was inserted that Murdock was to erect the engine ; on this occasion one of the adventurers said that ' two years ago M. might have

[1] Ibid. Boulton to Watt, 1788 Jan. 25.
[2] Ibid. Wilson to Watt, 1788 Mar. 5.
[3] Ibid. Gregory Watt to J. Watt, junior, 1798 May 12 and May 17.
[4] Ibid. Gregory Watt to Watt, 1798 Jan. 24.
[5] Ibid. James Watt, junior, to Boulton & Watt (in London), 1796 Jan. 2.

gone to the devil for anything the county cared, but that by the blessing of God their eyes were now opened ',[1] and even after he had left Cornwall he was pressed to come back to their assistance at North Downs and other mines.

Upon his arrival at Soho, Murdock set himself with great vigour to getting the boring-mill completed. When this had been done, it seems that it did not answer very well, and there was some little delay in settling the plan of a new mill. Murdock then felt himself at a loose end : ' he will begin to think he is losing his time here ', writes the younger Boulton ; ' I have endeavoured to occupy his attention with several other schemes which are going forward, but the construction of the boring-mill being his principal hobby-horse, his thoughts are continually recurring to this favourite subject.'[2]

Possibly as one means of occupying his attention about this time, Murdock, with Storey the foundry manager for his companion, was sent on a voyage of discovery to Leeds. Matthew Murray was turning out iron castings and smithwork far superior to the Soho productions, and it was desired to find out how he managed this.

Another project in which Murdock's aid was enlisted was the production of a smaller and cheaper engine than had been made hitherto at Soho. The young partners were convinced that there was a wide field for such engines, and Murdock appears to have been responsible, in the main, for the design of a self-contained engine, known as the ' bell-crank ' engine. It did not prove a success, but it embodied one of Murdock's most important inventions—the long D-slide valve. This is one of the inventions covered by his patent of 1799. Another is the use of a worm and wheel for driving the bar of a boring-mill ; this was applied at Soho both for boring and for large turning-lathes. In his specification he says that the screw ' I have commonly used has three threads, is 16 inches diameter, and the advance by one revolution is 6 inches '. Probably this was the size used for the boring-mill referred to above.

No doubt from time to time Murdock paid business and other visits to Cornwall. One such visit was in the early part of 1800, and he writes to James Watt, junior, that some of the Cornish mining people say they ' are going to make an offer that will be greatly for my benefit if I will remain in Cornwall ',[3] and he asks for directions as to the answer he is to make. It appears, too, that Murdock has been asked to recruit smiths for Soho, and in the same letter he states : ' I hoped to have sent three before this, but there is a report circulates here that you have starved Simon [Vivian] and his family and at this time I cannot prevail on any of them to go.'

Murdock's position at Soho was put on a definite basis in the year 1800 when he entered into an agreement to serve for five years at a salary of £300 per annum, with an allowance of 1 per cent. on all orders for Soho Foundry. This arrangement prevailed up to the year 1810 ; but he was then getting, in addition, a commission of 1½ per cent. on all the gas-lighting apparatus made, and his total income amounted to £684. He was in 1810 put on the footing of a partner at Soho Foundry, but in lieu of a share in the profits, he elected to take a salary of £1,000 a year, and this he received for the next twenty years, up to 1830, nine years before his death.

Murdock was now an old man ; he had reached the age of seventy-six, and his

[1] B. & W. Colln. Gregory Watt to James Watt, junior, 1797 Apl. 18.
[2] Ibid. M. R. Boulton to James Watt, junior, 1799 Jan. 21.
[3] Ibid. Wm. Murdock (Redruth) to James Watt, junior, 1800 Mar. 1.

services cannot have been of any serious value ; on the other hand, James Watt, junior, found that Murdock was drawing more from the concern than the partners were : ' the sum paid annually exceeding our own emoluments.' ' The engine business having for several years proved unprofitable, and his infirmities having so much increased, as to render him of little service, notice was given that our connection must terminate on 30 Sept^r, 1830, to which period the sum of £1,000 per ann was paid up.' [1]

Murdock's sons, William and John, and his grandson left Soho Foundry at about the same date. William died soon afterwards in Cornwall. Murdock himself lived to the age of eighty-five years ; he died at his house, Sycamore Hill, Handsworth, in 1839. He had previously occupied one of the houses adjacent to Soho Foundry.

The Boulton-Watt partnership, as we know, came to an end in 1800. It is beyond the scope of the present work to deal with the many interesting characters that figure on the Soho roll after that date, but space should be made for a few words on William Brunton.

WILLIAM BRUNTON

In 1796 Gilbert Hamilton was engaging men in the Glasgow district for Boulton and Watt. Among others he took on Brunton, who had been working under his (Brunton's) father, ' making iron and brass machinery ' for the New Lanark Mills. He came with a testimonial that he ' is an ingenious lad in his business, that he has received a virtuous education, & is accordingly both temperate and attentive '. Brunton very soon made his mark with Boulton and Watt. A little more than twelve months after he was engaged, we find John Southern reporting : ' William Brunton is a very valuable hand, and has more *gumtion* than most of his cotemporary engine erectors.' Less than five years after this he was made superintendent of the engine yard at Soho, a position that he retained until 1808, at a salary rising from £80 to £100 a year. He left Soho to go as engineer to Butterley Ironworks, where he made his celebrated steam horse. After seven years or so at Butterley, he returned to Birmingham to become a partner in and manager of the Eagle Foundry, and while there invented one of the first (if not the first) moving fire grate for furnaces. Later on he was engaged in South Wales.

[1] Doldowlod Papers : ' Register of the loss and   of Boulton, Watt & Co.'
death of Agents and clerks in the employment

XXII

RIVALS AND PIRATES. PROJECTORS
AND SCHEMERS. LITIGATION

RIVALS AND PIRATES

IN a consideration of the progress of the Soho firm and of Watt's work on
the steam engine, an important factor which must not be overlooked is that
of the men with whom they had to compete.

When, in 1775, the Boulton-Watt partnership was commenced, there were
a number of engineers engaged in building the Newcomen engine in London and
various parts of the country for public water-supply and for pumping in mines.
Except in Cornwall, where the Newcomen engine disappeared entirely within
a few years of the advent of the Watt engine, these men and their successors
continued at work during the life of Watt's patent and afterwards; indeed it
seems to be the case that at least as many atmospheric engines were put up in
the last quarter of the eighteenth century, in which Boulton and Watt were at
work, as in the preceding quarter.

Among these men we have Hadley in London, possibly a descendant of the
John Hadley who was associated with Sorocold in the reconstruction of the London
Bridge Waterworks about the year 1704, and had a wide reputation in connexion
with pump work in his day. In Cornwall there were Hornblower and Budge; in
Derbyshire, Francis Thompson; in Warwick and Staffordshire, Perrins. Then there
was John Curr of Sheffield, but it is not clear that he was at work as early as 1775.
He was the author of the first practical handbook on the construction of steam
engines and boilers : *The Coal Viewers and Engine Builders Practical Companion.*
This was published in 1797 ; it gives engravings of atmospheric engines, and
tables for the dimensions of all the parts proportioned to different sizes of cylinders,
and, as Farey says, ' it is a useful guide for those who require to construct such
engines, being a complete manual for their instruction.' [1]

But the name best known in connexion with the atmospheric engine in 1775 was
that of Smeaton. He, however, did not undertake the building of engines ; what
he did was to supply designs and directions for making and working them.

No doubt all along inventors had been at work seeking to improve the application of steam as a motive power, but as the Boulton and Watt engine gained
ground, and reports spread of the great sums the firm was receiving in the form
of royalties, still more inventors entered the race. All this was much to the annoyance of both Boulton and Watt. They did not complain, at least their correspondence does not indicate that they did, of the continued building of the old engine,
but they considered it quite an improper thing for any one else to seek to improve
it, and although they had a very low opinion of the value of these other projects

---

[1] Watt's 'Directions for erecting and working
the newly-invented Steam Engines. By Boulton
and Watt' was earlier; it was printed in 1779,
but for private circulation only.

and their originators, they were considerably disturbed. In 1780, when Watt was in Cornwall, report came of a scheme on foot in Manchester, possibly that of Joshua Wrigley, for driving cotton-mills by using a Savery engine to return water for a water-wheel ; Boulton refrained from communicating this to Watt, fearing it would upset him. Watt, however, got to hear of it and made inquiries, whereupon Boulton wrote :

' as to the Manchester schemes I suppose we may add them to the same catalogue in wch are enrolled all Mr. Hateleys, Mr. Wasbroughs, Mr. Matthews, Mr. Pintos, Mr. Jones, the Spaniards & the Truro mans & therefore I did not think it necessary to vex you with nonsense or divert your attention one moment from the great Cornish object.' [1]

This rivalry forced the hands of Watt in more than one direction. As we have seen, the appearance of the Hornblower compound engine led him to embark upon expansive working, and that of the crank engine of Wasbrough and Pickard upon the rotative engine, in both cases before he was ready to do so.

It was not till the last ten years of the eighteenth century, and of the life of Watt's patent, that this rivalry became a serious matter. The Cornish mine-owners, or the greater number of them, had been almost from the beginning very restive under the heavy tax that they had to pay to Boulton and Watt, and they now backed up Jonathan Hornblower and Bull in the production of rival engines. But in so far as the production of new engines was concerned, the competition was felt more particularly in respect of the rotative engine.

The rapid development of factory building, of the iron manufacture and coal-mining, and the consequent demand for power, led to a very considerable call for steam engines. Boulton and Watt benefited by this increased demand, and a great many of their engines were made ; in some cases, however, they were not in the position to supply within a reasonable time, and in others the power-users were averse to paying the heavy premiums demanded, but always there were engineers available who undertook to make the engines and, as they thought or stated, without infringing Watt's patent. It was, however, very difficult, perhaps impossible, for them to make any material improvement without infringing in some way or other. It was not merely that they could not use the separate condenser. Watt held that the patent covered every one of his ' principles', whether used together or separately, so, for instance, they were precluded from putting a cover to the cylinder and using steam instead of the atmosphere to press on the piston. In fact, the patent completely blocked any progress by other engineers. The result was that some of them, possibly they may have had honest doubts as to the validity of Watt's patent, boldly embodied some of his inventions in their engines.

Watt himself was not very confident that his specification would stand the test of an action at law, and was reluctant to commence one. At last, however, the infringements became too numerous and glaring to be ignored ; the customers of Boulton and Watt were beginning to question why they should continue to pay premiums for inventions which their competitors had the free use of, and the firm were forced to take legal action. In the following pages some account will be given of the rival engineers and their productions, of those who adhered to the old Newcomen form, as well as those who pirated one or other of the features of Watt's invention, and finally of the litigation that ensued.

[1] B. & W. Colln. Boulton to Watt, 1780 July 1.

JOHN SMEATON, F.R.S.

SMEATON had been interested in the steam engine from his youth, and his papers, according to a statement by John Farey,[1] disclose that as early as 1741 he had made sketches of an inverted engine, but Watt had been investigating the steam engine and in fact had invented the separate condenser some years before Smeaton took up the subject in practice. It was in 1769 that the experimental engine, upon which Smeaton made a great number of trials, was set to work at Austhorpe near Leeds. In the same year he examined the pumping-engines in the Newcastle district ; he found fifty-seven at work, and obtained a list of 100 that had been erected there at various times. In the following year he was in Cornwall on a similar errand, and then in 1772 the first engine made to his designs was set to work at Long Benton Colliery, Northumberland. It had a 52-inch cylinder, and a stroke of 7 ft. This was followed by an engine for Cronstadt, and by the Chace-water engine, 72-inch cylinder, 9 ft. stroke, and by other engines.

At first Smeaton thought the Watt engine too complicated, and that its construction made too great a demand for the engineering skill of the day. He was, however, highly pleased with the working of the Bow engine which he inspected in 1777. It will be remembered that his gratuity to the engine-man on this occasion led to the unfortunate result of the part wrecking of the engine. In 1778 he conducted a trial of the engine at Smethwick, and became so convinced of the superior efficiency of the Watt engine that he relinquished to Boulton and Watt the construction of an engine for the Hull Waterworks. It seems, however, that he continued to be concerned in the construction of atmospheric engines for collieries after this date.

Smeaton seems to have thought that the application of the separate condenser to the atmospheric engine would secure, possibly not all, but a considerable part of the superior efficiency of the Boulton and Watt engine, in a far simpler construction, and he applied for permission to try it. Watt, in his reply, sets out at some length his objection to his so doing, but goes on to say, in case these objections do not strike Smeaton with the same force, that : ' In consideration of the good will we bear you, we give you our free license for one engine to employ a condenser, or rather to condense the steam in such a separate vessel as you shall judge proper, and to apply the profits, if any accrue, to your own emolument. But we request that you will not give these profits in a free gift to your employers, as you have too generously bestowed your own improvements.' [2] Smeaton accordingly tried the combination, but found no sensible advantage, the result, as he thought, of the film of water left on the surface of the cylinder from the water lying above the piston.[3] The trial was made, in all probability, on his small experimental engine which had a 10-inch cylinder.

The improvements made by Smeaton in the Newcomen engine were directed to the better proportioning and construction of the parts ; they did not alter its mode of action in any way. He, however, recognized the loss of steam due to the great variation in the temperature of the cylinder and sought to mitigate it by covering the lower surface of the piston with wood ; he also cased his injection nozzle with a non-conductor in the form of a wrapping of tarred rope.

Smeaton was Watt's senior by twelve years ; he died in 1792.

[1] Manuscript note in an annotated copy of Farey's *Steam Engine* in the Patent Office Library.

[2] B. & W. Colln. : Letter Books. Watt to Smeaton, 1778 Jan. 17.
[3] Ibid. Boulton to Watt, 1781 July 28.

HUMPHREY GAINSBOROUGH

In the course of the actions at law brought by Boulton and Watt against infringers of Watt's first patent, the name of Humphrey Gainsborough was brought up as having forestalled Watt in the invention of the separate condenser. This was twenty years after Gainsborough's death, and the evidence, such as it was, carries no weight whatever, although, as we shall see, Gainsborough did invent something on the same lines as Watt's invention, but at a later date.

Humphrey Gainsborough was the brother of the painter, Thomas Gainsborough, R.A.; he was the minister at the Independent Chapel at Henley-on-Thames, and had a considerable reputation as an ingenious and skilful mechanician and engineer. One of his inventions was a tide-mill, for which he obtained a premium of £50 from the Society of Arts in 1761. R. L. Edgeworth, who lived for some years within a few miles of Henley, knew him well and wrote : ' I do not think that I have ever known a man of a more inventive genius. . . . He was besides an excellent and most accurate workman.' [1]

Edgworth, who later on made the acquaintance of Boulton and Watt, and inspected the Bedworth engine soon after its erection, however, says nothing to suggest that Gainsborough, at the time he knew him, had turned his attention to the steam engine ; indeed, we find Boulton writing :

' I can positively prove by Mr. Edgworth's evidence that in ye year 1769 he mentioned his intention of driving a wheel carriage to Gainsborough and that he found him totaly ignorant of the principles of the Com[m] engine, having at that time never seen one or even made any expts. upon steam or steam engines.' [2]

The earliest authoritative evidence that we have of Gainsborough's connexion with the steam engine is just before the Bill for the extension of Watt's patent was brought before Parliament in February 1775, and it may be that it was the notice that Watt's invention received in public at this time that caused him to direct his attention to the subject.

However that may be, on Feb. 6, 1775,[3] Gainsborough entered a petition for a patent for his invention of ' a steam engine upon a new construction much more useful to the public than the common steam engine by having much greater power and velocity '.[4]

No patent was granted on this petition and this was due primarily to the action of Watt, who filed a caveat against the grant. The effect of this proceeding was that the Solicitor-General could not allow the petition to go forward until he had satisfied himself that Gainsborough had an invention different from that of Watt, or until the caveat was withdrawn. Watt endeavoured to arrange a meeting in London to discuss the matter, and wrote . ' I am sorry to have had occasion to give you this trouble, but judged it better for both parties that the matter should be cleared up now than to be left to be the source of lawsuits afterwards, as might be the case if the inventions clash with one another, as I hope they do not.' However, Gainsborough was ill and unable to come to London, so he wrote, May 16, 1775 :

' Those who know both inventions have assured me that mine is totally different from

---

[1] *Memoirs of R. L. Edgeworth*, 1820, Vol. I, p. 158.
[2] B. & W. Colln. Boulton to Watt, 1781 July 28.

[3] Watt's petition was presented to the House of Commons, Feb. 23, 1775.
[4] State Papers, Domestic. Entry Book, vol. 265, p. 355.

your's. I must therefore leave you to act at your pleasure at the Patent Office, especially as it is impossible for me to be in town at present, and when God only knows, both I and Mrs. Gainsborough being very ill. As you have been ungenteel enough to give me unnecessary trouble, I am only sorry that I did not endeavour to hinder your Bill passing in any form, which I have good reason to believe would have been in my power. However wish you success so far as your invention can go, being well persuaded it will do me no harm, it having once been my own, but was for many reasons given up for that which I am now upon.' [1]

Gainsborough, it will be observed, refers to Watt's plan as one that he had himself thought of and given up for another that he considered better, so it is possible that he had been at work on the subject earlier than is suggested above, but it is clear that he did not realize the wide ambit of Watt's patent, and probably he was ignorant of the fact that it had been taken out six years before this.

After this letter Gainsborough's petition seems to have remained in abeyance until July of the following year when, upon a reminder from his solicitor, Boulton proposed a meeting at the office of the Solicitor-General. Again Gainsborough was unable to attend, but Boulton did meet him later on and seems to have made some proposition, the nature of which is not stated. Here our knowledge of the matter ends. No doubt Gainsborough was in failing health. He died August 23rd of the same year, 1776, but to the last he kept up his interest in engineering affairs, for we are told that he died suddenly near Henley ' while he was conversing with some gentlemen about the locks of the river, which he had constructed '.[2]

The only information that we have of the particular plan that Gainsborough had in view is that furnished by Jabez Hornblower in his account of the steam engine in Gregory's *Mechanics*, and this, it must be noted, appeared about thirty years after the death of Gainsborough :

' About the time that Mr. Watt was engaged in bringing forward the improvement of the engine, it occured to Mr. Gainsborough . . . that it would be a great improvement to condense the steam in a vessel distinct from the cylinder, where the vacuum was formed ; and he undertook a set of experiments to apply the principle he had established ; which he did by placing a small vessel by the side of the cylinder, which was to receive just so much steam from the boiler, as would discharge the air and condensing-water, in the same manner as was the practice from the cylinder itself in the Newcomenian method ; that is by the snifting valve and sinking pipe. In this manner he used no more steam than was just necessary for that particular purpose, which, at the instant of discharging, was entirely unconnected with the main cylinder ; so that the cylinder was kept constantly as hot as the steam could make it.'

Hornblower goes on to say that the model succeeded so well as to induce some of the Cornish mine adventurers to send their engineers to examine it and that their report was so favourable as to produce an intention of adopting it. This, however, was just after the passing of the Act of Parliament for the extension of Watt's patent. ' It was asserted by Mr. Gainsborough ', Hornblower proceeds to say, ' that the mode of condensing out of the cylinder was communicated to Mr. Watt by the officious folly of an acquaintance, who was fully informed of what Mr. Gainsborough had in hand.'

Obviously it would be wrong to attach importance to testimony such as this, so long after the event related. It is quite possible that Gainsborough may have said that Watt found out that he was at work on the subject in the manner referred to, and so have been enabled to hinder him in getting a patent. That Boulton and

[1] Doldowlod Papers. Gainsborough to Watt, 1775 May 16.   [2] Burn : *History of Henley on Thames*, 1861.

Watt had some knowledge of Gainsborough's plan seems quite clear from the fact that at the time they were in difficulties with the piston of the first Soho engine Boulton expressed the wish that they knew how Gainsborough packed his piston.[1]

At Gainsborough's death his engine model, among other things, came into the possession of his brother, the painter, who handed it over to his friend, Thicknesse, the author of the *Sketch of the Life and Paintings of Thomas Gainsborough*. Thicknesse, in turn, presented it to Fores, a bookseller in Piccadilly, who offered to sell it to Boulton and Watt. This was in 1785, nine years after the death of the inventor; Fores refers to it as ' the much improved and last finished model made by the late ingenious Mr. Gainsborough of Henley, for raising water by steam', and says that ' it is very much decayed & disordered by being exposed to the weather by his brother in Pall-mall'; he concludes his letter with the statement, ' I am told it will make any man's fortune who can carry it in to execution.'[2] Boulton and Watt did not buy the model, and its ultimate fate is not known.

THE HORNBLOWERS

The first engine ordered from Boulton and Watt for Cornwall was that for Tingtang mine, in November 1776. Jonathan Hornblower, the elder, was the engineer of this mine, and it was with him that Watt corresponded in reference to the engine-house, and the parts of the engine to be made on the spot.

Hornblower was then sixty years of age, and had been in Cornwall thirty-two years.[3] At the date of his arrival there were five or six engines at work in the county; at this period there were about sixty, and he had put up a good proportion of them; he was recognized as one of the leading engineers of the district, and no doubt knew as much of the construction and working of the Newcomen engine as any man. His experience and knowledge of practical requirements must have been of great assistance to Watt. Hornblower, it is clear, was anxious to understand fully, and to prepare to the best of his power what was required for the new engine, and Watt was equally desirous to give him the fullest information. The correspondence between them is altogether on a cordial footing. In one letter Watt, in sending a makeshift drawing, apologizes for doing so, saying : ' This method of copying drawings of engines whose dimensions vary from yours is irregular and would not be put in practice if I had a less opinion of the skill and experience of the person to whom I send them.'[4] On his first visit to Cornwall, in 1777, Watt reports to Boulton that Hornblower ' seems a very pleasant sort of old Presbyterian';[5] and again, that he ' seems a good sort of man and carries himself very fair but is *I hear* an unbelieving Thomas'.[6]

Hornblower had a large family, including several sons; of these it will be necessary to speak only of Jabez Carter, the eldest, Jethro, and Jonathan.[7]

Jethro may be dismissed in a few words; he was at this time thirty-one years of age and fully established in Cornwall as an engineer; he was engaged in the erection of some of the early Boulton and Watt engines in that county.

[1] Doldowlod Papers. Boulton to Watt [May, 1775].
[2] B. & W. Colln. Fores to Boulton and Watt, 1785 Mar. 11.
[3] His father, Joseph Hornblower, knew Newcomen and went to Cornwall in 1725 to erect a Newcomen engine near Truro.

[4] B. & W. Colln.: Letter Books. Watt to Hornblower, 1776 Dec. 18.
[5] Boulton Papers. Watt to Boulton, 1777 Aug. 9.
[6] Ibid. Watt to Boulton, 1777 Aug. 14.
[7] One of the younger sons was the inventor of the double-beat valve.

Jabez and Jonathan demand a more lengthy notice ; they were both thorns in the sides of Boulton and Watt, the one with a rotative engine, and the other with his compound engine. In the first place we will consider Jonathan.

Watt had conceived the idea of expansive working of steam in 1769, and the second Soho engine built in 1777 was designed for this mode of operation ; further, we have seen that in September 1781 (see p. 124) he had made a drawing of a double expansive engine, using steam first in one cylinder and then in another, but that he had contemplated using cylinders of the same size.   Quite independently Jonathan Hornblower the younger hit upon the plan of using steam in two cylinders, and he made his second cylinder of larger capacity than the first ; he built a number of such engines and is fairly considered to be the inventor of the compound engine.   From his own account[1] the invention originated in 1776, when he was twenty-three years old ; he made a small model and submitted it to his father.   The elder Hornblower considered that the covering in of the tops of the cylinders would present serious difficulties in practice, and the model was laid aside.   Soon afterwards Watt's engine made its appearance in Cornwall, and young Hornblower, seeing that it had a cylinder cover and stuffing-box, and that it worked satisfactorily, again took up his idea and made a larger model.   It appears that an attempt was made to work compound at Wheal Maid, and in 1781 Hornblower obtained a patent.   The first engine built was at Radstock Colliery, near Bath ; it had cylinders 19 in. and 24 in. diam., strokes 6 ft. and 8 ft. respectively.

At an early stage of the proceedings rumours of a competing engine reached Watt ; at first he was under the impression that it was a hot-air engine, but very soon he was in possession of the correct story, and there is no doubt that he was seriously annoyed and alarmed.   His letters on this matter, most of which have been published,[2] show Watt in a somewhat unpleasant light.

There were a number of preliminary troubles with the Radstock engine, but it seems to have been got to work satisfactorily towards the end of 1782.   At about the same time Boulton and Watt informed the proprietors of the colliery that the engine was an infringement of their patent and that they intended to bring an action against Hornblower.   However, no steps were taken in this direction, and a year later the Radstock people asked Boulton and Watt whether they had relinquished the idea ; they had refrained from paying Hornblower for the engine, and now wished to have the matter settled.   The reply they got was not satisfactory ; Watt would not commit himself one way or the other.

Hornblower seems to have been satisfied in his own mind that he was not infringing Watt's patent.   He considered that he was condensing in the cylinder, so had no separate condenser, and thought that the only point that could be in dispute was the use of steam to press upon the piston ; but this idea, he said, had been published long before Watt's time, there being in existence an engraving ' where the expansive power of steam is employed to impel the piston, on a vacuum which is made in the upper part of the cylinder, and where the piston is connected to an iron rod that moves thro' the cylinder lid '.[3]   Hornblower gives no particulars for identifying this engraving, and nothing which agreed with his description of it was brought forward in the litigation that took place later on in respect of Bull's and Jabez Hornblower's engines.

---

[1] Howard's *New Royal Cyclopaedia* (1788).
[2] In Muirhead : *Mech. Inv.*
[3] An address to the Lords, &c., in the Mines of Cornwall by Jonathan Hornblower and John Winwood. Penryn, May 1, 1788.

Watt, on the other hand, prepared an advertisement for insertion in the Bristol newspapers, in which he enumerates six features, ' all original inventions of J. Watt, and whoever makes, vends, or uses fire engines constructed upon any of these principles shall be prosecuted by Boulton & Watt, the patentees.' The features are : (1) Cylinder with closed top ; (2) Piston pressed down by steam ; (3) Steam case, or non-conducting casing to cylinder ; (4) Separate condenser ; (5) Air-pump ; (6) Piston kept tight by oil or grease.[1]

The Radstock engine, in the first instance, had a multitubular surface condenser in the lower end of the large cylinder, below the working space of the piston, but on one occasion the beam chains broke and the piston came down and destroyed the tubes, and the surface condenser was then replaced by a jet acting under a perforated false-bottom; see Plate XCVII. Later on, possibly after Hornblower had ceased to be concerned with it, the engine was fitted with a separate condenser. The condensing arrangement shown in Hornblower's account of his engine in 1788 [2] has the injection nozzle at one side of the bottom of the cylinder and the eduction pipe at the other, ' the water spurts across the cylinder into the mouth of the pipe . . . and in its passage condenses all the steam below the large piston.' Hornblower was now using an air-pump, and it is interesting to note that his stuffing-boxes were formed with steam chambers. His engines had the two cylinders side by side working on to one beam, so that one had a shorter stroke than the other. Watt's double-cylinder arrangement, it will be remembered, comprised two separate engines.

The Cornish mines did not rush forward with orders for Hornblower's engines, but in 1784 he was putting one up at Penryn, which, however, did not turn out a success. Another engine put up at Tincroft in 1791, after the initial troubles had been overcome, worked satisfactorily. Trevithick, then at the beginning of his career, was employed by the adventurers of the Tincroft mine to examine and report on the performance of this engine as compared with one of Boulton and Watt's. He arrived at the conclusion that the two engines did about the same duty. The success of the Tincroft engine led to other orders, and altogether nine or ten engines on Hornblower's plan were set up in Cornwall.[3] The largest of these engines had cylinders 45 in. and 53 in. diam. There was also another engine near Bath, besides that at Radstock. The Tincroft engine, as built, had the condensing arrangement as described by Hornblower in 1788, see above ; we do not know whether this was retained, whether it was used for the other engines set up in Cornwall, or whether, as in the case of the Radstock engine, a separate condenser was applied later on. In any case Watt held that this plan of condensing in the bottom of the cylinder was merely an evasion of his own invention, and Hornblower, it is clear, was using all the other features of which Watt considered that he had a monopoly, i.e. the cylinder with closed top ; piston pressed down by steam ; non-conducting casing to cylinder ; air-pump ; and piston kept tight by oil or grease. In the end the owners of the Hornblower engines in Cornwall paid up the premiums demanded by Boulton and Watt. Apparently an action at law had been commenced in respect of one of the engines, but the mine-owners gave in before the case came on for trial. This was subsequent to the actions against Bull and against Jabez Hornblower and Maberley.

[1] B. & W. Colln. 'Advt. put in the Bristol Papers, 1782.' Draft by Watt.
[2] Howard's *New Royal Cyclopaedia*.

[3] Tincroft (two engines), Swanpool, Lost-withiel, Tresavean, Baldue, Wherry Mine, Wheal Unity, and Wheal Pool.

Hornblower then was in the position that he had a patent which he could not work without infringing Watt's patent. Boulton and Watt had constantly maintained the position that they would not license other engineers to build engines including their inventions. Their patent by the prolongation granted by Parliament held good until the year 1800, by which time the fourteen-year protection accorded to Hornblower by his patent of 1781 would have expired. If, however, he could carry on the life of his patent after the expiration of Watt's term, it might be possible for him to reap some benefit from it. Accordingly, in 1792, he applied to Parliament to have his patent prolonged. Boulton and Watt opposed his Bill most vigorously, and Hornblower failed to get his Act.

Finding himself debarred from proceeding with the reciprocating engine, Jonathan Hornblower took up the subject of the rotary engine, upon which he took out two patents and spent several years' work. He died in 1815 at the age of sixty-two. It is stated that he amassed a considerable fortune as an engineer. In an obituary notice in the *Gentleman's Magazine* he is referred to as ' a very eminent engineer'. With the steam-pressures used by Hornblower there could have been no appreciable advantage in the use of two cylinders ; the year of his death saw his compound engine set up in Cornwall by Woolf, who, being in a position to use higher steam-pressures and consequently greater expansion, obtained a high degree of efficiency.

To turn now to Jabez Hornblower. At the time of Watt's first visit to Cornwall he had just returned from Holland, where he had been employed by the Dutch Government on a pumping-engine for drainage purposes. It is said that at one time he had made an inverted-cylinder engine, but his first occupation upon his return home was the construction of an instantaneous steam-generator, ' a dry boiler (similar to that Mr. Payne tried at Wednesbury & in Derbyshire) being kept in an hot state and a small stream of water being conveyed in through the boiler top. by means of a cog wheel disperses the water in a centrifugal motion round its sides '.[1] After the failure of this he was employed at Tingtang for a short period, but was turned off because, according to Watt, he would do his work well and consequently expensively.[2] Watt, at this time, considered Jabez to be very clever, a good engineer, and industrious, but, he says, he ' seems not to have the faculty of conciliating people's affections'. There was an idea that he might go to Holland again, and as he would now be able to take with him a knowledge of the construction and working of the Watt engine, which had not yet been patented in that country, it was a matter of policy to keep him from going there. Accordingly, Boulton and Watt offered him a job at Soho at a guinea a week (i.e. the same wages as Murdock was then receiving) with an allowance of three guineas for his expenses in travelling to that town. Jabez took this offer, and when in Birmingham he seems to have been an inmate of Boulton's house. In the early part of 1779 he was out erecting engines at Donnington Wood and Ketley, and it was at this time that he suggested the improvement in the disposition of the blowing-through nozzle that was adopted by Boulton and Watt (see p. 209). Jabez Hornblower's bad temper and ungenial disposition, however, were very much against him, he could not get along with the people for whom he was sent out to work, so in spite of the danger of his going to Holland, when he was sent to erect the engine at Penryndee, Carnarvonshire, in September 1779, he was told

[1] B. & W. Colln. Dudley to Watt, 1778   [2] Ibid. Watt to Boulton, 1778 Sept. 6.
Mar. 26.

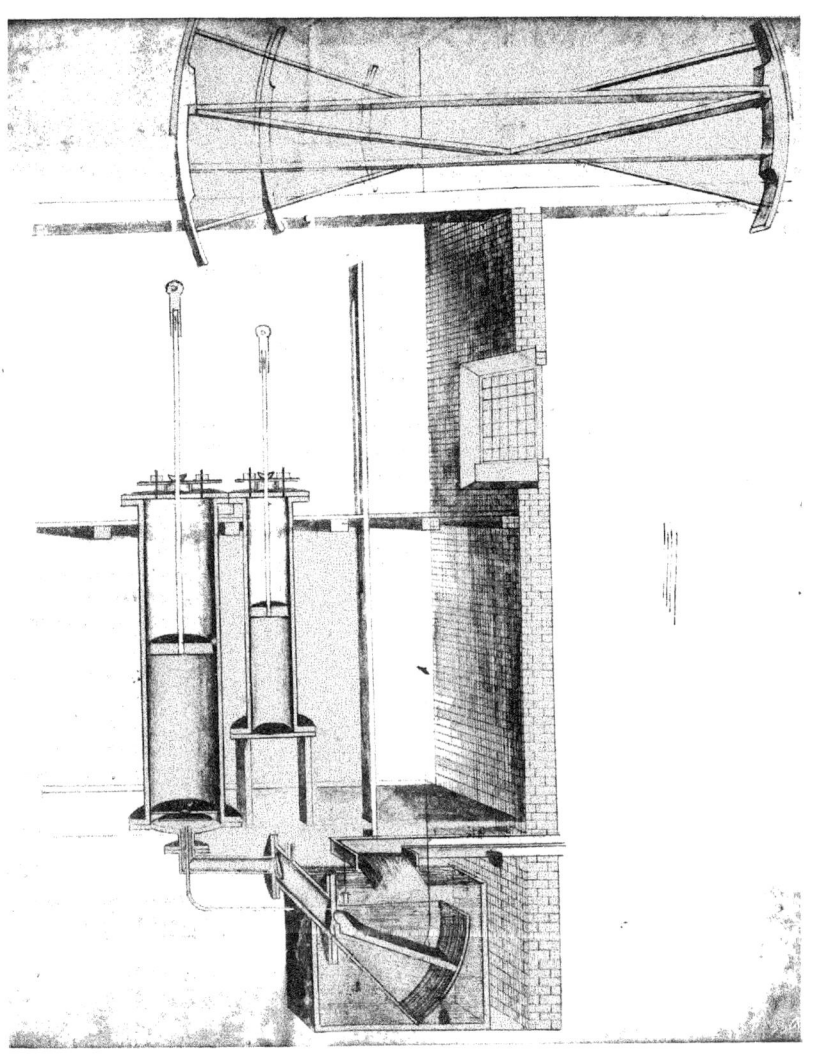

XCVII. HORNBLOWER'S COMPOUND ENGINE

that he must make his own arrangements as to pay with the proprietors of the mine and that he was not to consider himself any longer as on the staff of Boulton and Watt. In the spring of 1781 he was applying for the job of overlooking the erection of an engine at Coalbrookdale, but his bad reputation had spread abroad, and they would not have him there.[1]

After this he probably returned to Cornwall and worked with his brother Jonathan at the compound engine; at any rate in August 1784 he was back in Birmingham, recruiting engine-men and smiths. For the next eleven years the Boulton and Watt correspondence contains no reference to Jabez Hornblower, and it would seem that during this period he was engaged in civil engineering and building work. In a memoir given in Stuart's *Anecdotes of Steam Engines*, 1829, it is stated that: ' The improvements which forty years ago were made in the towns of Truro and Tewkesbury, and the erection of the Penitentiary at North-leach were entirely under his direction.'

However, by the year 1790 or soon after, Jabez had gravitated to London and set up in business as a maker of steam engines and other machinery. In 1795 his name reappears in the Boulton and Watt correspondence in conjunction with that of a partner, Maberley, as the maker of an engine which clearly was an infringe-ment of Watt's patent ; it had the separate condenser, closed cylinder top, and steam acting on the upper surface of the piston.[2] To describe it more particularly, the engine had two single-acting cylinders, the pistons of which were suspended by chains from a wheel on an overhead shaft, so that as one moved down the other moved up. The wheel had a plain periphery upon which the end of each chain was secured, and the overhead rocking-shaft carried an arm, the free end of which was coupled by a connecting-rod to a crank on a rotating shaft. There were two condensers and two air-pumps, the buckets of which were worked by chains from a small wheel on the overhead shaft. From an account, not very intelligible, by McMurdo, one of Boulton and Watt's erectors, it seems that there were valves in the pistons and that the cylinders were cross-connected, so that steam passed from one to the other, but each cylinder had a separate steam supply and a separate condenser and air-pump. McMurdo had examined the engine at Newbottle Colliery ; he said that it worked very well, was very manageable, and that he had heard no complaints about it. The design of this engine was based on a patent granted in 1791 to Isaac Manwaring, a saw-maker in Clerkenwell, for a ' pendulum steam engine with two or more cylinders to work immediately upon a wheel or wheels without a crank, long beam, or leaver, as now used '. Manwaring had a pair of open-topped vertical cylinders, and his piston-rods were formed with racks to engage on opposite sides of a wheel on an overhead shaft, the wheel being coupled to the shaft by ratchet-and-pawl arrangement. This patent had been acquired by Maberley, and when Hornblower joined him he set about adding his own improvements and modifications.

At the instance of Boulton and Watt injunctions were issued in January 1796 against Hornblower, Maberley, and Peareth, a coal-owner for whom one of the engines had been erected near Newcastle. Hornblower and Maberley sought to come to terms whereby they might be allowed to continue the manufacture of their engines, in fact to become licensees under Watt's patent for the remainder of its term, but the negotiations came to nothing and Boulton and Watt brought an

[1] Ibid.   Joseph Rathbone & Co. to Boulton & Watt, 1781 Apl. 16.

[2] There is a model of this engine at the Science Museum, South Kensington.

action against them for infringement. The case was finally decided in favour of Boulton and Watt in 1799. Under this reverse Hornblower's resources broke down and he seems to have been for a time in a debtor's prison. However, in the following year we find him taking out a patent for glazing calico ; this invention was tried in practice, but although it is said to have worked well it was not a financial success. Some years later Hornblower was engaged to design and erect a large malting and brewing establishment in Sweden ; he returned to England in 1813 and died in London in 1814.

Jabez Hornblower was the literary member of the family. In addition to the account of the steam engine that he contributed to Gregory's *Mechanics*, he wrote a number of papers and letters for *Nicholson's Journal* and other periodicals.

Arthur Woolf was for a time in the employ of Hornblower and Maberley. He put up the engines for winding at Charters Haugh and Newbottle in the Tyne district, and probably that at Meux's brewery in London, at which establishment he was employed as engineer from 1797 to 1806. Later on he was in business as an engine-builder, first in London and afterwards in Cornwall.

JOHN BUDGE

At the date of the introduction of the Boulton and Watt engine John Budge was one of the best-known engineers in Cornwall, in fact he shared with Jonathan Hornblower, the elder, the leading position. One of the Cornish correspondents of Boulton and Watt said of him : ' Though perhaps you may not find Mr. Budge inclined to talk very scientifically, even on the subject of engines, yet almost one half of those in this county are under his care, and were built by his direction, and he is esteemed one of the best practical engineers in it.' [1]

The occasion of this letter was the ordering of an engine for Wheal Union (Tregurtha Downs) and the dispatch of Budge, who it was intended should supervise its erection, to see the engines then at work in the neighbourhood of Birmingham. Budge came to Soho but did not appear impressed, and Watt feared that he had imbibed some prejudice against the engines from the fact that the Soho engine was in bad order and that they ' were obliged to work with steam 6 inches strong ', at the same time he admitted that Budge was at great pains to inform himself about the engine.[2]

In spite of this, it is clear that Budge had not grasped the principles of the new engine, for he returned to Cornwall unconvinced of its superiority, and suggested that Boulton and Watt were working with a slack bucket in the pump, so that there was only a show of working a large pump. In the end he declined the appointment to put up the engine. This was about the time of Watt's first visit to Cornwall. Budge seems to have kept out of his way for some time and Watt felt hurt at this. However, after a month or so had gone by, Budge did call and promised to read his recantation as soon as he was convinced and never to touch a common engine again. Watt had inspected five of Budge's engines and ' was far from seeing the wonders promised '.

In the following year, however, when he inspects some of the engines with Budge, he says they are very good ones and in good order. Budge about the same time went to see the Wheal Busy engine at work, and Watt says he praised

[1] B. & W. Colln. Edwards to Boulton, 1777 June 14.   [2] Ibid. Watt to Dudley, 1777 July 1.

the going of the engine and said that it was 'better than those he had seen of ours in the north'. Watt seems to have thought at the time that Budge intended to give up work, and he remarks that he is rich and has no family. Budge, however, continued in harness a few years longer ; in 1779 Watt writes in reference to the Wheal Chance engine that ' Mr. Bouge has built a very good house and has done everything in a masterly manner ', so that he exhibited his recantation in a practical manner by concerning himself with the erection of one of Watt's engines.

But no doubt he was getting old, had no particular reason for money-making, and wished to ease his burden, so we find him in 1780 inviting Murdock to enter into partnership with him and offering half of all his engines. Budge was the author of a plan of working engines that is spoken of as ' Stretching the Steam ', and he was the inventor of a boiler ; it was of the haystack form with a flue extending across it at an angle to the horizontal. It seems probable too that he is identical with the John Budge of Camborne who patented in 1772 a ' machine for raising metals, minerals, or other heavy materials . . . by means of a double scrole or barrel'. Moreover, we judge that he must have been the Mr. J. Budge of Camborne, who died in 1823, at the age of ninety-three years, a memoir of whom appears in the *Wesleyan Methodist Magazine* for 1824. According to this memoir he had been confined to his house for several years.

Both Boulton and Watt in their letters usually write the name of Budge as Bouge; probably this was a phonetic spelling of the name as they heard it pronounced in Cornwall. Smiles in *The Lives of Boulton and Watt* printed this name as ' Bonze ' ; the error was one of transcription from Watt's letters and was one that a copyist would readily fall into.

EDWARD BULL

ONE of the most familiar names in connexion with litigation in reference to Watt's patent is that of Edward Bull. Very little is known about the man ; in the legal proceedings he is contemptuously spoken of as a ' stoker'. As a matter of fact we first hear of him in 1779 as the engine-man at Bedworth, and getting a pay of 11s. a week. In that capacity he would no doubt have the boiler to attend to, but he had more to do than that, and must have been familiar with the engine. At any rate, in 1781 he was on the staff of Boulton and Watt and was one of the erectors sent to Cornwall that year. Up to the end of 1782 it is clear that he was one of the firm's erectors, and then we lose sight of him for some years. Probably as the demand for engines was supplied and fewer erectors were required, Bull took a situation as engine-man, for he seems to have remained in Cornwall, and to have become known to a good many of the people concerned in the mines. In 1791 he was in the position to have the erection of an engine or engines confided to him. He applied to Boulton and Watt for permission to make engines on their principle ; this was refused. However, he entered into a contract with the Wheal Rose adventurers. On behalf of Boulton and Watt these adventurers had been informed that the premium for an engine adapted to do the work they required would be £50 per month, increasing as the shaft was sunk deeper. ' Bull agreed to build the engine, furnish pumps & every article & keep them in repair & to work it to 45 fathoms for 42l. per month, to increase to 70l. in proportion to the water.' Wilson, who furnishes this information, supposes that he is backed by Martyn and says that Bull means to use a compound, but is not working with the Hornblowers. The Wheal Rose adventurers later on said that they did not buy an engine from Bull, but that they contracted with him to pump their water at a certain sum per gallon. In all, Bull seems to have put up ten engines.[1] One of the first was that at Balcoath mine which commenced work early in 1792. This engine is shown in Plate XCVIII, taken from a drawing in the Boulton and Watt Collection, probably made by Murdock. The characteristic feature of what became known as the Bull engine is that the cylinder was inverted and placed directly over the pump, the rod of which was coupled directly to the piston-rod. The drawing calls for no description ; it will be noted that the upper end of the cylinder is permanently open to the condenser, and that the air-pump and plug-rod are worked from a counterbalance lever coupled to the pump-rod.

The Bull arrangement was not a new one, nor does it appear that Bull claimed to have invented it ; possibly it was adopted in part with the idea of reducing the cost of construction, and in part to make the engine look as little as possible like the Watt engine. For there is little doubt that Bull was but an instrument in the hands of a section of the mining interest in Cornwall. There was some ground for thinking that Watt's patent would not stand the test in a court of law, and this section desired to have the matter brought to an issue. Bull himself was a poor man, and it is clear that without substantial backing he could not have entered upon a long and costly litigation. The Balcoath engine was made the subject of an action for infringement ; as this action will be dealt with later, here it will be sufficient to say that the jury had no hesitation in finding that Bull had infringed, but certain objections to the validity of the patent had to be remitted to another court, and a final decision was not arrived at until 1799. Bull, assisted by Richard Trevithick, continued to put up engines until 1795 when injunctions were served upon both, and this marks the end of his short career as an engine-builder ; indeed he did not live long after this, for he died before 1800.

RICHARD TREVITHICK

The life of Richard Trevithick has been written in considerable detail, but there is material in the Boulton and Watt Collection to which it is fitting that attention should be drawn ; in the main of course it concerns the relations of Trevithick to the firm. The earliest letter noted is one from Captain Thomas Gundry of Goldsithney in 1796, when Trevithick was twenty-five years of age. For a few years Trevithick had been assisting Edward Bull in the erection of his engines. Bull had in 1793 lost the first round in his fight with Boulton and Watt in connexion with the infringement of Watt's patent and the firm had caused an injunction to be served upon him restraining him from putting up any more of his engines. Trevithick was now desirous of employment under Boulton and Watt and Captain Gundry writes to them :

' I have taken the liberty to trouble you with this by desire of Richd Trevithick, Jr., who have been for some time past employed by Edwd. Bull in mechanism. He desires not to continue in opposition to you, and is ready to give up everything in this county, and be under your direction. If you should employ him, you will certainly find him possessed of good abilitys in mechanics, natural as well as acquired, and is of an honest and peaceable disposition, he would be glad to serve you either in Cornwall or Soho, *the latter place in particular*. If this step is taken I think the opposition in Cornwall would to a great measure subside. I would

[1] Balcoath, Wheal Rose, Wheal Treasury, Carzize Wood, Herland, Wheal Leids, Wheal Ann, Retallack, Pednandrea, and Ding Dong.

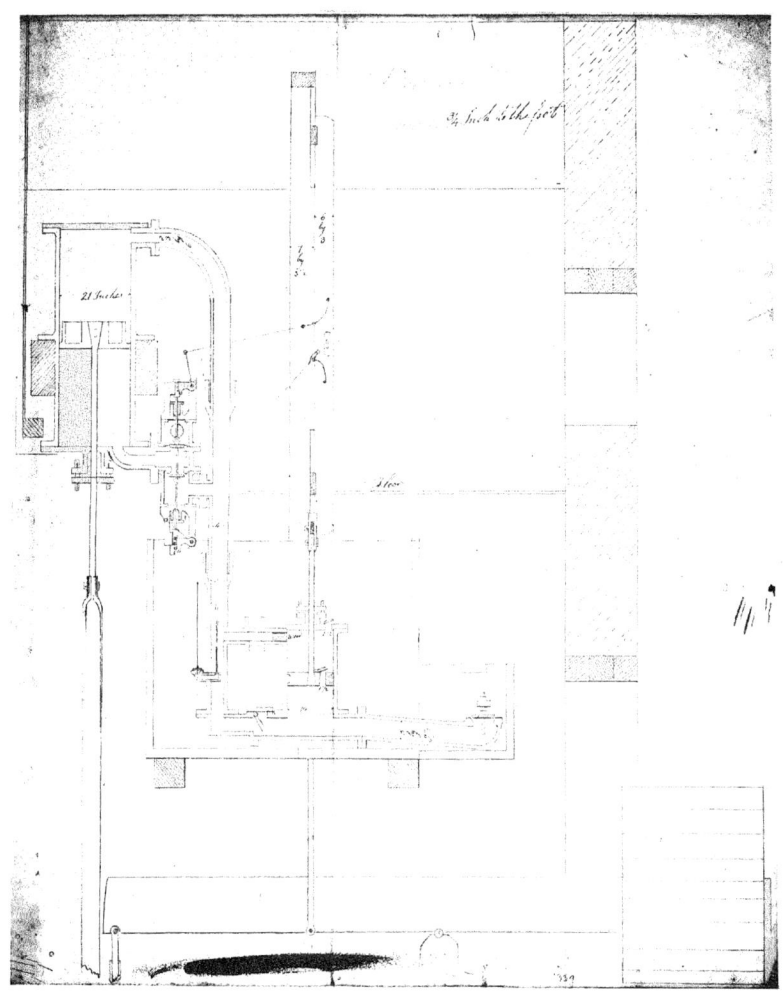

XCVIII. BULL'S ENGINE AT BALCOATH

esteem it a peculiar favour if you would take the matter into consideration and don't doubt but that any favour conferred will be gratefully acknowledged by him, as well as, yr. m⁰. obedᵗ. Servᵗ. Thos. Gundry.'

Gundry was a man whom Boulton and Watt held in great esteem, and no doubt they gave his letter careful consideration. Murdock, however (he had not yet left Cornwall), was against granting the request, and Trevithick was not engaged. Towards the end of the same year, 1796, both Bull and Trevithick were at Soho Foundry, apparently making a call upon another Cornishman, Vivian, who was one of the foremen there, with whom and Foreman, the clerk, they took dinner. Curiously enough, Trevithick who had evaded the service of an injunction in Cornwall was caught and had it served upon him at Birmingham. We have an account of the dinner in one of Southern's letters :

' Bull took his dinner quietly, but Trevithick walked backwards & forwards in the house like a mad man and firmly resisted all temptation to dinner, till the smell of a hot pie overcame his powers, on which he set to, & did pretty handsomely, but in such a manner as shewed him not quiet in mind. The rejoicing that was going forward at the very door, exhibited by fires and gunpowder work, not a little contributed to his *happiness* while he stay'd which was for a very short time indeed. They posted off towards Colebrookdale.' [1]

Southern in another letter is careful to state that neither Bull nor Trevithick was allowed to see the Foundry. The rejoicings referred to were due to the news of the result of the trial in the action against Hornblower and Maberley.

In the course of the following year Trevithick was making efforts to obtain the sanction of Boulton and Watt to his employment by the adventurers on certain mines in putting up engines under their patent, and he was anxious to contradict a rumour that Bull was in partnership with him. The firm saw that there were advantages to be gained by engaging him, but it seems that Trevithick refused the terms they offered. He was certainly engaged to put up an engine on Prince William Henry mine, for in December 1797 there is a letter from which it appears that the adventurers had engaged him, not knowing that the firm would have any objection, and they were now asking for the approval of the engagement. It will be understood that the mine-owners were free to erect the engine themselves, but that, under the agreements, Boulton and Watt had the right of veto in regard to the engineer in charge of the erection.

It was about this time that Trevithick took up work on the high-pressure engine. The explosion of the boiler of his engine at Greenwich is duly reported to Boulton and Watt by one of their London erectors, and then in 1804 another erector who had just returned from Wales ' reports that Trevithick & Homfray have quarrelled and that the former has taken himself off in dudgeon. The railway engine is in disgrace and the one sold for winding coals returned upon his hands, which Homfray says is owing to Lawson's interference.' [2]

MATTHEW WASBROUGH

The name of Matthew Wasbrough is identified with the application of the crank to the steam engine, a matter that has been dealt with in Chapter XIII. He was born at Bristol in 1753, and when he reached manhood he joined a brass-founding and clock-making business in which his father was a partner. In 1778,

---

[1] B. & W. Colln. Southern to Boulton & Watt, juniors, 1796 Dec. 22. [2] Ibid. J. Watt, junior, to Southern, 1804 Apl. 23.

when twenty-five years of age, he set about devising means for driving the lathes and other machines in the shop by a steam engine and for this purpose he invented a pawl-and-ratchet mechanism for converting the reciprocating motion of the beam into rotary motion. He was desirous of having a Boulton and Watt engine, but the firm were then full of work and were not anxious for his order, so Wasbrough was driven to use a Newcomen engine and in January 1779 he writes : ' The engine at our manufactory is nearly finished on the old plan for want of a conclusive agreement between us & I could not wait, likewise one more which is in hand, both of 2 foot dia^r.' He adds that he has orders for several other engines on hand.[1] At the beginning of April he writes that the engine is working much to his satis-faction, and again expresses his desire to come to terms with Boulton and Watt for the use of their engine upon which to apply his rotative mechanism.[2] At the same time he was making some copper eduction pipes for Boulton and Watt. These when delivered were found defective, which caused Watt to say: ' If M. Wasbrough makes no better engines than he does eduction pipes, he will soon be blown, it is the worst job you ever saw.'[3] Later on, Watt expressed himself in very emphatic terms as to Wasbrough's want of ability as an engineer : the general opinion was, however, that Wasbrough was a young engineer of great promise, a promise that was unfulfilled, for Wasbrough died, quite a young man, in the year 1781.

FRANCIS THOMPSON

THE Boulton and Watt engine at Gregory mine, Ashover, Derbyshire, was put up in 1780 by Francis Thompson; he had paid a visit to Soho in 1779 in reference to this engine, and he seems to have been concerned with the Yatestoop engine also. At this date Thompson had been in business as an engineer for some years and had built a number of Newcomen engines in Derbyshire, among them one at Gregory mine in 1774, when he was twenty-seven years of age. About the year 1785 his business as an engine-builder had fallen off, and he seems to have employed himself in experimenting with an air-engine. Two air-pumps worked by a water-wheel were used to set up a vacuum on one side of a piston in a cylinder, which had a valve to admit air to return the piston ; the piston was clothed with leather and covered with water.[4] Nothing came of this scheme, and in 1789 we find Thompson in conjunction with Smith, the ironfounder at Chesterfield who cast the cylinder of the Pentrich engine, referred to below, offering to supply a pumping-engine to the Commissioners of Middle Fen district.[5]

Farey gives an account of an atmospheric pumping-engine erected under-ground at the Yatestoop mine by Thompson in 1782 ; he mentions him as one of the principal makers of atmospheric rotative engines in the last decade of the eighteenth century, and states that he supplied not only Derbyshire, but Sheffield, Leeds, and other districts. One of his engines was applied to drive the first steam-mill for spinning worsted, near Nottingham, others for cotton mills at Macclesfield and Manchester. His engine was a double-acting beam engine with two cylinders

[1] B. & W. Colln. Wasbrough to Boulton & Watt, 1779 Jan. 30.
[2] Ibid. Wasbrough to Boulton & Watt, 1779 Apl. 3.
[3] Boulton Papers. Watt to Boulton, 1779 Oct. 4.

[4] B. & W. Colln. John Stratford, engine-man at Gregory mine, to Southern, 1785 Feb. 28, Apl. 8, and Aug. 12.
[5] Ibid. Robert Wild to Boulton & Watt, 1789 Sept. 3, Nov. 16 ; Rennie to Watt, 1789 Nov. 2.

in tandem, patented in 1792 ; it is said to have worked fairly well, but of course it was inferior in efficiency to Watt's engine, and Farey states that in 1797 Thompson put up two Boulton and Watt engines at a cutlery-grinding establishment in Sheffield.[1] A pumping-engine formerly at Pentrich, but now in the Science Museum, South Kensington, built by Thompson in 1791, has been described by Mr. W. T. Anderson in a paper which throws a good deal of light on Thompson's career.[2] He was one of the sons of a Stephen Thompson of Winster, and his name first appears in the books of the Ashover mine in 1770. He died in 1807, at the age of sixty-two.

ROBERT CAMERON

CAMERON started as a workman at Soho in the year 1780. It appears that he then had an idea of going to the West Indies as a millwright. Boulton and Watt thought he might be useful in erecting their engines out there and every opportunity was given him of thoroughly understanding their construction and action. It was he who fitted the swashplate rotative mechanism to the Soho 18-inch engine in 1781, a job that he carried through much to the satisfaction of Boulton. After this he was sent to Cornwall as an erector, and later on, in 1783, he was engaged in putting up the forge engine at Bradley Ironworks.

Before he came to Birmingham Cameron had caught the rotary-engine fever. He mentioned his scheme to Watt, who tried to turn him off it, telling him that the same idea had been in his mind years before and that he had made a model and, moreover, that in any case it could not be worked without infringing his patent, as it required a separate condenser.

Upon his return from Cornwall, Cameron still had the rotary engine in his mind and, finding that Watt would take no notice of his scheme, applied to Boulton for permission to make a model of it. This Boulton gave without consulting his partner, who did not learn of it until a few days after, when he thought it too late to interfere. The model when made would turn round, but it would do nothing else. In the meantime Watt had applied for another patent, the specification for which was dated July 3, 1782 ; it was written, Watt says, ' before the model was made, and delivered into Chancery about the time it was finished.' [3] This specification includes, *inter alia*, three forms of rotary engine. Boulton had by this time found an old and unfinished model of one of Watt's schemes, made in 1765 or 1766. When then Cameron asked to be allowed to go on with his rotary engine and effect further improvements, this model was displayed to him and he was told of the new patent. It was perhaps too much to expect Cameron to take this news very quietly—he said point-blank that Watt was not the inventor and, according to Watt, expressed himself in a very insolent manner to Boulton. Watt was for dismissing him forthwith, but Boulton, always reluctant to take this extreme step, made him certain offers, and allowed him to go on with his engine, and it would seem that he was allowed to work at it, possibly with intervals at other jobs, for some months, but at last Boulton put his foot down and wrote that Cameron was not to be allowed to begin another rotary engine (Feb. 15, 1783).

Cameron seems to have left Boulton and Watt in the latter part of 1783, at any rate in January 1784, when he took out his first patent for a steam- or fire-

[1] Farey: *Steam Engine*, 238, 422, 427, 508, 658, 661.
[2] *Trans. Inst. Min. Eng.*, Vol. LII, Part 4.
[3] Muirhead: *Mech. Inv.* Watt to Hamilton, 1782 Sept. 22.

engine, he was in London, living in Clerkenwell. Very soon he was trying to get his new invention taken up in Cornwall, for we find Boulton writing (Jan. 17, 1784) to Watt from Cornwall : ' Cameron is plagueing us here as much as Horn-blowers ever did.' As soon as it was possible to do so, Watt got a copy of Cameron's specification ; he failed to find therein anything that clashed with his own specifications, but thought that Cameron would be driven to use a separate condenser. Cameron himself protested that he used nothing of Watt's and did not intend to do so, and that he wished to proceed in amity with Boulton and Watt.[1]

Before the end of 1784 Cameron had got a moneyed man behind him in the person of Alexander Blair of Portland Place, London, and had established a work-shop at Green Dragon Wharf, Narrow Wall, Lambeth. Here he had his first engine running in January 1785 ; it was applied for driving his patent sawing-machinery for cutting up veneers. Later on in the year he put up an engine for a snuff mill ; this was found to be deficient in power, to make up for which the steam-pressure was raised until finally the boiler burst. In September Rennie writes to Watt that the engine has been set going again, he says that it ' has a double piston & division in the cylinder as usual, but, instead of forcing the condensed water out at the side of the cylinder, he has made the upper piston a pump bucket with valves. This piston pulls up the condensed steam to the top of the cylinder & there throws it off.' [2]

Cameron's next engine was for an oil mill at Battersea ; this was started in May 1786, and then, before the end of the following year, we learn that he had erected an engine at Newcastle and had an order for one for a paper mill.

By this time Cameron had taken out his third patent for methods of raising coals, ores, and water (Jan. 28, 1786), and a partnership had been formed including himself, Blair, and a man named Jeffries. The connexion was not a happy one ; it is clear that Cameron was a difficult man to work with, and by the middle of 1789 Blair and Jeffries were trying to induce him to withdraw from the concern, and were offering him the sum of £3,000 down, or an annual payment of £200 for the rest of his life. Without informing him of the position, Cameron approached Boulton to see if Boulton and Watt would take him up, and Boulton reports to his partner that Cameron ' would prefer being concerned with us than with his present partners'. It seems that Cameron had put up five coal-winding engines, and that one of these in the Newcastle district brought up 300 tons a day from a depth of

---

[1] In the correspondence between Boulton & Watt at this time Cameron is referred to under the name 'Wolf'. It was a fairly common practice for them to agree upon some disguised name, possibly as a precaution in case their letters fell into improper hands ; we have for instance 'Wasp', 'Hubble Bubble', and 'Six Sharp-Shafts'. The name 'Wolf' is applied to three different persons. First a man whose proper name was George Wolff, and with whom on one occasion Boulton dined at Clapham ; next, Arthur Woolf the Cornishman ; and then Cameron. There is little or no difficulty in distinguishing the man to whom the name is applied in each case.

[2] Farey has a fuller account of the construction of Cameron's engine. ' The cylinder is divided into two parts by a partition in the middle, so as to form two distinct cylinders, one above the other ; and there are two pistons fixed to an iron rod finely polished, and passing through the centre of the partition. The two pistons move both together, upwards and downwards ; each piston makes a stroke of 28 inches, and they are said to make 45 strokes per minute. The under part of the cylinder only has a communication with steam, and the upper part with air ; one of the pistons acts as an air-pump, making a complete vacuum every stroke ; the under part of the cylinder which receives the stream of injection is very hot, but the upper part is cool.' Farey : *Steam Engine*, 656 *n*.

95 fathoms. In this year 1789 he took out another patent, his last, for machines for raising coals, ore, and water in mines.

For the next few years we have no information either as to the man or the engines, indeed of Cameron we know nothing after 1789, but at the beginning of 1795 it is clear that Jeffries was the sole partner in the concern and was putting up engines near Newcastle ; but he was doing more than this, he was threatening legal action against mine-owners who were putting up atmospheric winding-engines with cranks, on the ground that they were infringements of Cameron's patent of 1784. Thomas Barnes, the agent for certain of the collieries, had arranged with Boulton and Watt for the erection of two winding-engines, and he at once wrote to the firm drawing attention to this advertisement and asking what position Cameron's patent really occupied. Later on we find Barnes informing Jeffries that Boulton and Watt are making two engines for him and have engaged to indemnify him for any claim that Jeffries might make to the exclusive right of drawing coals by the direct action of the steam engine without the intervention of a water-wheel. Jeffries upon this wrote a straightforward letter to Watt asking him for reasons for supposing that Cameron's patent was not valid and stating that in his opinion Watt was a better judge on such a matter than any lawyer could be. Watt in reply confined himself to stating that the application of the steam engine directly for winding coal could not be good subject-matter for a patent, since it had been done many years before, and that Cameron's patent would be valid for the particular machinery described, in so far as it was new at the date of the patent. Jeffries withdrew the action for infringement that he had begun, but he continued engine-making for a few years more, until 1798. It appears that to the last the engines were being made with two pistons and a central partition in the cylinder, as mentioned above.

JOSEPH HATELEY

At the time that Watt was putting up his experimental engine near Dr. Roebuck's house at Kinneil, Joseph Hateley was in the employ of the doctor and was working out an invention of his own. In 1768 he obtained a patent for ' a new fire engine with a boiler, both of a particular sort '. An interval of thirteen years passes before we hear anything more of him, then in 1781 Boulton writes to Watt that : ' Hateley from Scotland is going with Ld. Dunmore to Virginy, says that he & Somebody else in Scotland have invented an engine 3 times better than yours.' Hateley did not go to Virginia and a few weeks later we have Boulton again writing to Watt :

' When I was last at Broseley Wilkinson shewed me a letter from Joseph Hateley ye elder now in Scotland by which it appeared that he wanted Wilkinson to take him by the hand, saying that he had invented a new engine that would produce double the power of yours & that with one-fifteenth part of the fuel, & that it did not interfere with yours in any respect.'

Probably the invention was the ' rotary reciprocal fire engine ' for which Hateley, now described as of Dudley, Co. Worcester, engineer, obtained a patent in 1785. An engine made under this patent was put up in Cornwall, and Hateley's scheme attracted some attention in that county. In 1786 Boulton, then in London, informs Watt that : ' Mr. Edwards of Hayle and Mr. John Vivian are here and set out on Sunday to Birm^{gm}. & one part of their business is to see J. Hateley's engine, wch. now they say makes a great noise in Cornwall & that Gullett is erecting or going to erect one of them.' Hateley's engine at Gullett's mine was not a success, and it was replaced very soon by one of Boulton and Watt's construction.

We next hear of Hateley at Newcastle-upon-Tyne, whence in the year 1790 he petitioned for a patent for a ' pneumatic fire-engine on principles entirely new '. In 1792 he writes from London to Boulton about his new motor to replace the steam engine, and to Watt inviting him to come to London to see his engine at work. In the following year he was putting up an engine in Bristol, and Southern writes : ' I go to see an engine to-morrow erecting by Mr. Hateley of 24″ diam. 14 horses single engine with wonderful improvements.' Whether this was a ' pneumatic fire engine ' is not explained.

A few years after this, when the litigation arose about Watt's patent, Hateley put forward the statement that the invention of the separate condenser was made by Dr. Roebuck and not by Watt. Dr. Roebuck had been dead some years, but there was no difficulty in establishing the fact that he had invariably given all credit for the invention to Watt.

ADAM HESLOP

ADAM HESLOP of Ketley, Co. Salop, in 1790 obtained a patent for a ' new invented engine for lessening the consumption of steam and fuel in fire or steam engines '. He put up a number of engines, one of them in London for Henry Nock, a gunmaker ; this was at work in 1795.

Mention is made in the Boulton and Watt letters of seven or eight other engines by Heslop, viz. two at Coalbrookdale, two at Curwen's near Workington, and Lord Lonsdale is said to have had three or four. In regard to Curwen's engines there is a correspondence in 1798 as to infringement of Watt's patent ; Curwen pleads ignorance and says he is prepared to leave the matter to Boulton and Watt.

Southern, in a letter to James Watt, junior, describes the engines at Coalbrookdale as he found them in 1795 :

' We saw at the inclined planes two engines of Heslop's *invention*, but they have improved upon the construction since Mr. Boulton and I were at the Bank, by adding what they call an air pump (which I take to be merely a hot water pump) and putting a valve in the piston of what they call the cylinder, but which is now a very large air pump (& very unmanageable too we learnt). The *steam pipe*, as they call it, that goes from one cylinder to the other is covered as much as they can cover it, with a box of cold water, and may be called an eduction pipe. They have no snift that we could see, either real or sham. We did not see either of them work.'

One of Heslop's engines from Whitehaven is preserved at the Science Museum, South Kensington.

GEORGE MATTHEWS

GEORGE MATTHEWS of Broseley, Salop, ironmaster, in January 1781 obtained a patent for an ' engine to work by the power of fire and steam to be used in the iron and copper manufactories and for many other valuable purposes '. Later on in the year he put up an engine in London, but it does not seem to have been a success. In 1786, from a series of letters from John Rennie to Watt,[1] we find that Matthews is putting up another engine. He is now described as a founder

---

[1] B. & W. Colln. Rennie to Watt, 1786 July 26, Sept. 11 and 25.

on the Bankside, and ' is building a forge steam engine of his own improved !
construction at White Fryars Wharf'. When the engine was first started ' the
fly wheel by some means had catched hold of some part of the building and had
brought down a part of the house'. Then when it was repaired there was another
mischance soon after the start, for it broke the gudgeon. Rennie does not again
allude to this engine. It had a 30-inch cylinder, 6 ft. stroke, and worked a 500 lb.
hammer about 150 blows per minute.

NORWOOD

IN the year 1786 an engine was put up to drive a flour mill in London by a man
named Norwood. John Rennie in his letters to Watt mentions it on several
occasions.[1] He says it had a 30-inch cylinder, 2½ ft. stroke, and made 22 strokes
per minute, but he does not describe its construction ; it appears, however, that
the boiler was intended to be heated by a furnace which at the same time served
as a coke oven.[2] The engine was ready in February but could not be induced to
start ; then towards the end of March we read that ' Norwood has now after
trying again and again without success turned off all his engineers and employed
Simpson at Chelsea, who I suppose will make another Lambeth of it'. The
allusion here is to the fact that at this time there was another engine at Lambeth
that could not be made to work ; the terms of the letter suggest that Simpson
may have been its builder. However, by the end of May, Norwood's engine was
at work, but it drove but one pair of 4-foot millstones and ground seven bushels
of wheat per hour with 2½ bushels of coal. Apparently the boiler was now fired
in the ordinary way.

BOWSER

IN 1786 one Bowser undertook to erect a rotative engine to drive the bellows
at the ironworks of Foliot Scott & Co., Rotherhithe. ' Bowser was a tallow chandler
in London, but from his turn for schemes ruined himself and was a bankrupt
about 3 months ago.' [3] Scott had been tempted by the low price for which Bowser
offered to put up an engine, but very soon he began to doubt whether he would
not be landing himself in difficulties with Boulton and Watt over their patent
right, so, when Bowser had been on the job for a few months, the contract time
had elapsed, and the engine was still unfinished, Scott took the opportunity of
repudiating the bargain, told Bowser that he was only trifling and that he would
never make the engine answer, and obtained an order of the Court for the removal
of the engine. Bowser's engine was an atmospheric rotative engine ; his plan is
described in pretty clear language by John Rennie :

' The engine consists of two cylinders, one placed on each side of a wheel to which their
pistons has each a chain. Each cylinder has an outer and an inner bottom, the inner bottom
has the steam pipe to it & also a valve which occasionally opens and shuts a communication
with the space between the two bottoms. Into this space is inserted an injection and exhaustion
pipe which pipe communicates with an air pump. Thus the steam, being admitted into one
of the cylinders, the piston ascends. When it has ascended to its height, the steam is shut off
and the communication into the space between the bottoms is opened, which space the air

[1] Ibid. Rennie to Watt, 1786 Feb. 19, Mar.
1, 15, 25, 29, Apl. 6, May 31.
[2] General Conway at about this time had
a scheme for ' heating the boilers of steam

engines by coake ovens and selling the coakes '.
B. & W. Colln. Watt to Boulton, 1786 Jan. 15.
[3] Ibid. Rennie to Watt, 1786 May 31.

pump is supposed to have exhausted & the injection being thrown in at the same time the steam is condensed and falls down a perp^r. pipe into a trough or hot well so the atmosphere pushing on the piston forces it down. While the one piston descends the steam is to make the other ascend. The air pump, or condensing pump as he is pleased to call it, is a pump with a solid piston and a box at its bottom on which is two valves, one that opens upwards and the other downwards, i. e., the one that is on the eduction pipe opens upwards when the piston ascends, and when it descends that one is shut and the other one opens, by which the uncondensed steam or other fluid is blown or forced out. From the piston rod of each cylinder goes a connecting rod to a double throw crank, and as he supposed the one rod would turn the crank one half round while the other did it the other half.'[1]

What became of Bowser is not known. None of his schemes was patented. The idea of using a two-cylinder atmospheric engine for the production of rotary motion had been suggested by Dr. Falck in 1779, and some of the elements of Bowser's plan were utilized by Sturgess, Bateman and Sherratt, Thackeray, and Jabez Hornblower ; the idea of condensing in an extension of the steam cylinder had been used previously by Jonathan Hornblower in his compound engine, and it was employed by Symington.

WILLIAM SYMINGTON

SYMINGTON was one of those schemers who sought to obtain the benefit of the separate condenser, without infringing Watt's patent, by arranging the condenser and air-pump in the lower end of the cylinder itself, so that the condensation was not effected ' in vessels distinct from . . . the cylinders '.

In the Boulton and Watt letters we first meet with the name of Symington in connexion with the erection of the Torryburn and Wanlockhead engines, but this was the father of the inventor. The engine we hear of for the first time in 1789 when Rennie informs Watt that at Carron he had seen Symington's engine for ' Mr. Miller's navigation scheme '. The engine had two 18-inch cylinders, 2 ft. stroke ; it had been tried, but owing to some defects had been taken apart, in which state Rennie saw it. He adds that Meason (of Wanlockhead) had ordered one of Symington's engines with a 40-inch cylinder.[2] In 1792 we find Symington in London erecting engines for Walker, a colourman in Coldbath Fields, and for Hase, a brewer,[3] and by 1795 a number of his engines were at work, including one at Hunslet, Leeds, one at Limehouse, belonging to Trueman Harford, brewers, and one ' at the paper mill, Dockhead '. After this no doubt his operations were stopped by Boulton and Watt.

EDMUND CARTWRIGHT, D.D., F.R.S.

THE Reverend Edmund Cartwright, better known for his power loom and his wool-combing machine, was also a steam-engine inventor. He was at work on the subject in 1795 when he offered Boulton and Watt a share in a patent for a new steam engine.[4] It was not until 1797 that he applied for a patent, and then the proceedings were arrested at the Great Seal by a caveat by Watt,[5] but he did succeed in getting a patent in that year.

---

[1] B. & W. Colln. Rennie to. Watt, 1786 Sept. 15.
[2] Ibid. Rennie to Watt, 1789 Oct. 28.
[3] Ibid. Rennie to Watt, 1792 Jan. 3, 14,
Mar. 7.
[4] Ibid. Cartwright to Boulton & Watt, 1795 Feb. 25.
[5] Ibid. Cartwright to Watt, 1797 Aug. 27.

BATEMAN AND SHERRATT. THACKERAY 319

Cartwright's engine was single-acting without a beam ; the steam was introduced to the upper end of the cylinder, the lower end of which was permanently open to the condenser ; the equilibrium valve was fitted in the piston itself, and a surface condenser was used, but the noteworthy feature of the engine was the metallic piston and stuffing-box packing.

One of these engines—which Rennie reports ' will not answer'—had been put up at Wisbech in 1802.[1] This is, presumably, the same engine as that mentioned by Farey, who says it was a double-acting engine, that it would not perform the work it was appointed to do and was removed. Farey goes on to state that this engine was afterwards set up for the Duke of Bedford at his farm at Woburn, but that it proved a very defective engine and was broken up.[2]

SADLIER (SADLER)

On Garlick Hill, London, in 1796, a Mr. Sadlier set an engine to work for Sutton, Keen & Co., mustard makers. Rennie writes that it is reported ' that she answers exceedingly well. . . . I attempted to get a sight of her, but was foiled. Sadlier has been endeavouring to pervert the minds of the good citizens and to persuade them his engine is much better and cheaper than yours and uses only about half the coal ; this man should be looked after.'[3] In a subsequent letter Rennie says that the mustard firm ' have trumpetted the value of his invention with great industry'.[4] The engine is probably the same as that described in *Nicholson's Journal* in 1798 and there attributed to Mr. Sadler and said to have been patented. Farey gives a description based on Nicholson and adds that one of these engines ' was erected at Portsmouth Dockyard and was the first steam engine which was applied in public naval establishments'.[5] There are references to, and sketches of, Sadler's engine in the Sketch Book of William Reynolds, now preserved in the Science Museum, South Kensington. The indexes to patents show no grant at this date in the name of Sadler. In 1791 James Sadler of Oxford obtained a patent for a steam engine, but this was a rotary engine of the Barker's mill type with condensation.

BATEMAN AND SHERRATT. THACKERAY

The principal infringers of Watt's patent in Lancashire were Bateman and Sherratt of Salford. James Bateman is said to have commenced business as an ironfounder about the year 1770. In 1782 we find him writing to Boulton asking the terms for an engine for his foundry, sending a plan of the ground that he had vacant to receive it,[6] and soon afterwards inquiring about an engine for a cotton factory. Nothing came of these inquiries. Before 1794 Bateman had taken Sherratt into partnership. The firm seems, in the first instance, to have put up ordinary atmospheric engines, but in some of them they used Watt's parallel motion. Then they made two-cylinder atmospheric engines with separate condensers. A description and drawing of these engines will be found in Farey's

---

[1] Ibid. Rennie to Boulton, Watt & Co., 1802 Feb. 13.
[2] Farey: *Steam Engine*, 669.
[3] B. & W. Colln. Rennie to Boulton & Watt, 1796 Oct. 12.
[4] Ibid. Rennie to J. Watt, junior, 1796

Oct. 20.
[5] Farey: *Steam Engine*, 669, and *Trans. Newcomen Soc.*, III. 2.
[6] B. & W. Colln. Bateman to Boulton, 1782 June 5.

Steam Engine. The upper ends of the piston-rods of the two vertical cylinders were formed as racks to engage on opposite sides of a toothed wheel, on an overhead rocking shaft, so that they moved up and down alternately. The engine shaft on the level of the bottoms of the cylinders was driven by a crank and connecting-rod from an arm on the rocking shaft ; this shaft also carried a small toothed wheel which worked two air-pumps by means of racks.

It would seem that the first engine of this construction was set up at Thackeray's cotton mill at Garratt, Manchester, about 1794. It had 36-inch cylinders, 4 ft. stroke, and it was made to the instructions and plans of Thackeray and his engineer, Richard Bradley. Thackeray was concerned in at least one other engine, that at Stockdale's mill at Cark in Cartmel. Bateman and Sherratt had put up quite a number of them before the middle of the year 1796, when Boulton and Watt intervened with injunctions from the Court of Chancery for the immediate stopping of the engines. We read that they ' were driving a roaring trade and had at the time they were served with injunctions a great number of orders in hand '.[1] Bateman and Sherratt obeyed the injunctions and gave up making engines with separate condensers. As to the engines already at work, Boulton and Watt allowed them to continue, subject to the users paying up their premiums in full from the time the engines had been set to work.

Sherratt, it seems, made an atmospheric engine which he put into a barge on the Bridgewater Canal in 1794. There is a tracing of this engine at the Science Museum, South Kensington.

BOWLING IRONWORKS

In the year 1793 Boulton and Watt discovered that John Sturgess & Co., of the Bowling Ironworks, had made and were using engines that infringed the patent. Later on it was found that they had made engines for two other concerns. At the beginning of 1796 injunctions were served on Sturgess and his engineer, Barnett. After some correspondence it was agreed to pay Boulton and Watt premiums on their regular scale for all the engines, viz. a blowing-engine with a 50-inch cylinder, 6 ft. stroke, and a rolling-mill engine with a 44-inch cylinder, 5 ft. stroke, erected 1793, both at Bowling ; a 50-inch engine at the mill of Wilkinson, Holdforth & Co., Leeds, and a 31-inch engine belonging to Mr. Paley. Paley was a partner in the Bowling Works, and in Wilkinson, Holdforth & Co. Together, these premiums amounted to £670 per annum.

These engines were atmospheric engines, and all seem to have been of the same construction, with the condenser formed in the lower end of the cylinder under a false bottom fitted with a valve ; a passage from the bottom of the cylinder communicated with an air-pump ; they had parallel motions, and the valve gear was a close copy of that made at Soho.

LOWMOOR IRONWORKS

In 1795 Boulton and Watt heard of another infringing engine being under construction in the Bradford district. George Hawks, of Gateshead, had ordered an engine and it would seem that he came to Birmingham to see about it and on his return north had called at the Lowmoor Ironworks, then quite a new concern.[2]

---

[1] B. & W. Colln. M. R. Boulton to Southern, 1796 May 29.

[2] The first casting at Lowmoor was made in August 1791.

He informed the proprietors of his having ordered an engine and that he could not get it as soon as he wished on account of the time required for getting the cylinder made. The Lowmoor people at once offered to supply the cylinder and air-pump within a fortnight, and said that they were making a cylinder and air-pump on Boulton and Watt's construction for Messrs. Murray and Wood who, they understood, had the permission of Boulton and Watt to erect an engine on their plan, such permission having been granted for services rendered by Matthew Murray. What happened is not clear ; there is little doubt that the Lowmoor firm, Hird, Jarratt & Co., were acting in good faith, but that Murray had not been granted the permission referred to. Possibly Murray saw the Bowling people infringing with impunity and thought he might do the same.

MATTHEW MURRAY

It has been stated above that in 1795 a cylinder and air-pump were being made at the Lowmoor Ironworks for Murray and Wood, and that the proprietors were carrying out the order under the impression that that firm had permission from Boulton and Watt to erect an engine. It is not known whether the engine was actually built or not, but towards the end of 1797 Rennie draws the attention of the firm to the fact that Murray is building a 40 h.p. and a 16 h.p. engine and says : ' This man makes very free with your patents, would it not be well to look sharply after him.' [1] However, it does not appear that any action was taken against Murray. In 1802, after the expiry of Watt's patent, we find him undercutting the Soho firm in London to the extent of £400 on a 40 h.p. engine.[2] Later on, in 1803, Boulton, Watt & Co. took steps to revoke Murray's patent of 1801. To deal with this matter is rather beyond the scope of the present work, which does not aim at going later than the year 1800, still it may be desirable to say a few words about it. The matter really concerned the improvements in the steam engine made by Murdock. Murray's specification among other things covered some of these, and Boulton, Watt & Co. said that Murray had obtained a knowledge of their improvements by the seduction of their workmen, that those articles of the specification which had a claim to merit, and were apparently new, originated and were practised at their works some time before the date of the patent ; the other parts were impracticable or too trivial—whatever was useful was not new, whatever was new was not useful. Murray offered to grant them an exclusive licence to use his invention provided the proceedings against him were withdrawn. This offer was refused, and in the end Murray did not contest the revocation of the patent.

OTHER ENGINEERS

Some other engineers mentioned in the Boulton and Watt papers may be referred to very briefly. When the firm commenced supplying engines to London, Hadley seems to have been the engine-builder there. It was he who accompanied Smeaton on that visit to the Bow engine after which the engine-man got drunk on Smeaton's gratuity, with unfortunate results for the engine. He carried on business in Great Queen Street, Drury Lane, and Boulton refers to him

[1] B. & W. Colln. Rennie to M. R. Boulton, 1797 Nov. 8.

[2] Ibid. James Watt, junior, to Gregory Watt, 1802 Dec. 7.

as an ingenious smith and engineer. He died in 1779 or 1780, after which his business was carried on by his son. Both father and son seem to have done work for Boulton and Watt.

In 1784 a firm at Tipton, Bradley and Parker, were supplying materials for an engine to be erected by one Burnard at Hallen's Forge.

At the same time Banks and Onions at Broseley were building an engine with a crank ' without asking consent of the Snow Hill people '. This firm had set up a boring-mill on the same plan as that of Wilkinson ; they seem to have embarked upon engine-building with some vigour, and in 1789 they stated that they had as many engines on order as they could make in two years.

In 1785 the Reverend Mr. Derbyshire of Brierley, Staffordshire, was offering an engine that was to do great things ; it had a 36-inch cylinder and was to consume eleven tons of coal per week, but nothing more is said about it.[1] At the same time a man named Bond had put up an engine at Macclesfield, and another on his plan was being erected in the Potteries district for the purpose of grinding flint. This last was intended to raise 1,200 gallons of water per minute to work an overshot water-wheel of 26 ft. diam. ; the boiler was to be 6 ft. diam., the lower part of copper, and the upper part of cast iron, and the consumption of coal, slack, 80 lb. per hour. ' The expense is to be 240l. including a patent premium of 30l.' [1] Whether the engine was of the Savery or of the Newcomen type is not stated, and although reference is made to a patent premium there is in the official records no patent for this engine under the name of Bond, nor is there a patent for the Reverend Mr. Derbyshire's engine.

Another engine-maker, referred to by Southern in 1793 as Sowers, had put up an engine at Bristol which did not answer and was being replaced by one of Boulton and Watt's. Sowers was then putting up another engine on a different plan, and, as Southern was informed, was making use of a separate condenser and air-pump. In 1794 we read of engines in the textile factories of the West Riding by a man named Hurd. In 1795 David Weston erected an engine for Sutton Sharpe in King Street, Oxford Street, London. It had a separate condenser and air-pump, and the top of the cylinder was connected alternately to the boiler and to the condenser, while the bottom was in permanent communication with the condenser.

As the term of the patent was drawing to its end, naturally a number of people contemplated embarking upon engine-making. Rennie in 1800 says that Whitmore of Birmingham had been offering to supply engines to brewers at Bath, and that a man named Evans had put up an engine for one of them ' but it will not work '. Brodie and McNiven were going to start in Manchester and it was said that they had engaged Perrins, formerly one of Boulton and Watt's erectors.

PROJECTORS AND SCHEMERS

ALMOST from the beginning of the partnership Boulton and Watt were assailed, sometimes as individuals and sometimes as a firm, with the plans and schemes of projectors. Some of these had inventions for the production of motive power at a much cheaper rate than by Watt's engine, others wished to apply the engine to road vehicles or to boats.

One of the first of these projects noted is Simcock's invention for using an

[1] B. & W. Colln. Wedgwood to Watt, 1785 Jan. 13.

elastic vapour instead of steam. This is in 1777, and Watt in a letter to Boulton sets forth his ideas thereon. The idea of using air or some substitute for steam turns up several times. In 1785 Anthony Mitchell, the engine-man at Hawkesbury, writes to Watt that ' Mr. Barker says he has constructed a machine that is to be wrought with inflammable air, and that he can draw our water with half the quantity of coals that we consume. He intends drawing the coals out of the pit and also up the hills in the mine without the help of men. He says also that he can work a hammer in the Smithshop with the smoak that arises from the coals that is consumed in the forge.'

Then again, in 1796, John Cogdon Ragley, who is described as Clerk of the Works under James Wyatt at the Marquess of Hertford's, communicates his ideas for a road carriage to be worked by an air-engine, and describes the difficulties likely to be encountered.

A scheme for a fire-engine brought to the notice of Watt in 1780 was that of de Franchi, a Spaniard. He had made a model and said ' that having heated his machine with 1½ lb. of charcoal and 1 pennyworth of wood it worked about 40 minutes in what time 187 gallons of water was raised to the height of 9½ feet high, he said further that reckoning the time of its full work which he thinks to be about 30 minutes, the water elevated during this time to the above height (9½ feet) was 134 gallons . . . he has not yet taken any patent'.[1]

A Mr. Ewer, of Cheapside, brought forward a scheme for a new engine in 1781. Boulton refers to him as ' a grave young man ', and says that his plan infringes Watt's patent both in respect of using steam to press down the piston and in using a separate condenser.

A teacher of natural philosophy of Dundee, named Ware, called upon Boulton in 1789 ; he had invented improvements that would reduce both the cost of construction of the engine, and the cost of fuel by one-half.

About 1790 to 1793 a man named Yates, of Birmingham, seems to have had some engine scheme in hand. The matter comes forward in connexion with an application for employment ' by a poor man who alleges that he was enticed away from Sheffield about 2 years ago by fine promises carved & gilt by the *ingenious* Mr. Yates of our town late Banker & *engineer'*. It seems that this man had made several working models for Mr. Yates. He had also made on his own account a model of a rotative atmospheric engine without a beam and with the crank over the cylinder. This model Southern had seen working at the rate of 250 strokes per minute for a long time ; it had a cylinder 2⅛ in. diam. and the stroke was 3 in.[2]

In 1794 A. G. Eckhardt, F.R.S., who was concerned in a silk-painting manufactory in Sloane Street, writes that he has discovered a new construction of fire-engine and wants to submit it to the firm ; and in 1796 William Sellars, of Bristol, is about to take out a patent for working mills with less power by one-half than has been latterly in practice and offers them half the patent.

The sole use of a patent boiler is offered in 1790 by Roger Wearn, of Hayle, if Boulton and Watt will give him ' something handsom' for it.

Daniel Moore, of Wellingborough, in 1798 wants to submit an invention of ' a machine that will travel on the road without horses '. His machine will go up the steepest hill, and cannot possibly run back, but for want of power it moves

[1] Ibid. Magellan to Watt, 1780 Oct. 31, 1781 Jan. 13.   [2] Ibid. Southern to Lawson, 1793 June 4.

very slowly and he imagines that the power he is wanting may be supplied by steam.

It is known that the American steamboat inventor, James Rumsey, came to England and constructed a steamboat that was tried on the Thames. It appears that on his arrival in this country he called at Soho and discussed the subject with Boulton and Watt.[1] From a letter from Rennie to Watt we learn that in January 1790, ' The ship that is to be worked by steam now lies off Redriff Church nearly ready.'[2] Rumsey died before May 1794, when we find Boulton informing Watt that an American engineer has been sent over to take possession of his property ' which consists of a patent or two & some debts '.[3]

At the same time that Rumsey was at work, George Chase, of Oakingham, Berks., was writing to Boulton and Watt to know ' the expense of a commodious engine for driving a barge, the effect of which shall be equal to the draft of five horses. The length, breadth and hight respectively of the apparatus not if possible to exceed 8 or 9 feet and the stroke of the engine three feet.'[4]

In 1794 Robert Fulton, then residing in Manchester, comes upon the scene with an inquiry for the price and dimensions of a 3 or 4 h.p. engine to be placed in a boat, but nothing resulted from the inquiry at the time. It was not until 1804, after the retirement of Watt, that an engine was made to his order, and then it was a far more powerful one than he had in view in 1794.[5]

It may be as well to mention that Boulton himself had directed some attention to the question of propelling canal barges by steam before Watt came to Birmingham, and that Watt in correspondence with Dr. Small had suggested the use of the screw as the means of propulsion. Symington had been at work on the subject since 1788.

To conclude with a subject only remotely connected with the steam engine, we find James Watt, junior, in 1800 assisting with money a ' Mr. Dyer, the mineral paint inventor', and it seems to have been under consideration whether Boulton, Watt & Co. should enter into partnership in his business.

LITIGATION

HAVING taken a rapid view of the engines made by the competitors of Boulton and Watt we may consider briefly the proceedings in the actions brought in 1793 and the following years in connexion with the infringements of Watt's original patent. This patent, which it will be remembered had been granted in 1769 and had been prolonged by Parliament to the year 1800, was the only patent of Watt's that became the subject of an action at law.

For the first seventeen years Boulton and Watt had no real grievance in respect of infringement. It is true that they considered Jonathan Hornblower's compound engine to be within the patent, but, even if they were right in this, Hornblower's progress was so slow that they could not have suffered any material injury. In 1784 there seems to have been a definite intention to go to law with Hornblower and Winwood over the Radstock engine, but nothing was done at

[1] B. & W. Colln. Letter Books. Boulton & Watt to Handley, 1788 July 14.
[2] Ibid. Rennie to Watt, 1790 Jan. 25.
[3] Ibid. Boulton to Watt, 1794 May 15.
[4] Ibid. George Chase to Boulton & Watt,

1789 Aug. 2.
[5] The correspondence between Fulton and Boulton & Watt is given in Dickinson's *Robert Fulton*, 1913.

this time, nor indeed was any action ever brought to trial to test whether Horn-blower's engine was an infringement.

By the year 1792, however, the firm had both Hornblower and Bull troubling them in Cornwall; both engines were doing very well, and Watt writes from Truro : ' We have given Bull notice that we mean to bring an action against him next term (two engines have been ordered from him and one from Hornblower since we came here).' An action was commenced accordingly and the trial came on in June 1793 in the Court of Common Pleas before Lord Chief Justice Eyre and a special jury. Boulton and Watt had a fine array of witnesses, including De Luc, Herschel, Lind, Southern, Murdock, Rennie, and Ramsden.

Bull's engine was an inverted engine, but it had steam pressing upon the piston instead of the atmosphere, and it had a separate condenser, so the jury had no difficulty in finding that it infringed ; they also found that the specification contained an insufficient disclosure to enable the invention to be carried into effect. The verdict accordingly was for Boulton and Watt, but subject to the opinion of the court as to whether the patent was good in law. On this question there was considerable delay in settling the case to be argued, and it was not until February 1795 that it came on for hearing before Lord Chief Justice Eyre and Justices Rooke, Heath, and Buller. Judgement was reserved, and was not de-livered until May, and then it was indecisive—the judges were equally divided, for and against the validity of the patent.

In the meantime Boulton and Watt had taken action in the Court of Chancery for the issue of injunctions to Bull and others to restrain them from making and using the patented invention. In May 1794 they had a dozen suits of different kinds in Chancery.[1] It was the practice to grant injunctions on the ground of long possession notwithstanding doubts as to the validity of the patent ; the court gave credit to the validity of the patent until its invalidity had been regularly established by a proper proceeding in a court of law. Lord Rosslyn refused to dissolve the injunction against Bull, although the Court of Common Pleas had been equally divided upon the validity of the patent ; there had been long posses-sion and he declined to disturb the possession.[3]

The indecisive result of the trial in the Court of Common Pleas made it clear that there was reasonable ground for questioning the validity of Watt's patent and it was no doubt a contributory cause for the increased number of infringing engines about this time. As we have seen above, by the year 1796 quite a num-ber of rotative engines embodying features covered by the patent had been made by Hornblower and Maberley, Thackeray, Bateman and Sherratt, Sturgess, and Symington. Boulton and Watt went to the Court of Chancery and obtained injunctions against these people With the exception of Hornblower and Maberley all had submitted by June in that year. Hornblower and Maberley, however, were trying to get better terms than the others had accepted, but the negotiations with them fell through and Boulton and Watt decided to bring an action for infringement against them.

The trial of the cause Boulton and Watt versus Hornblower and Maberley came on in the Court of Common Pleas in December 1796 before Lord Chief Justice Eyre (who had tried the action Boulton and Watt against Bull) and a special jury. The result was a verdict for Boulton and Watt. The defendants then proceeded

[1] B. & W. Colln. Southern to Lawson, 1794   [2] Webster's *Patent Cases*, I. 282.
May 7.   [3] Ibid., 285. .

to have the case retried by means of a ' writ of error '. Here again there was a long delay, and it was not until 1799 that the case was finally decided in the Court of King's Bench. In this instance the judges, Lord Kenyon, Ashurst, Grove, and Lawrence, were all in agreement and affirmed the judgement of the Court of Common Pleas. It was still possible for Hornblower and Maberley to appeal to the House of Lords, but this they did not do, and now at last, in the year before the end of its term, the validity of Watt's patent was established.

In the Court of Chancery there were actions still pending, but these we need not pursue ; in the end all the infringers of the patent submitted and paid up the premiums demanded by Boulton and Watt.

The letters from Watt to his partner in reference to these proceedings bring out the fact that at the beginning of 1799 he was urging the advisability of applying to Parliament for an Act to explain and amend the Act of 1775. He even got so far as to draft a petition to the House of Commons for leave to bring in a Bill further extending the term of ' the parliamentary privilege for a reasonable time and for reviving ' the patents granted to him in 1782 and 1784 ' and extending the terms thereof in like manner and for removing doubts and difficulties concerning the same '.

In the course of the various proceedings a number of objections were brought forward against the patent. The grant was for ' a method of lessening the consumption of steam and fuel in fire engines', but the extending Act was for ' certain steam engines commonly called fire engines of his invention '. It was contended that these were not the same inventions and accordingly that the patent had not been prolonged. Then it was argued that a *method* of lessening the consumption of steam and fuel in fire-engines was not proper subject-matter for a patent under the Statute of Monopolies which restricts the grant of patents to ' any manner of new manufacture '. The same objection was taken on the specification itself, which states that the method ' consists of the following principles ' ; a principle was not a manner of manufacture and could not form a patentable invention. In the Court of Chancery it was argued that the grant of a second patent to Watt necessarily rendered void the first.

These are purely formal matters, although considerable time was spent in arguing some of them. More vital were the questions of novelty and the sufficiency of the specification. It was put forward that the real inventor was Dr. Roebuck, not Watt, that it was not new to use steam to press on a piston, that the air-pump was not new, nor the use of oil and grease on the piston, and that it was known to surround the cylinder with heated bodies and to condense without permitting water to enter the cylinder. As to the specification, objection was taken that it did not describe an engine ; an engine could not be made from it without experiment ; it gives no directions as to the form and proportions of the condenser ; does not say that condensation may be effected by injection, or how the water is to be removed. Moreover, it describes constructions that are impracticable.

As we have seen, the juries in both trials gave verdicts in favour of the patentees. They found that Watt was the inventor, that the invention was new and useful, had been infringed, and that the specification was of itself sufficient to enable a mechanic acquainted with the fire-engines previously in use to construct fire-engines producing the effect of lessening the consumption of fuel and steam upon the principle invented by Watt. In the action against Hornblower and Maberley the judges in the Court of King's Bench agreed to accept the finding of the jury on the sufficiency of the specification as conclusive.

One is inclined to think that the courts fully recognized the great value of the invention and wished to do all they could to support the patent, for it is very doubtful whether Watt's specification did come up to the standard that was called for even in those days. This is the view of John Farey, who was the great patent expert of the early part of the nineteenth century. Writing about twenty-five years after the trial, he says :

' According to the ordinary practice of the Courts of law in other cases, Mr. Watt's patent ought to have been annulled, for the insufficiency of the specification, which is a series of definitions of principles of action, without any description of the means of carrying them into effect. And it is certain that if the specification had not been supported by testimony of many scientific artists, who stated that it was sufficient in their opinion, and if the merit of Mr. Watt's engine had not been so universally allowed at the time of those trials as to have obtained a leaning in his favour, his right could not have been established as a mere question of law, according to the usual practice of the courts in other similar cases.' [1]

As already stated Watt himself was very doubtful about the specification. As long before as 1781 he had written to Boulton in reference to Arkwright's case : ' I don't like the precedent of setting aside patents through default of specification. I fear for our own. The specification is not perfect according to the rules lately laid down by the judges,' and it is curious to reflect that the defect arose mainly from his having taken the advice of his friends, Dr. Small and Boulton, in the preparation of the specification. He himself contemplated giving drawings and descriptions of the parts and had indeed begun the preparation of the drawings, but they dissuaded him, saying that his invention was for the broad idea, and not for any particular constructions. The specifications of his later patents are all very full.

[1] Farey : *Steam Engine*, 649.

XXIII

BOULTON AND WATT IN CORNWALL

CORNWALL was the most important field of the activities of Boulton and Watt, not so much in respect of the number of engines set up which, although considerable, seems not to have exceeded that for Lancashire, but in that the engines were for the most part of large size and the price of coal was high, so that the premiums and hence the revenue of the firm from that county amounted to a large sum ; moreover, it was the scene of the greatest struggle against Watt's patent.

The construction of the engines and their introduction into Cornwall have been dealt with in preceding chapters. The Boulton and Watt Collection affords material for an extended treatment of the Cornish affairs of the firm and of mining in Cornwall, but such a treatment would be of local interest only, and we limit ourselves here to a rapid survey of the relations of Boulton and Watt with the county generally.

The fact that in 1783, within six years of the erection of the first Boulton and Watt engine in Cornwall, Watt was able to say that there was but one Newcomen engine left in the county is sufficient evidence that the Cornishmen had not been slow to recognize the merits of the new engine. It was not only in regard to economy in fuel that it surpassed the old engine—for the same size it was also a more powerful machine. Whether the limit was imposed by difficulties of production, or to those of transport and handling, the Newcomen engine had attained the largest size that was possible in those days (Smeaton's Chacewater engine had a 72-inch cylinder, and there were at least four engines with 70-inch cylinders in the county), and the maximum depth of pumping seems to have been 80 fathoms. With the Watt engine the miner had at his hand a machine that could deal with more water and with greater depths. It is possible that this may have led to more speculative adventures and to an increased amount of unprofitable mining, so that the very merits of the engine may have been one of the sources of the trouble in regard to the payment of premiums to which we shall have to refer.

From September 1777, when their first engine was set to work, until the end of 1783 twenty-one engines by Boulton and Watt had been put up in Cornwall. They were all single-acting pumping-engines. The year 1784 saw the introduction of the double-acting engine and of the rotative engine, and from that date onwards we find engines of these types, as well as single-acting engines, being erected in the county. During the years 1784 to 1788 inclusive, eighteen engines were put up ; of these four were single-acting and eleven were double-acting pumping-engines, the remaining three were rotative engines for raising ore. No engines were put up in the years 1789 and 1790. Then from 1790 up to September 1796 we have five engines, two single-acting, two double-acting, and one rotative. From this date until the close of the connexion of Boulton and Watt with the county it is difficult to give an exact figure, but we shall not be far wrong if we put the number of engines erected at eleven. This gives a total of fifty-five Boulton and Watt engines put up in Cornwall.[1]

It will be seen that single-acting engines continued to be made throughout the period, and indeed they outnumber the double-acting engines by about two to one; moreover the list of single-acting engines shows a greater number of large cylinders than that of double-acting engines. The largest single-acting engine was a 64-inch, 9 ft. stroke, put up at Herland mine in 1792, and the largest double-acting engine was a 63-inch, 9 ft. stroke, put up at Wheal Maid in 1788, but whereas this was the only double-acting engine with a 63-inch cylinder, there were eight single-acting engines of that size. There were nineteen single-acting with cylinders 50 in. and above, and but four double-acting. The smallest single-acting engine had a 28-inch cylinder, and the smallest double-acting an 18-inch cylinder. The larger single-acting engines were of 9 ft. stroke, the smaller 8 ft. ; apart from the 63-inch, the larger double-acting engines were 8 ft. stroke and the smallest 4 ft.

In going through the records one is struck by the extent to which the engines were moved about from one mine to another ; the practice, however, was carried on in Cornwall before Watt appeared on the scene, and it still prevails. Of the twenty-one engines set up in our first period 1777-83, in 1796 but six were still on their original sites. Some of them had been moved three or four times. The first engine, Wheal Busy 30-inch, for instance, was first moved to Wheal Chance, Scorrier, then to Wheal Virgin in St. Hilary, next to Crane mine near Camborne ; then in 1794 it was at work on the United Mines, and in 1797 it was about to be erected at Wheal Susan near Godolphin.

Although it would seem that the Wheal Busy engine when twenty years old and after three removals was still thought in a sufficiently good condition to justify re-erection, some of the other early engines had not so long a life. Tingtang and Chacewater, both set to work the year after Wheal Busy, were reported in 1794, the one to be ' completely worn out ', and the other as ' now extinct and broken up '.

The subject of the performance or duty of the Boulton and Watt engines generally is dealt with in another chapter. The remarks which follow have special reference to the engines in Cornwall. Trials of the Newcomen engines at Poldice mine in August and September, 1778, showed that 7 million pounds of water were raised 1 foot by the consumption of 1 bushel of coal. Watt said that these engines were considered to be in good order, ' and among the best in the county, producing effects which might be assumed as a fair average of the effects of engines of that species.' [2] Some of the Mine Captains held that there were better engines in Cornwall, and it is possible that there were one or two doing as much as 10 millions. The Boulton and Watt engine, 63-inch, put up in place of the two Newcomen engines, gave a duty of 26 millions. The Polgooth engine in 1785 is credited with 29 millions, and the Ale and Cakes engine in 1786 with 31 millions. In 1791 Thomas Wilson, the agent for Boulton and Watt in Cornwall, sent in a report of the performance of thirteen engines ; the best are Poldice 58-inch double-acting (Oppy's), 31·5 millions, and Wheal Maid 63-inch double-acting,

[1] It would no doubt be possible to unearth in the Boulton & Watt Collection material which would show the exact number of engines, but the notes taken by the present writers do not furnish this; however, the figures given are certainly very near the mark.

[2] B. & W. Colln. Draft in Watt's handwriting headed ' Hints for answers to the *argumentum ad invidiam* '. Undated but written in 1793 or soon after.

29 millions ; the poorest, Poldice ' little-double' 24-inch, 13·5 millions, and Wheal Virgin East 56-inch single-acting, 15 millions.[1]

In the following year Wilson reported that the regular effect of the Poldice engine during the month of August had been 32·8 millions, and that a new engine at Wheal Butson, 36-inch single-acting, had, in a 12 hours' trial, attained 33 millions ; in a subsequent trial the duty shown was 28·5 millions.   This engine, he notes, is ' not encumbered with extra quantities of dry rods '.[2]

It will be seen that although duties of over 30 millions were attained, there were engines doing very much less than this.   The contrast between the two double-acting engines at Poldice in 1791 is very striking, the one 31·5 millions, and the other 13·5 millions.   Watt had noticed at an early date, in connexion with Newcomen engines, that some did far more work than others with the same nominal quantity of fuel, which, he says, ' is frequently owing to accidental circumstances, which cannot be discovered.   They are also more or less perfect according to the skill of the engineman, the goodness of the water, the quality of the coals, the measure of the coals, with other circumstances, such as the opportunity which can be allowed for repairs &c.' [3]   Being on the same mine these two engines would presumably use the same water and the same coal, but the boiler of the one may have been more efficient, and it may have had a more skilled engine-man than the other.   They were both erected in 1787.

Davies Gilbert and four others investigated the performance of the engines in Cornwall in 1798, and they found the average duty for twenty-three engines to be 17·6 millions.[4]   This was the year in which Murdock left Cornwall.   There seems to be a consensus of opinion that for some years after 1800, when the payment of premiums to Boulton and Watt ceased, the engines were, in general, neglected, and their performance fell off.   According to one authority, the average duty fell so low as 14 millions.   This may have been due, in part at any rate, to the absence of Murdock, but a more likely explanation is that the relief due to the cessation of payments to Boulton and Watt induced a feeling of slackness, and led the miners to think that there was no longer the same necessity for economy in the use of coal.

It is somewhat curious to find that there is considerable doubt as to the contents of the bushel used in Cornwall in Watt's time ; he took 84 pounds as the weight of a bushel of Newcastle coal, but so far as we are aware he gives no indication of the weight of a bushel of Swansea coal as measured in Cornwall.   From his silence on this point we may perhaps infer that he assumed there was no difference.   In the report of the trial of the Poldice engines in 1778 the coal consumption is expressed in weys, ' each wey being 64 Winchester coal bushels', and in 1830 Davies Gilbert in referring to this trial states that the bushel is ' supposed to weigh 84 lb.'   This seems fairly straightforward, but unfortunately by the earliest recorded test the weight of a Cornish bushel of coal was found to be 94 pounds ; the difference is a substantial one.   This test was made in 1835 (the year after the sale of coal by measure was made illegal) in connexion with an engine trial that is recorded in Leans's *Historical Statement of the ... steam engines in Cornwall*, 1839, and the authors in the same book give a summary table of the performance

[1] B. & W. Colln. Wilson to Boulton, 1791 Mar. 21.
[2] Ibid.   Wilson to Boulton & Watt, 1792 Sept. 20.
[3] Ibid.   Boulton & Watt to Mr. Daniel, 1779

Dec. (copy).
[4] *Phil. Trans.* 1830, Pt. 1, p. 122.   Davies Gilbert, President, ' On the progressive improvements made in the efficiency of steam engines in Cornwall '.

of the engines in Cornwall covering the years 1814 to 1835 inclusive, one of the columns of which is headed ' Coal in bushels of 94 lbs.', from which it is clear that they were not cognizant of any change in the size of the measure in that period ; had any change been made it could not fail to have come under their notice. (The imperial bushel had been established by an Act of Parliament in 1824, but the Winchester *coal* bushel was almost identical with it in capacity.) This takes us back to the year 1814, and we have next to consider whether there is any probability of an alteration having been made between that date and, say, the year 1800. First it may be remarked that the Swansea coal was a little heavier than the Newcastle coal. Edington (*A Treatise on the Coal Trade*) in 1813 states that a bushel of Wallsend coal weighs 84 to 85 pounds, and a bushel of Swansea coal 85 to 86 pounds ; this was by the London measurement. A difference of this order does not affect the question, we still have to account for the difference between 94 and 85—nine pounds. It is not easy to see why the Cornish people should have made any change in their measures, and it is reasonable to suppose that the revenue officers would oppose a change by way of increase in the size of the measures upon which the coal duty was rated. The writers are inclined to the belief that no change was made, that the Cornish people, while they may have thought they were using the Winchester coal bushel, were actually using a measure one-ninth larger all along, and that the bushel used in Watt's time held 94 pounds of coal just as it did in 1835.

At the outset the royalty or premium for the use of the Boulton and Watt engines was based on the value of the coal saved by the new engine as compared with the Newcomen engine it displaced, both doing the same work ; Boulton and Watt taking one-third of the saving. Obviously the simple case where a Newcomen engine was replaced by a Boulton and Watt engine to do exactly the same work was rare, if it ever occurred in Cornwall. In the ordinary course the new engine was required to raise more water, or to pump from greater depths. It was necessary then to compare the actual coal consumption of the Boulton and Watt engine, not with that of the engine it displaced, but with that of a Newcomen engine capable of doing the same work, and to keep in view that although in the main such variations as might occur in the amount of work done corresponded with the number of strokes made by the engine, it was necessary also to be able to deal with changes in the size of the pumps and in the height of the lift.

The coal consumption was readily ascertainable, for it was the regular practice to measure all coal used by steam engines. There was in these days a customs duty on sea-borne coal, but a drawback was allowed on the coal consumed by pumping-engines in the tin and copper mines of Cornwall, and to obtain this drawback the engine-man was required to put in a sworn statement of the quantity used. The work done was calculated from the diameter of the pumps, the height to which the water was raised, and the number of strokes made by the engine. This last factor was determined by a counter fixed to the engine beam.

To arrive at the size and coal consumption of the supposititious Newcomen engine was not so simple a matter ; it involved a good deal of calculation, and it will be of interest to consider, in as brief a manner as possible, how the problem was attacked, and for this purpose we shall make use of the printed pamphlet ' Proposals to the Adventurers in . . . By Boulton and Watt '.

Watt based his calculations on the performance of two engines at Poldice mine ; one had a 66-inch cylinder and raised water from a depth of 33 fathoms, the other had a 60-inch cylinder and raised from 25 fathoms ; the pumps of both engines were 17 in. diam. From the mine account books it appeared that :

' The quantity of coals consumed by both engines during the sixty-one days of August and September [1778] was two hundred and forty-two weys of sixty-four bushels each ; but it being alleged that part of the coals then used were not of so good a quality as usual, it was agreed to deduct twenty-two weys on that account, and to fix the consumption of both engines for these two months at two hundred and twenty weys.'

' It was agreed by all who examined the going of the engines, that six strokes per minute was fully their average rate of working, when the water was kept in fork, and that five feet six inches was also fully the average length of their stroke.'

. . . ' it appeared both by the testimony and opinions of those persons who had experience of engines in that state, and also from calculation, that an engine with a sixty inch cylinder loaded only to one pound per square inch of piston, going at the rate of six strokes per minute of five feet six inches long each, would consume in thirty days 1920 bushels.'

' It was agreed, to facilitate calculation, that the *increase* of consumption under other loads should be proportioned to the loads in the manner which shall be explained.'

' Also it was agreed, for the same reason, that the consumption should be considered as proportioned to the number of strokes per minute.'

Watt's first step was to express the consumption of the two engines in terms of a single engine. He found that a 60-inch engine, loaded to 5·4 pounds per square inch (the mean of the loads on the two engines), making 6 strokes per minute each of 5·5 ft. long, would consume in 30 days, 3,131 bushels. Then deducting the agreed consumption for a load of 1 pound per inch, i.e. 1,920 bushels, from 3,131 he obtained 1,211 bushels for the quantity necessary to raise the load from 1 pound to 5·4 pounds. This gave him 280 bushels for the additional quantity necessary monthly for each pound on the inch increase of load after the first pound. Thus the consumption with a load of 5 pounds on the square inch would be 1,920 +(280×4) = 3,040 bushels.

For Newcomen engines of other sizes the consumption was assumed to vary directly with the area of the cylinder, and the length and number of the strokes per minute. In comparing the Newcomen engine with his own, Watt took the piston load of seven pounds per square inch for the one, and 10½ pounds for the other. Thus the Tingtang 52-inch Boulton and Watt engine was equal in power to a Newcomen engine with a 63¾-inch cylinder, 7¾ ft. stroke.

From the figures for the 60-inch engine a tabular statement was prepared for the imaginary Newcomen engine showing for loads, advancing by one pound, from one to seven pounds per square inch, the total load on the piston and the coal consumed in thirty days, when going at the rate of six strokes per minute. To the table for the Tingtang engine is appended the following constants : ' differential number 406', ' multiplier 0·1272', ' add 2646'.[1] The differential number represented the bushels required for each increase of one pound load per square inch above the first. The multiplier was this last number divided by the area of the piston 406÷3,192 = 0·1272 ; it represented the consumption in bushels per pound of the total load on the piston above the total load corresponding to one pound per square inch. The number 2,646 was got by taking the differential number 406 from 3,052, the number of bushels corresponding to a load of one pound

[1] These figures are taken from two memorandum books in the Boulton Papers : one, marked on the cover ' Calculations & Performance of Cornish Engines, 1778', is in the handwriting of Watt ; the other is Boulton's private copy of the monthly returns for the engines in Cornwall from 1779 to 1782. They are based on the assumption that the 60-inch engine consumed 3,027 bushels in 30 days, not 3,131 as given above.

per inch ; it represented consumption for a zero load and was used merely to facilitate calculation.

To ascertain the savings in a given month by the Tingtang engine, first from the diameter of the pump and the height of the lift, the weight in pounds of the column of water was ascertained, this represented the total load on the piston ; if the figure agreed with one of the entries in the table, the consumption was found at once ; if not, it was multiplied by 0·1272 and to the product was added 2,646. The sum represented the number of bushels that the engine would consume going at the rate of six strokes per minute for thirty days. The number of strokes recorded by the counter is now to be divided by 8,640 (24 × 60 × 6) to get the number of days the engine must have worked at six strokes a minute to make that number of strokes, and then as thirty is to this number of days so is the consumption found above to the number of bushels that would be consumed by a Newcomen engine in making the same number of strokes under the same load. The actual consumption of the Watt engine had been kept account of at the mine ; this was deducted from the calculated consumption to show the total saving in bushels. For each engine there was kept a book, the pages of which were ruled into columns for, date, counter reading, depth of shaft, diameter of pump, load in pounds, consumption of the equivalent Newcomen engine, real consumption, total savings, B. & W's third part, price of the coal, amount claimed by Boulton and Watt. The agreement entered into by the Mine Adventurers for the use of the Boulton and Watt engine embodied the basis on which the payments were to be made, i.e. it specified the size of the equivalent Newcomen engine and contained a table showing its coal consumption at various loads.

The arrangement was quite a fair one, but it had its drawbacks ; one, mentioned by Watt, was that those months in which there was most water to be pumped, and consequently the highest premiums had to be paid, were also those in which the mines were least productive ; another was that it was not possible for the adventurers to know what payments they were liable for until the end of each month, so they could not plan out their expenditure in advance. Then the procedure in arriving at the amount of the saving was complicated, far too much so, one would imagine, to be grasped by the ordinary mine captain of those days ; possibly too there were suspicions as to the reliability of the counters, for there is no doubt that in several instances their records were wrong.

The result of all this was that a feeling arose adverse to this method of assessing the premiums, and certain of the mines approached Boulton and Watt with the request that it should be replaced by a fixed payment for each engine according to its size.

Somewhat reluctantly Watt, who had taken a good deal of trouble and spent some time over the calculations and tables, acceded to this and a scale based on the size of the cylinder was determined upon, and then we find in Cornwall two systems in use, payment by the counter, and payment according to a fixed rate. Whether the first system continued in use until the expiry of the patent we have not been able to determine. The rate agreed upon for the five engines for the Consolidated Mines in 1780 was £63 per month for an engine with a 63-inch cylinder and 9 ft. stroke, and for other sizes in proportion to the area of the piston and the length of the stroke. This rate, however, was soon reduced to £55 per month, and in 1784, in connexion with the engines on North Downs Mines, mention is made of a scale according to which the amount of the annual premium in pounds was equal to one-sixth of the square of the diameter of the cylinder in inches for a single-acting engine, and one-third for a double-acting engine. In the case of North Downs a new element was introduced in fixing the rate—the load per square inch on the engine piston. The adventurers in these mines were contemplating sinking to greater depths and proposed to put down engines larger than were required for the existing workings, and they considered it unfair that they should be charged the full rate while the engines were not carrying their full load. .

We have not discovered a statement of the total amount received by Boulton and Watt from Cornwall on account of their premiums, but it seems, from a document headed ' Statement of the Commission paid Thos. Wilson at sundry times by Boulton and Watt ',[1] that from the beginning of 1781 until Sept. 30, 1800, the sum of £139,400 was collected by Wilson. His commission at 2½ per cent. amounted to £3,485. For the twelve years 1780 to 1791 inclusive there is a statement showing the annual receipts ; the best year is 1786 with £9,281, the worst 1780 with £2,598, and the average is £6,330. The total for the twelve years amounts to very nearly £76,000.[2]

These were very considerable payments for that time and Boulton realized that they constituted a tax on the mining industry of Cornwall ; he calculated that for the year 1787 this tax amounted to about 1½ guineas per ton on the copper and about 2 guineas per ton on the tin produced.[3] The period, however, was one of considerable difficulty in Cornwall ; for the year in question the ore sold produced £211,780 9s. 4d., but the total cost of working the mines had been £238,944 17s. 4d., so that there had been a loss of £27,164 8s. 0d.[4] Still, in these circumstances, even had Boulton and Watt waived their premiums altogether, the relief would not have been perceptible ; they amounted to but a small percentage on the cost of working.

However, there is ample evidence that Boulton and Watt at various times and to meet conditions of bad trade, or exceptionally heavy expenditure, did make very considerable abatements in the premiums agreed upon. For instance, it was calculated that up to 1794 they had given up in premiums to the adventurers in Poldice mine nearly £13,000.

Nevertheless, it will not be a matter of surprise that the Cornishmen felt aggrieved at these heavy payments ; the last Newcomen engine had now disappeared from the county and there was nothing to remind them of the immense benefits they were deriving from the use of Watt's invention, and that the payments were measures of these benefits. Moreover, certain sections of them had from an early period considerable doubt, more or less well founded, as to the validity of Watt's patent. Possibly had the aggregate payments been, say, one-half of what they actually were, matters would have gone on smoothly, for there can be no doubt that Watt was right in saying :

' The truth is that had the savings made by their invention been only ¼ of the fuel formerly used, instead of ¾, and B. & W. had taken only 1/3 of the savings, there would have been no complaint, although the mines would have saved a comparatively small sum. It is the magnitude of the sums which were justly due to B. & W. which have excited all this envy and malevo-

[1] B. & W. Colln.
[2] Ibid.
[3] Ibid.   Boulton to Watt, 1787 Oct. 3. Boulton states that the total quantity of copper and tin got in Cornwall from Aug. 1, 1786, to Aug. 1, 1787, amounted to 4,768 tons

and 693 tons respectively, and that the premiums the firm had received for the first six months amounted to £4,120, or at the rate of £8,240 per annum.
[4] Ibid.   Boulton : ' General State of the Cornish Copper Mines, 1787.'

lence. If they had gained little or been loosers by their services, they would have been more highly praised, though they would have merited it less.' [1]

Years went by and the Cornishmen bore what they considered a great grievance with protests and grumbling, then in the latter half of 1793 matters came to a head and payment of the premiums was refused by the United Mines, the Consolidated Mines, Wheal Crenver, Wheal Treasury, Wheal Gons, Poldice, Herland, and Godolphin. The United Mines had for the first two engines agreed to pay by the counter ; at the end of 1782 a change was made to a fixed monthly payment of £83 6s. 8d. ; at the beginning of 1785, when a third engine was erected, the premium was advanced to £108 6s. 8d. a month ; again when in July 1787 a fourth engine was set to work the sum was raised to £133 6s. 8d. This sum the adventurers continue to pay to January 1791, when they refused to make the full payment and tendered £83 6s. 8d., which sum was accepted by the agent of Boulton and Watt ' on account '. As from January 1792 Boulton and Watt offered to accept two-thirds of the agreed premiums, an indulgence that they were making to other mines in the same neighbourhood. Of this offer the United Mines adventurers took no notice and in January 1794 they were withholding payment, they said, until Boulton and Watt's patent was established.

The Consolidated Mines had five engines put up in 1782 for which they agreed to pay £208 6s. 8d. per month or £2,500 per annum. In 1787 in consideration of the bad state of the mine Boulton and Watt abated this sum, by one-third, to £1,667. This continued to be paid until May 1793 although in the meantime a 63-inch double-acting engine had been put up.

The main engine at Wheal Crenver was a 60-inch set to work in 1787 ; the adventurers had paid the agreed premium £49 17s. monthly up to the end of 1791 ; they had also a 36-inch auxiliary engine which they had bought from Wheal Treasury. The Wheal Gons engine was the 63-inch put up at Dolcoath in 1781 ; at Dolcoath the payment had been based on the counter. Wheal Gons agreed to pay £55 monthly. The Poldice mines had, to begin with, two 63-inch engines ; in December 1782 Boulton and Watt offered to accept £1,500 per annum as the premium on the two ; in 1785 a third 63-inch engine was set to work and the annual payment was increased to £2,000 ; in the next year the adventurers determined to erect two new engines, double-acting, and Boulton and Watt agreed to give up the premiums until the mines should have repaid £8,000 proposed to be expended on them ; no premium was paid until 1791, and then it was agreed to make a reduction of one-third ; the premium so reduced was paid up to October 1793.

In September 1796 there were 33 Boulton and Watt engines in existence in Cornwall, out of 44 erected, and of the 33, 29 were at work, most of them regularly, a few occasionally only. Of the 29 at work, litigation was then proceeding in respect of 12, viz. the engines of the United Mines (4), of the Consolidated Mines (4), Wheal Crenver (2), Wheal Gons (1), and Wheal Treasury (1). In the interval since the beginning of 1794 Poldice and Herland had come to terms and Godolphin seems to have ceased working.

The culmination of the discontent in Cornwall in 1793 coincided with the more or less successful introduction of the Hornblower and the Bull engines. Jonathan Hornblower had been working at the compound engine for some years,

[1] B. & W. Colln. ' Hints for answers to the *argumentum ad invidiam*.'

but the first engine of his in Cornwall that was continued in work was that started at Tincroft mine early in 1791 ; from that date until the beginning of 1794 he put up eight other engines. Edward Bull built an engine at Balcoath which was started about the beginning of 1792, and in the next few years he put up nine others. Boulton and Watt brought an action against Bull in respect of the Balcoath engine for infringement of their patent. Hornblower and Bull, their engines, and the litigation are dealt with in Chapter XXII. It will be sufficient here to say that the action against Bull was indecisive; it was found that he had infringed the patent, but on the question whether the patent itself was good in law the judges could not agree. It was not until 1799 that the validity of Watt's patent was finally decided, and then it was on another action, that against Jabez Hornblower and Maberley. Boulton and Watt had in the meantime obtained injunctions from the Court of Chancery against Bull and the users of his engines, to restrain them from making and using their invention. Action was taken also against the users of the Hornblower engine and those users of the Boulton and Watt engine who continued to withhold their premiums. In the end all these came to terms. The final settlement of the claims was entrusted to young Boulton, now a member of the firm, who spent some time in Cornwall conducting the negotiations in person. He is said to have secured about £30,000. The final payments, however, were not made until some years after the patent had come to an end. The document referred to above, ' Statement of Commission paid Thos. Wilson &c.', shows that from September 1801 to November 1803 a sum of over £1,100 was received on account of engines built by Hornblower.

The course of events in the last few years of the connexion of Boulton and Watt with Cornwall is difficult to follow. This is due, on the one hand, to the repudiation of the Watt patent by certain of the mine-owners, and the erection of infringing engines by Trevithick the younger, Hornblower, and Bull, and on the other hand to the fact that Murdock was putting up engines for which he himself made the drawings, so that of these engines the Soho portfolios contain no representations. No doubt this was done with the knowledge and approval of the firm, and after the regular agreements had been entered into, but as the portfolios of drawings in the Boulton and Watt Collection have been the recognized source of information about the engines put up in Cornwall and elsewhere, it will be appreciated that the absence of these drawings makes it difficult to state the exact number of engines put up.

Richard Michell, then the engineer of the United Mines, informs Boulton and Watt in 1797 that the adventurers have a 63-inch cylinder for Ale and Cakes already cast at Coalbrookdale and says : ' I am ordered to make drawings for nozels for it to work double if wanted. I had much rather the [sic] were to be had at Soho, for in that case it would save me a deal of trouble, but I can make no kind of objections to anything of this sort, for Bull and Trevithick are at hand.' [1] In another letter he states that the 63-inch cylinder is now on the mine and that he is ordered to make drawings for another engine of the same size for Poldory, ' if I refuse the erection Trevithick or Bull will snap at it '.[2]

In this last letter Michell mentions that Trevithick is erecting an engine at Ding Dong mine, for which the nozzle had been cast at Hayle. It seems that Trevithick aimed at constructing this engine in such a way as to avoid Watt's

---

[1] B. & W. Colln. Richard Michell to Boulton & Watt, 1797 Apl. 27.

[2] Ibid. Richard Michell to Boulton & Watt, 1797 Nov. 23.

patent, for later on he altered it and we find him writing to the firm to inform them of this and offering to pay the premium.   In the same letter Michell mentions also an engine by Bull at Pednandrea which he describes as atmospheric with a column of water instead of an air-pump ; this engine he says Murdock is converting into a Watt engine.   Murdock was in the same year engaged upon engines for Trescowe, Wheal Jewel West, and New Roskear or Prince William Henry. For the Wheal Jewel West engine a 40-inch cylinder made for a Bull engine was used, and the condenser and air-pump were cast at Hayle.   For the New Roskear engine Murdock was making the drawings in the first weeks in April.   The nozzles were ordered from Soho, to be made the same as those for the Wheal Butson engine, the cylinder from a foundry near Bristol, and the air-pump and condenser from Perran Foundry.[1]   Here it may not be amiss, at the risk of repetition, to say that none of the pumping-engines was at any time actually made and erected by the firm.   The mine adventurers entered into an agreement to pay certain royalties ; Boulton and Watt supplied drawings of the engine and instructions for putting it up.   It was a condition that the valve nozzles were to be obtained from Soho, and in the ordinary way this was the only part of the engine made there ; the cylinder and air-pump came from Wilkinson's foundry at Bersham near Wrexham, as also did, usually, most of the other castings of any size.   The wrought-iron work, except the piston-rod, was made on or near the mine.   The mine adventurers paid for all this material themselves, and they paid the wages of the erectors. It was not incumbent upon them to employ Murdock or any of the Soho erectors ; they might appoint any engineer to take charge of the erection subject to the approval of Boulton and Watt.

After 1795, when the Bersham Ironworks came to an end, Boulton and Watt ceased to exercise any control on the place of production of the cylinders, and the mine-owners ordered their material where they thought fit, from Coalbrookdale, Bristol, and, for the smaller cylinders, the Cornish foundries, Hayle and Perran.

The Watt engine called for greater skill and care in its construction and management than did the Newcomen, so we find the introduction of the new engine reacting upon the craftsmen and engine-men.   Richard Michell wrote, about 1793 :

' Prior to the introduction of these engines into this county there was scarce anything scientific in the best engineering, but as soon as these engines were erected, scientific knowledge began to spread to the astonishment of the publick, so much that now there are few counties in England who can outdo us in practical engineers, as miners, or as handy craft men.   Our blacksmiths knew nothing, comparatively speaking, of the proper method of work[s] iron & steel, of making screw taps and plates for cutting these large adjusting screws so very usefull and necessary in mining, yet so little thought on by the Cornish as by the inhabitants of the Pellew Islands.   Likewise the method of boring & turn[s] cast and wrought iron.   I never remember to hear of but one hole bored in cast iron before B. & Watt came and pointed out the way.   As to cutting brass work it was done in the clumsiest way & manner possible, but now we have both brass & iron cast almost equal to any of the best foundries in England.   They were altogether in the dark with regard to making boilers capable of holding water and steam, and as well with respect to their form &c.   Whereas now we have them compleatly finish'd and tight.   There was scarce ever an engine house built sufficiently strong till they pointed out the method.   And as to the pump work, bobs, &c. every person must acknowledge they have made vast improvements on them.' [2]

Michell was a confirmed adherent of Boulton and Watt, and beyond doubt

[1] Ibid.   Gregory Watt (Redruth) to J. Watt, junior, 1797 Apl. 24.

[2] Ibid., ' Copy of a letter from Rich[d] Michell to a friend.'

his picture is too highly coloured. The Cornish craftsmen were not so unskilful when Watt made his first visit to the county, nor was the progress made so great, as he suggests. From the beginning the smithwork about the engines had been done locally, and Watt in 1777, while he remarks on the low price charged for it, thinks it ' is little inferior to our own, if any '. ' The engine carpentry and stone work', however, he found ' very rough'. As to foundry work, Cornwall probably was not more backward than many other parts of the country ; it certainly had two small foundries within its borders.

There can be no question that the last quarter of the eighteenth century witnessed in Cornwall, as in other parts of England, decided progress in mechanical engineering and iron-founding. For instance, in 1797 we find air-pumps, condensers, and valve nozzles being made locally, at Perran and Hayle. The cylinders, at any rate for the larger engines, continued to be supplied from outside the county. Some years had to go by before the great days of Cornish iron-founding and engine-making set in.

The quotation from Michell's statement, given above, ends with the remark : ' And as to the pump work, bobs, &c. every person must acknowledge they have made vast improvements on them.' It seems clear that by ' they' the writer meant Boulton and Watt, but while no doubt considerable improvements were effected during the period between 1777 and 1800 there is no distinct evidence that the firm ever did very much in regard to the pumps themselves, at least in Cornwall. Doubtless the substitution of a main pump-rod, descending the shaft and working the separate lifts by offsets, for a series of separate rods brought up to the top of the shaft as, clearly, was due to Watt, as the arrangement of bobs for working two sets of lifts by the double-acting engine. The pumps in use when Watt first went to Cornwall were bucket pumps, and, although it does seem that the force or plunger pump was coming into use, bucket pumps remained in general use until after the year 1800. The date of the first introduction of force pumps into the Cornish mines is uncertain ;[1] we have seen that in 1782, when the idea of the double-acting engine was being developed, Boulton suggested the combination of force and lift pumps. Watt took no notice of this suggestion, but it is clear that within a few years ' plungers or forcing pumps' were at work in Cornwall. In 1786, in connexion with an order for a large double-acting engine for Wheal Virgin, Boulton writes from Cornwall :

' Captn. Paul (& indeed every one of the Capt^{ns}.) are very fond of the plungers or forceing pumps. 1st. because they save ball^e bobs totaly with their friction & vis inertia. 2. they save $\frac{3}{4}$ of the leather. 3. save time & stopages as they are never out of order. 4. saves the ware of the pumps from the iron rods wch go down them. 5. saves the iron rods themselves. 6. saves much of the engine power as they evidently work quicker, and from all I can learn of the pitmen & everybody about the engines it is certainly better than ball^e bobs.'[2]

The terms of this letter certainly indicate that plunger pumps were in use as early as 1786 ; such a statement as ' they are never out of order ' can hardly have been made except from experience ; possibly, however, that experience had not been long enough to test the construction thoroughly and later on defects may have shown themselves the surmounting of which took some time, for it was not

---

[1] Desaguliers has a figure of a plunger pump which he says is ' often used in the Engine to raise Water by Fire ', but he does not suggest that it was in use in the mines of Cornwall.

(J. T. Desaguliers: *A Course of Experimental Philosophy*, Vol. II, pl. 15, 3rd edition, 1763).
[2] B. & W. Colln. Boulton to Watt, 1786 Sept. 17.

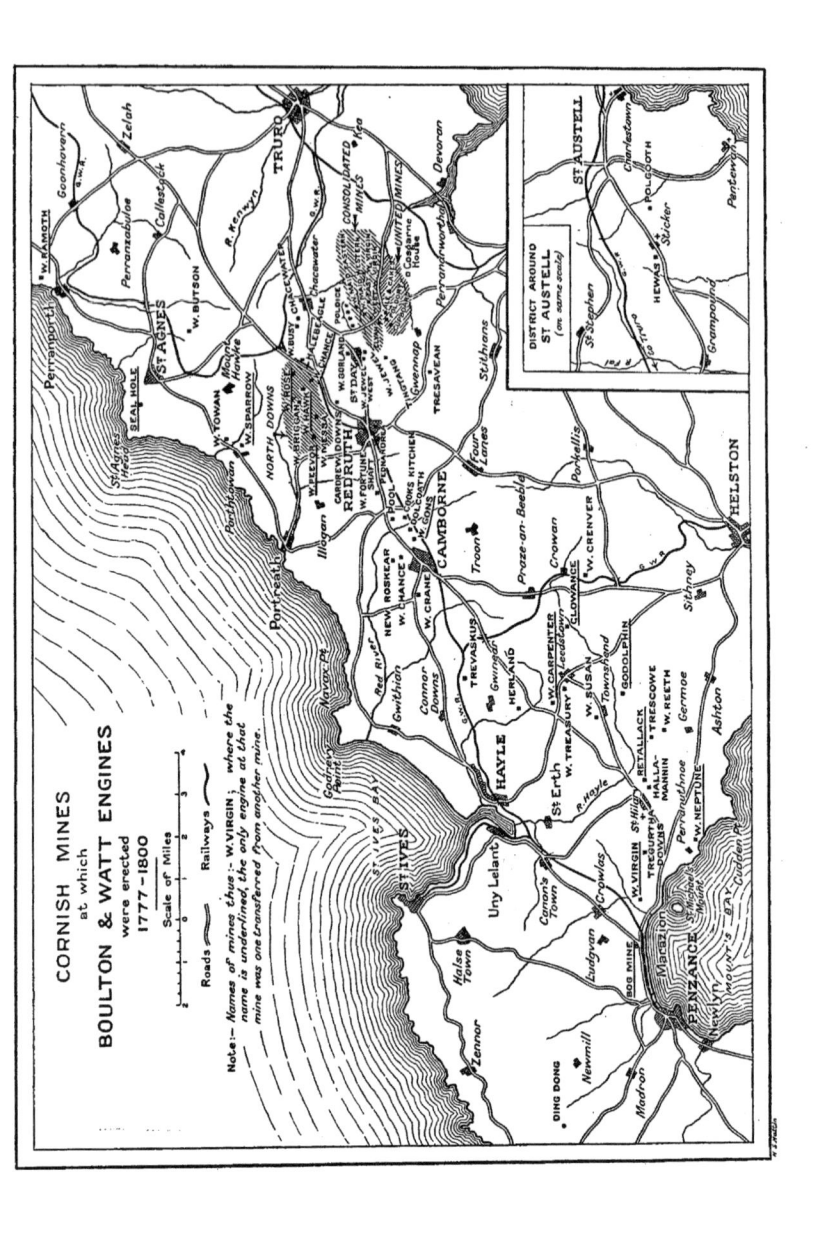

CORNISH MINES
at which
BOULTON & WATT ENGINES
were erected
1777-1800

Scale of Miles

Roads   Railways

Note:- Names of mines thus:- W.VIRGIN; where the
name is underlined, the only engine at that
mine was one transferred from another mine.

DISTRICT AROUND
ST AUSTELL
(on same scale)

ST AUSTELL
St Stephen   Charlestown
Polgooth   HEWAS   St Blazey
Scorrier   Grampound   Pentewan

Goonhavern   Zelah
Perranzabuloe   Callestock
W. PENMOTH   TRURO
Perranporth   R. Kenwyn
ST AGNES   W. BUTSON   R. Allen
SEAL HOLE   St Agnes Head
Mount Hawke   CONSOLIDATED MINES
W. TOWAN   Chacewater   UNITED MINES
W. SPARROW   CARHARRACK   Devoran
NORTH DOWNS   POLDICE
W. VIRGIN   W. FORTUNE   Perranarworthal
Illogan   REDRUTH   St DAY   Gwennap
Portreath   POOL   Tresavean
Hayle river   KING'S KITCHEN   Stithians
NEW ROSKEAR   CAMBORNE   Four Lanes
W. CRANE   Troon   Praze-an-Beeble
W. CHANCE   W. CARPENTER   Porkellis   HELSTON
Red River   Gwennap   W. CRENVER
Gwithian   TREVASKUS   HERLAND   Crowan
Cannon Downs   GODOLPHIN   Sithney
HAYLE   W. SUSAN   Townsend   Ashton
St Erth   W. TREASURY   TRESCOWE
ST IVES   R. Hayle   RETALLACK   W. REETH
ST IVES BAY   HALLA-MANNIN   Germoe
Uny Lelant   W. VIRGIN St. HILL   Perranuthnoe
Carnon Town   TRECURTH   W. NEPTUNE
Crowlas   TOWEN   Marazion
Halse Town   BOG MINE   PENZANCE
Ludgvan   Newlyn   MOUNT'S BAY   Mousehole
Zennor   DING DONG   Newmill   Madron

until after the year 1800 that it can be said that the plunger pump came into extensive use.

The table which follows shows the distribution of the Boulton and Watt engines in Cornwall up to the year 1800. It is arranged alphabetically in accordance with the names of the mines, with dates of erection or starting, size of cylinder, kind of engine, and remarks on the history of the engine after its first erection, and the like. Up to the year 1796 the table is based on definite records in the Boulton and Watt Collection. From that date on to 1800 the historical material available is not so explicit, and it is possible that some minor changes may have taken place which have not been noted.

The sites of many of the mines are difficult to identify at the present day. Two sources of this difficulty are that, in course of time, new names are given to old mines, and old names to new mines. It is believed that the accompanying map is substantially correct.

BOULTON AND WATT ENGINES ON CORNISH MINES

* indicates the mines which had only engines moved from other mines in Cornwall.
S.A. = single-acting ; D.A. = double-acting ; R. = rotative.

Erected or Started	Mine	Cylinder diameter in inches and kind	Remarks
	*Bog Mine	36 S.A.	Between Marazion and Penzance. Wheal Butson engine moved here. Working in 1794.
1777	Wheal Busy (Chacewater)	30 S.A.	The first B. & W. engine put up in Cornwall. It was put up for temporary use and ceased working in September 1778. It was first removed to Wheal Chance (Scorrier), from there to Wheal Virgin (St. Hilary), thence again to Wheal Crane (Camborne). In 1794 it was working at the United Mines in the same house as the Poldory 48-inch engine ; but it was thrown out soon after this, and in 1797 had been bought by the Adventurers of Wheal Susan, near Godolphin.
1791	Wheal Butson (St. Agnes)	36 S.A.	Removed to Bog Mine and working there in 1794.
	*Wheal Carpenter	14¾ D.A.R.	This was the Wheal Crane engine. See Wheal Crane below.
1794	Cardrew Downs	20 D.A.R.	This seems to have been a second-hand engine got from outside the county. It was at work drawing water in 1794 ; in 1796 it was idle.
1781	*Castle Adit	63 S.A.	Tregurtha Downs engine moved here in 1781. Referred to also as Kestal Adit. Site not identified.
1778	Chacewater	63 S.A.	This was a reconstruction of Smeaton's 72-inch engine. It worked from 1778 to 1784. In 1794 it was reported as ' now extinct and broken up '.
	Do.	14¾ D.A,R.	This was the engine from Wheal Crane. It was used at Chacewater as a whim engine ; sold before 1794.
1780	Wheal Chance or Higher Rosewarne (Camborne)	63 S.A.	Worked about two years. Removed to Polgooth where it was working in 1794.
	*Wheal Chance (Scorrier)	30 S.A.	Wheal Busy engine moved here before 1782. See Wheal Busy.
	Do.	52 S.A.	This was the Tingtang engine. See Tingtang.
	*Clowance	20 D.A.	This was the engine originally erected at Wheal Reeth. In 1794 it is reported that ' it is let to the Advrs. to pay 10 guineas per month as long as they use it and to deliver it in good condition when done with '.
	Consolidated Mines :		
1782	Elvan	58 S.A.	This was still at work in 1796. It had been altered by Bull by the addition of another cylinder (that of Wheal Maid 50-inch) but had been restored to the original construction.
1782	Old Wheal Maid (Wheal Maid Eastern)	58 S.A.	Worked up to September 1793. In 1794 it was reported to be in excellent order, and was for sale.
1782	New Wheal Maid (Wheal Maid Western)	50 S.A.	This engine was removed to work by the side of Elvan (see above). In 1794 it was ' destroyed excepting the cylinder, which is good for little '.
1782	East Wheal Virgin	56 S.A.	Worked up to September 1793. In 1794 it was for sale, but is said to have been in very bad order.
1782	West Wheal Virgin	52 S.A.	In 1796 this engine was being worked occasionally as an auxiliary to Wheal Maid 63 D.A. (see below).
1784	Wheal Maid Rotative	18 S.A.R.	In 1794 it was reported that this engine had been applied to winding ores upon Wheal Maid until September 1793, when it had been sold for a colliery in Glamorganshire. According to the Life of Trevithick, it was brought back from Wales to Herland and afterwards moved to Dolcoath.

Erected or Started	Mine	Cylinder diameter in inches and kind	Remarks
	Consolidated Mines (contd.) :		
1788	Wheal Maid, Double	63 D.A.	This is referred to also as Wheal Virgin, Double. Still at work in 1796.
	Wheal Virgin, Little Double	20 D.A.	This ' worked at drawing water 100 fathoms underground '. It was sold to the Wheal Jewel Adventurers in 1792.
1793	Cooks Kitchen	36 S.A.	Still at work in 1796.
1784	Wheal Crane	14¾ D.A.R.	This was applied to pumping water underground. It was removed to Chacewater for winding ores, and afterwards to Wheal Carpenter where it was used for ' drawing water with a rotative motion and fly to regulate it '. Then in 1797 it was ' bought by the United Mines for drawing up ores &c.'
	Do.	30 S.A.	Wheal Busy engine moved here from Wheal Virgin, St. Hilary. See Wheal Busy above.
1782	Wheal Crenver	48 S.A.	Removed to Godolphin in 1786.
1786	Do.	60 S.A.	Still at work in 1794.
	Do.	36 S.A.	Wheal Treasury 36-inch engine moved here from Wheal Gons and put to work by the side of the 60-inch engine above, the engines being independent but together working one set of pumps and being supplied from one boiler.
1797	Ding Dong		This engine was put up by Trevithick, apparently as an atmospheric engine, but he altered it to a B. & W. double-acting engine and offered to pay the premium.
1781	Dolcoath	63 S.A.	This was sold to Wheal Gons in 1789 and was working there in 1794.
	Wheal Fortune	36 S.A.	Working in 1796. Site uncertain.
	*Godolphin	48 S.A.	Wheal Crenver 48-inch engine moved here ; stopped in 1794.
	*Wheal Gons	36 S.A.	Wheal Treasury 36-inch engine moved here, afterwards to Wheal Crenver.
	Do.	63 S.A.	Dolcoath 63-inch engine working here 1789–96.
	Wheal Gorland	24 D.A.	Working single in 1796.
1778	Hallamannin	40 S.A.	Removed to Retallack and thence to Wheal Treasury.
1787	Do.	60 D.A.	Working there until 1794, then removed to Wheal Ramoth, and afterwards to Herland.
1792	Herland	64 S.A.	This was built of the materials of Poldice No. 3 engine, with a new cylinder.
1797	Do.	60 D.A.	Hallamannin 60-inch engine moved here from Wheal Ramoth.
1798	Do.		A whim engine supplied by B. & W., and an 18-inch single-acting engine ordered from them.
1794	Hewas (Huyas)	45 D.A.	Working in 1796.
1792	Wheal Jewel	20 D.A.R.	Bought from the Consolidated Mines and used for ' drawing water with a rotative motion '. It was at work in 1796.
1798	Do.		A new engine on order.
1796	Wheal Jewel West	20 D.A.	An engine of this size was on the mine in 1796.
	Do.	40 S.A.	This was Hallamannin 40-inch engine moved from Wheal Treasury. It was at work in 1797.
1798	Do.	40	In 1797 Murdock was at work upon an engine for Wheal Jewel West, using a cylinder that had been made for a Bull engine at Herland.
	Kestal. See Castle Adit.		
	*Wheal Let.	52 S.A.	This mine was a part of Wheal Chance (Scorrier) ; the Tingtang engine was moved here from Wheal Chance.
1786	Wheal Mount	20 D.A.	Used for drawing water. Site not identified. Sold to Mr. Gullet of Beerferris Mine, Devonshire. Destroyed by fire before 1794.
	*Wheal Neptune	20 D.A.	Wheal Reeth engine moved here.
1797	New Roskear or Prince William Henry.	36 S.A.	

Erected or Started	Mine	Cylinder diameter in inches and kind	Remarks
1786	North Downs : Wheal Messa or Lemons	42 D.A.	Still at work in 1794.
1785	Wheal Fortune or Briggan	45 D.A.	Still at work in 1796.
1796	Halebeagle	52 D.A.	This was an exceptional engine. The cylinder was directly over the shaft and the piston-rod was connected to the pump-rod by a cross-head and side-rods.
	Do.	52 S.A.	Tingtang engine moved here from Wheal Let. It was lying idle, being completely worn out, in 1794.
	Do.	45 S.A.	This was the Trevaskus engine ; it was brought here to take the place of the Tingtang engine before 1794.
	*Wheal Rose	63 S.A.	This was the Tregurtha Downs engine moved here from Castle Adit. It was working on Wheal Rose, 1794–6.
1794	*Wheal Peevor	28 S.A.	The Tresavean engine was erected here in 1794.
	*Wheal Hawk		This mine is mentioned by Boulton in 1787, but it is not clear what engine they had. In 1796 there is mention of a ' 42-inch double acting, Wheal Hawk, say Lemons ', so possibly the Wheal Messa engine had been moved here.
	Pednandrea		This was an engine put up by Bull, atmospheric with a column of water instead of an air-pump. It was altered by Murdock in 1797.
1779	Poldice No. 1. Eastern engine	63 S.A.	This was at work until 1794. It was sold to Wheal Treasury in 1796 and afterwards to Wheal Sparrow.
1782	No. 2 Middle or Western Engine	63 S.A.	At work in 1794. Idle in 1796.
1785	No. 3	63 S.A.	This engine was replaced in 1787 by Nos. 4 and 5, and was removed to Chacewater, but the cylinder was resold to Poldice where it was used to replace the worn-out cylinder of engine No. 2. The remaining materials were sold to Herland.
1787	No. 4 Little Double	24 D.A.	Applied to drawing water, at work in 1794.
1787	No. 5 Oppy's	58 D.A.	At work in 1796.
1783	Polgooth	63 S.A.	This was the Wheal Chance engine. At work in 1796.
1786	Do.	58 S.A.	In 1794 this engine is reported to be ' kept in good condition but is only used very rarely in winter '. In 1796 it is said that Coxshead 58-inch engine is used occasionally. This seems to be the same engine.
1781	Pool	60 S.A.	In 1794 it was reported that this engine ' was removed to Chacewater, and after laying idle several years, was lately sold to the United Mines, where it is shortly to be erected in place of Poldory 48-inch and Wheal Busy 30-inch '.
1787	Prince George Mine	20 D.A.R.	Applied to drawing water. Removed to Seal Hole. Site not identified.
	*Wheal Ramoth	60 S.A.	Hallamannin engine working here, 1796–7.
1786	Wheal Reeth	20 D.A.	Applied to drawing water. Removed to Wheal Neptune.
	*Retallack	40 S.A.	Hallamannin 40-inch moved here.
	*Seal Hole	20 D.A.	Prince George engine was moved to Seal Hole, ' where B. & W. undertook to work it at a monthly premium, but the bad quality of the water corroded the iron vessels so much that they were glad to pay a considerable sum to be off their bargain, and lost in all about 1000l. The engine by agreement continues to work without paying any premium.' This account was written in 1794.
1797	*Wheal Sparrow	63 S.A.	Poldice Eastern engine re-erected here by Bull.
1797	*Wheal Susan near Godolphin	30 S.A.	Wheal Busy 30-inch engine bought from the United Mines.
1778	Tingtang	52 S.A.	In 1794 it was reported that this engine had been ' bought by Fenton & Co. and erected at Wheal Chance (Scorrier), then sold to John Vivian and erected upon Wheal Let another part of the same mine of Wheal Chance, which was afterwards transferred to Jno. Williams & Partners. It was again removed to Halebeagle, part of North Downs, and is now lying idle, being completely worn out.'

Erected or Started	Mine	Cylinder diameter in inches and kind	Remarks
1785	Wheal Towan	18 D.A.	This worked upon the mine for four years, drawing water underground. It was afterwards sold out of Cornwall, and made rotative for drawing coal.
1780	Wheal Treasury	36 S.A.	Removed to Wheal Gons and thence to Wheal Crenver.
	Do.	40 S.A.	Hallamannin 40-inch engine moved here from Retallack. In 1797 this engine is reported to be working at Wheal Jewel West.
	Do.	63 S.A.	Poldice No. 1 engine moved here in 1796 ; sold to Wheal Sparrow in 1797. In 1797 a new engine seems to have been erected on this mine, but particulars are lacking.
1778	Tregurtha Downs (Wheal Union)	63 S.A.	The engine was removed in 1781 to Castle Adit and afterwards to Wheal Rose on North Downs where it was working in 1794.
1780	Tresavean	28 S.A.	In 1794 it was reported that, ' This mine had been for several years extinct, and the engine lying idle upon it, but it is at present erecting upon Wheal Peever, part of North Downs mine.'
1797	Trescowe		Put up by Murdock. No other information.
1782	Trevaskus	45 S.A.	Removed to Halebeagle, North Downs, before 1794.
	United Mines :		
1780	Ale & Cakes	58 S.A.	Still working in 1796.
1780	Poldory	48 S.A.	Still working in 1794 with Wheal Busy 30-inch as an auxiliary. Soon after, these engines were replaced by Pool 60-inch engine.
	Do.	60 S.A.	This was the Pool engine. See above. It was working in 1796.
1797	Do.	14¾ D.A.R.	Wheal Crane engine bought from Wheal Carpenter for drawing ores.
1784	New Poldory (Cupboard)	45	Working in 1796.
1787	New Ale & Cakes	45 D.A.	Working in 1796.
1801		14¾ D.A.R.	Towards the end of 1800 the United Mines ordered ' a 6 h.p. engine and drawing apparatus '. In 1797 a 63 D.A. engine was being put up on Ale and Cakes without the consent of B. & W. Its erection, and also the working of Wheal Crane engine, *above*, was stopped by Injunction.
	*Wheal Virgin (St. Hilary)	30 S.A.	Wheal Busy engine moved here from Wheal Chance (Scorrier), 1782.

XXIV

THE ENGINE PREMIUMS OR ROYALTIES.
THE PARTNERSHIPS

PREMIUMS

AT the outset the royalty, or premium as they termed it, that Boulton and Watt charged for the use of engines made according to Watt's patent was based upon the saving of coal effected as compared with an atmospheric engine. Their position is set forth very clearly in some of the letters. For instance Watt, writing to Jonathan Hornblower, the elder, in 1776, says : ' Our profits arise not from making the engine, but from a certain proportion of the savings in fuel which we make over any common engine that raises the same quantity of water to the same height. The proportion of the savings we ask is one-third part to be paid to us annually for twenty-five years, or if our employers choose it they may purchase up our part at ten years price in ready money.' [1] Boulton in a letter written about two years later puts the matter at greater length :

' We do not at present turn our eyes toward collieries for great profits, but to places where coals are at a high price, and there we take annually for the term of our Act of Parl^t a sum equal to 1/3rd of the value of the coal when deliver'd at the engine which evidently appears to be saved by our engine when compared with the common fire engine. There are many savings made by the proprietors of our engines over and above the 2/3rds of the savings of fuel. In Cornwall we take down two large cylinders and erect one of ours of equal or even smaller diameter in their place which doth more work than the former two and in general will go 20 fathom deeper with the very same pumps, and our one doth not burn more than half the coals of one of the former two, hence the erection of one large engine is saved and consequently the annual support of it, with engineers, fire men, repairs &c. which amount in Cornwall to between £500 or £600 p. annum ; if there is but ¼th of the coals burnt, there is less than ¼th of the boiler burnt, our cylinders so are not liable [to] corrosion by the vitriolic waters of mines as the common ones are, because no water is ever admitted into ours, but they are always dry and constantly polished by a greasy piston. Our engine is much more under command than a common one, as it will work quietly and oeconomically under any load from 1 lb on the square inch to eleven pounds, and if work'd by a man of common sense the buckets and clacks in the pumps may be more favor'd by it. In many collieries the proprietors value their engine coals at nothing, consequently we are obliged to deviate in such cases from our rate of taking 1/3rd of the value of the fuel sav'd because 1/3rd of 0 is 0 and our time is too valuable to work at that price even if it be interlarded with honor.'

Boulton goes on to say :

' We will make drawings for your self of the whole, and for the founding, the smiths, for every minute part, all which you will probably get the best terms you can. We will send you an experienced person (whose wages you must pay) to assist and direct your own people to erect it, which they cannot do as it ought to be done in the first engine without such assistance, although we shall furnish you with all necessary directions in writing, and we will guarantee the good performance of the engine conformable to our agreement. Such part of the foundry

---

[1] B. & W. Colln. Watt to Hornblower, 1776 Oct. 17.

work as relate to the pump we leave you at liberty to take where you please, but such parts as relate to the engine end of the beam must be cast at Bersham in Wales, as there is no other proper apparatus in Britain for producing the parts with that truth and exactness we require.' [1]

Boulton's idea in the first instance had been somewhat different to that set out above. In April 1775 he informs Watt that he had ventured to say to one of the Cornish visitors to Soho :

' That we will undertake and contract to make an engine or engines capable of doing any quantity of work that shall be requested and described, for as little money as com^n engines will cost that are capable of doing as much work, & we will guarantee them to do that work with half the expense of fuel that com^n ones will require ; provided we are allowed a sum that shall be equal to its further savings over and above the said half.' [2]

This letter seems to have led to a discussion of the subject between the partners and within two months, as the correspondence relating to the Bow engine shows, the plan of taking one-third of the savings had been decided upon.[3] This was the system that was applied in connexion with the London Waterworks' engines, and with the first engines in Cornwall. In Cornwall, however, it was soon replaced, to a large extent, by a fixed rate based on the size of the engine ; this and some other matters concerning the assessment of the premiums in that county are discussed in Chapter XXIII.

With the rotative engine the system of payment by savings was not applicable, and the other plan of a fixed rate was adopted. Whitbread, one of the first London users, paid £63 per annum for an engine with a 24-inch cylinder, 6 ft. stroke. Calvert and Barclay and Perkins paid the same premiums for engines of the same size. Very soon, however, a rate per horse-power was fixed upon, viz. six guineas per horse-power per annum for engines in the London district, and £5 for engines in the country. All the premiums, it will be understood, were payable so long as Watt's original patent remained in force.

As has been explained elsewhere, at the outset and for many years only a small part of the engine was made at Soho and supplied by Boulton and Watt. As time went on the practice grew, particularly in connexion with rotative engines, for the customers to order all the engine materials through Boulton and Watt. Then as the period of the patent was drawing to a close, in 1795, the firm decided not to make the premium a separate charge, but to make the price of the engine cover the premiums for the remaining years of the patent.

THE PARTNERSHIPS

In view of what has been stated above, that only a small part of the engine was made at Soho, it is clear that the manufacturing profits cannot have been large and, in fact, the main source of the income of the firm was the patent royalties or premiums. We can now proceed to the consideration of how this income was divided between Boulton and Watt. The circumstances of the formation of the partnership have been dealt with in Chapter IV ; Boulton had two-thirds and Watt one-third share in the concern. After payment of all expenses, including a salary

[1] Patent Office Library. Boulton to Morris, 1778 Dec. 28. Transcript.
[2] Doldowlod Papers. Boulton to Watt, 1775 Apl. 24. It will be noted that the comparison is to be made with the ' common ' engine. Boulton & Watt declined to grant their usual terms if the comparison was to be made with one of Smeaton's engines. (B. & W. Colln. Watt to Smeaton, 1778 Jan. 17.)
[3] In the case of this particular engine, however, the premiums were commuted for a sum of £150.

of £330 to Watt, the net profit was to be divided in the proportion of two to one. Boulton was to pay all preliminary expenses, the cost of obtaining the Act of Parliament for extending the term of the invention, experiments, &c., and the partnership was to endure for the extended term of the patent, till June 1800.

In so far as relates to pumping-engines this arrangement held good until the end, but when the rotative engine was introduced it was agreed that the profits on it should be equally divided.

About the year 1795, when the term of the patent was drawing to an end, steps were taken to develop the manufacturing side of the business and considerable extensions were made in the engine shops at Soho Manufactory, and Soho Foundry was commenced ; it was after this that the production of complete engines was undertaken.

An explicit statement of the reason for this change is not. forthcoming. In so far as the business was one of collecting royalties, that would obviously come to an end with the patent. Now by this time the name of Boulton and Watt and the fame of the engine had become known very widely; inquiries were coming in to Soho from all parts of the country and from abroad. The partners must have experienced a certain feeling of regret at the idea that the reputation so built up would be of no further advantage to them. Moreover their sons were growing up, and no doubt it seemed to them that this was the opportunity for a change in the Soho policy; the younger men should start as engine-makers with the advantage of the reputation of the old firm. Whether this be the correct explanation or not, in October 1794 a new firm was formed under the style of Boulton, Watt & Sons, Soho, and composed of Boulton, his son Matthew Robinson, and Watt and his sons James Watt, junior, and Gregory, then a student at the College of Glasgow.[1] The original firm continued in existence, but it handed over to the new firm ' the future manufactory and profits of steam engines made by them'. Five years later Watt gave up, to his sons, the small share he had retained for himself in the new concern, which now became Boulton, Watt & Co., Soho.

At first the new firm had been owned by the Boultons and the Watts in the same proportion as the partnership of the original firm, but at the end of the year 1800 the concern was equally divided, both as regards pumping and rotative engines, between the Boultons and the Watts ; Boulton, his son, and the two young Watts each holding a share of one-fourth. When Gregory Watt died, in 1804, James Watt, junior, succeeded to his share, and when Boulton died, 1809, his son became the owner of his share and the firm now consisted of James Watt, junior, and Matthew Robinson Boulton. In the following year, 1810, Southern was made a partner, or perhaps it would be more correct to say was given a percentage on the profits. He died in 1815.

So far we have been dealing only with the engine works at Soho Manufactory. We have now to consider Soho Foundry. The land upon which the foundry was built was bought by Watt in 1795, but he had no share in the concern, the partners were Boulton, his son, and the two young Watts; they held equal shares. In 1801 Watt conveyed the land to them for the price he paid for it. The funds for erecting the buildings and machinery were advanced by Boulton and Watt ; they amounted to over £21,000. The debt with interest at 5 per cent. had been entirely repaid from the profits of the concern in August 1812. The Soho Foundry was carried

---

[1] Doldowlod Papers. ' Short memorandums of the different firms under which Mr. Boulton and Mr. Watt and their sons have carried on business.' [By James Watt, junior.]

on under the style ' Boulton, Watt & Co. Soho Foundry'. Gregory Watt's share was merged in that of James Watt, junior, and Boulton's in that of his son, as in the case of the firm at Soho Manufactory. When Southern was made a partner, a partnership in the Foundry was offered to Murdock, but he preferred to have an increased salary. He was offered and accepted £1,000 a year.

The memorandum by James Watt, junior, upon which the preceding statements are based furnishes us also with particulars of the ' London Banking House' and of the Copying Company.

The Copying Company—James Watt & Co.—had three partners, Boulton, Watt, and Keir, each holding a one-third share. In 1794, when the copying patent expired, Keir retired, and from then forwards the partners were James Watt, junior, and Matthew Robinson Boulton, who held equal shares.

Boulton banked with William Matthews of Green Lettice Lane, in the City of London, and the account of Boulton and Watt was kept with him till his death. The business was continued by his widow until her death in 1802, when it was taken up by M. & R. Boulton, J. & G. Watt & Co. who appointed as their agents the two clerks who had been employed by Mrs. Matthews. Upon the death of Gregory Watt and of Boulton, the firm became M. R. Boulton, J. Watt & Co. The books of this concern were closed finally in 1833 when an account was opened at the Bank of England.

XXV

PERFORMANCE OF ENGINES.
EXPERIMENTS AND TRIALS

FOR present-day purposes the particulars that have come down to us in regard to the performances of steam engines in the time of Watt are not of great value. It is very rare indeed that we have exact figures for the consumption of steam, or of boiler-feed water. The factor that appealed to the user of the engine was the consumption of coal, and this at once brings in the question of boiler efficiency, but even as to the consumption of coal the figures are of doubtful value. Now and again we are given the weight in pounds, and sometimes in hundredweights, but here we have to be on our guard, for the hundredweight was sometimes 112 and at other times 120 lb. In general, the absence of suitable weighing-apparatus led to coal being sold by bulk, so that usually the consumption is given in chaldrons, weys, or bushels.[1] In Cornwall there was a long wey of 72 bushels and a short wey of 64. The capacity of the bushel varied in different localities, and the weight of coal in a bushel, even from the same coalfield, differed widely. Boulton found a bushel of Pontop coal to weigh 91 lb., whereas the usual figure for coal from the same district, the Tyne, was 84 lb. No doubt a part of this variation may be accounted for by variation in the size to which the coal was broken.

The calorific value of the coal one does not expect to find stated at this date; there are, however, statements comparing the evaporative power of one coal with another, but necessarily they are rough approximations.

The statement of the work done in pumping is in terms of net work, the quantity of water raised to the given height, and even this was determined by calculation from the length and number of the strokes and the diameter of the pumps. There was no means for ensuring that the engine stroke was invariably of full length, and defects in the pump buckets might cause the actual delivery to fall far short of that obtained by calculation. This last point was fully recognized, and on the first appearance of the Boulton and Watt engines in Cornwall some of the Cornishmen, unable to credit the results, did not hesitate to say that they were working without leather on the buckets and making holes in the clacks, in order to deceive strangers. To record the variations in the length of stroke, Watt devised an apparatus called a 'stroke measurer'; this seems to have been made but not to have been used to any great extent, and its construction is not known.

However, assuming the pump to be in good order and all the strokes of full length, the calculated delivery was not very much in excess of the quantity obtained

---

[1] Wyatt invented his bridge weighing-machine in 1743, and there were two public weighing-machines in use in Birmingham within a few years. This explains how it was that Boulton and Watt were able to buy coal by weight in 1774.

by measurement,[1] and at any rate it was quite a fair basis for the comparison of engines of about the same size working similar pumping-sets, and using the same coal. We say similar pumping-sets advisedly because the conditions of pumping and the arrangement of the pumps varied a good deal, and the friction losses occasioned in lifting a given quantity of water one foot high might be very different in different situations.

Although it cannot be said that these differences were lost sight of, the performance of an engine was always given, as stated above, in terms of net work, and the weight of water on the pump piston divided by the area of the engine piston in square inches was termed 'the load'. Builders of atmospheric engines had by Watt's time settled down to a 'load' of from 7 to 8 lb. on the square inch. An engine could be made to work with a greater load, and engines were frequently found working with a less load. The load of 7 to 8 lb. represents, say, half the pressure of the atmosphere, but, as the engines in addition to the 'load' had to overcome their own friction and that of the pumps, the degree of vacuum under the piston must have been considerably greater than half the atmospheric pressure, possibly 21 inches is a fair average figure. A better vacuum, and thus a greater load on the piston, could have been attained by pouring in more injection water, but only at the expense of a greater proportional loss of steam by condensation against the cooled cylinder wall. The load of 7 to 8 lb. was what experience had shown to be the most efficient compromise.

The effect of Watt's invention was to allow a better vacuum to be obtained without undue cooling of the cylinder. This meant, on the one hand, that the engine piston could take a bigger load, and, on the other, that the work would be done with less fuel. Watt in the first instance seems to have over-estimated the load that he could put on his piston. He started with $12\frac{1}{2}$ lb. on the square inch on the piston of the Soho 18-inch engine. When he designed Bloomfield engine he based his calculation on 11 lb., but in consequence of some mistake as to the depth from which the water was to be drawn, the engine when at work was loaded to $11\frac{3}{4}$ lb., which was considered too much.

Within a few years Watt settled down to a standard load of 10 lb., but many of his pumping engines were worked at 9 lb. and even less, and in the rotative engines the effective pressure was assumed at 7 lb. The diminished load does not necessarily imply a poor vacuum, or a defective engine ; with the lower load the engine could move more quickly and perform a greater number of strokes per minute. One of the most efficient engines in Cornwall, i. e. of Watt's construction, was loaded to 9 lb.

The scientific value of the figures obtained in these early trials may not be great, but at the time they served a useful purpose as a comparison of Watt's engines among themselves, and with the atmospheric engine, and they produced a spirit of emulation in those in charge of the engines. Accordingly some particulars of the nature of the experiments and trials made by Watt and by Boulton cannot fail to be of interest.

In the Boulton and Watt Collection at Birmingham is a manuscript book that has been alluded to on previous pages. This is in the handwriting of Watt and has for its title : ' Experiments on ye first engine at Soho, 18 inch cylinder.' This book contains a record of a series of trials made with different steam pressures, with variations in the firegrate and flues of the boiler, and alterations in the engine

[1] Trials made in Cornwall showed that the actual discharge was 0·917 of the calculated discharge.

itself, mainly in the piston. The engine, it will be recalled, had a surface condenser, and was employed to pump back water for a water-wheel, so that it was working against a constant load. The first part of the book is concerned with the engine as it came from Kinneil with the original block-tin cylinder; the second part, with the engine after a cast-iron cylinder had been put in.

In the first experiments the engine beam is pivoted in such a position that the load on the piston amounted to 12½ lb. on the square inch; later on the position of the fulcrum was so altered that the load became 11 lb., and still later it was restored to its original position. With these loads Watt had to use steam at higher pressure than was his practice in after years. In the first recorded experiment, with a load of 12½ lb., it was 9 in. of mercury, then when the fulcrum of the beam had been shifted and the load was 11 lb., it was found that the engine went very well with steam at 5 in. and tolerably with steam at 3 in. pressure. Subsequently we find him using pressures of 6 and 8 in. of mercury. The trials were directed in the main to determining the consumption of coal and the corresponding number of strokes made. The coal was Wednesbury coal, a hundredweight of which was 120 lb., and the strokes were numbered by a counter worked from the engine beam.

In one experiment Watt observed the time taken in getting up steam from the lighting of the fire. The water began to boil in 2 hours 55 minutes, a pressure of 3 in. was attained 15½ minutes later, and then in a further 7 minutes the pressure rose to 9 in. The engine was run for 1¾ hours at the rate, very nearly, of 13 strokes per minute. No feed was introduced into the boiler during the trial and the consumption of water was calculated from the difference in the water levels in the boiler, and its area at those levels.[1] However, in some of the experiments the feed was measured by a pan of known capacity.

After the iron cylinder had been substituted for the block-tin one Watt in one experiment, calculating from the measured consumption of feed water, determined the volume of steam used per stroke. He found it to amount to 11 cubic ft., whereas the content of the cylinder was but 8·8 cubic ft., so that the loss by condensation in the cylinder and possibly by water carried over by the steam was quite pronounced. It was, however, recognized that the experiment was quite a rough one and Watt judged that the real consumption was less than the figure arrived at. At any rate it was not very long before he was able to boast that he could work with no more steam than corresponded to the contents of the cylinder.

The last trial recorded in the book (the end pages of which have been lost) is dated June 16, 1775. Further experiments carried on by Boulton in the same month are mentioned in his correspondence with Watt. We find steam pressures of from 9 to 12 in. of mercury being used, and observations taken of the temperature of the water entering and leaving the surface condenser. In one case the water enters at 66° and leaves at 80°, in another it enters at 68° and leaves at 80°. Boulton fitted a mercurial gauge to the condenser and was interested in noting the variation in the vacuum during each stroke, from 25 to 28 or 28½ in.

Early in 1776 the amount of condensation in the jacket is noted. In 1,800 strokes made in 2 hours 24 minutes and using 10 cubic feet of feed water, the water

[1] The evaporation was found to be 11 cubic feet for 120 pounds of coal. A few years later Watt recorded 7 cubic feet for 56 pounds. Swansea coal he then said would do 11 cubic feet for 79 pounds. On another occasion he stated that 120 pounds of Wednesbury coal, when good, are not equal to one bushel of Newcastle.

condensed between the inner and outer cylinders amounted to 23 lb. This was after the cylinder had been provided with a non-conducting covering. Previous experiments, without the covering, gave 90 lb. on one occasion and 70 on another, condensed in the same time, from which Boulton deduced that the covering had reduced the loss two-thirds.[1]

The result of these trials with the Soho engine was a statement in October 1775 that the engine raised 16,000 cubic feet of water 24 ft. high with the consumption of 1 cwt. of the slack of Wednesbury coal. At the beginning of 1777 it was stated that it would raise 20,000 cubic feet with the same coal consumption, and a little later Boulton put the figure somewhat higher still.

Other experiments made at Soho that may be mentioned are a set in 1781 on expansive working, and a set in 1783 to determine the loss in friction between the beam and the shaft in rotative engines. In the expansion experiments the engine was first tried with a cut-off of one-fourth, and then with a cut-off of one-fifth, and observations were taken of the number of strokes, made by a given weight of water. The result was rather in favour of the higher expansion ; the average of three experiments in each case gives 2,640 strokes for the four expansions, and 2,836 for the five expansions. For the friction experiments the experimental forge engine was tried, first with the flywheel and rotative mechanism coupled to to the beam, and then without the rotative motion, but in each case the engine was working a pump directly from the beam. It was found that at twenty strokes per minute with the rotative motion the consumption of coal was 0·28 cwt. per hour, and without 0·235 cwt. The boiler evaporation was the same in the two cases, 11 cubic feet per cwt.

We may now turn to an interesting series of experiments on London water-works engines. In a trial made in April 1775 on the atmospheric pumping-engine at York Buildings, Watt measured the amount of injection water used, its temperature and that of the hot-well. From these data he proceeded to work out the volume of steam sent into the cylinder per stroke. He found it to be 206 cubic feet ; now the content of the cylinder itself, allowing for the clearance space, was 126 cubic feet ; accordingly 40 per cent. of the steam was wasted. It will be well to set out Watt's calculation as he wrote it. ' Heat of injection cistern 46°. Hot Well 170°. 54 strokes took 75,600 inches of injection water × heat gained 124°=9,374,400 ÷ heat lost by steam=842°=11,133⅔ ÷ 54 strokes=206 per stroke. The stroke was 9·2 ft. long, but allow for empty spaces 9·8, then the contents of cyl^r=126 cubic feet.' [2] It will be apparent that this calculation, as written, does not give an answer in terms of cubic feet of steam ; it is in the same terms as the measurement of the injection water, cubic inches of water. We have here one of Watt's short-cuts ; for ordinary purposes he reckoned that a cubic inch of water would form a cubic foot of steam. The number 842 also demands explanation. At this date the latent heat of steam was taken at 800° ; to this he added 212 and deducted 170. Six years after this his own experiments gave the latent heat at 948° or 950°, and the figure we find him using subsequently is 960°. The temperatures are given on the Fahrenheit scale throughout.

[1] Boulton formulated a rate for the amount of jacket condensation : ' when cylinders are cloathed & ye air 50 degrees, there will be one pound of water per hour condensed for every 4 sqr. feet & when *not* cloathed there will be one pound condensed for every □ foot per hour.' Boulton to Watt, 1779 Oct. 18 (B. & W. Colln.).

[2] Boulton Papers. Watt to Boulton, 1780 Apl. 11, sending a copy of the record.

In April 1777 Boulton made a trial of the atmospheric engine at Shadwell Waterworks which is noteworthy in that he determined by measurement, in cisterns, the quantity of injection water, of the hot-well discharge, and of the water actually pumped during a limited number of strokes.[1] He found the engine doing better than he thought it should do. It must be understood that the object of the trial was to obtain the performance of the engine for comparison with that of the Boulton and Watt engine that was to replace it, and thus to determine the saving of coal effected by the latter and the amount of royalty payable to Boulton and Watt. Having measured the hot-well discharge for eight strokes of the engine, Boulton says 'by calculation ye engine took less steam than ever I knew one take'. As a check he proceeded to measure the injection water, but 'in both cases the engine appeared to be better than I could wish it'. 'I was not satisfyd', he says, 'with this too well doing,' so he next measured the pump delivery for twelve strokes to see whether it agreed with the calculated discharge. He gives us no figures for comparison, but evidently he found no serious discrepancy, for he concludes with the remark that the actual quantity of water will always be a little less than the amount obtained by calculation. In spite of Boulton's forebodings the Boulton and Watt engine when erected showed a substantial saving. It made the same number of strokes with one-third the coal.

Soon after the Shadwell experiment we find Boulton at the Chelsea Waterworks measuring the quantity of hot water discharged from one of the engines. He does not give the figures but notes that the temperature of the injection water was 57° and that of the hot well 157° in one experiment, and 158½° in another.[2] In the definitive trials of the Chelsea old engines made in May 1778 observations were taken only of the consumption of coal and the numbers of strokes made, but here we do learn the weight of the bushel of coal they were using; it was 84 lb. No doubt the coal was from the Newcastle district. The Boulton and Watt engine was tested in July 1779, and the result was that the new engine with the same quantity of coal would do between three and four times as much work as the old. Watt writing in 1780 said that this engine raised 520,127 cubic feet of water one foot high per bushel of coal.[3]

Upon his arrival in Cornwall in 1777 Watt made observations of the temperature and quantity of the hot-well discharge in some of the engines found there, the highest temperature noted is 166°, the lowest 150°, and in 1779 he was using temperature observations to determine the consumption of steam in the Birmingham Navigation engine at Smethwick for the purpose of comparing the engine with and without the upper working-gear in use.

In the Soho experiments the performance is expressed in cubic feet of water and hundredweights of coal. For the London engines and in Cornwall the result is expressed in cubic feet of water raised one foot high by the expenditure of one bushel of coal. As early as 1782 Watt had advanced a step and was converting the cubic feet into pounds and expressing the result in millions of pounds raised one foot high. This new form for expressing the performance did not come into general use at once (we find the duty expressed in cubic feet after 1790), but ultimately it did entirely displace the old form.

The performance of the Soho engine stated above amounts for the 1775 figures

---

[1] B. & W. Colln. Boulton to Watt, 1777 Apl. 17.
[2] Ibid. Boulton to Watt, 1777 May 7.
[3] Doldowlod Papers. Watt to Hamilton, 1780 Jan. 1.

to 24 million pounds raised one foot with 1 cwt. of Wednesbury coal; for the 1777 figures to 30 millions. The figures are certainly too high.   Smeaton, who had had previous experience in testing pumping engines, and was no doubt perfectly independent, tested the Birmingham Navigation engine at Smethwick in 1778, soon after its erection.   The result (it was highly satisfactory in Smeaton's opinion) did not come up to the 30 million credited to the Soho engine, but Watt some years afterwards expressed the view that even this was too high, and thought that some mistake had been made in the measurement of the height to which the water was raised ; as he says, Smeaton's figures ' represent an effect equal to 26½ million pounds raised one foot high per 120 lbs. of coal, and 120 pounds of Wednesbury coal, when good, are not equal to one bushel of Newcastle'.   At this date, 1792, Watt had gained considerable experience in engine trials, and some of the very best engines in Cornwall were doing 32 and 33 millions, while others were doing very much lower duties.   The whole question of engine performances must have perplexed Watt a good deal.   The Chelsea engine in 1779 is credited with a duty of 32½ millions, upon a trial that seems to have been conducted with great care, and then in 1790 we have Boulton, when it is a question of another waterworks engine in London (for the Borough Waterworks), saying that he wants to put up the best engine possible and thinks it ought to do 25·2 millions, quite as if it was a matter they could not be sure of, and it is clear that he was convinced that this result could not be attained without great care.   The best results in Cornwall recorded in the Boulton and Watt papers are Poldice (in 1779) 26·5 millions, Polgooth (1785) 29·2, Ale and Cakes (1786) 31·3, Poldice (1792) 32, and Wheal Butson (1792) 33 millions.   In 1798 the average duty of twenty-three engines in the county is given as 17·6 millions.[1]

THE HORSE-POWER UNIT

So long as the engines were applied only for raising water the rating of their power was a comparatively simple matter; the diameter of the cylinder, the length of stroke, and the number of strokes per minute was sufficient information for the men concerned, and the performance expressed in terms of the quantity of water raised to a stated height in proportion to the consumption of coal was exactly what they required.

When the rotative engine came forward different conditions arose.   It was of little use to tell a customer the size of the engine cylinder; he had certain machinery to be driven and it was for the engineer to determine the size of the engine and to provide sufficient power.   In some cases Watt had for his guidance the information that a water wheel using a certain quantity of water at a certain head would do the work ; in others that it could be done by a certain number of horses.   He found it convenient to express the capacities of his engines in terms of the number of horses, working at the same time, that would do the same work. It was a comparison that would apply directly in many cases, and generally it would give a sufficiently good idea of the power available.

The idea of comparing the power of an engine with that of horses was not a new one.   Savery, in introducing his engine to the public in *The Miner's Friend*,

[1] *Phil. Trans.* 1830, 122.   Davies Gilbert: ' On the progressive improvements made in the efficiency of steam engines in Cornwall.'   More detailed information as to the performance of the engines in Cornwall will be found in Chapter XXIII.

had pointed out that one of his engines that would raise as much water as two horses working together at the same time was really equivalent to the ten or twelve horses that would require to be kept to maintain the work in constant operation. Again Smeaton in reference to his Cronstadt engine, 1775, states that it will raise 27,300 tons of water in 24 hours to the height of 53 feet, which, he says, is ' equal to the labour of 400 horses '.[1] This is perhaps the first time in which we get a numerical comparison between the power of a steam engine and that of horses, but it is clear that Smeaton had in view the total number of animals required, not the number in use at one time, and in the absence of any indication as to the number of relays that he reckoned upon the figures are of little value. Indeed his object was to give a striking illustration of the great power of the machine ; no idea of rating the sizes of engines was in his mind.

The formulation of a unit of power impresses one as an important step in mechanical science ; not that Watt had the interests of science in view or imagined that he was defining a unit which was to become universal ; he was concerned with facilitating and regularizing his own work as an engine designer and builder, and it was not until many years later that an intimation was published that his ' horse-power ' had a definite meaning, and what that meaning was. Boulton and Watt did not send out, broadcast, lists of engines giving cylinder diameter, &c., corresponding to different horse-powers. As late as 1792 we find Watt giving Southern instructions about the reply to be made to a certain inquiry for an engine ; he is to give prices, but not the diameter of the cylinder, ' as we wish to give them no information we can avoid until we know if it is likely to be an order.' But, setting all this aside, it is a matter of considerable interest to know the steps by which Watt arrived at his unit, and fortunately we have contemporary material to assist us in the investigation of this point in his ' Blotting and Calculation Book 1782 & 1783 '. In this book under the date August 1782 we read :

' Mr. Worthington of Manchester wants a mill to grind and rasp logwood and to drive a calendar. The power for all which is computed to be about that of 12 horses. Mr. Wriggley, his millwright, says a mill-horse walks in 24 feet dia^r and makes $2\frac{1}{2}$ turns p. minute. $2\frac{1}{2}$ turns= 60 y^{ds} p. minute, say at the rate of 180^{lb} p. horse.' Watt then proceeds to work out the size of the engine : ' 60 yds × 3 = 180 × 180 pounds = 32,400 ÷ 120 feet of piston's motion = 270 lbs × 12 horses = 3240^{lbs} load of cyl^r, which at 5^{lb} p. inch = 29 inch cyl^r, 6 feet stroke, 20 p. minute.'

Here, then, we have Watt's first value for the horse-power unit, 32,400 lb. raised one foot high per minute. How he arrived at the figure for the pull of the horse— 180 lb. when moving at the rate of 60 yds. a minute, or 2 miles per hour—does not appear. There is nothing to suggest that it was determined by experiment, and it is a good deal lower than the figure published some years earlier by Desaguliers, who states that ' horses one with another, draw about 200 lbs ', when walking about $2\frac{1}{2}$ miles an hour and working 8 hours a day, ' up out of a well over a single pulley or roller (made to have as little friction as possible).' [2] Although Watt does not set it down in his book, it is quite likely that it was given him, together with the rate of walking, by the millwright Wriggley. The men concerned in setting up horse-gins for bringing up coal, ores, and water in mines must have had fairly definite ideas as to what a horse could be expected to do. The low

---

[1] Smeaton's *Reports*, II. 360.
[2] Desaguliers : *A Course of Experimental Philosophy*, I. 251, 1763 (3rd ed.). It should be noted that under the conditions stated the horse

is not performing continuous work ; he travels from the well, pulling up the load, and then back free from the load.

figures for the pressure on the piston—5 lb. per square inch—may have been assumed partly as a safeguard against any error on the wrong side in the estimate of the pull of the horse, but no doubt the main reason was that Watt did not know what the friction losses between the beam and the shaft would amount to. At this period we are still in the very early days of the rotative engine ; Watt's specification for the patent of 1781 had been completed only six months before.

The next entry in the ' Blotting and Calculation Book' relating to horse-power is in respect of an inquiry for an engine to work two pairs of millstones. Here again we have the figure 32,400. After this, in September 1783, comes an entry in which, without any warning or indication of the change, the figure becomes 33,000. Watt had been furnished with particulars of an engine put up by Wasborough in a flour mill at Southampton, and he sets out to calculate the size of an engine on his own plan that would do the same work. The cylinder, he says, ' Contains 1010 sq^r inches area which at 6^{lb} p. inch $= 6060^{lb} \times 8$ feet stroke $= 48,480 \times 14$ strokes, $678,720 \div 33,000$ (power of horse) gives 20 horses $=$ power of engine $=$ grinding and dressing 20 bushels coarse flour p. hour—the engine $=$ to a 30 inch of our construction.'

Farther on in the same book we find : ' Blackfryars corn mill engine. It is required to work 18 pairs of stones and each pair to grind 6 bushels p. hour and reckoning each bushel p. hour $=$ to one horse and each horse $= 33,000^{lb}$ 1 foot high p. minute.' Here we have the full statement, 33,000 lb. one foot high per minute. This note is not dated, but it must have been made before Oct. 22, 1783, for on that day Watt had nearly finished the drawings of the engine. The Blackfriars mill was the establishment afterwards known as the Albion Mills.

Within a few years it became the common practice at Soho to refer to engines by their horse-power, or rather by ' horses', 14-horse engine, 20-horse engine, &c.

The horse-power unit is defined for the first time in print in the *Edinburgh Review* for January 1809, where we are told that Boulton and Watt, from the result of experiments made with the strong horses employed by the brewers in London, have assumed ' as a standard of a horse's power, a force able to raise thirty-three thousand lib. one foot high in a minute '. The occasion is a review of the second edition of Gregory's *Mechanics*, and, in particular, a statement therein that : ' What is called the horse's power, is of so fluctuating and indefinite a nature, that it is perfectly ridiculous to assume it as a common measure by which the force of steam engines and other machines should be appreciated.' The reviewer considers that the assumption made by Boulton and Watt could not have been unknown to Gregory, but he does not suggest how he was to have acquired the knowledge.

Five years after the *Edinburgh Review* article, i. e. in 1814, Watt himself penned an account of the origin of the horse-power standard :

' When Boulton and Watt set about the introduction of the rotative steam-engines, to give motion to mill-work, they felt the necessity of adopting some mode of describing the power, which should be easily understood by the persons who were likely to use them. Horses being the power then generally employed to move the machinery in the great breweries and distilleries of the metropolis, where these engines first came into demand, the power of a mill-horse was considered by them to afford an obvious and concise standard of comparison, and one sufficiently definite for the purpose in view. A horse going at the rate of 2½ miles an hour raises a weight of 150 lbs. by a rope passing over a pulley, which is equal to the raising 33,000 pounds one foot high in a minute. This was considered the horse's power ; but in calculating the size of the engines, it was judged advisable to make a very ample allowance for the probable case of their not being kept in the best order, and therefore the load was only assumed at about 7 lbs. on the square inch of the piston, although the engines work well to 10 lbs on the inch, exclusive of their own friction.' [1]

In the book in which this account appears Watt gives the dimensions of three particular sizes of engine. The 40 horse-power has a cylinder 31½ in. diam., a stroke of 7 ft., and makes 17½ strokes per minute. The 20 horse-power, a 23¾-inch cylinder, 5 ft. stroke, and 21½ strokes per minute. The 10 horse-power, a 17½-inch cylinder, 4 ft. stroke, and 25 strokes per minute. The piston speeds work out to 245, 215, and 200 ft. per minute. The ' load' is 7 lb. per square inch, or rather less than this.

These statements as to the origin of the horse-power unit written, the one twenty-six years, and the other thirty-one years after Watt had fixed the value, do not quite tally with the account derived from the contemporary notes. This, perhaps, is not very surprising ; moreover there are other instances in which the account of an event given by Watt in his old age is not in accord with the contemporary record. The Edinburgh Reviewer, it will be noticed, speaks of experiments made with strong horses, and in such terms as to suggest that they had been made by Boulton and Watt. The terms used in Watt's own account, however, do not seem to bear this out ; rather they suggest that the figures given were already known and accepted, although if this was the case it is difficult to see what Gregory had to cavil at. Still it is quite possible that Watt may have made experiments and obtained the figures mentioned. On the other hand the change from the original figure 32,400 to 33,000 may have been made merely with a desire to get a round number for facility in calculation ; in doing this he would naturally take the higher number, for it would have been much against the interests of the firm to supply an engine that would not do the work it was ordered for. Another point is that the first orders for engines for London breweries are subsequent in date to the notes quoted above. We have the figure 33,000 in September 1783, but Goodwyn's engine was not ordered until April 1784, and Whitbread's later still in that year. Still there may have been inquiries at an earlier date, Barclay and Perkins, certainly, were asking about an engine in February 1783 ; but there is no indication that any action was taken on this inquiry and it was not until 1786 that they had an engine.

With the expiry, simultaneously in the year 1800, of the extended term of Watt's patent of 1769, and of the Boulton and Watt partnership, the task of considering Watt's activities in connexion with the steam engine comes to an end. Indeed he had ceased to interest himself in the subject some years before this.

The work that a Boulton and Watt engine would perform at about this time is summed up by Boulton in the following terms :

' one bushel (84 lbs) of Newcastle or Swansey coal.
(1) will raise 30 million lbs. of water 1 foot high.
(2) will grind & dress 10, or 11, or 12 bushels of wheat, according to the state of it.
(3) will turn 1000 or more cotton spinning spindles per hour.
(4) will roll & slit 4 cwt. of bar iron into small nailor's rods.
(5) will do as much work per hour as ten horses.' [2]

Thus it will be seen that the efficiency of the engine had not been materially

---

[1] Robison : *Steam and Steam Engines*, 1818, 145 *n.*
[2] B. & W. Colln. Boulton to James Watt, junior, 1796 Nov. 28.

increased for a number of years. As compared with the Newcomen engine, Watt had, at one bound, increased the amount of water raised by the consumption of a given weight of coal, say, four times. After that he made no changes in principle, no improvements in the direction of securing increased economy of steam. The advantage of his engine over the earlier engine was so great that possibly there was little incentive for him to do so ; in any case the most obvious course for securing increased efficiency, i. e. expansive working, he was precluded from taking effective advantage of by his rooted objection to the use of steam at pressures of more than a pound or two above the atmosphere. He had in his first patent covered a non-condensing engine, so would at that time have contemplated using steam at higher pressures, but his experience in practice of the difficulty of securing tight joints, and his fear of boiler explosions, led him to consider the use of steam at high pressure as little short of a crime.

The removal of the restraint imposed by his patent allowed other men to set to work on the steam engine, and some of these were prepared to deal with the difficulties of higher pressure. One of the first results was the introduction of the non-condensing engine into practice. There was a loss of efficiency, but in many cases that was a matter of no great moment, and to compensate for it there was a great reduction in the weight of the engine, and a great gain in compactness ; very soon this form of engine was applied to railway locomotives. Another result was that the Cornish engineers were free to enter upon that course of improvement in the pumping-engine that resulted in the highly efficient ' Cornish ' engine. Watt himself lived to see the duty figures brought up to 50 millions, and this mainly by the use of higher pressures and expansive working. The engine was still the Watt engine.

APPENDIX I

BIBLIOGRAPHY CONCERNING JAMES WATT

GENERAL NOTES

IT is a matter for regret that James Watt, although so ready with his pen, as evidenced by his voluminous letter-writing, wrote practically nothing for publication ; what he did write might be said to have been almost exclusively for professional or business purposes. He had some thoughts in 1773 of writing a book on ' the elements of the theory of steam engines', in order to advertise the engine.[1] Again, in 1778 it was stated : [2] ' The result of these experiments (i. e. on the volume of steam) he intends to lay before the publick, in a treatise upon that subject.' Nothing came of these, however.

After his retirement from business, at the solicitations of his friends, his thoughts reverted to the subject of writing a history of his improvements in the steam engine, and there is evidence that about 1808 he collected and annotated a good deal of matter for the purpose. His disinclination to court publicity, and his advancing age led to the project being shelved. The occasion of the publication of the collected works of his old friend Robison, however, stirred Watt to action in 1814, but the result was rather annotations on the former's work than an original monograph such as we, at the present day, should have prized.[3]

A very valuable source of information about Watt personally is to be found in the documentary matter preserved by the present representative of the family. This matter covers the whole of his life and comprises letters, letter-books, note-books, journals, accounts, reports, memoranda, patent specifications, law cases, drawings, and books on the steam engine. Of the letters alone, there are several thousands. (Referred to in this volume as the Doldowlod Papers.) Speaking broadly, this is the material used by Muirhead in his biographical works on Watt mentioned below.

An equally valuable source of information, more particularly from the technical point of view, for the period 1775–1800 and, indeed, later still, is the Boulton and Watt Collection. This material constitutes the records of the firm, and comprises upwards of 10,000 drawings with a very large number of documents such as correspondence and letter-books, together with printed books, models, and plant. When the effects of the firm were dispersed in 1895 these objects were purchased by the late Mr. George Tangye, who presented them in 1911 to the City of Birmingham ; they are now housed in the Free Reference Library. (Referred to in this volume as the B. & W. Colln.) This matter has not been utilized hitherto to any extent and has consequently been drawn on extensively for the present volume.

Another collection comprising journals, deeds, ledgers, letters, printed matter, and other documents, in the main complementary to the two collections named above, was in the possession of Lionel B. C. L. Muirhead, Esq., a relative of the biographer of Watt, till September 1921 when it was handed over to the care of the Free Reference Library, Birmingham (referred to in this volume as the Muirhead Papers).

About a thousand letters from Watt to Matthew Boulton, together with a few to Dr. William Small, covering the period 1768 till his death, along with a large amount of documentary matter concerning Boulton's business undertakings, were preserved by the Boulton family at the family seat, Tew Park, Oxon., till last year when they were removed to the Assay Office, Birmingham. (Referred to in this volume as the Boulton Papers.) This material was used both by Muirhead and Smiles.

---

[1] Watt to Small, 1773 Aug. 17. Muirhead : *Mech. Inv.*, II. 59.

[2] Pryce : *Mineralogia Cornubiensis*, 309.
[3] See p. 365.

APPENDIX I

The Birmingham Free Reference Library also possesses the following papers:

Letters from James WATT *to John* SOUTHERN, 1789 to 1815. In 1 vol. 4to. (Referred to in this volume as the Southern Papers.)

Collection of original letters, &c., relating to WATT and BOULTON and SOHO (1761 et seq.) formed by Samuel TIMMINS. 1 vol. fol.

Letters from R. CARRUTHERS, five in number, to W. C. Aitken, relating to Burns and James WATT, 1851–69.

Letters from Samuel SMILES to W. C. Aitken relating to James WATT and James KEIR, 1861–70. (In a collection of papers of W. C. Aitken.) 8vo.

Extracts relating to BOULTON, WATT, and EGINTON, 1808–95 (*c.* 1895), by G. N. OSBORNE. 1 vol. 4to.

Other bibliographical matter is arranged below chronologically, explanatory notes being appended to any item where this course seems to be needed. Some sifting has been effected, and mere compilations of matter already published have been excluded.

DETAILED NOTES

MACKELL, Robert, and WATT, James.
An account of the navigable Canal, proposed to be cut from the River Clyde to the River Carron, as surveyed by Robert Mackell and James Watt.
London, 1767. 18 pp., 4to. 1 engraved map.
This was a scheme for joining the rivers Forth and Clyde by a canal via the Loch Lomond passage. As eventually carried out by Brindley and Smeaton, the direct route for the Canal was adopted.

WATT, James.
A scheme for making a navigable Canal from the City of Glasgow to the Monkland Coalierys.
Glasgow, [1769]. 12 pp., 4to.
This was the Monkland Canal, the first part of which was carried out entirely by Watt.

WATT, James.
Report concerning the Harbour of Port-Glasgow, made to the Magistrates of Glasgow, by James Watt, Engineer, and submitted to the consideration of the Merchants.
Glasgow, Aug. 9, 1771. 8 pp., 4to.

WATT, James.
A Report to the Honorable His Majesty's Commissioners for managing the Annexed States in Scotland concerning the Isthmusses of Tarbert and Crinan.
Glasgow, Dec. 21, 1772. 107 pp., 4to.
Only known to the Editors in the manuscript state. One copy, no part of which is in Watt's handwriting, is preserved in the British Museum. (Add. MS. 9059.)

WATT, James.
An account of the Scheme for rendering navigable the Rivers Forth and Devon with estimates of the Expence by James Watt.
Edinburgh, MDCCLXXIV. 4to.

WATT, James.
An account of JAMES WATT'S *Improvements upon the* Steam *or* Fire Engine.
(London 1774.) 8 pp., fcp. 4to.
This was the statement that Watt distributed to Members of Parliament when the Bill for extending his patent was before the House, as evidenced by his note on one copy in the Boulton and Watt Collection: ' Delivered to the members of the House of Commons, 1774–5.'

PARLIAMENT, Houses of.
An Act for vesting in James Watt, Engineer, his executors, administrators and assigns the sole use and property of certain steam engines commonly called fire engines, of his invention, for a limited time. 15 Geo. III, c. lxi, pp. 1587–94.
London (1775), fol.
Prior to its passage into law, it was printed as a Bill. It was reprinted subsequently in several forms.

BIBLIOGRAPHY CONCERNING WATT 361

PRYCE, William.
> Mineralogia Cornubiensis.
> London, 1778, fol.
>> In the Appendix, p. 308, a full description of and historical notes on Watt's engine are given, but no drawing. The information was supplied by Watt (Boulton to Henderson, 1777 Aug. 13) and is the first public notice of the engine in print. Both Boulton and Watt were subscribers to the book.

BOULTON and WATT.
> Proposals to the adventurers in . . . by Boulton and Watt.
> (Birmingham, 1778.) 8 pp., 4to.
>> A form to be filled up for any one wishing to erect their engines. The experiments at Poldice on the coal consumption of the existing fire-engine are described. One paragraph reads :
>> (3) ' The profit which we require for such licence drawings and instructions is to be such sums . . . as shall be equal in value to a *third part* of the *savings in fuel.*'

WATT, James.
> Directions for erecting and working the newly-invented steam engines by Boulton and Watt.
> (Birmingham, 1779.) iv + 24 + 16 pp., 6 plates, 12mo.
>> Reproduced in full in Appendix II. Comprises ' General Directions for building the engine house ' ; ' Directions for putting the engine together ' ; ' Directions for working the engine '; ' Additional directions ' and ' Explanation of plates '. These directions are detailed and precise, clearly revealing Watt's hand. The booklet was meant for private circulation among the firm's engine erectors and clients. It is obviously a compilation, e. g. the ' additional directions ' are separately paged and the plates are numbered X to XII. This is known to be the case, as separate portions exist in manuscript in the Boulton and Watt Collection. Five of the copper plates are still preserved there. The date is fixed by entries in Watt's journals, see Chap. XX, showing when he was at work upon it.

PRIESTLEY, Joseph, LL.D.
> Experiments and observations relating to various branches of Natural Philosophy.
> Birmingham, 1781. 8vo.
>> The second volume has on p. 388 an appendix Number III, entitled ' Observations on this volume with which I was favoured by Mr. Watt '. This is one page of notes which reveal Watt's acquaintance with the chemistry of the day.

[BOULTON and WATT.]
> Directions relating to the Engine.
> [Birmingham, 1784.] 1 sh., post.
>> Reproduced in full in Appendix III. The sheet was obviously meant to be hung up on the wall of the engine house.

[WATT, James.]
> Remarks on a Government Paper entitled Iron Trade, England and Ireland.
> (London, 1784), 3 pp., fcp.
>> Several copies of this paper exist in the Boulton and Watt Collection, but there is no definite evidence that Watt wrote it ; the subject matter would point rather to Boulton or Wilkinson as the author. The occasion was a proposal by Mr. Pitt to impose additional duties on manufactures exported to Ireland from Gt. Britain.
>> There is another paper in existence, entitled ' An Answer to the Treasury paper on the Iron Trade of England and Ireland ', whose composition much more savours of the hand of Watt.

WATT, James.
> Thoughts on the Constituent Parts of Water and of Dephlogisticated Air ; with an account of some Experiments on that Subject. In a letter from Mr. James Watt, Engineer, to Mr. De Luc, F.R.S.
> Read Apl. 29, 1784 (before the Royal Society). Printed in *Phil. Trans.*, LXXIV, pp. 329–53.
> Sequel to the Thoughts on the Constituent Parts of Water and Dephlogisticated Air : in a subsequent letter from Mr. James Watt, Engineer, to Mr. De Luc, F.R.S.
> Read May 6, 1784. Printed in same vol. of *Transactions* : pp. 354–7.
>> The original letters are still in the possession of the Royal Society (Guard Book No. 74). These letters gave rise to a bitter controversy between the supporters of Watt and Cavendish as to which of them had the prior claim to the discovery of the composition of water. The letters, with much additional matter, were subsequently reprinted in volume form (see below, p. 367).

DALRYMPLE, Sir John.

Address and proposals from Sir John Dalrymple on the subject of the Coal, Tar and Iron Branches of the trade.

London and Edinburgh, 1784.   15 pp., 8vo.

Sir John was one of the Barons of Exchequer in Scotland.   Writing to Matthew Boulton on Aug. 6, 1784, he encloses a copy of what he calls ' a little pamphlet ' in one part of which he says : ' I have done justice to your and Mr. Watt's great improvements upon the fire Engine.'

WATT, James.

Heads of a Bill to explain and amend the laws relative to Letters Patent and grants of privilege for new Inventions.

B. & W. Colln.   MS. 15 pages, 4to.

Probably drafted in 1785 or 1786 when there was some talk by opulent manufacturers of combining to attack patents just as Arkwright's had been treated.   Watt says : [1]
' A pursuance of such decisions as have been given lately in several cases must at length drive men of invention to take shelter in countries where their ingenuity will be protected.'
Boulton writing to De Luc was still stronger :
' Some late decisions against the validity of certain patents have raised the spirits of the illiberal, sordid, unjust, ungenerous and inventionless misers who prey upon the vitals of the ingenious, and make haste to seize upon what their laborious and often costly application has produced.'

[BOULTON and WATT.]

Short Statement, on the part of Messrs. Boulton and Watt, in Opposition to Mr. Jonathan Hornblower's Application to Parliament for an Act to prolong the Term of his Patent.

Birmingham, 1792.   1 sheet fcp., folded.

Hornblower was only trying to do for his compound engine patent what Watt and Boulton had done for the separate condenser patent.   It appears from the Boulton MS. that Watt got thoroughly alarmed at the competition threatened by Hornblower, and it is said that they ' lobbied ' actively against the prolongation of the patent with the result that it was not granted.

[BOULTON and WATT.]

Observations on the part of Messrs. Boulton and Watt re Hornblower's steam engine bill.

[Birmingham], Apl. 17, 1792.   1 p., fcp.

WILSON, Thomas.

A comparative statement of the effects of Messrs. Boulton and Watt's Steam Engines with Newcommen's and Mr. Hornblower's.   Addressed to the lords of, and adventurers in, mines in Cornwall.

Truro, 1792.   25 pp., 8vo.

WILSON, Thomas.

An address to the Mining Interest of Cornwall on the subject of Messrs. Boulton and Watt's and Mr. Hornblower's engines.

Truro, 1793.   22 pp., 8vo.

Addressed similarly to the last.

Wilson was the agent in Cornwall of Boulton and Watt, and there can be little doubt that the above two pamphlets were prepared by him at the firm's instigation, or possibly by Watt himself, to counteract the Hornblower competition.

COURT OF COMMON PLEAS.

The Special Case in the cause Boulton and Watt against Bull, in the Court of Common Pleas, with the arguments of the Judges thereon ; and an Appendix of matters referred to.

London, 1795.   3 pp., fcp.

Quite probably edited by Watt.

The Special Case in the cause Boulton and Watt against Bull, in the Court of Common Pleas, with the arguments of the Judges thereon ; and an Appendix of matters referred to.

London, 1795.   8 pp., 8vo.

This is merely a reprint of the preceding in handier format.   Contains the shorthand report of the case.   The Appendix is a list of patents in which *methods* of doing something were specified.

[1] Boulton Papers : Watt to Boulton, 1786 Mar. 19.

BEDDOES, Thomas, M.D., and WATT, James.
Considerations on the medicinal use of factitious airs, and the manner of obtaining them in large quantities. In two parts. Part I by Thomas Beddoes, M.D. Part II by James Watt, Esq.
Bristol, 1794. 8vo. Separately paged, Part I, 48 pp., and Part II, 34 pp.
The sub-title of Part II is : Description of an Air Apparatus ; with hints respecting the use and properties of different elastic fluids. By James Watt, Esq.
Consists of a series of letters of various dates from June 17 to Oct. 2, 1794, giving descriptions and drawings of the apparatus. Under date July 14th, Watt mentions that at Dr. Beddoes's desire : ' Boulton and Watt have agreed to manufacture these machines for the Public.'

BEDDOES, Thomas, M.D., and WATT, James.
Considerations on the medicinal use, and on the production of factitious airs Part I by Thomas Beddoes, M.D. Part II by James Watt, Engineer. Edition the second, to which are added communications from Doctors Carmichael, Darwin, Ewart, . . . and others.
Bristol, 1795. 8vo. Separately paged, Part I has 172 pp., Part II has 40 pp., 5 copper-plates, and two tables.
In Part I there is a note that ' at Mr. Chippendale's, *Salisbury Court, Fleet Street, London*, Mr. Watt's Air apparatus may be seen '.
Part II has the sub-title, ' Description of a pneumatic apparatus with directions for procuring the factitious airs, by James Watt, engineer.' The preface dated January 1795 says : ' The Author has also availed himself of this opportunity to methodize and elucidate his description in a manner which the former hasty publication would not admit of.'
Both second and third editions were printed also at Birmingham. ' Edition the third, corrected and enlarged ', appeared in 1796. Paged continuously 222 pp., 5 pl.

BEDDOES, Thomas, M.D., and WATT, James.
Considerations on. the medicinal use and production of factitious airs. By Thomas Beddoes, M.D., and James Watt, Engineer. Part III.
Bristol, 1795. x + 121 pp., 3 pl., 8vo.
Contains letters from physicians and others as to the treatment, among them being three from Watt (pp. 34, 36, and 105).
A ' Second Edition, corrected and enlarged ' appeared the following year.
Bristol, 1796. xx + 178 pp., 3 pl., 8vo.

BEDDOES, Thomas, M.D., and WATT, James.
Medical cases and speculations ; including Parts IV & V, of considerations on the medicinal powers, and the production of factitious airs by Thomas Beddoes, M.D., and James Watt, Engineer.
Bristol, 1796. 8vo. Part IV, xv + 168 pp. Part V separately paged, 96 + (25 to 42 bis), 2 plates Nos. 4 and 5, is entitled :
' Supplement to the description of a pneumatic apparatus, for preparing factitious airs ; containing a description of a simplified apparatus and of a portable apparatus. By James Watt, Engineer.'
42 pp. of this is devoted to a description of the apparatus. A price list is given. The descriptive portion of Part V was reprinted with the same title as a separate pamphlet, evidently to push the sale of the apparatus.
Birmingham, 1796. 48 pp., 8vo.
The reason why Watt was led to the study of this branch of medicinal treatment, viz. inhalation of oxygen, &c., was that his youngest son Gregory, by his second wife, a lad of great promise, suffered from consumption. In fact, consumption was in the family, for a daughter, Jessy, had died of this complaint in 1794. In spite of all the care expended on Gregory he, too, died of the disease in 1804, while staying in Devon on account of his health ; he was buried in Exeter.
Thomas Beddoes was the well-known and enterprising physician who founded the Pneumatic Institution.

WATT, James, junior.
Directions for using the patent portable Copying Machines. Invented & made by James Watt and Company of Soho, near Birmingham.
[Birmingham, 1795.] 18 pp., 12mo.
Some copies have as a frontispiece a copperplate engraving of the desk. The pamphlet states that machines are to be had from Richardson and Harrison, Leadenhall St.

Directions for using the patent portable Copying Machines invented and made by James Watt & Company of Soho, near Birmingham.

Birmingham, 1813. 16 pp., 12mo.

With the exception that some of the introductory matter is deleted, and a few mistakes corrected, this is only a reprint of the preceding. Reprints of this were brought out in 1818 and 1830.

Manière de se servir de la Machine Portative à Copier. Inventée & patentée par Jacques Watt & Co., de Soho, près de Birmingham.

Birmingham, 1805. 16 pp., 12mo.

A French translation of the pamphlet became necessary owing to the fact that a sale for the copying machine had been created abroad. This is possibly not the earliest edition.

Manière de se servir de la machine à copier. Inventée & patentée par Jacques Watt & Co., à Soho près de Birmingham.

Birmingham, 1807. 24 pp., 12mo, 1 plate.

Practically a reprint of the preceding. A further reprint was brought out in 1818, 16 pp., 12mo.

Anweisung zum Gebrauch der tragbaren Kopir-Maschinen erfunden und verfertigt von James Watt und Comp. zu Soho bei Birmingham.

Birmingham (? 1805). 24 pp., 12mo.

Probably prepared at the same time as the French translation. The agent for the machines was G. H. Busch, Hamburg.

Watt patented the press copying system, and the machine therefore in 1781. A separate firm, James Watt & Co., was formed to exploit the invention, and their letter book, the first in which the system now so wide-spread was used, is preserved in the Boulton and Watt Collection. The sale of the machine was hindered by the prejudice of bankers and business people generally, and it was not till the patent had wellnigh run its course that this opposition gave way before the merits of the system. It was at this time that James Watt, the younger, saw that the firm would not reap where they had sown unless the manufacture of the machines was taken up more vigorously, as another firm was already making machines. He brought out a portable form of the machine, combined with a writing-case and, as we learn from the Boulton correspondence [Boulton Papers] which fixes the date, compiled the above ' Directions ' in order to push the invention. It is not unlikely that Watt may have revised the manuscript.

PRONY, Gaspard Clair François Marie Riche de.

Nouvelle Architecture Hydraulique.

Paris, 1796. 2 vols. 4to, plates.

The double-acting engine described is stated (vol. i, p. 571) to be that constructed by MM. Périer Frères in 1790, based on a model to the scale of 1 in. to a foot made by M. le Chevalier de Bettancourt as a result of a visit to England in 1788. It is stated to have been ' entièrement de l'invention ' of that gentleman, but this claim was afterwards withdrawn and it was admitted that the information was taken from Watt's engines.

The drawings are therefore the earliest published representing the Watt engine.

BRAMAH, Joseph.

A Letter to the Rt. Hon. Sir James Eyre, Lord Chief Justice of the Common Pleas: on the subject of the cause, Boulton and Watt v. Hornblower and Maberly, for infringement of Mr. Watt's patent for an improvement on the Steam Engine, by Joseph Bramah, Engineer.

London, 1797. 90 pp., 8vo.

Substance of matter prepared by him as witness for the defence in the above trial.

ROBISON, John, M.A., LL.D., F.R.S.E.

Articles ' Steam ' and ' Steam Engine ' in the *Encyclopaedia Britannica*, 3rd edition, vol. XVI.

Edinburgh, 1797. 4to, 3 plates.

WATT, James.

Specification of an invention of certain new improvements upon Steam or Fire Engines for raising water and other mechanical purposes, and certain new pieces of mechanism applicable to the same by James Watt, Engineer.

[London, 1798.] 4to, 19 pp. + 7 pp. references, 2 pl.

Simply a transcript of the patent specification of Mar. 12, 1782 (No. 1321). It is not obvious why it was printed, as the patent had then expired—probably it was intended for private circulation.

COURT OF COMMON PLEAS.

The arguments of the Judges in two Causes relating to the Letters Patent granted to James Watt, Engineer, for his method of lessening the consumption of steam and fuel in Fire Engines. Taken in shorthand by Mr. Gurney. In two parts: with an Appendix. London, 1799. 104 pp., 8vo.

The first part is a transcript of Watt's first patent specification, Jan. 5, 1769 (No. 913), and of the Act for extending the patent for twenty-five years from May 22, 1775.
The second part contains the reports of the causes Boulton and Watt *v.* Bull, and the same plaintiffs *v.* Hornblower and Maberley. As to the first of these two, the pamphlet merely reprints the matter given earlier, dated 1795.

ROBISON, John, LL.D., F.R.S.E.

Two articles, 'Steam' and 'Steam Engines', written for the *Encyclopaedia Britannica* . . . with notes and additions, by James Watt. . . . And a letter on 'Some Properties of Steam' by the late John Southern, Esq. Edinburgh, 1818. 184 pp., 8vo., 8 plates.

This is a reprint of the article above, with much additional matter comprising corrections in the text, footnotes, and an appendix from Watt's own pen, together with a covering letter to Sir David Brewster from Watt, dated Heathfield, May 1814.
The reason which had moved Watt to take all this trouble was his great regard for the memory of his old friend, Robison, an edition of whose works (see below) was then being prepared under the editorship of Sir David Brewster. This is explained in a pencil note on the copy now in the British Museum, which Watt sent inscribed 'To the Right Hon. Sir Jos. Banks, Bart., G.C.B., etc., etc. from his obliged servant James Watt.' The note reads : 'The book of which this forms a part will not be published till winter, and until that time Sir Joseph is respectfully requested to keep this in his own possession, June, 1818.'
In the covering letter to Brewster, which, by the way, was also printed in *Edin. Phil. Journ.*, 1820, II, by permission of James Watt, junior, Watt says : 'I have not attempted to render Dr. Robison's memoir a complete history of the Steam-engine ; nor have I even given a *detailed* account of my own improvements upon it. The former would have been an undertaking beyond my present powers, and the latter must have exceeded the limits of a commentary upon my friend's work. I have therefore confined myself to correcting such part as appeared necessary, and to adding such matter as he had not an opportunity of knowing.'
Watt takes the opportunity of correcting the statement that he had been the *pupil* of Dr. Black at Glasgow, and that he had owed the improvements on the steam engine to Black's instruction. The importance of these memoranda, for we can call them nothing more, is that they are really all we have at first hand from Watt himself upon his inventions.
The letter from Southern is addressed to Watt under date March 1814, and gives the result of his and William Creighton's experiments and calculations on the temperature and pressure of steam, figures which remained the authority till superseded by the more precise figures established by Rankine. One or two letters passed between Watt and Southern about this time. For example, Watt writes [1] : 'I am at work on the Proffrs acct of my invention in the Dictionary, which I find will be a difficult thing to correct leaving any of the Proffrs words.'

FAREY, John.

Article 'Steam Engine' in Rees's *Cyclopaedia*, vol. XXXIV. London, 1819. 4to, plates.

The fullest practical account of the steam engine that had appeared up to that date. Written subsequently to 1816 with information obtained from Watt. The plates alluded to are dated 1812 to 1818. The author subsequently expanded the matter into book form (see below).

HACHETTE, Jean Nicolas Pierre.

Notice sur la vie et les travaux de James Watt. *Bulletin de la Société d'Encouragement*, CLXXXII, August 1819.

Very brief—Hachette had just been in England and had received from James Watt, junior, a bust by Chantrey for the Société d'Encouragement.

JEFFREY, Hon. Francis, Lord.

Character of Mr. Watt. Obituary eulogy in *Scotsman* newspaper. Edinburgh, Sept. 4, 1819. This was reprinted in *Edinburgh Magazine*, September 1819, 203.

---

[1] Southern Papers : Watt to Southern, 1813 Dec. 20.

PLAYFAIR, William.
    Original memoirs of Eminent Persons. The late James Watt, Esq., F.R.S., &c., &c., communicated by Mr. Wm. Playfair.
    Monthly Magazine or British Register, XLVIII, 1819, pp. 230–9.
    Playfair was in the employ of the firm from 1777 to 1782 and therefore had inside knowledge of this period.

PLAYFAIR, William.
    Memoir of James Watt, F.R.S.
    New Monthly Magazine and Universal Register, 1819, vol. xii, p. 576.
    In a prefatory note the Editor states that it was written by Mr. Playfair and remarks that he had disposed of a copy to the rival magazine (i. e. the above) ; the memoirs are, however, quite different in matter.

WATT, James.
    Thirteen letters from the late James Watt, Esq., to James Lind.
    Monthly Magazine or British Register, L, 1820, p. 239.
    These cover the period 1764–99 and deal with Watt's scientific pursuits.

ROBISON, John, LL.D.
    A system of mechanical philosophy. With notes by (Sir) David Brewster.
    Edinburgh, 1822. 5 vols., 8vo.
    The first part of Vol. II consists of the matter given above under date 1818, in fact the pagination, even to the title-page, is identical ; obviously the type had been kept standing.

[WATT, James, Jun., and JEFFREY, Francis, Lord].
    Article ' James Watt ' in the supplement to the fourth, fifth, and sixth editions of the *Encyclopaedia Britannica*, vol. VI, pp. 778–85.
    Edinburgh, 1824. 4to.
    The first satisfactory biographical memoir of Watt. Three-fourths is from the pen of James Watt, jun. ; the rest of it is taken up by an eulogy on Watt from the pen of Lord Jeffrey, being, in fact, a reprint of his *Scotsman* article (*see ante*).

[WATT, James, jun., and JEFFREY, Francis, Lord].
    Memoir of James Watt, F.R.S., L. & E., from the supplement to the *Encyclopaedia Britannica*, vol. VI.
    [Edinburgh, 1824.] 32 pp., 8vo, privately printed.
    Consists of above matter paged up in book form.

[TURNER, Charles Hampden, *Chairman*.]
    Proceedings of the public meeting held at Freemasons' Hall [London] on the 18th June, 1824, for erecting a monument to the late James Watt.
    London, 1824. 96 pp., 8vo. A large paper copy was also issued.
    As a result of this meeting, a public subscription was started, and the outcome was the Chantrey statue in Henry VII's Chapel, Westminster Abbey.

FAREY, John.
    A treatise on the Steam Engine, historical, practical, and descriptive.
    London, 1827. 4to.
    An expansion in volume form of the article in the *Encyclopaedia Britannica* mentioned above. Contains a large amount of first-hand detailed information about the Watt engines.

STUART, Robert. (Pseudonym of Robert Meikleham.)
    Historical and Descriptive Anecdotes of Steam Engines, and of their Inventors and Improvers.
    London, 1829. 2 vols, 16mo, portraits and plates

ARAGO, Dominique François Jean.
    Éloge historique de James Watt.
    Mémoires de l'Académie Royale des Sciences de l'Institut de France, vol. XVII, pp. lxi–clxxxviii.
    Paris, 1840. 127 pp., 4to.
    Watt occupied the distinguished position of being one of the eight foreign Associates of the Academy,

an honour which was bestowed upon him in 1814; he had been a corresponding Member of the Institute since 1808. In 1833, by direction of the Academy, M. Arago, the Perpetual Secretary, was charged with the task of writing a memoir of Watt. For this purpose he came to this country, and gleaned much information at first hand, notably from James Watt, junior, and Lord Brougham. The result was this eulogy, which was read at a public meeting of the Institute on Dec. 8, 1834. The Éloge is a valuable contribution to the biography of Watt.

To the Éloge is appended a translation of an historical account of the discovery of water by Lord Brougham who, before transmitting the manuscript to Arago, submitted it to James Watt, junior. The latter added a number of notes, which Lord Brougham considered so valuable that he asked Arago to retain them in printing. This he did, and there is a note to that effect on p. clxxxviii. It may be remarked that the above matter was printed off by itself with new pagination, pp. 128, but without a title-page. Copies were presented to his friends by James Watt, jun.

ARAGO, D. F. J.
    Biographical Memoir of James Watt, one of the Eight Foreign Associates of the Academy of Sciences, by M. Arago, Perpetual Secretary.
    Edinburgh New Philosophical Journal, XXVII. October 1839, pp. 221–324. 12mo.

    A translation made by Hyde Clarke of the Éloge mentioned above. To it is appended Arago's essay 'On Machinery considered in Relation to the Prosperity of the Working Classes', which did not appear in the *Mémoires*. This essay is intended to show that inventions such as those of Watt are not detrimental to the interests of workers, as was then maintained by some.

ARAGO, D. F. J.
    Life of James Watt, to which are subjoined, Memoir on Machinery considered in relation to the prosperity of the working classes, by M. Arago; and Historical Account of Discovery of the Composition of Water, by Lord Brougham.
    Edinburgh, October 1839. 142 pp., 12mo.

    This is the matter from the *Edin. New Phil. Journ.* above, in book form. Its early appearance was apparently due to the enterprise of the Editor and proprietors of that Journal.
    Although not mentioned on the title-page, the volume contains on p. 125 the *Eulogium of James Watt* by Lord Jeffrey (from the *Encyclopaedia Britannica* already cited).
    This Life of Watt ran into a second and a third edition the same year (1839); the latter was illustrated with engravings and ran to 222 pp.

ARAGO, D. F. J.
    Historical Éloge of James Watt. Translated from the French, with additional notes and an appendix, by James Patrick Muirhead, Esq., M.A.
    London and Edinburgh, November 1839. ix + 261 pp., 8vo, portrait. A large paper copy was also issued.

    Although this contains the same matter as the preceding book, the translation is different and in some respects of greater merit. The Appendix includes: the Article on the Composition of Water by Lord Brougham; the Eulogy by Jeffrey; the Memoir on Machinery by Arago, and a reprint of the proceedings of the Meeting for Erecting a Monument to Watt, already cited *ante*.

WILLIAMSON, George.
    Letters respecting the Watt Family.
    Greenock, 1840. 69 pp., 8vo. Privately printed.
    Practically all the matter is incorporated in the author's 'Memorials' (see *infra*).

BROUGHAM, Henry, Lord, F.R.S.
    Lives of Men of Letters and Science who flourished in the time of George III.
    London, 1845, 8vo., steel engraved portrait.
    The biography of Watt occupies pp. 353–401 and includes an appendix on the discovery of the theory of the composition of water.

MUIRHEAD, James Patrick, M.A., F.R.S.E.
    Correspondence of the late James Watt on his Discovery of the theory of the Composition of Water, with a letter from his son. Edited with introductory remarks and an appendix by James Patrick Muirhead, Esq., F.R.S.E.
    London, 1846. 264 pp., 8vo, portrait. A large paper copy was also issued.

    As we have said above, acrimonious discussion arose between the partisans of Watt and Cavendish as to their respective claims to priority in the discovery of the composition of water. The writer, being a relative of Watt, is naturally biased. It is now conceded that while Watt was the first to adduce reasoned arguments to show that water was not an element, Cavendish independently supplied the experimental data on which accurate knowledge alone could be founded.

ENGLISH, Henry.

    Mining Almanack.

    London, 1849.   8vo.

English was the Editor of the *Mining Journal*. On p. 301 there is a ' Life of James Watt ' by the Editor and on p. 302 we read : ' An éloge has been published by M. Arago, translations of which have been written by Mr. Muirhead and Mr. Hyde Clark.'

ANON. (Religious Tract Society.)

    James Watt, and the Steam Engine.

    London, 1852.   192 pp., 16mo.

[MURRAY, Thomas, LL.D., Secretary.]

    Inauguration of the Statue of James Watt in connexion with the Watt Institution and Edinburgh School of Arts.

    Edinburgh, 1854.   23 pp., 8vo.

In 1824 a public meeting was held in Edinburgh for the purpose of erecting a memorial in honour of Watt. The original idea was to erect a building ' for the accommodation of the Edinburgh School of Arts, whereby the memory of Watt may be for ever connected with the promotion, among a class to which he himself originally belonged, of those mechanical arts from which his own usefulness and glory arose '.

Eventually the fund was devoted to the purchase in 1852 of the premises that had been long used for the purposes of the School of Art, when the name was changed to the ' Watt Institution and Edinburgh School of Arts '. It was felt that, in addition, something distinctive should be done to mark this concatenation : accordingly a statue was erected in front of the School (see p. 87). The School is now merged in the Heriot-Watt College.

MUIRHEAD, James Patrick, M.A., F.R.S.E.

    The Origin and Progress of the Mechanical Inventions of James Watt illustrated by his correspondence with his friends and the specifications of his patents.

    London, 1854.   3 vols., 8vo., portrait.   A large paper copy was also issued.

Based almost entirely on letters in the Doldowlod Papers, it is undoubtedly the richest mine of information about Watt.

The first volume is mainly taken up with an account, not sufficiently discriminating perhaps, of Watt's life. The second is devoted to transcripts of his letters commencing in 1765, and continuing till the year before his death. Possibly the letters have been chosen too much for their social and literary rather than for their scientific interest, but on the other hand such a choice appeals to the widest circle. The third volume gives the patent specifications and reprints of the Patent Cases already referred to.

WILLIAMSON, George.

    Memorials of the lineage, early life, education, and development of the genius of James Watt.

    Printed for the Watt Club.

    (Greenock), 1856.   4to.

Authoritative on Watt's ancestry and early life. There are two portraits : one, frontispiece, an engraving of the Henning portrait, 1803, the other, facing p. 120, a lithograph of the oil painting by Partridge (see p. 83).

COMMISSIONERS OF PATENTS.

    A. D. 1769, No. 913.   Specification of James Watt.   Steam Engines, &c.

New invented method of lessening the Consumption of Steam and Fuel in Fire Engines. Dated Jan. 5, 1769. It is important to note that the patent only covered England, Wales, and the Colonies.

    London, 1855.   3 pp., 4to.

    A. D. 1769.   No. 913*.   Extension of Patent of James Watt.   Steam Engines, &c.

    London, 1857.   6 pp., 4to.

    An Act for vesting in James Watt, Engineer, his executors administrators and assigns, the sole use and property of certain Steam Engines commonly called Fire Engines, of his Invention, described in the said Act, throughout His Majesty's Dominions, for a limited time (22nd May, 1775).

A point to be noted is that this Act extended the Patent to Scotland.

    A. D. 1780.   No. 1244.   Specification of James Watt.   Copying Letters, &c.

A new method of copying letters and other writings expeditiously. Dated Feb. 14, 1780.

    London, 1856.   4 pp., 4to.

COMMISSIONERS OF PATENTS (*continued*).

A. D. 1781. No. 1306. Specification of James Watt. Steam Engines.

Certain new methods of applying the vibrating or reciprocating motion of steam or fire engines to produce a continued rotative or circular motion round an axis or centre, and thereby to give motion to the wheels of mills or other machines. Dated Oct. 25, 1781.

London, 1855. 9 pp., 4to, plates.

A. D. 1782. No. 1321. Specification of James Watt. Steam Engines.

Certain new improvements upon steam or fire engines for raising water, and other mechanical purposes, and certain new pieces of mechanism applicable to the same. Dated Mar. 12, 1782.

London, 1855. 16 pp., 4to, plates.

A. D. 1784. No. 1432. Specification of James Watt. Fire and Steam Engines, &c.

Certain new improvements on fire and steam engines, and upon machines worked or moved by the same. Dated Apl. 28, 1784.

London, 1855. 14 pp., 4to, plates.

A. D. 1785. No. 1485. Specification of James Watt. Furnaces and Fireplaces.

Certain newly improved methods of constructing furnaces or fireplaces for heating, boiling, or evaporating of water and other liquids, which are applicable to steam engines and other purposes; and also for heating, melting, and smelting of metals and their ores, whereby greater effects are produced from the fuel, and the smoke is in a great measure prevented or consumed. Dated June 14, 1785.

London, 1854. 4 pp., 4to. plates.

SCHIMMELPENNINCK, Mary Anne.

Life of, edited by Christiana C. Hankin.

London, 1858. 8vo.

Reminiscent character sketches written in 1856 of Matthew Boulton, James Watt, and the family of the latter as they were in 1788–90.

MUIRHEAD, James Patrick, M.A., F.R.S.E.

Life of James Watt, with selections from his correspondence.

London, 1859. 8vo.

This is in substance the memoir from the author's *Mechanical Inventions* referred to above. It ran into a second revised edition, 1859.

HART, Robert.

Reminiscences of James Watt. *Trans. Glasgow Archaeological Society*, Pt. I, 1.

Glasgow, 1859. 8vo.

Valuable recollections of Watt in his old age, i. e. in 1813 or 1814.

SMILES, Samuel, LL.D.

Lives of Boulton and Watt, principally from the original Soho MSS. comprising also a History of the Invention and Introduction of the Steam Engine.

London, 1865. xvi + 521 pp., 8vo in 1 or 2 vols.

Attractively written, avoids the introduction of technical detail, but is not always accurate. This ran into a second edition the year following.

COMMISSIONERS OF PATENTS.

Contributions to the History of the Steam Engine being two deeds relating to the Erection by Messrs. Boulton and Watt of Steam Engines on the United Mines at Gwennap, Cornwall, and at Werneth Colliery, near Oldham, Lancashire, from the originals in the Patent Office Library.

London, 1872. 16 pp., 8vo.

The date of these deeds is 1779 and 1799 respectively. The title should read : ' . . . Werneth Colliery in the parish of Prestwich, Lancashire . . .'

TIMMINS, Samuel

James Watt, from *Trans. Archaeol. Section of the Birmingham and Midland Inst.*, 1872.

Birmingham, 1873. 4to.

SMILES, Samuel, LL.D.
Lives of the Engineers. The Steam Engine—Boulton and Watt. New and revised edition.
London, 1878. Plates, 12mo.
Vol. IV of the 'Lives' slightly abridged from the author's larger work, *ante*.

COWPER, Edward A., M.I.Mech.E.
On the Inventions of James Watt and his models preserved at Handsworth and South Kensington.
Excerpt *Proc. Inst. Mech. Eng.*, 1883. Pp. 599–631, and plates 55–87.
London, 1883. 8vo.
Written by a capable engineer who had made a study of the subject; perhaps the best description extant of Watt's inventions.

TANGYE, [Sir] R[ichard] and G[eorge].
James Watt and William Murdock.
Birmingham (1888). 11 pp., 2 prints, 12mo.
Comprises: 'The Earliest Locomotive in England' and 'James Watt's Garret'.

BARR, Archibald, LL.D., D.Sc.
James Watt and the Application of Science to the Mechanical Arts: An Address.
Glasgow, 1889. 27 pp., 8vo.

PREECE, Sir William H., C.B., F.R.S., M.Inst.C.E.
Watt and the Measurement of Power. Watt Anniversary Lecture delivered before the Greenock Philosophical Society, 1897.
Greenock, 1897. 13 pp., 8vo., 2 figs.

THORPE, Sir Thomas E., D.Sc., LL.D., F.R.S.
James Watt and the Discovery of the Composition of Water. Watt Anniversary Lecture delivered before the Greenock Philosophical Society, 1898.
Greenock, 1898. 19 pp., 8vo.
A full, impartial, and judicious summing up, by a chemist of world-wide fame, of the long drawn-out controversy among the partisans of Watt, Cavendish, and Lavoisier, as to what were their respective shares in this great discovery.

BRAMWELL, Sir Frederick, Bart., LL.D., F.R.S., M.Inst.C.E.
Article, 'James Watt' in *Dictionary of National Biography*, vol. lx, pp. 51–62. Also paged up separately as a booklet of 24 pp.
London, 1899. 8vo.
The best short biography extant, by an engineer of ripe experience.

BECK, Theodor.
Beiträge zur Geschichte des Maschinenbaues.
Berlin, 1900. 582 pp., 8vo.

THOMSON, William, Baron Kelvin, P.R.S.
James Watt, an Oration delivered in the University of Glasgow on the commemoration of its ninth Jubilee.
Glasgow, 1901. 22 pp., 8vo.
Deals in the main with Watt's connexion with the College of Glasgow.

THORPE, Sir Thomas E., F.R.S., &c.
Essays in Historical Chemistry.
London, 1902. 8vo.
Essay V, pp. 98–122, is devoted to 'James Watt and the Discovery of the Composition of Water'.
This is a reprint of the Watt Anniversary Lecture noted above.
This Essay was not, of course, in the author's earlier volume under the same title published in 1894.
A third edition appeared in 1911.

JACKS, William, LL.D., D.L.
James Watt.
Glasgow, 1901. 215 pp., portrait, 12mo.
Repeats some misstatements of former writers, and is not sufficiently critical.

HELE-SHAW, H. S., F.R.S., M.Inst.C.E., M.I.Mech.E.

James Watt, Inventor. Watt Anniversary Lecture delivered before the Greenock Philosophical Society, 1902.
Greenock, 1902. 28 pp., 4 plates, 8vo.

PEMBERTON, T. Edgar.

James Watt of Soho and Heathfield. Annals of Industry and Genius.
Birmingham, 1905. 233 pp., portrait, 3 plates, 12mo.
The Pemberton family occupied Heathfield from 1857 to 1876. Gives matter relative to Watt's connexion with the house and estate, his associates at Soho, and with the Lunar Society. From 1876 to 1920 Heathfield was occupied by the late Mr. George Tangye.

CARNEGIE, Andrew.

James Watt.
London, 1905. 240 pp., 8vo.
Painstaking and eulogistic, but brings out no new facts.

CARNEGIE, Andrew.

James Watt. (Famous Scots Series.)
Edinburgh and London [1905]. 164 pp., 12mo.
Condensed from the preceding.

MATSCHOSS, Conrad, Ph.D., Dipl.Eng.

Die Entwicklung der Dampfmaschine.
Berlin, 1908. 2 vols., 4to.
Appraises Watt's experiments and discoveries, pp. 339–72. Comments upon his influence on the progress of machine design.

BURSTALL, Henry Frederic William, M.A., M.Inst.C.E.

Nine Famous Birmingham Men. Lectures delivered in the University, edited by J. H. Muirhead, LL.D.
Birmingham, 1909. Portraits, 8vo.
The Fourth Lecture, pp. 109–30 is devoted to James Watt, by Prof. Burstall.

FOX, Howard, F.G.S.

Boulton and Watt.
Reprinted from the Report of the Royal Cornwall Polytechnic Society for 1909.
Penryn, 1910. 20 pp., 8vo.
Letters, or extracts therefrom, received by Thomas Wilson, agent of Boulton and Watt in Cornwall, during the period 1794–1802.

CAIRD, Robert, LL.D., F.R.S.E., M.Inst.C.E.

James Watt's Contribution to the Advancement of Engineering. Watt Anniversary Lecture delivered before the Greenock Philosophical Society, 1910.
Greenock, 1910. 23 pp., 8vo.

CORMACK, J. D., C.M.G., D.Sc., M.Inst.C.E.

In the Days of Watt . . . Watt Anniversary Lecture delivered before the Greenock Philosophical Society, 1915.
[Greenock, 1915.] 17 pp., 8vo.

DICKINSON, Henry Winram, M.I.Mech.E.

Some unpublished letters of James Watt. Excerpt *Proc. Inst. Mech. Eng.*, pp. 487–534.
London, 1913. 8vo.
Annotated transcripts of letters from the Boulton and Watt Collection and from the Boulton Papers, bringing out points of technical interest.

GRANT, John W

James Watt and the Steam Age.
London, 1917. pp. 223, portrait, 8vo.
A gossiping sketch.

BOARD OF EDUCATION.

Catalogue of Mechanical Engineering Collection in the Science Museum, South Kensington. Pt. I, 6th edition, with a supplement containing illustrations.

London, 1919. 8vo.

Includes technical description of original Watt models preserved at South Kensington.

HENDERSON, H., *Librarian.*

James Watt Centenary Exhibition [at the Watt Monument, Greenock], Sept. 4, 1919. Greenock, 1919. 19 pp., portrait, 8vo.

WATT CENTENARY COMMITTEE.

James Watt Centenary Commemoration, Sept. 16–20, 1919. Souvenir Guide Book of Special Exhibit in the Art Gallery, Birmingham.

Birmingham, 1919. 60 pp., 8vo.

BOARD OF EDUCATION, Science Museum.

Catalogue of Watt Centenary Exhibition.

London, 1919. 45 pp., portrait, 8vo.

A Collection of portraits, drawings, holograph letters, original models, books, &c.

GALLOWAY, T. Lindsay, M.A., F.G.S.

James Watt.

Proc. Roy. Phil. Soc. Glasgow, vol. L, 154.

Glasgow, 1921. 16 pp., 8vo.

FLEMING, James Arnold, O.B.E., F.R.S.E., F.S.A., Scot., F.C.S.

Scottish Pottery.

Glasgow, 1923. 8vo.

Gives *inter alia* an account of Watt's connexion with pottery manufacture—a hitherto little-known activity of his.

LORD, John, B.A.

Capital and Steam-Power, 1750–1800.

London, 1923. 8vo.

Deals with the economics of the introduction of the steam engine into industry. Uses the history of the firm of Boulton and Watt as the basis of his thesis. Employs the same material very largely, although from a different angle, as do the authors of the present volume, in conjunction with which it should be read.

MARSHALL, Thomas Humphrey, M.A.

James Watt (1736–1819). Roadmaker Series.

London and Boston, 1925. 8vo.

Racily written—effectively uses the known material, including the preceding volume, well set in relationship to contemporary events.

JENKIN, A. K. Hamilton, M.A. B.Litt.

Boulton and Watt in Cornwall. Excerpt Rept. Roy. Cornwall Polytechnic Soc., N.S., vol. V. 1926.

Camborne, 1926. 8vo.

Extracts are given from a collection of about 1,000 letters from the firm to Thomas Wilson, their agent. The letters are now the property of the Society and are preserved in their Library.

APPENDIX II

DIRECTIONS FOR ERECTING AND WORKING THE ENGINES, 1779

DIRECTIONS

FOR

ERECTING and WORKING

THE NEWLY-INVENTED

STEAM ENGINES.

BY

BOULTON and WATT.

CONTENTS
OF THE
DIRECTIONS

GENERAL DIRECTIONS

For Building the Engine Houſe.

I. HAVING fixed upon the proper situation of the pump in the pit, from its centre meaſure out the distance to the centre of the cylinder, that is the length of the working beam, or great lever, and the half breadth of each of the great chains, as ſhewn by the drawing. Then from the centre of the cylinder set off all the other dimenſions of the house, including the thickness of the walls, and dig out the whole ground included (to the depth of the bottom of the cellar) so that the bottom of the cylinder may ſtand on a level with the natural ground of the place, or lower, if convenient, for the less height the house has above ground, so much the firmer will it be. The foundations of the walls muſt be laid at least two feet lower than the bottom of the cellar, unless the foundation be firm rock, and care muſt be taken to leave a ſmall open drain into the pit quite through the lowest part of the foundation of the lever wall, to let off any water that may accidentally be ſpilt in the engine house, or may naturally come into the cellar. If the foundation at that depth does not prove good, you muſt either go down to a better, if in your reach, or make it good by a platform of wood or piles, or both.

II. The foundation of the lever wall muſt be carried down lower than the bottom of the ſpace left under the condenſer ciſtern, (to get at the ſcrews which fix the condenſer) and two ſhort walls muſt be built to carry the beams under the condenſer ciſtern. Two other ſlight walls ſhould be built one on each ſide, at a little diſtance from that ciſtern, to keep the earth from it, which would otherwise cause it to rot.

III. Within the house, low walls muſt be firmly built to carry the lower cylinder beams, so as to leave ſufficient room to come at the holding-down ſcrews, as ſhewn in the drawing, and the ends of these beams muſt also be lodged in the wall, but the platform is not to be built on them until the house is otherwise finished.

IV. The lever wall muſt be built in the firmeſt manner, and run solid course by course with thin lime mortar, and care muſt be taken that the lime has not been long ſlacked.——If the house be built of ſtone, let the ſtones be long and large, and let many headers be laid through the wall ; it ſhould also be a rule, that every ſtone be laid on the broadeſt bed it has, and never ſet on it's edge.——A course or two above the lintel of the door which leads to the condenſer, build in the wall two parallel flat thin bars of iron equally diſtant from each other, and from the outside and inside of the wall, and reaching the whole breadth of the lever walls. About a foot higher in the wall, lay at every four feet of the breadth of the front, other bars of the ſame kind at right angles to the former course, and reaching quite through the thickness of the wall, and at each front corner lay a long bar, in the middle of the ſide walls, and reaching quite through the front wall. If these bars are 10 or 12 feet long it will be ſufficient.——When the house is built up nearly to the bottom of the opening under the great beam, another double course of bars are to be built in, as has been directed.

V. At the level of the upper cylinder beams, holes muſt be left in the walls for their ends, with room to move them laterally, so that the cylinder may be got in, and ſmaller holes muſt be left quite through the walls, for the introduction of iron bars ; which being firmly faſtened to the cylinder beams at one end, and ſcrewed at the other or outer end, will ſerve by their going through both the front and back walls, to bind the house more firmly together.

VI. The ſpring beams, or iron bars faſtened firmly to them, muſt reach quite through this back wall, and be keyed or ſcrewed up tight, and they muſt be firmly faſtened to the lever wall on each ſide, either by iron bars, firm pieces of wood, or long ſtrong ſtones reaching far back into the wall ; they muſt also be bedded ſolidly, and the ſides of the opening built in the firmeſt manner with wood or ſtone. The ſpring beams muſt always be laid 8 inches on each ſide diſtant from the working beam, to give room for the ſide arches.

VII. The house being finished, a wooden platform, of 2½ inch plank, is to be laid on the lower cylinder beams, and the centre of the cylinder being accurately marked on it, four holes are to be bored through the cylinder beams, for the holding down screws, and four boxes, about seven or eight inches square, and as long as the stone platform is to be deep, are to be placed perpendicularly over them. Then the stone or brick platform is to be built up to the level of the cylinder's bottom, as shewn in the drawing; it must be composed of the heaviest materials which can readily and cheaply be procured. A very solid pillar of stone or brick-work, laid in the best lime mortar, must be carried up directly under the cylinder, and must be, at least, of the diameter of the outside of the flanches; the rest of the platform may be filled up with the heavy materials, bedded solidly in a mortar of clay and sand, and well beat into their places, so as never to settle or yield.

VIII. The lever, or great working beam, is best when composed of one single log of seasoned oak; where that cannot be obtained two may be used, or four, or more; but the fewer logs it is composed of, so much the more durable will the lever be, or of so much smaller scantling may it be made. This beam is to be fashioned and mounted as in the drawing; the diagonal stays which are fastened to the arches and to the lower log, or lower edge of the beam, are to prevent the logs from sliding on one another, by the difference of the direction in which the chains act upon them when the end of the lever is up or down. These stays are to be let into the side of the beam, that the other diagonal braces may pass over them.—The diagonal braces which reach from the top of the king post to the lower edge of the beam, are intended to prevent the logs from bending or sliding on one another; they are fastened to the beam at their lower end by means of a strong square bar of iron, screwed at both ends, which passes through the beam, and serves to bind it together laterally; and they must no where else have any fixture to the beam; their screw at the top of the king post must be tightened from time to time, as required.

The gudgeon is to be placed on the top of the beam, and is not to be at all let into it, only the corners of the log may be taken off to fit the saddle-plate, and to prevent the saddle-plate from sliding on the beam; one or two pieces of hard wood, about five inches broad and a foot long, by three inches thick, may be let into the upper side of the beam, one inch deep, with their ends butting up against the saddle-plate. They must be spiked down in their places, and both them and the saddle-plate must be laid in a bed of tar and tallow mixed and used boiling hot, which will prevent the wood from rotting under them.—A clamp of oak, four inches thick, and from four to six feet long, must be spiked on the lower side of the beam; this clamp must be rounded on the edges, as shewn in the drawing; its use is to prevent the beam straps from hurting or weakening the beam in that critical place.—These beam straps must not be made out of thick bars or lumps of iron, but must be made up of a number of thin or small bars welded together, and they and all the other iron-work of the beam ought to be made of iron of the best quality; all the big pieces should be made up of smaller or thin bars in the way I have mentioned.—Upon no account whatsoever let any holes be bored through the beam near the gudgeon, nor any thing else be done which may weaken it there.

IX. The arches for the plug-tree and condenser pumps should be screwed to the beam by screw bolts, which should pass through the joints of the logs of the beam, if it be composed of more logs than one, and one bolt may generally pass above the beam; these bolts also serve to keep the beam together laterally; these arches should be made with a shoulder of two inches projection to rest on the upper-side of the beam.—The tails of the martingales of the plug-tree, and the condenser pumps, must also be secured by bolts passing through the beam in some joint, if it can conveniently be done.—The lower end of the king post should have a hollow in it to fit the gudgeon, but care should be taken that it rest upon the gudgeon, and not upon the saddle-plate.—It should be contrived that the tails of two of the great martingales should rest on the middle of each of the two logs which compose the thickness of the beam; that is, when the beam consists of four or more logs. The martingale screws should be strong, and should go down through the beam, as it is them that principally keep the beam together in the direction of its depth; near these screws must be placed the keys, or pieces of hard dry wood, which being half let into each log of the beam, prevent the logs from sliding one upon the other; these keys should never be above two inches thick, that is, one inch let into each log; they may be made in three pieces, the two outside pieces dove-tail ways, and the middle one tapering, by driving up which they are made to jam themselves in their mortoise. Or they may be made of one piece six inches broad at one end, and five at the other, so that by driving the whole in, it may check the sliding of the logs; if there are more sets of logs than two in the depth, the keys must be placed alternately on different sides of the martingale screws; care must be taken in placing the chains for the plug-tree and condenser pumps, that all the heads of the chain bolts be next the beam, and that they be far enough off not to rub on the diagonal stays, or any other thing.

X. The great chains must be made, according to the drawing, of the very best iron, and the martingales must be placed so that the adjusting screws may lye parallel to the arches, and the upper surface of the head of the martingale be at right angles to them. The holes in the martingales should be quite easy for the adjusting screws, and a washer, thinned about the outer edges, should be put under the nuts. There should be a sufficient length of chain to reach one link lower than the under end of the arch of the beam.

XI. The cap and cross bar for the piston rod should be made exactly according to the drawing, firm work of good iron; the mortoise in the cap should be made exactly to suit the mortoise in the piston rod, and the cutter or fore lock to fit them both exactly; and this cutter, above all things, should be the very best of iron, as the whole depends on it; there is always a sufficient size given it in the drawing, so that if it should fail it must be the fault of the iron or workmanship. This cutter must be kept in its place by two cross cutters, and these again by a thong of leather past through some holes in them.

XII. It will seldom happen that the plug-tree can be hung directly under its arch; you are to place them exactly in the places fixed by the drawings; that of the arch will always be found right in the general section; but the place of the plug-tree and guide posts must be taken by measuring from the nozzle, in the drawing of the working gear; a strong iron bracket with a stay must be fastened to the top of the plug-tree, in such a direction that the point of the bracket may come directly under the arch. There must be a hole in the point of the bracket, to receive the end of an iron rod reaching down from the chain, and the end of this rod must be screwed for five or six inches, and have a nut on the lower side of the bracket to adjust the height of the plug-tree by.

XIII. There should be placed upon the spring beams over the cylinder, two uprights, connected at top by a strong cape piece. These uprights serve to support a windlass with a wheel and pinion, by means of which, and a pair of tackle pully blocks hung to the cape piece, it will be easy to lift and put the cylinder, &c. in their places, and after the engine is completed, it makes it easy for the engine-men to raise the cylinder lid to pack the piston without other assistance.—The barrel of the windlass may be of oak about 6 inches diameter, and must have a square gudgeon of iron drove quite through it, on one end of which the toothed wheel is fixed; the gudgeon may be from $1\frac{1}{2}$ square to $2\frac{1}{2}$ inch square, according to the size of the engine, and the wheel about 2 feet diameter, driven by a pinion of 5 inches diameter; but these may be larger or lesser according to the weights to be commonly raised.—It is necessary to mention to those who are disposed to look on a wheel and pinion windlass as a superfluous expence, that there is no trusting to windlasses wrought by bars, and that many bad accidents have happened through the use of them, which obliges us absolutely to condemn them for this purpose.

XIV. The springs to receive and in some degree save the blow when the engine comes down too suddenly, are best made of a piece of square dry elastic timber, reaching from the plummer blocks to nine or ten inches beyond the catch pins, their size must be suited to that of the engine from six inches square to twelve or fourteen. The end next the catch pin must be sloped off on the under side for four or six feet in length, according to the size of the engine, so that their points may be one inch distant from the spring beams to which they must be bolted down by a screw bolt at the end of the sloped part, and another at the end next the plummer blocks.— The part of these springs which are struck by the catch pins, should be covered by a plate of iron, and that again by a piece of strong leather, to prevent the clattering noise they might otherwise make.

XV. The utmost attention to dimensions ought to be observed in constructing the masonry of the building, particularly in regard to heights, mistakes in them are of the worst consequences.

XVI. The condenser cistern ought to be made of the best Dantzick three inch deal plank if they can be got. If they are not to be readily got, any other good red deal or oak may be employed, but whatever kind of wood you use, be sure to cut off all the sap wood, otherwise the cistern will soon become useless. The best way of putting the cistern together, is by means of long screw bolts of iron, about $\frac{3}{4}$ square, put through the planks edgeways from top to bottom of the cistern ; these screws may be 18 inches distant from one another.—The bottom may be put together in the same manner with screws, and then fixed down upon the beam or beams represented in the drawing, and supported by so many more smaller beams as may be necessary. If the cistern is not more than seven feet long, no uprights on the outside are necessary, only one about six inches square in each corner in the inside, and in no cistern ought there to be any uprights on the side next the wall. The joints of the planks should be plain joints, and put together on a strip of coarse flannel soaked with a mixture of tar and tallow equal parts, used warm, or upon bullrushes. A large cock or a brass valve should be fixed in the bottom of the cistern to let off the water occasionally, and a notch about four inches deep, and 18 inches broad, with a trough fitted to it should be made in the upper edge of the cistern, to convey away the waste water.—If surface water cannot be found to supply the injection, a small pump should be fixed to bring up water from the main pump head into this cistern.— In case the water from the pit is good, and is raised to the surface, the main pump may deliver it directly into the cistern, but if the water be subject to be muddy, or mixed with sand, &c. it will be best to put it into another cistern to deposit some of that matter first.—If the pit water be vitriolic or encrusting water, it becomes necessary to use every means to procure better water, otherwise it will destroy the condenser, &c.

XVII. In making the boiler you should use rivets between 5-8ths and 3-4ths inch diameter. In the bottom and sides the heads of the rivets should be large and placed next the fire, or on the outside, and in the boiler top the heads should be on the inside. The Rivets should be placed at two inches distant from the centre of one rivet to the centre of the other, and their centres should be about one inch distant from the edge of the plate.—The edges of the plates should be evenly cut to a line, both outside and inside. It is impossible to make a boiler top truly tight which is done otherwise. After the boiler is all put together, the edges of the plates should be thickened up, and made close by a blunt chissel about $\frac{1}{4}$ inch thick in the edge impelled by a hammer of three or more pounds weight, one man holding and moving the chissel gradually, while another strikes. All the joints above water should be wetted with a solution of sal armoniac in water, or rather in urine, which by rusting them will help to make them steam tight. After the boiler is set, it may be dryed by a small fire under it, and every joint and rivet above water painted over with thin putty, made with whiting and linseed oil, applied with a brush.—A gentle fire must be continued until the putty becomes quite hard so as scarcely to be capable of being scratched off by the thumb nail, but care must be taken not to burn the putty, nor to leave off fire until it become dry.

XVIII. In building the brick work of the boiler-setting, no lime must be used where the fire or flame comes, but a mortar made of loam or sand and clay ; but lime mortar should be used towards the outside. Pieces of old cart tire or other such like pieces of iron, may be laid under the chime of the boiler, between it and the bricks, which will prevent its being so soon burnt out there. The brick work which covers the boiler top, should be laid in the best lime, which will not hurt it there but will preserve it ; the mortar should be used thin, and the boiler top well plaistered with it, which will conduce greatly to tightness, if done some time before the engine be set to work.—If your lime be not of that species which stands water, it will be well to mix some Dutch or Italian terrass, or pan scratch from the salt works with it, but in any case the lime should be newly slacked. In carrying up the brick work round the flues, long pieces of rolled iron should be built in two or three courses to prevent the brick work from splitting.——Four holes at convenient places should be made into the flues, large enough to admit a boy to go in to clean them.—One of them may be over the fire door, and another right behind the damper in the back side of the chimney. This last may be as high as the flues themselves are. These holes when not in use are to be built up with nine inch brick work, and made perfectly air tight. Immediately above the brick work of the boiler-setting, a hole must be left in the chimney on the side next the boiler. This hole must be as wide as the chimney, and one foot or 18 inches high, and must have a sliding door fitted to it, to open it more or less at pleasure ; the use of it is to moderate the draught of the chimney, and to prevent the flame being drawn up it before it has acted sufficiently on the boiler. A groove must be left in the brick work for the damper to move up and down in easily, which should fit flat to the face of it.—The damper may be made to move easily up and down by means of a beam or a wheel, with a counterpoise equal to the weight of the damper. The best form of a fire door is two feet long and one foot high, inside measure, to have two leaves made of boiler plates hinged on the two sides, and over lapping one another about an inch in the middle. The scantling of the frame may be three inches broad by two inches thick.

XIX. The guage pipes may be fixed into the boiler top in some convenient place ; the lower end of the longest should reach within 6 inches of the top of the flues, and the shortest should be 4 inches above it.—The feed pipe should reach two feet under the surface of the water in the boiler, and should have a valve at its lower end, to prevent the water being ever forced up through it by the steam.—Its upper end should rise seven feet higher than the surface of the water in the boiler. It should be supplied with water by a pipe from the top of the hot-water pump regulated by a cock near the feed pipe.

XX. If you have not land water that will naturally run into the condenser cistern, you must make a pool somewhere in the neighbourhood to receive the water from the hot-water pump, and reserve it for supplying the boiler and condenser cistern when the engine stands still on any occasion.—This pool may be at least 40 feet long and 20 feet wide to hold 3 feet deep of water, and pipes or troughs must be laid from its bottom to the boiler feed pipe and to the cistern. That at the feedpipe must have a cock on purpose.—It is only meant that this pool be simply dug in the earth and lined with turf, puddled, or otherwise made water tight.——If no ground within a reasonable distance be high enough for the water to run from the bottom of the pool into the boiler, then a pool may be made on lower ground and a hand pump fixed up to supply the boiler and cistern ; but this ought to be avoided if possible.

DIRECTIONS
For putting the ENGINE together.

XXI. HAVING put the working beam together, and fastened the gudgeon to it, rest it on the plummer blocks ; but do not fasten these blocks until the Cylinder is fixed.

XXII. Level the top of the stone platform, and lay the outer bottom of the cylinder down in its place, truly level, and corresponding to the holding down screw boxes.

XXIII. Apply the inner bottom upon the outer one, and set its upper joint level, by wedging betwixt it and the outer bottom if it requires it ;—then cut out segments of pasteboard, such as is used for the boards of books, (not such as is composed of paper pasted together,) let these segments be of such thicknesses as the different parts of the joint may require (if it be more open in some places than in others.) Soak these pasteboard segments in warm water until they become quite soft, then lay them upon boards to dry, and when quite dry put them into a flat pan with a quantity of drying LINTSEED oil ; warm the oil until the pasteboard ceases to emit bubbles of air, but take care not to heat the oil much hotter than boiling water, otherwise it will harden or burn the pasteboard. Anoint the segments on both sides with thin putty made with fine whiting and some of the lintseed oil ; let the whiting be very dry, otherwise it will be difficult to mix with the oil, and N.B. that white lead will not answer in place of it.

You must as much as you can avoid using more than one thickness of pasteboard, and the segments should be a little broader than the flanch, with all the holes cut out by a chissel, but not quite so large as the holes in the iron. The segments should also be thinned at the ends where they overlap each other, so that they may form a circle of pasteboard of an uniform thickness.

XXIV. Lift up the inner bottom, and lay your segments regularly round upon the flanch of the outer bottom, then place the inner bottom upon them, taking care at the same time to put a proper thickness of pasteboard in the joint under the pipe which proceeds from the inner bottom.

In like manner prepare pasteboards for the joint between the inner bottom and the cylinder, and proceed as has been directed for the other joint.

XXV. Having the cylinder ready suspended, lower it down in its place, so that the square pipe at its upper end be exactly over the pipe of the inner bottom ; then with a square taper piece of iron, of a proper size, thrust into each hole, enlarge the holes in the pasteboard, so as to admit the screws.—Put in the screws and screw up the joint gradually all round, and do not screw up one side faster than another, otherwise you will be apt to crack the flanch of the cylinder or bottom : or else will make a bad joint. No screws are to be put through the cylinder flanch over the pipe, therefore that part of the joint ought to be made with the utmost care, and the pasteboard ought to be a trifle thicker there. The general thickness of the pasteboard for these joints, ought to be 3-16ths of an inch.

XXVI. Put in the holding down screws, which ought to have screws and nuts at both ends, then set the cylinder truly upright, which is done by putting a piece of wood across at bottom, and another at top, and marking upon both the centre of the cylinder at their respective places ; then hang a plummet from the upper centre, and examine if the LINE be in the centre below ; if it be not, you must wedge under the outer bottom, until you bring the line to hang truly in the axis of the cylinder.——The holding down screws should be screwed tight, so as to keep the cylinder in its true position, after which the screws of the joint must be again screwed up, then taken out one by one, and lapped round with a rope yarn and some putty, both under the head of the screw and under the nut, so that each screw may be air tight of itself.

XXVII. Carefully scrape, or rather scour, the rust from the sides and bottom of the cylinder; clean it well out and grease the sides with tallow.—Hang on the chains and piston-rod cap in their places—put the piston-rod into the cylinder ;—suspend the piston by two half links fastened to one of the crosses, and lower it down upon the piston rod ; but previous to this, the rod should be tried into the piston, and if the hollow and convex cones do not fit one another, they must be made to do so, by chisselling and filing the cone of the rod. A lead ring an inch square exactly fitting the inside circumference of the cylinder, should be laid upon the small rim of the inner bottom, to save it in case of dropping the piston at any time ; and an iron gland an inch thick should be screwed across the base of the cone of the piston rod, by means of two screws coming through the bottom of the piston and screwed into the gland. The points of these screws should be cut off so that they may not strike the bottom when the piston strikes the ring.

XXVIII. The piston being lowered down upon the rod, the lid of the cylinder is to be laid on, without the stuffing box. The end of the working beam is to be lowered down, and the piston rod cap put on the rod and forelocked fast.—The beam is to be raised, and the lid also, and an examination made whether the piston have dropped truly down to its place upon the rod. If so, the lid or cover is to be let down, and by lowering and raising the beam and piston, you will perceive whether the rod always moves up and down truly in the axis of the cylinder. It must be made to do so by shifting the plummer blocks out or in, or by shifting the martingales to one side or to the other.

The utmost care should be taken that the plummer blocks be placed both of one height ; and after the beam has been some days in place, it should be examined if the gudgeon lie truly horizontal, as otherwise it will cause a most disagreeable motion in the piston rod.

XXIX. Caulk the joint round the pipe of the inner bottom between it and the pipe of the outer bottom, with rope yarn or oakum as hard drove in as possible. Screw the nozzle to the pipe of the inner bottom, making the joint as has been directed, and with the utmost care, so that the nozzle shall hang a quarter or half an inch lower at the point than at the joint, that any water condensed in it may run to the exhaustion pipe.——Put a strong wooden prop from the ground to the lower side of the nozzle, right under the perpendicular steam pipe. Care should be taken that the inside of the bottom of the nozzle, be even with, or rather lower than the inside of the bottom of the pipe which comes from the inner bottom of the cylinder; so that no water may lodge.

XXX. Put on the steam case, screwing the pannels together with a few screws. If found to be too short, it may be lengthened by means of a lead flanch put in the middle joint with a thickness of paste-board on each side of it, but if found too narrow and the deficiency upon being divided equally among all the joints amounts to more than ¼ inch to each joint, at the inner side, then a bar of iron is to be prepared of such breadth as will make up the whole deficience, and as thick as it can be put between the screw holes, in the perpendicular flanches of the steam case, and the rings on the cylinder ; this bar is to be put into a joint of the steam case on the backside of the cylinder, and made tight by caulking or by pasteboard. Remember to put the middle of a pannell, opposite to the perpendicular steam pipe.

XXXI. When you have found that the steam case is of a proper diameter and length, or have adjusted it as has been directed, it is to be made tight. Make the joint between the pannels behind the perpendicular pipe, and the upper and under rings of the cylinder, by applying a proper thickness of pasteboard and putty or soft ropeing, upon the cylinder rings before you put up these pannels ; or, if you perceive that the joint will admit of it, you may wind a soft rope, slackly twisted, once or twice all round the cylinder rings ; then screw the perpendicular joints of the pannels together, (putting in all the screws) until the insides of the joints are quite close, or as close as they can admit of ; afterwards take oakum, mixed with some putty, made with thick linseed oil ; or a soft rope covered with putty, and with a caulking chissel drive it forcibly into the joint, and continue caulking in a little at a time, until you have filled the joint quite to the outside of the flanches. Remember to put oakum or soft rope yarn under the head and nut of each of the screws, as you put it in, and don't force the screws too much, lest you break the flanches ; rather trust to the caulking. In like manner you are to make tight by caulking, the joints between the steam case and the upper and under rings, using a crooked chissel for the more conveniently getting at the under one.

XXXII. Put on the upper part of the lower nozzle and make its joint. Set on the perpendicular steam pipe, and try the upper nozzle to its place ; if the pipe prove too short, lead flanches, of a proper thickness, must be introduced equally above and below, to make up the length ; but where-ever lead flanches are used, where hot steam comes at them, it is necessary to put a thickness of pasteboard, with putty on each side of them, and the lead should be free from tin. These lead flanches should be a little larger all round than the iron flanches, that their edges may be riveted up afterwards, when any leaks are perceived. If the pipe proves a little too long, the upper or top nozzle may be raised a little higher than its natural joint, provided the over length does not exceed an inch. The round flanch of the perpendicular steam pipe goes uppermost. Four round holes are to be drilled in the top of the upper part of the lower nozzle, corresponding to four holes in the flanch of the perpendicular pipe, and they are to be screwed together by screws, with heads within the nozzle. Five screws may, in like manner, be put in the flanch above.

XXXIII. The cross pipe is to be put on, and its joint made. The boiler steam pipe is to be screwed to one end of it, and the other shut by a plate. If any of the joints are not of a proper angle, fill them up with lead.

XXXIV. The steam is to be communicated from some convenient place of the cross pipe to the steam case, by means of a copper pipe with thin copper flanches, fixed to the cross pipe and to the steam case, by small pierced glands, with a square hole in each end, to admit the square necks of two screws, which being screwed at both ends, one end must be screwed into the cast iron, first tapped for that purpose, and the other with a nut serves to keep on the gland. Another similar but smaller pipe must be fixed to the very lowest part of the steam case, bent over the flanches, and inserted into the perpendicular part of the outer bottom to fill it also with steam. In some convenient part of the outer bottom, as low down as may be, is to be fixed a waste pipe, to let out the condensed water. This waste pipe must reach down about five or six feet, and be bent upwards a little at the lower end, and shut by a valve loaded with a proper weight, which will open whenever the elasticity of the steam and the weight of the pillar of water in the pipe are able to overcome the weight which shuts the valve.

XXXV. The condenser is now to be put in its place in its cistern according to the drawing. Its joints may be put together with pasteboard soaked in oil, as directed, and putty, firmly screwed up and caulked afterwards : or, any where under water plates of lead may be used, about $\frac{1}{4}$ inch thick, well fitted to the joints and puttied on both sides ; after these joints made with lead, are well screwed up, and the condenser warm by fire or steam, the edges of the lead which had been left projecting a little, must be rivetted up inside and outside. A soft rope about half an inch diameter, coiled round and round until it covers the flanch, and well puttied, may be used, in default of pasteboard or lead ; but either of the two former are preferable, and in every case caulking or rivetting should be used.

XXXVI. If the clack of the hot-water pump has two valves, and is not sent ready fitted, the beating or fixed part must be chiseled and filed truly flat. The pivots or axis of the valves must be from ¾ to an inch diameter, according to the size of the engine, the flat part of the iron of the valve about ¼ inch thick.—The copper facing 1-6th inch, and the iron plate under it also 1-6th inch. After the two iron plates and the copper facing are firmly rivetted together, they are to be heated red hot, laid on their place, a short piece of end wood set above them, and beat down by some blows of a sledge hammer.—The pieces of iron the pivots move in, are to be fixed by means of pins of iron half inch or ¾ square, screwed into the cast iron of the clack, passing through a square hole in the pivot pieces, and forelocked above by spring cutters. Every one of these parts ought to be made very secure and firm.——A guard to prevent these valves from over opening, is to be fixed in the hot-water pump, according to the drawing, this guard may be about an inch thick, and should not touch the edges of the valves, but catch them on the flat part behind.—The cast iron face of the eduction pipe foot is also to be made flat for the valve there to beat against.—The pivots of this valve should be one inch diameter. The thickness of the iron and copper the same as for the others.—The ends of the valve should be one quarter of an inch clear of the sides, and one half inch clear of the bottom of the place it plays in.—The pivots should be sunk into the cast-iron of the sides until their lower edge be within one quarter of an inch of the opening of the beating part.—They should have one inch of hold of the iron at each end, and have no play in that direction.—In the lid or clack door for this valve there should be a groove for the axis of the valve, so that it may not touch it when the lid is screwed on.— The pivots should not be confined close against the beating part, but should have a quarter of an inch of play in that direction, as the air makes its escape partly at the hinge.—The valves of the air and hot water pump buckets are to be fitted in the same manner, remembering to make the pivots proportionable to the size of the valves.

XXXVII. The condenser being fixed in the cistern at the height below the nozzle, and distance from the centre of the gudgeon or of the cylinder, shewn by the drawings, and so that the middle between the centres of the pumps shall be directly under the middle of the working beam, and the line between these centres at right angles to the beam, the copper eduction pipe is to be fitted to its place. It is to be screwed to the flanch of the short pipe under the nozzle by means of a loose flanch of hammered or cast iron applied on the underside of the copper flanch of the pipe. The outside diameter of the loose flanch should be the same as that of the flanch on the nozzle, and its inside diameter should be one inch more than the outside diameter of the bent copper pipe, and should have its inner angle taken off a little on the side next the copper flanch, lest it should cut that flanch, or crack the soldering.——If the loose flanch be made of hammered iron, it should be ¾ of an inch thick, and the holes should be drilled and not punched ; and in the same way you are to proceed with the joint at the foot of the eduction pipe.

XXXVIII. Having carefully tinned the inside of the upper end of the wide or perpendicular part of the eduction pipe, and also the outside of the brass ring which goes within it, the ring is to be put into its place, and being heated, the joint is to be run with fluid tin solder, after which four or more holes may be drilled through both the copper and the brass, and some copper rivets put in them.——The spiggot and fosset joints are to be secured as follows: An iron ring three or four inches broad, and half an inch thick, is to be put red hot on the outside of the fosset part, so that by its contraction in cooling it may grasp it firmly.——The spiggot part is then to be put into it and made tight by caulking in soft roping and putty. The proper width of a joint for caulking is 3-16ths of an inch at the wide or open end, and drawing quite close at the inner end ; but will answer although a little wider or narrower.

The joint of the bent part of the eduction part to the perpendicular part at the brass ring, is also to be done by caulking.

When the engine is set to work, if any of the spiggot and fosset joints shew a disposition to slide or move, that may be cured by putting screw hoops round both the spiggot and the fosset part near the joint, and pulling the joint together, by means of two screws connecting the screw hoops.

Care must be taken in putting the eduction pipe together, to keep the brased joint upwards, so that if any defects appear they may be cured by tin solder.

When the eduction pipe is all put together, a hole is to be cut for the short fosset pipe of the injection, as shewn in the drawing. This hole must be cut, so as to fit the outside of the fosset accurately.——The fosset pipe should point up the eduction pipe in such manner that the injection water may strike the upper side of the eduction pipe within about two feet of the nozzle ; but care should be taken that it do not spout too low, otherwise it may, by the bent pipe, be reflected up against the exhaustion regulator, which will be very hurtful.—The fosset pipe being adjusted to its proper position, and the knee of the eduction pipe tinned round the hole, the fosset is to be fixed in its place, either by a strong body of plummer's solder, or by a copper bosse or case run full of the same solder heated to a dull red heat.——The upper edge of the inner end of the fosset should only go half an inch within the eduction pipe, and the nozzle of the injection not quite so far.——The injection pipe being set in its true position, the joints soldered with plummer's solder, and its valve soldered on, a hole is to be cut for the blowing pipe fosset at or about the level of the valve of the injection, but not lower, otherwise the engine will blow at the injection, and heat the cistern.——The fosset for the blowing pipe is to be fixed by soldering, or by a bosse, as directed for the injection, and its inner end ought not to go more than one or two inches within the eduction pipe, according to the diameter of that pipe.——The blowing pipe may then be put together, and its valve soldered on, taking care that the pipe be of such a length that its valve may be 6 inches under the surface of the water in the cistern.——Care must be taken that the stems of both the blowing and injection valves stand truly perpendicular when fixed in their places.—The injection and blowing pipes are to be fixed in their fossets by caulking as directed. Care must be taken that no tin from any of the solderings be left in the pipes.

XXXIX. The condenser pumps must be fixed down by means of screws passing through the bottom of the cistern, and the beam under it, as shewn in the drawings ; and it must be remembered, that its disposition to rise is very powerful, and that if it has any play it will be sure to spoil the joints of the eduction pipe, or perhaps break it. The hot water pump must have a strong prop under it, and be tied down as well as the other. A beam of deal, nine or ten inches square, must be put across the cistern, near the air pump rod, to support a pair of shears or uprights, for a pump brake for that pump, to examine the tightness of the joints by. This brake must have an arch and a chain with a hook, to join it to the chain of the air pump rod, when in use. The buckets of the condenser pumps must be surrounded by a plaited rope made of rope yarn, of such breadth as to fill easily the interstice between the bucket and the pump barrel. A pudding link chain, four foot long, must be fastened to the top of the sliding-rod of the air pump, and to the other part of that rod which reaches up to the working-beam ; its use is to suffer the engine to work without unloosing the hook of the pump brake, when trying experiments on the tightness of the engine.

XL. The stuffing box of the air pump must be packed with a small soft rope wrapt round the rod, and forced down into the box pretty tight, but so that the rod may move easily. A flat round piece of wood about 1½ inch thick, fitted easy to the inside of the box, and to the outside of the rod ; must be put above the stuffing and screwed down by the gland. There need be no screws put to hold down that side of the air pump lid which is over the connecting box ; those on each side of the box are sufficient, if care be taken in making the joints. In like manner the lid or clack door of the lower valve of the eduction pipe foot needs only two screws, one at each end. In the bottom of the air pump must be placed a ring of hammered iron, with three or four feet, for the bucket of the pump to rest upon when at its lowest, i. e. when the lower edge of the packing of the bucket is within one inch of the under end of the working barrel ; this ring must be fixed so that it may not turn round and come in the way of the lower valve of eduction pipe. An upright, six inches square, must be fixed from the bottom of the cistern, near the injection, to screw that pipe to ; and its upper end must be fastened to the beam which carries the lever, and the end of the working gear of the injection. This upright should be fixed firmly, and the injection pipe should be fastened to it, by a stirrup with screwed ends, grasping the neck of the valve, going through the upright, and having nuts behind it. Any motion in the injection pipe will be apt to loosen or crack the joints of it, therefore it must be firmly fastened.

XLI. Guards must be fixed over the injection and blowing valves, to prevent their over opening ; for the knob of the spindle which stops them, by the bridge of the valves, is not to be trusted, and may therefore be cut off, which will give the convenience of taking out the fly part of the valve at pleasure. An S hook of iron is to be fitted into the eye of the valve, so as to have no motion there, and the rod which pulls it open is to have a hole in its lower end, for the upper end of the S to play easily ; if it be allowed to have motion in the eye of the valve, it will soon wear it out. Guards are also to be fixed over the valves on the air pump lid, to prevent their over opening ; these guards may be fixed by means of two of the screws which fasten on the lid.

XLII. The guide posts, or Y posts, of the plug frame, are to be fixed exactly according to the drawing sent for that purpose ; and the cross swords which slide in the guide posts should be of oak or beech, two inches thick and eight or nine broad. The plug-tree itself should be of hard, straight grained, seasoned oak, the holes $1\frac{1}{4}$ inch diameter, bored off both sides by a centre bit ; for if you bore them by an augre, they will be apt to break out into one another ; care should be taken to bore a sufficient length of the plug. The opening horns, or arches of the Y shafts, which act upon the levers of the regulators, must be bent exactly to the curves of the FULL SIZE DRAWINGS sent for them. This is best done by taking a piece of soft iron, an inch broad and three sixteenths thick, and bending it cold until its hollow side exactly fit the drawing, and by applying this mould to the arch, while red hot, you can set it truly into form. These moulds should be carefully laid up, lest, by any accident, the arches should require repairs. To fix the Y shafts, make the levers of both regulator spindles truly horrizontal, and so long as just to reach to their proper places on the Y shafts, and then the lower side of the exhaustion lever, and the upper side of the stream lever will point to the axis or centres of their respective Y shafts. The coupling brasses for the Y shaft pivots or gudgeons, are to be fixed one inch from the inside of the guide posts, and the centres of the pivots are to lye exactly in the line of the inner side of the rabbits or grooves, in which the swords move, (as drawn). A piece of wood, with a slit in it three inches wide and about three feet long, having holes in it, like an old-fashioned plug-tree, must be placed to receive the opening horn and lever of the stream regulator, and by means of wooden pegs, one inch diameter, put through its holes, and saddles of leather laid above them, regulate the opening of the steam regulator. To prevent shaking and noise, the lower end of this piece of wood should rest on the ground, in the floor of the cellar. The lower end of the guide posts must be fixed upon sills parallel to the working beam ; otherwise the weight of the exhaustion would fall upon them and shake them every stroke. The floor over the eduction pipe should be easily moveable, that the pipe may be come at. There should be a window in the door which leads to the condenser, to give light to the plug-frame. The weight which hangs to the detent of the exhaustion, and serves to raise the arch and open that regulator should be of lead cast on the rod, and square pieces of lead, with a notch in them, to admit the rod, may be laid on, if the weight proves too light. Some oakum should be laid between these saddles to prevent noise ; a box, eighteen inches square, and two feet deep, should be fixed about the blowing pipe, to prevent the hot water from mixing with the cold in the cistern, but there should be a few holes in the bottom of this box to suffer the water to go out below ; this box should rise six inches above water.

XLIII. Care should be taken that both the regulators fall into their seat without touching sooner on one side than the other ; and if the copper cones, under the regulators, be not already rivetted or screwed to them, it should be done before you begin, but avoid bending the valves in so doing ; some threads of oakum, well puttied, should be lapped round the necks of the regulator spindles, beyound the shoulders, to keep them steam and air tight ; but this should be done in such manner as not to prevent the spindles from going quite home to their shoulders, otherwise the regulators cannot fall right in their places.

XLIV. The brass of the cylinder stuffing-box should be fixed in its place, and the upper or thin edge of it set out against the sides of the iron part. When the piston rod plays truly up and down in the axis of the cylinder, put on the stuffing-box, and screw it down by its flanch ; then pack the box with soft rope yarn, wrapt round the rod, until you have nearly filled the box, then take a collar of deal wood, two inches thick, made easy for the rod and for the box ; divide it in two by its diameter, lay it on the top of the stuffing, and apply the gland above it ; as you go on with the packing, melt some grease and pour amongst it, and when finished, screw down the gland moderately tight.

XLV. The cylinder lid is to have no screw holes over the square pipe ; its joint is to be made with pasteboard, puttied on the lower side but not on the upper side, and the lid being greased with tallow, the pasteboard will not stick to it, but will lye in its place when the lid is raised. Two long iron rods, with hooks at their lower end, should be hung to eye bolts in the spring beams, so that when the lid is raised about three feet from the cylinder, these hooks may be put into two opposite screw holes, to support the lid at that height while you pack the piston.

XLVI. To pack the piston, take sixty commonsized WHITE or untarred rope yarns, and with them plait a gasket or flat rope, as close and firm as possible, tapering for 18 inches at each end, and long enough to go round the piston, and overlap for that length ; coil this rope the thin way as hard as you can, lay it on an iron plate, and beat it with a sledge hammer until its breadth answers its place ; put it in and beat it down with a wooden driver and a hand-mallet ; pour some melted tallow all round ; then pack in a layer of white oakum, half an inch thick, then another rope, then more oakum, so that the whole packing may have the depth of about four inches, or only three inches if the engine be a small one. Cast segments of a circle of lead, about 12 inches long, three inches deep, and $1\frac{1}{4}$ inch thick, fitted to the circle of the piston, and cut down square at both ends ; lay them round upon the packing as close as they can lye to one another without jamming, and screw down the piston springs upon them ; the piston springs should be bent downwards at the end next the piston rod, and a little mortoise should be cut in the cast iron there, for the bent down point of each of them to lodge in, which will prevent their coming forwards to touch the cylinder. Previous to the piston being put into the cylinder, the hollows among the crosses should be quite filled up with solid pieces of deal wood, put in radius fashion. The packing of the piston should be beat solid, but not too hard, otherwise it will create so great a friction as to hinder the easy going of the engine. Abundance of tallow should be allowed it, especially at first ; the quantity required will be less as the cylinder grows smooth.

XLVII. The joints being all made, the regulator valves in their places, and their covers screwed on, but no water in the condenser cistern, admit steam, and when the cylinder and steam case are thoroughly warmed, screw up the nuts of all your screws, and caulk the pasteboard or oakum of such joints as may require it, with a caulking chissel, until you find that every thing about the cylinder is perfectly staunch ; then pour three or four feet deep of water into the hot-water pump ; stake down the injection and blowing valves, and also those on the air pump lid, then let the steam into the condenser, which will shew the defects or leaks, if there be any.

XLVIII. Screw on the steam gauge to the steam case near the nozzle, and behind the engine-man's place, pour as much mercury into it as will half fill the open leg ; put a float on it, broad at bottom, but very slender in the stem ; cut the float or index off close to the end of the open tube, and fix a scale to it, reckoning every half inch the float rises equal to an augmentation of the elasticity of the steam, corresponding to the supporting a column of mercury an inch high, because the surface has sunk as much in the one leg as it has risen in the other.—— Solder a small copper fosset pipe, to fit the copper communicating tube of the barometer, into the eduction pipe, 12 inches under the fosset of the blowing valve, and on the opposite side of the eduction pipe ; place the barometer in the door way to the condenser on the further side from the plugtree, so that the engine man may see it when at his station ; join the copper tube to it, by pouring melted sealing-wax into the copper cup at top, fill the short leg of the barometer with mercury, within four or five inches of its top, and put a light float in it, long enough to reach to the top of its frame.

XLIX. Fill the condenser cistern, shut the lower regulator, and there being no steam in the cylinder or its communication with the boiler being cut off, take off the bonnet or cover of the exhaustion regulator, shut that regulator, and work the air pump by means of the brake. If then you find that air enters by the regulator, pour some water on it, and continue pumping until you have raised the barometer, i. e. sunk its float to 27 or 28 inches ; leave off pumping, and observe if the vacuum continues good, or is a long time in being destroyed. If it loses fast, seek for the leaks which must be somewhere in the eduction pipe, and will make a noise if touched with a wet hand ; (observe if the condenser moves by the pumping, and secure it.) After having cured these leaks, you may try the tightness of the cylinder, by staking the working beam, so that the piston cannot descend ; then taking the cover off the cylinder, open the exhaustion regulator, and shut the steam regulator; on beginning to pump, you will perceive if the piston be tight, if it is not, it may be beat a little, and some water being thrown upon it, and on the steam regulator, whatever air enters, must be by leaks, which must be sought for and cured by screwing or caulking in oakum.——N. B. A critical tightness in the piston cannot be obtained until the engine has gone a few days, without beating it too hard, to permit the engine to move easily.——When you can detect no more leaks in this way, the steam must be admitted, and the same examination made as before.

L. The piston chain should be so adjusted, that it may descend within one inch of the lead ring at bottom when the springs are pressed down by the catch pins, and that, when it is at its highest its upper edge may be level with the square opening at top, so that no water may lodge there, but may run down the perpendicular pipe ; and the engine should always be made to work full stroke, otherwise it will spoil the cylinder.—A collar of soft rope should be lapt round the piston rod under the lid to prevent the piston striking it if it should rise with a jump. And if the cap of the piston rod does not touch the gland of the stuffing-box when the catch pin have pressed down the springs above, a collar of iron must be fitted on the rod to make up the deficience, and to help to save the blow if the chains should give way and the piston fall ; for, though it should break the cylinder lid, that is a much smaller damage than the bottom would be, as it may be clasped or otherwise mended.

LI. There ought to be cleets or strong brackets of wood firmly bolted to the dry pump rods, and beams put across the pit at proper distances to receive them in case of the accident of their breaking.

LII. After the engine has been set a-going, and has gone a few hours, the holding-down screws should be screwed tight, and so from time to time as they become slack ; and in like manner all the other screws about the cylinder or nozzles should be screwed up as they slacken, and the joints caulked and puttied where they require it.

Directions for Working the Engine.

LIII. IT being necessary that the uses of the several regulators be thoroughly understood by those who attend the engine, we shall begin by describing them.

In the lower nozzle or regulator box are two regulating valves. When the upper one is opened, it admits the steam, from the perpendicular steam pipe into the cylinder below the piston, and thereby permits the piston to ascend, or in the engine man's phrase, allows the engine TO GO OUT OF THE HOUSE ; this regulator we call the STEAM REGULATOR. The lower regulator, which is placed in the bottom of the nozzle or regulator box, when open suffers the steam to pass from the cylinder into the air pump of the condenser, and thereby a vacuum is produced in the cylinder ; this valve is called the EXHAUSTION REGULATOR.—There is a third regulating valve, called the TOP REGULATOR, placed in the cross pipe at the upper end of the perpendicular steam pipe, which serves to proportion the Quantity of steam to be admitted from the boiler, to the load of the Engine ; so that when the load is less than ten pounds and a half on the inch, the steam in the upper part of the cylinder, which presses on the piston, may be less dense or weaker than the steam in the boiler, and consequently a smaller quantity may be employed to do the work than what is required when the engine is fully loaded. This regulation may be effected in two ways ; either by opening the top regulator fully, at the beginning of the stroke, and shutting it before the piston arrives at the bottom ; or by opening it so far as just to give the piston a sufficient velocity, and keeping it open until the end of the stroke.

LIV. The engine being supposed in motion, the operation of these valves will be as follows, when the piston is at the bottom of the cylinder, and the exhaustion regulator is shut, if the steam regulator be opened, the steam will pass through the perpendicular steam pipe and that regulator from the part of the cylinder above the piston into the part below it, and the steam thereby becoming equally strong or dense above the piston and below it, will give no resistance to the ascent of the piston which will therefore be pulled up by the superior weight at the pump end of the working beam.

When the piston is come to the upper end of the cylinder, the steam regulator must be shut, the exhaustion regulator opened fully, and at the same instant the top regulator opened so far as to admit the proper quantity of steam, (the degree of this opening must be determined by experience) : The steam contained below the piston will then rush from the cylinder through the exhaustion regulator into the vacuum or empty space in the eduction pipe, where it will meet the jet or stream of injection water which will instantly condense or reduce it to water, and thereby exhaust or empty the cylinder of steam.

The steam in the upper part of the cylinder being no longer ballanced by steam below the piston will press upon it by its elasticity, and it will begin its motion downwards ; as the piston moves downwards the steam in the upper part of the cylinder will become less dense than that in the boiler, which will therefore enter the upper part of the cylinder by the opening of the top regulator, and will maintain the steam in that part of the cylinder in a proper degree of density or strength to give the necessary velocity to the piston, and to press it to the bottom of the cylinder ; but if the engine be underloaded it will be necessary to shut the top regulator a little before the piston is at the end of its stroke. It has been observed that the precise time at which the top regulator should be shut must be determined by experience, no certain rule can be given, because it depends upon the degree to which it is opened, and upon the load of the engine at the time ; but it must always be shut sooner than the exhaustion regulator, which is kept open to the end of the stroke.

The injection valve should be opened a little before the exhaustion regulator, that the exhaustion pipe and the water remaining from the last stroke may be cold when the steam enters, by which means the condensation will be performed more suddenly ; and the injection should be shut very soon after the piston begins to descend, observing however to let it play so long that the degree of vacuum shewn by the barometer may be greater in the latter part of the stroke than in the beginning of it. The opening or adjutage of the injection pipe should be proportioned to the load of the engine, so that the proper quantity of water may enter in about one second of time, and as the load encreases, the opening must be enlarged.

LV. The eduction pipe serves to convey the injection water and condensed steam to the foot of the air pump of the condenser ; the injection pipe enters it at its knee, and spouts along the horizontal part of it ; and from its side issues the blowing pipe, the use of which is to empty the eduction pipe of air and water when the engine is put in motion after it has been stopt at any time. At the bottom of the eduction pipe is a hinged valve or clack which permits the water and air to pass into the air pump, but prevents it from returning. This valve should be very tight ; it is called the valve of the eduction pipe foot.

LVI. The AIR PUMP is the lowermost and widest pump of the condenser. When the steam enters the eduction pipe it spoils the vacuum for an instant, and then presses upon the water in the lower part of the eduction pipe, and forces a part of it into the air pump ; as the piston of the cylinder descends, the bucket of the air pump ascends, carries up along with it the hot water which was above it, and leaves a vacuum under it, into which the remaining injection water enters ; first because it stands higher in the eduction pipe than in the air pump ; and secondly because the vacuum in the eduction pipe is not quite so complete as in the air pump.

The water raised by the air pump bucket passes through the clack of the hot-water pump into the vacuum produced by the rising of the bucket of that pump, which is raised at the same time with the bucket of the air pump, and no part of it will come out at the valves on the lid or cover of the air pump, unless the bucket of the hot-water pump is not tight, or an overplus quantity of water enters the eduction pipe or condenser by leaks ; for if there be a sufficient empty space left by the bucket of the hot water pump it is evident that the water will rush into it, and fill it before it can open the valves on the lid, which are kept shut by the pressure of the atmosphere so long as there is any degree of vacuum in the upper part of the air pump, or that part of the hot water pump which communicates with it. When the air pump bucket descends, it leaves a vacuum behind it, because the water is retained by the hot water pump ; and the water in the lower part of the air pump passes through the valves of the bucket, which lifts it up the next stroke as before.

The hot water pump raises the water high enough to let it run into the boiler by the feed pipe, or into a reservoir to be cooled, and so to serve the purpose of injection a second time.

LVII. The barometer serves to shew the degree to which the cylinder is exhausted of air and steam ; it consists of a longer and a shorter tube of iron, both of one diameter and truly bored, and joined together at bottom by a bent iron pipe ; it should be fixed up perpendicular, and should be filled with mercury until it stands 18 inches deep in the shorter or open leg ; a light float of wood something like a gun-rammer should be put into the short leg, and cut off even with the top of the scale when the engine is at rest, and the eduction pipe filled with air ; the scale is divided into half inches, which correspond to inches on a common barometer, because for every half inch the mercury rises in the long leg, it falls half an inch in the short leg, which, added together, make one inch difference of height ; a pipe from the top of the long leg is joined to the eduction pipe, below the blowing valve, for were it fixed higher, steam might come through it and loosen the cement that connects the pipe and the barometer.——When the mercury in the common barometer stands at 30 inches, it should stand at 28½ inches in this barometer, if your engine be in order, or in proportion at other heights.

The steam gauge is a similar instrument, in which the steam presses up a column of mercury proportioned to its elasticity. When an engine is underloaded it ought to be wrought with steam able to support one inch of mercury, and when full loaded it ought not to exceed two inches ; but if the engine be loaded to more than ten pounds and a half on the square inch of the piston, the strength of the steam must be increased accordingly.

It is never adviseable to work with a strong steam where it can be avoided, as it increases the leakages of the boiler and joints of the steam case, and answers no good end.

LVIII. A very important article is the proper packing of the piston, directions for doing which have been already given, (XLVI) but as that part may not come into the engine-man's hands, it is proper to repeat it here : Take sixty white or untarred rope yarns, and with them plait a gasket or flat rope, as close and firm as possible, tapering for eighteen inches at each end, and long enough to go round the piston and overlap for that length ; coil this rope the thin way as hard as you can, lay it on an iron plate and beat it with a sledge hammer until its breadth answers its place ; put it in and beat it down with a wooden driver and a hand mallet ; pour some melted tallow all round ; then pack in a layer of white oakum, about half an inch thick, then another rope and more oakum, so that the whole packing may have the depth of four inches, or only three inches if the engine be a small one ; soak the whole well with melted tallow, and after having beat the packing moderately, lay on the piston leads ; put on the springs and screw them down. In a new engine the piston must be examined after about twelve hours going, and be beat a little and fresh greased ; but you must be careful not to pack or beat it too hard ; otherwise it will create so much friction as almost to stop the engine.

LIX. The buckets of the air and hot water pumps are to be packed with a flat rope, wrapt round them edge ways ; and the ends of these gaskets should be made fast by being drawn through holes made in the buckets, for that purpose, and secured there by wooden pegs hard drove in.——The gaskets should be well smeared with tallow before the buckets are put in, and they should not fit the pumps too tight, as their sticking is very troublesome, especially at first.

The stuffing boxes of the cylinder and air pump are to be packed by wrapping a soft rope round the rod, and beating it in until it nearly fills the stuffing box, remembering to soak it well with tallow as you go on ; above this rope lay on the wooden collar, and screw the gland down upon it moderately tight.

LX. To set the engine a going, raise the steam until the index of the steam gauge comes to three inches on the scale ; when the outer cylinder is fully warmed, and steam issues freely on opening the small valve at the bottom of the syphon or waste pipe, which discharges the condensed water from the outer bottom, open all the regulators ; the steam will then forcibly blow out the air or water contained in the eduction pipe, by the blowing valve, but cannot immediately take place of the air in the cylinder itself ; to get quit of it, after you have blown the engine a few minutes, shut the steam regulator, the cold water of the condenser cistern will condense some of the steam contained in the eduction pipe, and its place will be supplied by some of the air from the cylinder ; open the steam regulator and blow out that air ; and repeat the operation until you judge the cylinder to be cleared of air ; when that is the case, shut all the regulators and observe if the barometer shews that there is any vacuum in the eduction pipe ; when the barometer gauge has sunk three inches, open the injection a very little, and shut it again immediately ; if this produces any considerable degree of vacuum, open the exhaustion regulator a very little way, and the injection at the same time, if the engine does not commence its motion, it must be blown again and the same operation repeated until it moves ; if the engine be very lightly loaded, or if there is no water in the pumps, you must be very nimble and shut the exhaustion and top regulators, so soon as it begins to move quickly, otherwise it will make its stroke with great violence, and perhaps do some mischief. To prevent which, open the top and exhaustion regulators, only a little way, and put pegs in the plug-tree, so that they may be sure to shut these regulators long before the piston comes to the bottom.

If there is much unbalanced weight on the pump end, you must also take care to put a peg in the ladder which guards the steam regulator lever, so as to allow that regulator to open only a little way, and so to lessen the passage for the steam, when it enters to fill the cylinder, otherwise the rods, &c. at the pump end may descend too fast and be prejudicial ; if you find after a few strokes, that the engine goes out too slow, the steam regulator may be opened wider. In order to regulate the opening of the exhaustion regulator, you should have pieces of board of various thicknesses, to put under the weight which pulls it open, by means of which it may be made to open more or less at pleasure, and the top regulator may be managed in the same manner.

LXI. Should the engine work with too great violence on account of its being underloaded, you may correct it by giving the top regulator a lesser opening, and shutting it at such a part of the stroke as will just give the piston sufficient force to come to the bottom. Whenever the top regulator is used, the exhaustion regulator should be thrown fully open every stroke, in order to give a free exit to the steam, on which a great part of the good effects of the top regulator depends.

The engine should always be made to work full stroke, that is until the catch-pins come within half an inch of the springs on each end, which is easily managed by an attention to the pegs. Care must be taken, that the piston rise high enough in the cylinder when the engine is at rest, to spill over into the perpendicular steam pipe any water which may be condensed above it ; for if any water remain there, or in any other part of the cylinder while it is working, it will very much encrease the consumption of steam. When the engine is to be stopt, shut the injection and secure it, put a peg in the plug-tree to prevent the exhaustion regulator from opening, and take out the peg on the other side, so as to allow the steam regulator to open and to remain open ; otherwise you may have a partial vacuum in the cylinder, and it may be filled with water from the injection or leakages, which is a troublesome accident.—The top regulator should also be open while the engine stands.

When an engine is in tolerable good order it will bear to stand ten minutes, and go to work again without blowing afresh, and though it has stood two or three hours, if there has been any steam issuing from the boiler, and no air has been admitted into the cylinder, it will generally go off with once blowing for about a minute.

LXII. If you find, after following the above directions, that the engine does not go to work, shut the exhaustion regulator, and give some injection, if it then makes no vacuum, it is likely there are air leaks about the eduction pipe ; if it does make a vacuum, which remains but a short time, it may be owing either to air or water leaks, these may be distinguished by blowing as before, and shutting the lower regulator for about a minute, without giving any injection. If upon opening it again, it throws out a good deal of water at the blowing pipe before it blows steam, it is certain that it either has some leak in the condenser under water, or that the injection or blowing valve does not shut close, every joint should be examined, and also the valve at the foot of the eduction pipe.

If after blowing as before, you find that immediately on opening the exhaustion regulator, a quantity of air is thrown out at the blowing valve, the leak is in the eduction pipe some where between the surface of the water in the cistern and the nozzle. The particular place of these leaks may be found, by emptying the cistern of water, putting three or four feet deep of water into the hot water pump, and staking down the blowing and injection valves with those on the air pump lid ; then if steam is admitted into the eduction pipe, it will come out at the leaks and point them out.—If not found out in this way, apply the brake to the air pump, taking care first to put some water on its bucket, and then by working that pump by hand, you will probably on an attentive examination observe where air goes in, which may be known more distinctly by wetting the place suspected.

If upon shutting the lower regulator and making a vacuum in the exhaustion pipe by pumping, or by injection, you find that vacuum continues good for a considerable time, then the fault does not lie in the eduction pipe, but in the nozzle or joint of the cylinder bottom, where it must be sought for.

In these examinations by pumping it is proper to take off the bonnet or cover of the exhaustion regulator, and to examine if air enters at that regulator, if it does, and only in small quantity, throw some water on the regulator while you are examining the eduction pipe; and when the leak is suspected to be in the bottom joint of the cylinder, or in the lower nozzle, you must throw some water on the steam regulator and also on the piston, then by pumping and strict examination you will soon find where the air enters. When you are examining the tightness of the piston by pumping, you must stake the beam, so that the piston may not descend.

LXIII. If in course of working, you do not find the vacuum keep good, and the engine goes sluggishly, or stops and requires to be blowed frequently, you must examine whether an uncommon quantity of air or water issues at the hot water pump, or if any comes out at the valves on the air pump lid; if the quantity of air is great, the engine has some air leak, and if the quantity of water be great, and is rather cooler than usual, it proceeds from a water leak in the condenser; if the quantity of water be great, and at the same time very hot, it proceeds from a bad piston, or from the steam regulator not shutting close.

The engine will also go badly if the air pump or water pump buckets or clacks strip the water, that is let it pass by them; you will know if this be the case with the water pump bucket, by observing whether the water follows down after it at the return of the stroke, and leaves a part of the pump empty; if it does not, either the bucket strips the water, or the engine receives water in some way which it ought not.

LXIV. Attention ought to be given to feeding the boiler in a regular manner, that it may not be spoiled, nor steam wanted. When there is too much water in the boiler, the engine will not work regular, and if there is too little, the sides of the boiler will be burnt by the flame in the fiues.——If by accident it should at any time run a little too low, the feed should be augmented, so as to fill it gradually; for if you run in too much at once, you will check the steam and stop the engine; but if it be run very low, stop the engine, open the puppet clack, and fill the boiler from the pool or reservoir if you have one; otherwise fill it by working the air pump, having first staked down the valves on its cover, and opened the injection valve.— In working the engine the steam ought to be strong enough to make the index of the steam guage stand half an inch high at least, otherwise air will enter at the joints of the boiler, &c. and spoil the vacuum, so as to cause a good deal of trouble to get quit of it again. Therefore if you perceive the steam guage to be lower, stop the engine until it rises again. By a little attention, you will find the proper opening of the feeding cock for any rate of working.

LXV. Let all the coals employed to feed the fire, be thoroughly watered just before they are thrown on, as that will prevent their being swept into the fiues by the draught of the chimney.

The fire should be kept of an equal thickness and free from open places or holes, which are extremely prejudicial, and should be filled up as soon as they appear; if the fire grows foul and wants air by clinkers collecting on the bars, they must be got out with a poker, but the fire should be as little disturbed in that operation as possible, and the greatest care taken not to make any coals or coaks fall through, which are not thoroughly consumed; it is very common for a fourth of the whole coals to be wasted in this manner, by mere carelessness. When the fire is newly made, the damper should be raised a little, so as to let off the smoke freely, but should be let down to its proper place so soon as the smoke is gone off. The air door in the chimney should be always open more or less; it prevents the flame from being sucked up the chimney, and very considerably increases the effect of the coals. Once a month, the boiler and fiues ought to be cleaned, or oftener if the water be very subject to incrust the boiler. Every morning the ashes ought to be taken out, the engine house swept clean, and a view taken of every part of the engine, to see that nothing be working out of its place, or want oiling. Particular attention ought to be paid to the bolts and cutters of the great chains and piston rod, so that none of them get loose.

LXVI. Once every week let the top of the cylinder be taken off, and also the springs and leads of the piston; let the packing be beat down moderately, with the driver and mallet, and fresh oakum, or a gasket added when necessary. For every foot the cylinder is in diameter, pour two pounds of melted tallow on the packing, before you put in the leads, and for two or three hours after you have added the tallow keep the piston from rising quite to the top of the cylinder, by laying two pieces of wood three inches thick on the outside springs, that the tallow may not be spilt off before it has time to soak into the packing. At the same time you pack the piston, you should examine the state of the condenser, and rectify any thing you find amiss ; and while these things are doing the pitwork should not be neglected, that one stoppage may serve for all.

LXVII. The regulator valves should be examined from time to time, and a little fresh oakum should be lapt about the necks of their spindles to keep them air and steam tight. The stuffing-boxes also should be minded, and no steam suffered to escape any where ; its escaping is a mark of slovenliness, and a material injury both in extra-consumption of coals, and in the destruction of the iron and wood-work.

An engine, when in good order, ought to be capable of going so slow as one stroke in ten minutes, and so fast as ten strokes in one minute ; and if it does not fulfil these conditions, somewhat is amiss that can be remedied.

The hot water should issue of the heat of 96 degrees of Fahrenheit's thermometer, that is blood warm, when the engine is in excellent order, and should never exceed the heat of 110 degrees, unless when the injection or cold water is hotter than 70 degrees, and in that case the vacuum will not be good.

LXVIII. At the end of the horizontal steam pipe next the boiler is fixt the STEAM REGU-LATOR, the use of which is to shut off the steam while any thing is doing about the top regulator, or other parts connected with it. It may also be used to stop the communication with one boiler, while another is in use.

LXIX. At the first setting an engine to work, it frequently happens that there is a difficulty in procuring a sufficient quantity of cold water for condensation, and it also frequently happens that there is something or other amiss, which may occasion the engine to be a long time in setting to work, and by the repeated blowing, the water in the cistern gets too hot to serve for condensation. In such cases a great deal of trouble may be saved by exhausting the air from the cylinder by working the air pump by the brake, having first opened the exhaustion regulator and shut the steam one.—And in any case when the engine does not go readily to work by blowing, and the quantity of injection water is limited, it is best to set on by pumping, and even to assist the engine for a stroke or two by the same means, if it be fully loaded.—As the bucket of the air pump ascends, you must hook the chain of the pump break to a lower part of the pump chain, by which means you can keep pumping until the engine has made its full stroke.

LXX. To make putty for making or repairing the joints. Take whiting, or chalk finely powdered, dry it on an iron plate, or in a ladle, until all the moisture be exhaled ; then mix it with raw lintseed oil, and beat or grind it well, adding more oil or whiting, until it be of the consistence of thick paint, and perfectly free from lumps or inequalities.

For some purposes, where the putty is wanted to dry and to be very sticky, use painter's drying oil, which is made by boiling the oil with a small quantity of litharge or red lead.

Where the putty is wanted to continue always soft, mix about two ounces of butter, or common sallet oil with each pound or pint of the lintseed oil : This soft putty is principally useful in the caulked joints of the eduction pipe, above water. N.B. White lead will not answer in place of the whiting.

No wet cloaths should be suffered to be laid on the cylinder, boiler or steam pipes, and every part containing steam should be guarded as much as possible from the influence of cold air or water.

The proper grease for the piston and cylinder stuffing box is melted tallow, and for the chains, gudgeons, &c. common Spanish olive oil (called sallet oil) which for some uses may be thickened by dissolving some tallow or butter in it, by means of heat.—Lintseed oil should never be used as grease, as it dries and creates more friction than would have been without it.——Hogs lard, or train oil, if applied any where about the cylinder, or where it is hot, will thicken like lintseed oil.——When the oil or grease about the great chains, or any of the working parts, grows clotted or very thick, it should be scraped off before new grease is added.

ADDITIONAL DIRECTIONS.

The Numbers denote the Paragraph to which they correspond.

VI. AS the whole weight of the great beam, and also of the power to be exerted, is supported by the plummer blocks, care must be taken that they stand firmly on the spring beams, and that the latter be well supported from the lever wall.   To do which, wherever the building is made of bricks, or of indifferent stone work, form the bottom of the opening, under the beam, of three planks of oak, or of the best deal, six or eight inches thick, and twelve or fourteen inches wide.   These planks must reach at least four feet into the walls at each side of the opening, one of them must be laid in the line of the outside of the wall, another in the line of the inside of the wall, and the third, which should be the strongest, in the middle, right under the gudgeon.   Upon these planks, at each side of the opening place three others of the same dimensions upright ; let their upper ends reach to the upperside of the spring beams, and let the spring beams be let into the uprights, so that only two inches of their thickness shall project beyond the face of the spring beams, and that the remaining four inches of the thickness of the uprights shall form a shoulder under the spring beams, which will support them firmly under the sides which are next the beam, where it is most necessary ; for were the insides of the spring beams or plummer blocks to give way to the pressure, and the outsides to be supported, the gudgeon would rest on its points, and by the leverage it would gain thereby, might be broken.   The lower ends of these six uprights may have small tenants to fit mortices in the sills, which will prevent their slipping.

IX. *Page* 3. The holes through the great beam for the screw bolts of the martingale tails should be quite easy for them, otherwise the screws will be broken if the logs of the beam come to slide upon one another.   The keys to prevent the logs from sliding upon one another, are best made of pieces of very dry and hard oak, two inches thick, six or seven inches broad at one end, and four or five inches broad at the other end ; their length being suited to the thickness of the beam.

XVI. *Page* 5. In large engines, where the condenser pumps are consequently heavy, it is found proper to make the bottom of the condenser cistern of planks five inches thick.

XVIII. *Page* 6. An improvement has lately been made in the covering boiler tops.   The setting being built up to nine inches above the flues as usual, a course of horse or cow dung, three inches thick, and well beat, is applied to the boiler top ; on the outside of that is laid some good lime mortar, about an inch in thickness, to which is applied a course of bricks flatwise, with their ends upwards ; on the outside of that another course of bricks (also laid in good mortar) in the same position, but so as to break joint with the first course ; in which manner the covering is carried on until the whole top is covered, taking care to leave an opening for the man hole : every flanch may be thus covered, and when well done, it effectually makes the top steam tight, and also defends it from cold and rain, so that a boiler house is not necessary.   The mortar employed must be such as stands water.

XIX. *Page* 6. The valve put into the boiler feed-pipe, to prevent boiling over, is best fixt in its upper end, so that it may be taken out when any material is wanted to be introduced into the boiler by the steam pipe.   The proper valve for this purpose is one of the kind which are used for the injection, and blowing pipe, which must be put into the feeding pipe, in an inverted position.

XXIII. *Page* 7. Instead of using painters drying oil to make the joints with, take good raw or unboiled lintseed oil, put it in an iron pot, place it over a gentle fire, (out of doors, but protected from rain) let it be watched as it heats, as it is very liable to boil over ; when it boils make the fire more moderate, but continue to heat the oil, until upon dropping some of it upon a cold stone or piece of iron, you find it is, when cold, of the thickness of thick tar or treacle. The pasteboards for the joints are to be soaked in this oil warm, or painted over with it, and laid in a hot place to suck it up ; and it is also to be used to make the putty with.

XXVII. *Page* 8. Instead of putting a gland across the bottom of the piston rod, to prevent it from dropping, it is better to drill two opposite holes through the cone of the piston, and one inch each into the cone of the rod ; two iron pins put into these holes will effectually keep the rod in its place. There should be a groove about a quarter of an inch deep, and half an inch wide, cut round the base of the cone on the rod below these pins, which grooves being lapped round with rope yarn and putty, will serve to prevent steam from getting through the piston by the sides of the pins. To make it more easy to get these pins out, they should have flat tails bent upwards, so as to lie close against the outside of the cone of the piston when the pins are in their places ; and to secure them there, mortices must be cut in the wood which fills the hollows of the piston, to which must be fitted wooden wedges, made *very* tapering, by driving which down, the tails of the pins will be prest against the cone, and the tapering form of the wedges will make it easy to dislodge them when the pins are wanted to be taken out. It is necessary to observe, that the pins should be fitted tight into the holes in the piston cone, and that the holes into which their points enter in the cone of the rod, should be made easy for them, otherwise they might prevent the one cone from being pulled far enough into the other.

XXIX. *Page 8.* The oakum with which the joints are caulked, should be well smeared with the strong or thick boiled oil, mentioned in these additional directions. If the under side of the pipe of the inner bottom does not fit close to the lower edge of the opening made for it in the outer bottom, that is to say, if the space left there for pasteboard or caulking be wider than one quarter of an inch, a piece of hammered iron an inch and half broad must be forged of such thickness as to fill up the space so as to make it tight by the help of a thickness of pasteboard above it, and another below it. Lead ought not to be used in these cases, as its expansion and contraction by heat and cold, are too great. Instead of putting a prop from the nozle to the ground, it is found better to put a balance beam off sideways under the floor, with a short upright having a flat end to take a broad bearing under the nozle. The weight of the balance should not support more than two thirds of the weight of the nozle.

XXXI. *Page 9.* The lower ring on the cylinder, to which the steam case is fixed, is some-times made with a projecting flanch, on which the steam case rests, and the joint is then made tight by caulking between the flanch of the steam case, and that on the ring.

XXXII. To avoid the inconvenience of the perpendicular steam pipes occasionally proving too short, they are now made without any flanch at the lower end, and a socket is cast upon the nozle to fit them, in which they are to be made tight by caulking.

The weight of the upper nozle must be supported by a prop from the cross piece between the cylinder beams. And if the boiler steam pipe be very long, and consequently heavy, part of its weight should be supported by a balance beam near the wall of the house.

XXXV. *Page 10.* The best way of making the standing joints of the condenser, is by means of rings of lead a quarter of an inch thick, as broad as the flanches, and pierced for all the screws. They may either have putty made with the thick oil put on each side of them, or, for greater security, they may be covered with Russia duck and putty. In other respects proceed as directed in xxxv. The soft rope does not answer well.

XXXVI. *Page 10.* In addition to this article, see the explanation of plate xiv.

XXXVIII. *Page 11.* Where the joints of the eduction pipe are made with flanches, they are to be fixed together by means of strong flat rings of iron, put on each side of them, as has been directed for that at the nozle and the joints are to be made tight by pasteboard and putty ; for, on account of its expansion, lead will not answer where it is subjected to be alter-nately hot and cold.

XXXIX. *Page 12.* The hot water pump must be fixed down by means of two long bars of iron with screwed ends, which go through the bottom of the cistern, and extend upwards through two of the holes of the lower flanch of the hot water pump. One of these bars is shown in the drawing of the front of the engine house.

XLII. *Page 13.* The guide posts may be fixed upon a sill passing from one to the other ; and the best way of fixing the weight of the exhaustion regulator is to make it in the form of a saddle, moveable at discretion, upon a beam centred at the further guide post, so that the beam may fall flat upon the sill when at lowest, and the saddle will produce the effect of a greater or lesser weight, according as you place it farther from the centre or nearer to it.

The door of the condenser may be converted into a window, and a seat for the engine man, as soon as the condenser and eduction pipe are fixed.

XLV. Some people use a plaited rope to make the joint of the cylinder lid, which is a bad practice ; for though a plaited rope may make a joint apparently steam tight, yet it has been found by experience, that such joints are not air tight ; but when, by the working of the top regulator, a partial vacuum is produced in the upper part of the cylinder ; they permit some air to enter imperceptibly, and without noise, which in course passes to the condenser ; and by persons that are not aware of this circumstance, may be thought to enter at some air leak in another place. *We therefore recommend that this joint be always made with pasteboard and putty ; and that a strict attention be paid to the tightness of the stuffing box, wherever the top regulator is used.*

XLVI. *Page* 23. The proper quantity of tallow to grease the piston, is two pounds every week for every foot the cylinder is in diameter. But where opportunity can be obtained of adding it more frequently, the whole quantity ought not to be added at once, but divided according to your opportunities. When the top regulator is used, if the tallow is put into a flat funnel which ought to be made to surround the piston rod above the cylinder stuffing box, it will be gradually sucked in without the trouble of taking off the lid.

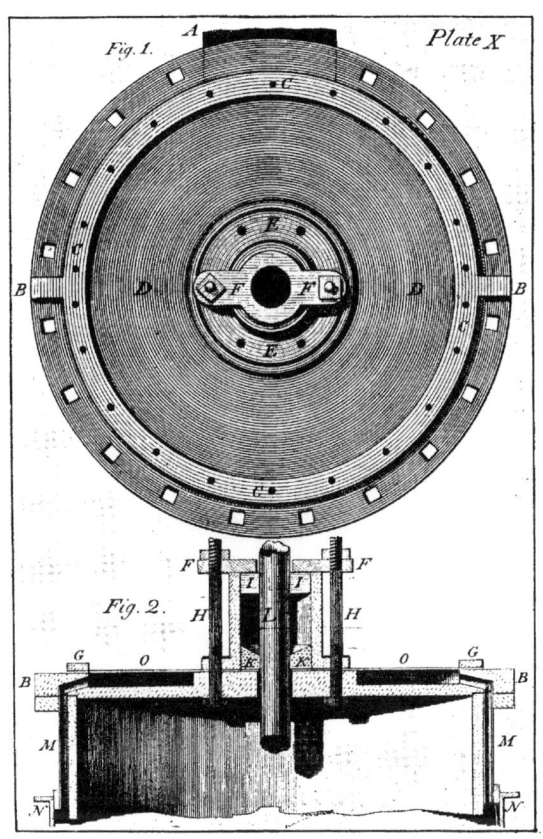

XCIX. Directions for erecting and working the newly invented
Steam Engines [1779]

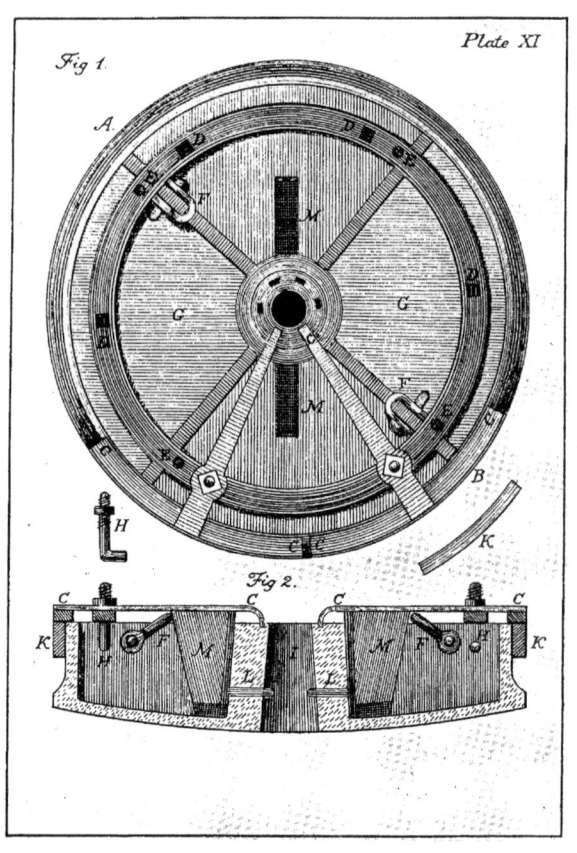

Plate XI

Fig 1.

Fig 2.

C. Directions for erecting and working the newly invented
Steam Engines [1779]

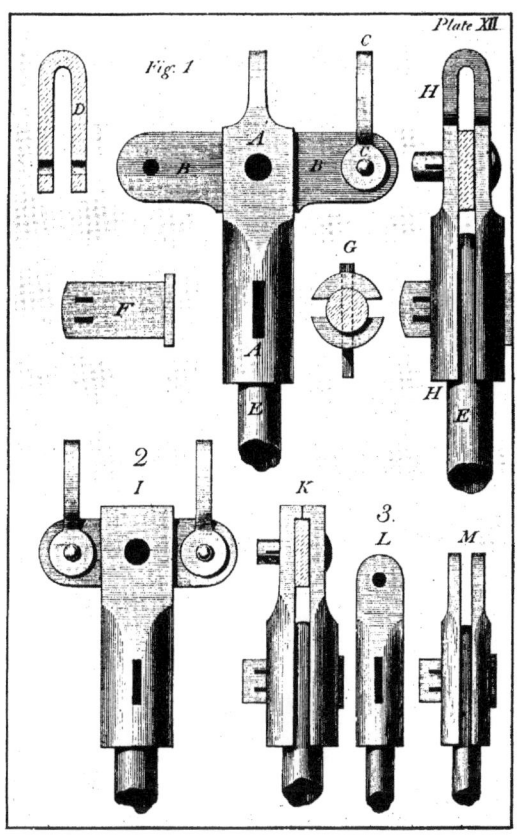

Plate XII.

CI. Directions for erecting and working the newly invented
Steam Engines [1779]

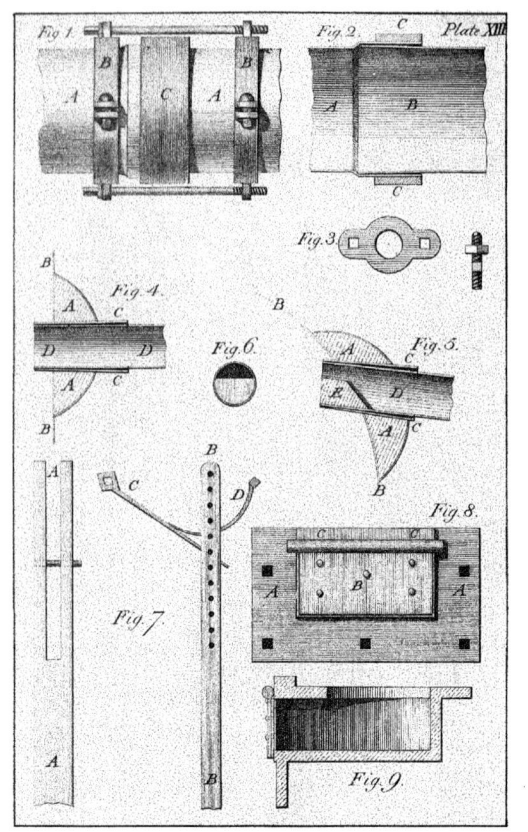

CII. Directions for erecting and working the newly invented
Steam Engines [1779]

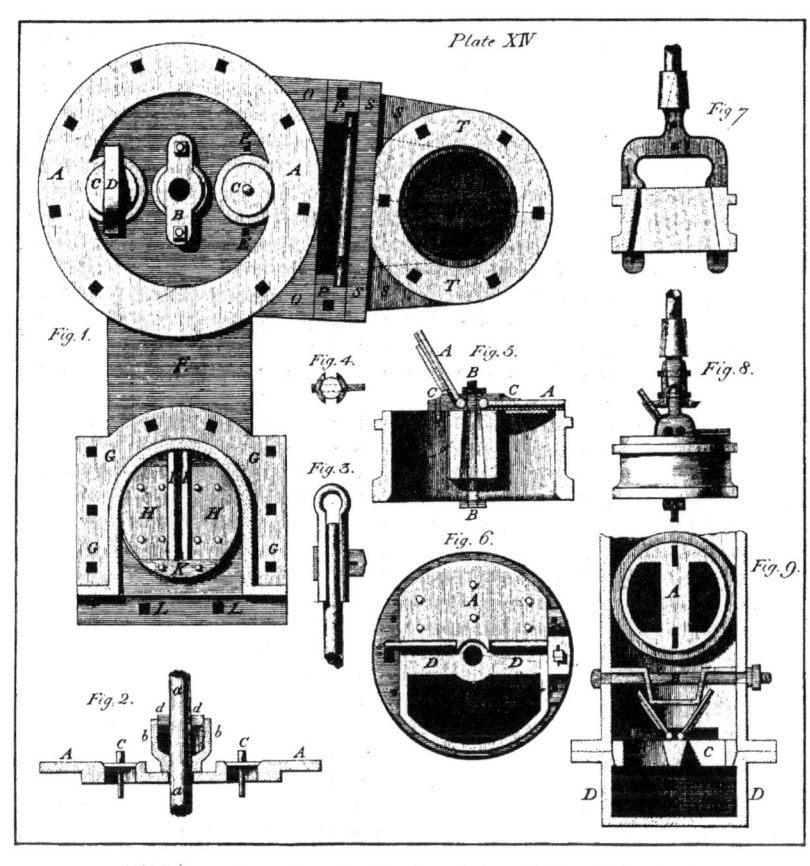

Plate XIV

Fig. 1.

Fig. 2.

Fig. 3.

Fig. 4.

Fig. 5.

Fig. 6.

Fig. 7

Fig. 8.

Fig. 9.

CIII. Directions for erecting and working the newly invented Steam Engines [1779]

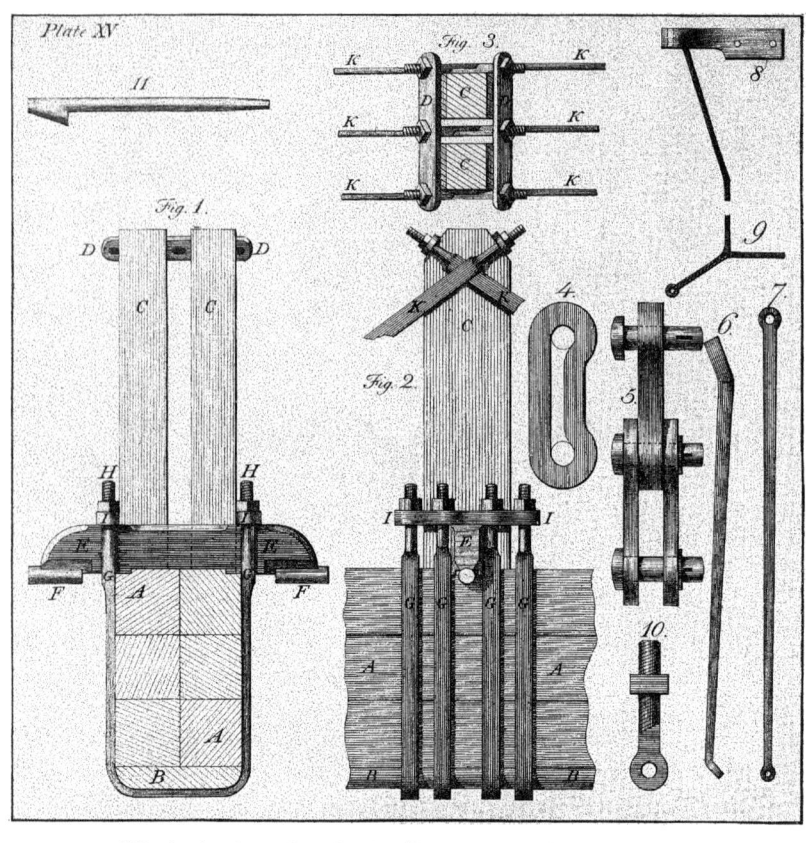

CIV. Directions for erecting and working the newly invented Steam Engines [1779]

APPENDIX III

DIRECTIONS FOR WORKING
ROTATIVE ENGINES c. 1784

DIRECTIONS RELATING TO THE ENGINE

I. Everything to be kept as clean as possible.

II. When the parts are oiled or greased, the old scurf must be taken off as much as possible, before the fresh grease or oil is applied.

III. The working gear ⎫
IV. The parallel motion ⎬ Should be oiled every twelve hours of working.
V. The rotative motion ⎭

VI. The regulator valves should be examined and cleaned once a week, if the Engine goes night and day ; or once a fortnight if it goes in the day time only.

VII. Care must be taken to put the valves in with the same side towards you, as they were taken out with.

VIII. The spindles of the valves should be examined when the valves are, and the lapping repaired if necessary.

IX. The piston should be packt once a week, and *seventy* [1] ounces of tallow melted each time, and poured upon the packing before the upper gasket is put in ; but if the packing be wholly new, as much tallow should be put in as will soak up ; and so much packing should be put in as will keep the under side of the ring from touching the upper side of the piston, when the ring is screwed down.

X. When the piston is packt, the sides of the top part of the cylinder should be scraped to take off the old grease, putty, dirt, &c. that is generally retained there ; and the greatest care should be taken that there remains no dirt of any sort, either on the piston, or in the passage to the upper valve.

XI. In packing the stuffing box of the cylinder, so much packing must be put in as to prevent the gland from touching the top of the stuffing box ; otherwise there will be no certainty that the box is tight ; in which case, either air will pass through it, and injure the vacuum, or much tallow will be wasted, and will not cure the evil.

XII. About 2 *ounces* [1] of tallow may be put in the cup of the stuffing box once in six hours, which will be enough, when the box is tight, to keep the packing elastic.

XIII. The air pump bucket should be examined every month, and new packt if it require it.

XIV. The brasses of the parallel motion should be frequently examined, and the cutters driven up *with a small hammer*, two or three times a day ; as any shake there causes noise and bad work, and *vibrations* in the millwork ; and the same attention should be paid to the brasses of the connecting link, for the same reason.

XV. The brasses in the connecting link must be packt up with thin bits of iron or copper, from time to time as they wear, so as to prevent the teeth from bottoming, or the edges of the rims upon the sides of the wheels from touching in any part of their revolution.

XVI. The brasses at the top of the connecting rod should be examined, and any shake that may arise prevented by cuttering them up.

XVII. The sliders that shut the hands of the working gear, should be so adjusted as just to let the catch slip into its place at each end of the stroke ; and the catch should be pressed by a *weak* spring to make it act quickly.

[1] These words are written in by hand to suit the particular size of engine involved.

XVIII. The catch should be so trimmed as to disengage as finely as possible, that is as near the end of the stroke as can be : Much of the smoothness of going depends upon that circumstance.

XIX. The boiler should be cleaned at least once a month ; but if the water be muddy or scurfy, more frequently ; as it will otherwise not only be liable to destruction by burning, but will likewise require more coals : *Two evils to be carefully guarded against.*

XX. The *water* in the *boiler* should be kept as *nearly* of the *same height* as possible ; as carelessness in this point may cause the most sudden destruction of the boiler, and the consequent stoppage of the works (where there is not a spare one ready to supply its place).

XXI. The flues round the boiler should be cleaned once in six weeks.

XXII. The coals when of that sort which does not cake together, should not be of the small kind, but of the lumpy, and the quantity thrown in at a time should be rather small than much, and more frequent the times.

XXIII. The coals, if they do not cake, should not be heaped upon the grate, but should be of an uniform thickness, 6 or 8 inches thick.

XXIV. The weight upon the safety valve should never be so much as to raise the index in the steam gage, in any case, more than three proper inches, or 6 upon the gage, as too great a strain will be brought upon the boiler unless this circumstance be attended to.

One of the copies of this sheet of directions in the Boulton and Watt Collection has a series of facetious notes, evidently written by one of the young men in the drawing office. We reproduce these notes as an early example of engineering humour, and have numbered them to correspond to the paragraphs against which they are written.

II.   ' The engine man to butter his bread with what comes off.'

VI.   ' And once a year if not going at all.'

VII.   ' And the same side uppermost.'

X.   ' Don't leave your hat, coat or shoes in the cylinder.'

XIII.   ' Old wigs and dog's tails are better than hemp.'

XVII.   ' A brick end tied on with string is an excellent substitute.'

XVIII.   ' And much black grease.'

XX.   ' And hasty general holiday.'

XXI.   ' And the man to have a quart of ale for the job.'

XXII.   ' The engineman to be allowed his pockets full every day for taking home.'

XXIII.   ' Lumps more than this thickness to be broken in two.'

XXIV.   ' Blow down the gage cocks to find if water is right height.'

APPENDIX IV

ACCOUNT OF THE WATT CENTENARY COMMEMORATION

IT has been mentioned already in the Preface that this Volume owes its inception to the Committee that was set up to commemorate the centenary of James Watt's death. It is necessary, therefore, to give, however briefly, a history of the movement and what it achieved.

The idea that the centenary was a proper occasion for a commemoration had occurred to the minds of several persons both at home and abroad, and schemes were being thought out when the Great War broke out and distracted the attention of every one. Undeterred by this fact that when the Armistice was signed in November 1918 the time available before the centenary—August 19, 1919—was all too short, there were yet enthusiastic spirits who determined that the occasion should not pass unnoticed.

A few engineers in the city of Birmingham, appropriately enough, were the first to take steps in the matter, and it is with their activities that we are here concerned. It should be mentioned, however, that in other places, notably Scotland, the idea was taken up with energy, and that important results were the outcome.

The Birmingham engineers issued invitations to persons in the Midlands likely to be interested, and the conference was held on February 27, 1919, at the Birmingham Chamber of Commerce. A provisional committee was elected with Mr. (now Sir) William Mills as Chairman. After much discussion a draft scheme for the fitting commemoration of the centenary, embodying the principal suggestions put forward at the meeting, was drawn up.

Briefly the scheme comprised :

1. The collection of funds for the erection of a Watt Memorial Building in Birmingham, as being the place where Watt brought his inventions to fruition, and where he spent the greater part of his life.

2. The provision of endowment at Birmingham University for a Watt Chair in some branch of engineering, particularly connected with the trend of Watt's inventions.

3. The publication of a Memorial Volume.

With this scheme in hand a meeting was held in the Council Chamber of the Town Hall, Birmingham, presided over by the Lord Mayor, Sir David Brooks, and supported by upwards of 200 gentlemen representing the engineering profession, scientific institutions, and the public services of the country. A resolution adopting the scheme, broadened however to assume an international character and to include a week of commemoration meetings, was moved by Sir Oliver Lodge, seconded by Sir William Mills, and carried unanimously. A representative committee was thereupon set up to carry out the scheme. Sir William Mills was elected Chairman and Mr. R. B. Askquith Ellis the Hon. Sec. of this body. The Birmingham Chamber of Commerce generously placed office accommodation at the disposal of the Committee.

At a meeting on May 20 several sub-committees were set up to further the hundred and one activities that such an important occasion involves. Space does not admit of giving the names of these gentlemen, and they were many, who gave time and energy voluntarily to the prosecution of the work.

COMMEMORATION PROCEEDINGS

Other meetings prevented the Commemoration being held on the actual anniversary of Watt's death, and September was decided upon, especially as it gave a few weeks longer for the preparations.

FIRST DAY'S PROCEEDINGS

The proceedings opened on Tuesday, September 16, in the morning with a civic welcome by the Lord Mayor at the University Buildings in Edmund Street to an assembly of guests from all parts, among whom were descendants of both of the Watt and of the Boulton families.

A number of papers specially written for the occasion, dealing with different aspects of the life-work of James Watt and its world-wide results, were presented. Prof. H. F. W. Burstall, M.A., who was first called upon, took as his subject ' The Rise of Engineering Manufacture '.[1]

At the conclusion of the address a telegram from the Rektor and Senate of the Norges Tekniske Höjskole, Trondhjem, was received asking the Committee to accept from that body an illuminated address that had previously been forwarded.

The Lord Mayor vacated the chair, which was taken by Sir William Ashley, Vice-Principal of the University. He introduced the next speaker, Dr. H. S. Hele-Shaw, who discoursed upon ' James Watt and Invention '.

In the afternoon an impressive memorial service was held in the Parish Church of St. Mary, Handsworth, where Watt's remains are deposited, and where the famous statue of him by Chantrey is enshrined in the Memorial Chapel. Special prayers were offered by the Rector, Canon H. G. Daniell Bainbridge, M.A., and an anthem, composed for the occasion by Dr. A. J. Silver on the words ' Let us now praise famous men ', was sung.

An impressive pause ensued while Major J. M. Gibson Watt, representing the family, and the Lord Mayor with Sir William Mills on behalf of the Committee, laid laurel wreaths on the plinth of the Watt Statue.

Following this ceremony, Canon E. W. Barnes, M.A., F.R.S. (subsequently Bishop of Birmingham), pronounced a scholarly and sympathetic eulogy on Watt, which was listened to with very close attention. At the conclusion of the service, an opportunity was afforded to every one present to inspect the famous monument.

The guests then made their way to Heathfield, Watt's old home hard by, where, by kind permission of Mr. George Tangye, a garden party was held. An opportunity was taken to secure a photograph of those present with the old house as background. During the course of the afternoon those of the guests who wished to do so were given the opportunity of climbing the back staircase to take a peep at the garret workshop over the kitchen where, after his retirement from active business, Watt spent many an hour engaged in mechanical pursuits.

In the evening a reception was given by the Lord Mayor and Lady Mayoress in the Council House, Birmingham. The adjoining Art Gallery was thrown open, and in the gallery of the Industrial Hall a Loan Collection of Boulton and Watt relics had been brought together. A printed catalogue of the objects, prepared by Mr. H. C. James-Carrington, was available.

SECOND DAY'S PROCEEDINGS

On Wednesday, September 17, the proceedings opened in the morning by a gathering at the University Buildings, Edmund Street, under the chairmanship of Mr. C. H. Wordingham, C.B.E., to hear a lecture by Sir Oliver Lodge on a subject that he has made specially his own —' New Sources of Energy '. A joint paper by Emeritus Professor Archibald Barr, D.Sc., and Regius Professor J. D. Cormack, C.M.G., D.Sc., of Glasgow University, on the ' Model of the Newcomen Engine repaired by Watt ' was also read.

During the interval between these lectures, representatives from Australia, France, Japan, and the United States were welcomed.

The afternoon of Wednesday was devoted to visits of inspection to some of the Boulton and Watt engines that still remain in the Birmingham district; for this a service of motor-cars had been arranged.

The oldest engine is that of the Smethwick pumping station of the Birmingham Canal Navigations, 1777; this is now at Ocker Hill, Tipton, whither it was removed in 1898 and re-erected in working order. Parties of visitors were shown round by Mr. A. W. Willet, Chief

---

[1] Space does not admit of giving a report of this and succeeding papers and addresses, but an account will be found in the engineering press at the time, cf. *Engineer*, 1919, II. 277, 309, and *Engineering*, 1919, II. 385, 415.

Engineer to the Navigations, and his staff. The engine was meanwhile worked under steam and with the aid of an original Boulton and Watt indicator, diagrams were taken at intervals by Prof. Burstall, assisted by students from the Engineering Dept. of the University.

Another engine, that belonging to the Warwick and Birmingham Canal Co., 1796, at Bordesley, was shown by permission of the directors. Its duty, till 1884, when its place was taken by fresh plant, was to pump water from the Saltley Canal into the higher reaches ; the cylinder is 46 in. diam. by 8 ft. stroke, and the pressure was 10 lb. per sq. in.

The third engine, that at Lawley Street, the property of Birmingham Canal Navigations, dates from 1817 and is still in use.

In the evening of Wednesday the Commemoration Dinner was held in the Grand Hotel. Guests to the number of upwards of 300 were received by the Lord Mayor and Lady Mayoress, supported by Major and Mrs. Gibson Watt. After the loyal toast had been duly honoured, the Hon. Sec. read out telegrams from engineering institutions and learned societies wishing success to the Commemoration. The oratory that followed reached a high level, and not least in the speech of the American Ambassador, the Hon. J. W. Davis, when proposing the City of Birmingham.

THIRD DAY'S PROCEEDINGS

The first visit on Thursday, September 18, was to Soho Foundry, now in the possession of Messrs. W. & T. Avery, Ltd., where the visitors were welcomed by the Managing Director, Mr. G. C. Vyle, and his staff. The parts of the works associated with Watt, Boulton, and Murdock were shown : i. e. the gate house ; the old mint with its barred windows (now a smithy) ; a wall planing machine ; an old lathe capable of taking work 26 ft. diam. ; the house where Murdock lived till he removed to Sycamore Hill in 1817 ; the early gas holder associated with his introduction of coal gas lighting ; and lastly some underground passages and chambers the use of which is not fully understood.

At noon a Degree Congregation of the University of Birmingham was held in the Great Hall of the Birmingham and Midland Institute. Degrees of Doctor of Laws *honoris causa* were conferred upon :

His Excellency the Hon. John William Davis, Ambassador of the United States of America to the Court of St. James's.

Emeritus Professor Archibald Barr, M.A., University of Glasgow.

Engineer Vice-Admiral Sir George Goodwin, K.C.B., R.N., Engineer-in-Chief of the Fleet.

Prof. Auguste Courelle Edmond Rateau of the Académie des Sciences, France.

Sir George Thomas Beilby, F.R.S., Director of Fuel Research in Great Britain, representing the Royal Society.

William Cuthbert Blackett, C.B.E., President of the Institute of Mining Engineers.

Frederick William Lanchester, M.Inst.C.E.

The afternoon of Thursday was devoted to visits to the engines alluded to above and to excursions as far afield as Stratford-on-Avon.

On Saturday afternoon a procession and demonstration organized by a Workers' Committee to pay the Craftsman's tribute to the genius of James Watt took place. The workmen marched through the principal streets of Birmingham to the Botanic Gardens, Edgbaston. Pride of place at the head of the procession was given to a model, half full size, of Watt's rotative engine, 1788, known as the 'Lap' engine.[1] This model had been built in the remarkably short space of four weeks by co-operation among the pattern makers, members of their trade society employed by several well-known firms. The model was carried on a lorry, and excited great interest from the fact that it was shown in motion. Following the model came the banners of the principal trade unions, and these again were succeeded by trade exhibits sent by a number of prominent Birmingham engineering firms.

Arrived at the Botanic Gardens, the lorry with the model was drawn up in an open space. Using the lorry as a platform, Eng. Comm. Edgar C. Smith, O.B.E., R.N., delivered a lecture on ' The Development of the Steam Engine ' with special reference to Watt's inventions. In the afternoon, in the Exhibition Hall, Prof. Burstall lectured on ' The Life and Work of James

[1] See p. 221.

Watt '. In the evening Prof. F. C. Lea, M.Inst.C.E., lectured in the same hall on ' James Watt and his Influence on Everyday Life '.

The Loan Collection of Boulton and Watt relics at the Art Gallery was opened to the general public on September 17, and remained on view for several weeks.

SUBSEQUENT PROCEEDINGS

The Commemoration over, attention was concentrated by the Committee on their other aims. The Memorial Hall and the Watt Chair of Engineering, if established in a fitting manner, would require, it was thought, about £100,000, and this was the sum aimed at. For eighteen months a staff, with Mr. J. F. Chambers as Secretary, was maintained, carrying on propaganda for funds; but owing to the economic situation that supervened in the country, it became evident that nothing like the desired sum would be obtained, so that early in 1921 the Committee decided to suspend its activities. Only about £11,000 was in hand. The circumstances were laid before the subscribers, and in conformity with the replies received, the sum of £5,000, most of which had in fact been earmarked for that purpose, was handed over to the University of Birmingham, and has been used to establish a James Watt Fellowship in Engineering. After setting aside a sum for the Memorial Volume, which needs no mention as these presents witness, it was decided to place the balance in trust until such time as the funds could be applied to the fulfilment of the purpose for which they were subscribed ; to secure the funds a trust deed has been executed. The following, of whom the first named acts as Chairman, were elected Trustees :

> Sir William Mills.
> Sir Gilbert Barling, Bart., C.B.
> R. A. Chattock, M.I.E.E.
> John W. Hall, M.Inst.C.E.
> J. D. Watson, M.Inst.C.E.
> R. B. Askquith Ellis, A.M.I.Mech.E.

Mr. Ellis acted as Honorary Secretary to the Trustees but resigned his trusteeship in 1925, and the vacancy so created has not yet been filled. For the time being Mr. Hall is acting as Honorary Secretary, and can be addressed at 25 Temple Row, Birmingham.

INDEX

PRINTED IN ENGLAND AT THE
UNIVERSITY PRESS, OXFORD
BY JOHN JOHNSON
PRINTER TO THE UNIVERSITY

SD - #0015 - 081121 - C0 - 229/152/32 - PB - 9780265913901 - Gloss Lamination